PESTICIDE FACT HANDBOOK

PESTICIDE
FACT
HANDBOOK

Volume 2

U.S. Environmental Protection Agency

NOYES DATA CORPORATION
Park Ridge, New Jersey, U.S.A.

Published in the United States of America by
Noyes Data Corporation
Mill Road, Park Ridge, New Jersey 07656

10 9 8 7 6 5 4 3 2 1

Library of Congress Cataloging-in-Publication Data

(Revised for vol. 2)
Pesticide fact handbook.

    Includes indexes.
    1. Pesticides--Handbooks, manuals, etc.
I. United States. Environmental Protection Agency.
SB951.P396   1988   632'.95'0212   87-31528
ISBN 0-8155-1145-0 (v. 1)
ISBN 0-8155-1239-2 (v. 2)

Foreword

Volume 2 of the *Pesticide Fact Handbook* contains 87 Pesticide Fact Sheets released by the U.S. Environmental Protection Agency, and announced in the *Federal Register*, from January 1988 through December 1989.

Each individual pesticide listing includes a description of the chemical use patterns and formulations, scientific findings, a summary of the Agency's regulatory position/rationale, and a summary of major data gaps. The Fact Sheets in Volume 2 cover more than 430 trade-named pesticides.

The Fact Sheets are issued if one of the following regulatory actions occurs: (1) a Registration Standard has been issued, (2) a significantly different use pattern has been registered, (3) a new chemical is registered, or (4) a Special Review determination document has been issued.

Fact Sheets have been prepared for Registration Standards issued since June 1982, and for new chemicals and for chemicals with significantly changed use patterns registered since January 1984. Fact Sheets have also been issued for Special Review final determinations since June 1983.

Noyes has reproduced these Fact Sheets in Volume 2 directly from copies of EPA original material and, because of their expected usefulness, bound them in a durable hard cover edition at a fraction of their cost if purchased separately ($11.00 per fact sheet, or $957.00).

Volume 1 of the *Pesticide Fact Handbook* was published by Noyes in 1988. It contains 130 Fact Sheets, covering more than 550 trade-named pesticides, released by the EPA through December 1987. (If purchased separately, these 130 Fact Sheets would cost $1430.)

The table of contents is organized alphabetically and provides easy access to the information contained in the book. A Glossary and a Numerical List of Pesticide Fact Sheets, as well as Indexes of Common Names, Generic Names, and Trade Names, can be found at the end of the book.

Also included at the end of this volume, as a source of reference for the reader, are the alphabetical and numerical lists of fact sheets for the first volume.

Advanced composition and production methods developed by Noyes Data Corporation are employed to bring this durably bound book to you in a minimum of time. Special techniques are used to close the gap between "manuscript" and "completed book." Due to this method of publication, certain portions of the book may be less legible than desired.

NOTICE

The information in this book was prepared by the U.S. Environmental Protection Agency. On this basis the Publisher assumes no responsibility nor liability for errors or any consequences arising from the use of the information contained herein. Mention of trade names or commercial products does not constitute endorsement or recommendation for use by the Agency or the Publisher.

Final determination of the suitability of any information or procedure for use contemplated by any user, and the manner of that use, is the sole responsibility of the user. The reader is warned that caution must always be exercised when dealing with pesticides and pesticide residues, and expert advice should be sought at all times.

Contents

ACEPHATE

Reason for Issuance: Issuance of a Registration Standard
Date Issued: October 1987
Fact Sheet Number: 140

1. Description of Chemical

   Generic Name: O,S-dimethyl acetylphosphoramidothioate
   Common Name: Acephate
   Trade Name: Orthene
   EPA Shaughnessy Code: 103301
   Chemical Abstracts Service (CAS) Number: 30560-19-1
   Year of Initial Registration: 1974
   Pesticide Type: Insecticide
   Chemical Family: Organophosphate
   U.S. and Foreign Producers: Chevron Chemical Co. (U.S.A.)

2. Use Patterns and Formulations

   Application Sites: Agricultural crops; ornamentals (field grown, greenhouse, and home garden); lawns and turf; pasture and rangeland; forestry; indoor homeowner use on houseplants; and commercial applicator use in residential and commercial buildings including food processing establishments.

   Types and Methods of Application: Aerial; ground; direct injection into tree trunks; dip treatment (ornamentals); soil incorporated; and sprinklers.

   Type of Formulations: Granular, pressurized liquid, soluble concentrates (both liquids and solids), and cartridge.

3. Science Findings

   Summary Science Statement: Acephate has a relatively low acute toxicity to laboratory animals through the oral, dermal, and inhalation routes of exposure. Based on the available evidence, i.e., findings from the mouse oncogenicity study, and the mutagenicity assays, the Agency has classified the chemical as a category C carcinogen (a possible human carcinogen). The mouse oncogenicity study indicated a statistically significant increase in the proportion of liver adenomas/carcinomas and hyperplastic nodules occurred only in the high dose (1000 ppm) females and only at the time of terminal sacrifice. The EPA Guidelines for carcinogenic

1

risk assessment (FR September 24, 1986) were followed for the evaluation and the classification of the oncogenic effect of acephate.
Following the guidance set forth in the EPA guidelines, the mouse
oncogenic response was considered as "limited evidence."

The available data are not sufficient to enable the Agency to
accurately assess the potential risk to humans from this oncogenic
effect resulting from exposure to acephate. The data gaps include
residue reduction studies, exposure studies, usage data, a dermal
penetration study, a glove permeability study, and reentry data.

The available rat reproduction study showed reproductive effects at
50.0 ppm, the lowest dose tested. A new rat reproduction study is
needed to determine the no-observable-effect level (NOEL) for these
effects and to enable the Agency to assess the potential risk to
humans resulting from exposure to acephate.

Methylthioacetate (MTA) occurs as an impurity in the current
registered technical material. The available data suggest that the
MTA, despite its generally low acute toxicity, may pose a hazard to
the optic tract and pituitary gland in rabbits and other mammals at
low doses. Data were not provided to demonstrate a NOEL for lesions
at these target organs. Since visual impairment is inherently
difficult to diagnose in animals, it is possible that this effect
occurred in other studies but was not detected. In addition, a
mutagenic effect was seen in the mouse lymphoma assay in the activated
system. Due to the insufficiency of the submitted data to explain
the toxic and mutagenic potential of MTA, the Agency requires that
additional studies be performed.

Methamidophos, the cholinesterase-inhibiting metabolite of acephate,
is also an insecticide in its own right, and as such, was assessed
under a separate Registration Standard issued for the chemical in
September 1982. Several of the data gaps identified in that standard
have been fulfilled. It is highly toxic by both the oral and dermal
routes (Toxicity Category I). Results of two oncogenicity studies
show that methamidophos was not oncogenic in rats at dose levels of
2,6,18 and 54 ppm nor in mice at dose levels of 1,5,and 25 ppm. The
available teratogenicity studies show that it is not teratogenic to
rats or rabbits. The chemical was negative for acute delayed
neurotoxicity in the submitted study on hens. The lowest effect
level (LEL) for cholinesterase inhibition activity was determined
to be 2 ppm (0.05 mg/kg/day) in both the 1-year dog study and the
2-year rat study.

Data gaps for methamidophos include a rat reproduction study and
mutagenicity studies.

Chemical Characteristics

Physical State: Solid
Color: White
Odor: Strong, pungent, mercaptan-type
Boiling Point: N/A

Melting Point: 82-89 °C (97% technical)
Flammability: N/A
Solubility in Water: High solubility (65%)

Toxicology Characteristics:

ACEPHATE:

- o Acute Oral — Rat: 945 mg/kg (male); 866 mg/kg (female)
     Toxicity Category III

- o Acute Dermal — Rabbit: > 10,000 mg/kg (male)
     Toxicity Category III

- o Acute Inhalation — Rat: > 61.7 mg/kg (male and female)
     Toxicity Category IV

- o Acute Delayed Neurotoxicity — Hen: Negative at 785 mg/kg of
     body weight

- o Mouse Oncogenicity: Female mice fed 1000 ppm of technical
     acephate (highest dose tested) had a statistically
     significant higher incidence of hepatocellular carcinomas
     (15.8%) and hyperplastic nodules (19.7%) than did the
     controls.

- o Rat Oncogenicity: Not oncogenic to male and female rats
     under the conditions of the study; highest dose tested
     was 700 ppm (35 mg/kg).

- o Rat Chronic Feeding: LEL = 5 ppm (0.25 mg/kg) based on the
     inhibition of cholinesterase activity in plasma, RBC,
     and brain.

- o Dog Chronic Feeding: NOEL = 30 ppm (0.75 mg/kg) based on
     the inhibition of plasma, RBC, and brain cholinesterase
     activity. NOEL = > 100 ppm (2.5 mg/kg) for systemic
     toxicity.

- o Rabbit Teratogenicity: Not fetotoxic or teratogenic at
     10 mg/kg (highest dose tested).

- o Rat Teratogenicity: Not teratogenic at 200 mg/kg (highest
     dose tested).

- o Mutagenicity: The available studies indicate that acephate
     can induce gene mutations, DNA repair, and sister
     chromatid exchanges. However, in vivo studies did not
     indicate that these effects and structural chromosome
     aberrations are produced at a detectable level in an
     intact mammalian system.

o Rat Reproduction: Various reproductive effects (low
pregnancy rate, high loss of total litters, high fetal
losses, decreased size and weight of total litters, and
decreased number of live fetuses) were observed at the
lowest dose level tested, which was 50.0 ppm of technical
acephate (93% acephate).

METHYLTHIOACETATE (MTA):

o Acute Dermal - Rabbit: 1720-2820 mg/kg; Toxicity Category II-III.
Clinical signs included irreversible absence/diminution of
pupillary light reflex and apparent blindness.

o Acute Inhalation - Rat: 3.47 mg/L; Toxicity Category III.

o Skin Irritation - Rabbit: 2.6 Primary Irritation Score;
Toxicity Category III.

o Skin Sensitization - Guinea Pig: Nonsensitizing and
nonirritating; dose level tested was 0.3 ml (0.3 g).

o Eye Irritation - Rabbit: Toxicity Category III; dose level
tested was 0.1 mL of 93.5% MTA.

o Mutagenicity - Mouse Lymphoma Assay: Mutagenic to lymphoma
cells in the activated system but not in the nonactivated
system; levels tested were 1-10,000 ug/ml (activated)
and 10-5000 ug/ml (nonactivated).

Physiological and Biochemical Behavioral Characteristics:

Translocation: The available plant metabolism studies show
that acephate residues are readily absorbed by the roots
and translocated throughout the plant. However, data show
that acephate does not accumulate in carrot plants rotated
in acephate-treated soil or in fish, daphnia, or diatoms.

Mechanism of Pesticidal Action: Acephate is a contact and
systemic insecticide. As an organophosphate, acephate
exerts its toxic action by inhibiting certain important
enzymes of the nervous system (cholinesterase).

Metabolism and Persistence in Plants and Animals: The metabolism
of acephate in plants and animals is adequately understood.
Available data show that the residues in or on plants
resulting from acephate use may be largely or wholly intact
acephate and its metabolite, methamidophos. Available
animal metabolism data show that most of the radiolabeled
material is rapidly eliminated from the body and that a
majority of the material is excreted in the urine.

Methamidophos is not the major metabolite in ruminants. About 80 percent of the radiolabeled material in the urine was associated with unchanged acephate and less than 10 percent with the metabolite O,S-dimethylphosphorothioate. Most of the methamidophos formed is probably eliminated and excreted in the urine as O,S-dimethylphosphorothioate.

Environmental Characteristics:

Due to its rapid leaching behavior, acephate has the potential for ground water contamination. Available data are insufficient to fully assess this potential. Pertinent data (mobility, photodegradation, metabolism, and dissipation) are being required under the Acephate Registration Standard on an accelerated basis.

Available soil metabolism studies show that acephate dissipates rapidly with half-lives of < 3 and 6 days in aerobic and anaerobic soils, respectively. The major metabolite was CO_2 in both types of soil. The available leaching data include a soil thin-layer chromatography (TLC) study and a soil column study. Results of these studies indicate that acephate is mobile in most soils but that aged acephate residues (excluding acephate and its degradate methamidophos) are immobile in sandy loam soil. Apparently most of the applied acephate and the degradate methamidophos degrade to immobile compounds in 20 days.

Ecological Characteristics:

o Avian Oral Acute Toxicity: 350 mg/kg (mallard) and 140 mg/kg (pheasant)

o Avian Dietary Toxicity: > 5000 ppm (mallard) and 1280 ppm (bobwhite)

o Fish Acute Toxicity: > 1000 ppm (rainbow trout) and > 1000 ppm (bluegill sunfish)

o Freshwater Invertebrate Acute Toxicity: > 1000 ppm (Chironomus) and > 100 (Gammarus)

o Avian Reproduction: NOEL = > 5 ppm but < 20 ppm for mallard and NOEL = > 20 ppm but < 80 ppm for bobwhite.

o Honey Bee Acute Toxicity: 1.2 ug/bee.

Based on these studies, acephate is moderately toxic to avian species, practically nontoxic to freshwater fish and freshwater invertebrates, and highly toxic to honey bees.

However, acephate's metabolite, methamidophos, has been shown to
be very toxic to birds. Therefore, additional testing (residue
monitoring studies) are being requested to complete a hazard
assessment for the multiple-application, high-use rate field
crops. Appropriate labeling for the protection of endangered
species determined to be in jeopardy from use of acephate on
forests, range and pastureland, soybeans, and cotton have been
developed by the Agency and were imposed under PR Notices 87-4
and 87-5.

Tolerance Assessment:

Refer to Attachment A for the list of currently established tolerances
for acephate as well as the tolerance changes to be initiated by the
Agency.

To achieve compatibility with the maximum residue levels of the
Codex Alimentarius Commission, the following revisions in 40 CFR
180.108, 21 CFR 561.20, 40 CFR 180.315, 21 CFR 561.277, and 21 CFR
193.10 are to be initiated by the Agency.

o   40 CFR 180.108 and 21 CFR 561.20

The acephate tolerances currently established under these
sections are to be expressed in terms of only acephate per
se, with references to 40 CFR 180.315 and 21 CFR 561.277
indicating that tolerance for the metabolite methamidophos
are also in effect.

o   40 CFR 180.315 and 21 CFR 561.277

The methamidophos tolerances currently established under
these sections are to be divided into parts (a) and
(b) where (a) includes (1) tolerances reflecting uses of
methamidophos and (2) tolerances where both acephate and
methamidophos formulations are used on the same crop and
(b) includes tolerances reflecting uses of acephate formulations alone, i.e., residues of methamidophos resulting
from the metabolism of acephate.

o   21 CFR 193.10

These food additive tolerances reflecting crack and crevice
treatment in food-handling facilities are to be expressed
in terms of only acephate per se, i.e., based on the available data. No residues of the metabolite methamidophos are
expected to occur (< 0.001 ppm) in or on these foods.

Also, such a change in the residue definition would require deletion
of the paragraph (d)(8) of 40 CFR 180.3, which states that methamidophos residues may not exceed the higher of the two tolerances established for the use of acephate or methamidophos as a pesticide.

The tolerance for acephate residues in milk has been found likely
to be exceeded if maximum levels of spent mint hay or grass hay are
included in the dairy animal diet. Labeling restrictions prohibiting the feeding of spent mint hay and grass hay to dairy animals
are being imposed under this Standard for the use of acephate on
pasture, rangeland, peppermint, and spearmint. If the registrant
elects to submit additional residue data to support a lower tolerance for grass and grass hay, or to support a longer pregrazing or
preharvest interval for dairy animals, the labeling restriction
for use on pasture and rangeland would be imposed as an interim
precautionary measure pending submittal and evaluation of these
data.

Available data are not sufficient to conduct a full tolerance
assessment. Data gaps exist for magnitude of residue studies,
residue storage stability studies, a dairy cattle feeding study, a
rat reproduction study, and a rat feeding study.

4. Summary of Regulatory Position And Required Unique Labeling

Registration of current registered uses of acephate is to be continued.
Additional data to allow the Agency to better define the dietary,
occupational, and domestic exposure risks from the registered uses of
the chemical are being required. Once the Agency has evaluated these
data, it will determine whether the chemical should be placed in
Special Review or returned to the normal registration process.
Pending submittal and evaluation of these data, no additional
tolerances, including temporary tolerances, will be established for
acephate and no new uses will be registered, i.e., uses that would
result in an increase in the current exposure to humans or in new
exposure to humans.

As interim measures to reduce exposure pending submittal and
evaluation of the additional studies specified above, the following
restrictions are being imposed or, in the case of the last restriction
concerning domestic use, continued: A reentry interval of 24 hours
for fieldworkers; the use of protective clothing, including
chemical-resistant gloves, long-sleeved shirts and long-legged
trousers, shoes and socks by mixer/loaders, applicators, and early
reentry workers who may be exposed to treated plant surfaces within
24 hours of acephate application; dairy animal feeding restrictions
as described above under Tolerance Assessment; and, for domestic
use, the restriction not to allow children or pets on treated
surfaces until sprays have dried.

The 24-hour reentry interval is being imposed for the use of acephate
on agricultural crops, commercially grown ornamentals, in commercial
or governmental forestry seed production, and in greenhouses.

As described above under <u>Ecological Characteristics</u>, the Agency
has imposed labeling restrictions for the protection of endangered
species determined to be in jeopardy from use of acephate.

5.   <u>Summary of Major Data Gaps</u>

<u>Toxicology</u>	<u>Date Due</u>
- Acephate	
Rat Reproduction	39 Months
21-Day Inhalation	6 Months (Protocol)
Rat Feeding Study	6 Months (Protocol)
- Methylthioacetate (MTA)	
Acute Oral (Rat and Rabbit)	9 Months
Acute Dermal (Final Report - Rabbit)	9 Months
Acute Inhalation	9 Months
90-Day Dermal (Rabbit)	15 Months
Mutagenicity	12 Months

<u>Environmental Safety</u>

Avian Residue Monitoring	6 Months (Protocol)

<u>Environmental Fate</u>

Soil Photodegradation	9 Months
Anaerobic Aquatic Metabolism	27 Months
Adsorption/Desorption	12 Months
Soil Dissipation - Field	27 Months
Irrigated Crop	39 Months
Confined Rotational Crop	39 Months
Spray Drift	6 Months

<u>Exposure</u>

Applicator (Outdoor and Indoor)	6 Months (Protocol)
Indoor Inhabitants	6 Months (Protocol)
Glove Permeability	6 Months (Protocol)

<u>Residue Chemistry</u>

Storage Stability	24 Months
Magnitude of Residues	24 Months
Dairy Cattle Feeding	18 Months
Tobacco Residue	24 Months

Benefits

    Usage   6 Months
    Use-Related Exposure   6 Months

6. Contact Person at EPA

    William H. Miller (PM 16)
    Insecticide-Rodenticide Branch (TS-767C)
    401 M Street SW.
    Washington, DC 20460.

DISCLAIMER: The information presented in this Chemical Information Fact Sheet is for informational purposes only and may not be used to fulfill data requirements for pesticide registration and reregistration.

Attachment A

Tolerance Changes Reflecting the Recommended Change in Residue Definition

Commodity	Acephate*		Methamidophos	
	Established Tolerance (ppm)a	Recommended Tolerance (ppm)b	Established Tolerance (ppm)c	Recommended Tolerance (ppm)c
Beans (succulent and dry forms)	3(1)	3	—	1
Brussels sprouts	3(0.5)	3	1	1
Cauliflower	2(0.5)	2	1	1
Celery	10(1)	10	1	1
Cottonseed	2	2	0.1	0.5
Cranberries	0.5(0.1)	0.5	—	0.1
Eggs	0.1	0.1	—	0.01
Fat, meat, and meat byproductsd	0.1	0.1	—	0.01
Grass (pasture and range)	15	15	—	3
Grass hay	15	15	—	3
Lettuce (head)	10(1)	10	1e	1
Milk	0.1	0.1	—	0.05
Mint hay	15(1)	15	—	1
Peanuts	0.2	0.2	—	0.1
Peanuts hulls	5	5	—	1.5
Peppers	4(1)	4	1f	1
Soybeans	1	1	—	0.2
Cottonseed hullsg	8	8	—	0.1
Cottonseed mealg	4	4	—	2.5
Soybean mealg	4	4	—	2
Processed foodsh	0.02	0.02	—	—

a Expressed in terms of combined residues of acephate and methamidophos. If specified, limits of methamidophos are given parenthetically.

b Expressed in terms of acephate per se only.

c Expressed in terms of only methamidophos per se.

d Included are cattle, goats, hogs, horses, poultry, and sheep.

e The methamidophos tolerance covers all types of lettuce (head and leaf).

f The methamidophos tolerance covers all types of peppers.

g Feed additive tolerances in 21 CFR 561.20 (acephate) and recommendations for inclusion under 21 CFR 561.277 (methamidophos).

h Food additive tolerance reflecting spot and crack and crevice treatment of food areas of food handling establishments (21 CFR 193.10). There is no expectation of methamidophos residues in such foods (residues were nondetectable, < 0.001 ppm, in all cases).

* This table does not include the tolerance of acephate in/on macadamia nuts (.05 ppm), which was established subsequent to completion of the evaluation needed for this document. Acephate is not currently registered for use on macadamia nuts.

ALACHLOR

Reason for Issuance: Special Review
Date Issued: December 14, 1987
Fact Sheet Number: 97.1

1. Description of the Chemical

   Chemical name: 2-chloro-2'-6'-diethyl-N-(methoxymethyl)-
                     acetanilide

   Common name: Alachlor

   Trade names: Lasso®, Pillarzo®, Alanex®

   EPA Shaughnessy number: 090501

   Chemical abstracts service (CAS) number: 15972-60-8

   Year of initial registration: 1969

   Pesticide type: Herbicide

   Chemical Family: Acetanilide

2. Use Patterns and Formulations

   Application sites: preemergent use on field corn (including
                        sweet corn, popcorn), soybeans,
                        peanuts, dry beans, lima beans, green peas,
                        cotton, grain sorghum, sunflowers, ornamental
                        plants.

   Types of formulations: emulsifiable concentrate, granular,
                           microencapsulate

   Types and methods of application: ground or aerial methods

   Application rates: rate and frequency vary according to
                       site application; typically 1 to 4 pounds
                       of active ingredient per acre.

3. Science Findings:

    Physical and Chemical Properties

Color	--	White
Melting Point	--	40 to 41°C
Specific Gravity	--	1.133 (25/15.6°C)
Solubility	--	Soluble in ether, acetone, benzene, alcohol (unspecified) and ethyl acetate; slightly soluble in hexane; solubility in water - 240 ppm

    Physical and Chemical Properties:

Octanol/Water Partition Coefficient	--	434
Stability	--	Stable (first detectable heat evolution at 105°C)
Appearance at room temperature	--	White, crystalline solid at 23°C

    Tolerance Assessment: Tolerances were established in 40 CFR 180.249 for residues of alachlor and its metabolites.

    Toxicology Summary:

    Acute toxicity - Technical alachlor is not an acutely toxic product by any route of exposure. The acute oral LD_{50} in rats is .93 g/kg (Category III), the acute dermal LD_{50} in rabbits is 13.3 g/kg (Category III), and alachlor does not cause significant eye or skin irritation in rabbits (Category IV).

    Chronic Toxicity - Alachlor is oncogenic in both mice and rats. In mice, alachlor causes a statistically significant increase in lung bronchioalveolar tumors in females at 260 mg/kg/day (Highest Dose Tested). In rats, alachlor causes statistically significant increases at 42 mg/kg/day and above in nasal turbinate and stomach tumors in both sexes and thyroid follicular tumors in males. The following No Observed Effect Levels (NOELS) have been established for non-oncogenic effects: 1 mg/kg/day for liver and kidney effects; 2.5 mg/kg/day for uveal degeneration syndrome (UDS) of the eyes; 10 mg/kg/day for reproductive effects to kidneys of offspring. No birth defects were seen in highest dose tested for rats (400 mg/kg /day), but an additional teratogenicity study in rabbits is pending completion in 1988.

4.   Summary of Agency's Regulatory Position

The Agency position is that current registrations for
the use of alachlor on agricultural and nonfood crops will
be allowed to continue provided the following conditions and
label modifications are met to reduce applicator exposure:

a.   Reclassification as restricted use pesticide.

The application of alachlor is restricted to use by
certified applicators or persons under their direct
supervision.

b.   Continue use of tumor warning on label.

The following tumor warning statement imposed in the
alachlor Registration Standard must remain on the label:

"The use of this product may be hazardous to
your health. This product contains alachlor
which has been determined to cause tumors
in laboratory animals."

c.   Require the use of mechanical transfer devices by
all mixer/loaders and/or applicators who treat
300 acres or more annually with alachlor.

d.   Reinstate aerial application on the alachlor label
with the following additional label restriction:

"Human flaggers prohibited. Aerial application
may be performed using mechanical flaggers only."

The Agency has determined that the upper bound U.S. dietary
risk from alachlor residues on food crops is in the range of
10^{-6} (approximately one increased tumor case per million
persons exposed) based on the actual percent of crops treated with
alachlor. The Agency believes this risk is reasonable given the
benefits of continued use.

The Agency has determined that the risks associated with
alachlor exposure through ground water cannot be adequately
assessed at this time. The true extent of alachlor occurrence
in ground water is not known and cannot be properly estimated
for most areas of the country. Further monitoring will be
necessary considering the large volume, multiregional use of
alachlor, and the lack of statistically representative data from
the available studies. The Agency's final evaluation of the

potential for alachlor contamination of ground water is being
deferred, pending receipt of required monitoring data being
generated by the registrant under an EPA-approved protocol.
This study is scheduled for completion in late 1989, and the
results will be evaluated by EPA in early 1990.

Existing data on alachlor residues in surface water indicate
that the risk from drinking water sources supplied by surface water
will generally not exceed a range of 10^{-6}. The Agency believes
this level of risk is reasonable given the benefits of continued
use of alachlor products, and is not proposing regulatory action
under FIFRA on alachlor residues in surface water. The Agency
plans to promulgate regulations establishing a Maximum Contaminant
Level (MCL) for alachlor under the Safe Drinking Water Act in
the near future. These regulations would require the treatment
of drinking water which contains alachlor residues in excess of
the MCL, thereby maintaining the level of risk from exposure at
reasonable levels.

5.   Contact person:

James V. Roelofs
Review Manager
Special Review Branch
Registration Division (TS-767C)
Office of Pesticide Programs
U.S. Environmental Protection Agency
Washington, D.C.   20460
(703)-557-0064

DISCLAIMER: The information in this Chemical Information
Sheet is for informational purposes only, and
may not be used to fulfill data requirements
for pesticide registration or reregistration.

ALDICARB

Reason for Issuance: Special Review PD 2/3
Date Issued: June 22, 1988
Fact Sheet Number: 19.1

DESCRIPTION OF CHEMICAL

Common Name: Aldicarb

Chemical Name: 2-methyl-2-(methylthio)propionaldehyde
0-(methylcarbamoyl)oxime

Class Description: Member of the carbamate family

Trade Name: Temik

EPA Shaughnessy Code: 098301

Chemical Abstracts Service (CAS) Number: 116-06-3

Year of Initial Registration: 1970

Pesticide Type: Insecticide, acaricide, nematicide

U.S. and Foreign Producer: Rhone-Poulenc (formerly Union
Carbide Agricultural Chemical Co.)

USE PATTERNS AND FORMULATIONS

Aldicarb is currently registered for use only on cotton,
potatoes, citrus, peanuts, soybeans, sugar beets, pecans,
tobacco, sweet potatoes, ornamentals, seed alfalfa, grain
sorghum, dry beans, and sugar cane.

Types and Methods of Application: Soil incorporated.

Application Rates: 0.3 - 10.0 lbs. active ingredient.

Types of Formulation: Granular formulation (15%, 10%, and
5%). Also as a granular in a mixture with the fungicides
pentachloronitrobenzene and 5-ethoxy-3-(trichloromethyl)-1,2,4-
thiadiazole.

SCIENCE FINDINGS

Chemical Characteristics: Technical aldicarb is a white crystalline solid with a melting point of 98-100 C (pure material).   Under normal conditions, aldicarb is a heat-sensitive, inherently unstable chemical and must be stabilized to obtain a practical shelf-life.

Toxicological Characteristics:

Aldicarb is a carbamate pesticide which causes cholin-esterase (ChE) inhibition at very low exposure levels.   It is highly toxic by the oral, dermal, and inhalation routes of exposure (Toxicity Category I).   The oral LD_{50} value for technical aldicarb is 0.9 mg/kg and 1.0 mg/kg for male and female rates, respectively.   The acute dermal LD_{50} for aldicarb in rats is 3.0 mg for males and 2.5 mg for females.   Rats, mice and guinea pigs were exposed to aldicarb, finely ground, mixed with talc, and dispersed in the air at a concentration of 200 mg/m^3 for five minutes; all animals died.   At a lower concentration (6.7 mg/M^3) a 15 minute exposure was not lethal; however, 5 of 6 animals died during a 30 minute exposure. Exposure of rats for eight hours to air that had passed over technical or granular aldicarb produced no mortality.   Aldicarb applied to the eyes of rabbits at 100 mg of dry powder caused ChE effects and lethality.

The toxicity data base for aldicarb is complete. The toxicity data base includes a 2-year rat feeding/oncogenicity study which was negative for oncogenic effects at the no-observed-effect-level (NOEL) of 0.3 mg/kg bw/day; a 100-day dog feeding study and a 2-year dog feeding study with NOELs of 0.7 and 0.1 mg/kg bw/day, respectively, for effects other than cholinesterase inhibition (highest levels tested (HLT)); an 18-month mouse feeding/oncogenicity study with a NOEL of 0.7 mg/kg bw/day which was negative for oncogenic effects at the levels tested (0.1, 0.3, and 0.7 mg/kg bw/day); a 2-year mouse onco-genicity study which was negative for oncogenic effects; a 6-month rat feeding study using aldicarb sulfoxide with a NOEL of 0.125 mg/kg bw/day for ChE inhibition; a 3-generation rat reproduction study with a 0.7 mg/kg bw/day NOEL; a rat teratology study which was negative for teratogenic effects at 0.5 mg/kg bw/day (HLT); a hen neurotoxicity study which was negative at up to 4.5 mg/kg bw/day; a mutagenicity study utilizing the rat hepatocyte primary culture/DNA repair test which was negative for mutagenic effects at 10,000 ug/well; and a mutagenicity test utilizing an in vivo chromosome aberration analysis in Chinese hamster ovary cells which was negative for mutagenic effects at 500 ug/ml.

Physiological and Biochemical Behavioral Characteristics:

Aldicarb and its metabolites are absorbed by plants from the soil and translocated into the roots, stems, leaves, and fruit. The available data indicate that the metabolism of aldicarb in plants and small animals is similar.

Aldicarb is metabolized rapidly by oxidation to the sulfoxide metabolite and followed by a slower oxidation to the sulfone metabolite, which is 25 times less acutely toxic than aldicarb. Both metabolites are subsequently hydrolyzed and degraded further to yield less toxic entities. Available studies demonstrate that the administration of aldicarb to a lactating ruminant results in the rapid metabolism and elimination of the material. No residues of the parent compound and little, if any, residues of aldicarb sulfoxide or aldicarb sulfone are found in the tissues and milk. The predominant residue detected in tissues and milk is aldicarb sulfone nitrile.

Environmental Characteristics:

Sufficient data are available to assess the environmental fate of aldicarb. From the available data, aldicarb has been determined to be mobile in fine to coarse textured soils, even including those soils with high organic matter content, and has been found to reach ground water. Aldicarb is not expected to move horizontally from a bare, sloping field. Therefore, accumulation of aldicarb in aquatic nontarget organisms is expected to be minimal. This is further supported by an octanol/water partition coefficient of 5 and an ecological magnification value of 42.

Ecological Effects:

Aldicarb is highly toxic to mammals, birds, estuarine/marine and freshwater organisms. LC_{50} values for the bluegill sunfish and rainbow trout have been reported as 50 ug/liter and 560 ug/liter, respectively. A LC_{50} of 410.7 ug/liter was reported for Daphnia magna. Studies on the toxicity of aldicarb to the mallard duck and the bobwhite quail indicate LD_{50} values of 1.0 and 2.0 mg/kg, respectively.

Limited exposure to mammals is expected from a dietary standpoint. However, data from field studies and the use history of aldicarb provide sufficient information to suggest that application of this pesticide may result in some mortality, with possible local population reductions of some avian species. Whether these effects are excessive, long-lasting, or likely to diminish wildlife resources cannot be stated with any degree of certainty. Therefore, additional field studies have been required to further quantify the impact on avian and small mammal populations. Field study results will be submitted in April 1988.

Aldicarb has also been found to pose a threat to the endangered Attwaters Greater Prairie Chicken, living in or near aldicarb-treated fields. Accordingly, all aldicarb products are required to bear labeling restrictions prohibiting the use of the product in the Texas counties of Aransas, Austin, Brazoria, Colorado, Galveston, Goliad, Harris, Refugio, and Victoria if this species is located in or immediately adjacent to the treatment area.

Tolerance Assessment:

The Agency is in the process of reassessing the existing tolerances for aldicarb. Processing studies for coffee and potatoes have been submitted and are acceptable. A large animal metabolism study has been submitted to the Agency and satisfies the data requirement. A completed study of aldicarb residues on soybean processing fractions is to be submitted by August 1988. The requirement to submit a study analyzing aldicarb residues on treated cotton forage has been satisfied with a label restriction prohibiting the feeding of treated forage to livestock.

Exposure Incidents:

In 1979, aldicarb residues were found in drinking water wells located near aldicarb treated potato fields in Suffolk County, Long Island, New York, at levels greater than 200 parts per billion (ppb). Subsequently, aldicarb has been detected in ground water in 48 counties within 15 other States at levels up to 515 ppb. In all, the Agency has evaluated over 35,000 ground water samples of which 32% were positive for residues of aldicarb. The Agency's Office of Drinking Water (ODW) has established a Health Advisory level (HA) of 10 ppb for residues of aldicarb in drinking water.

The Pesticide Incident Monitoring System (PIMS) reports on aldicarb, from 1966 through 1982, contained 165 incidents

associated with human injury. Most of the human incidents
alleged that aldicarb was the cause of the problem, but there
was insufficient evidence to support such a conclusion. Those
incidents involving confirmed aldicarb poisonings appeared to be
the result of failure to use label recommended safety equipment
while applying aldicarb. Other incidents resulted from accidental spillage, ingestion of aldicarb, or consumption of food
commodities improperly treated with aldicarb.

The largest documented episode of foodborne pesticide
poisoning in North American history occurred in July 1985 from
aldicarb-contaminated California watermelons. More than a
thousand probable cases were reported from California, Oregon,
Washington, Alaska, Idaho, Nevada, Arizona and Canada. The
spectrum of illness attributed to aldicarb ranged from mild to
severe and included cases of grand mal seizures, cardiac
arrhythmias, severe dehydration, bronchospasms, and at least two
stillbirths occurring shortly after maternal illness. The
prompt embargo of watermelons on July 4, 1985 abruptly terminated
the major portion of the outbreak and reported illnesses
occurring after the implementation of the watermelon certification program were far fewer and milder in comparison to
earlier cases. Contamination of the watermelons ranged up to 3.3
ppm of aldicarb sulfoxide (ASO), a metabolite of aldicarb.
Clinical signs occurred from exposures to dosages estimated to be
as low as 0.0026 mg/kg ASO.

SUMMARY OF REGULATORY POSITION AND RATIONALE

a. Dietary Exposure to Treated Food Commodities

The Agency has recently received the final results of a
National Food Survey which monitored raw agricultural commodities
for residues of aldicarb in the market place. After these data
have been evaluated, the dietary exposure from consuming treated
food commodities will be estimated, and a risk assessment will
be conducted. The Agency may propose further regulatory action
depending on the results of this study.

b. Dietary Exposure to Contaminated Ground Water

The Agency has concluded that there are unacceptable risks
to persons consuming drinking water that is contaminated with
aldicarb at levels greater than the HA of 10 ppb due to a reduced
margin of safety for ChE inhibition.

The Agency cannot identify all specific areas of the nation where aldicarb residues exceed the HA, or the number of people who would be exposed to these high levels of contamination. However, the Agency can predict certain areas of the nation where the ground water supplies have a relatively high vulnerability to aldicarb contamination due to the hydrogeology and/or agronomic practices found in that area. Additionally, the Agency can predict certain areas which would have a medium vulnerability to contamination, although the vulnerability within some of these areas could vary greatly with some areas being much more vulnerable.

It is the Agency's presumption that the risks posed by aldicarb contamination of ground water above the HA in current or potential drinking waters will likely be more significant, in almost all cases, than any local benefit derived from aldicarb's continued use. Consequently, the Agency is proposing to regulate the use of aldicarb in order to eliminate or prevent contamination of ground water at levels above the HA. As a basic level of protection for all areas where aldicarb is used, the Agency is proposing a number of restrictions on the label. Specifically, no use of aldicarb would be permitted within 300-feet of a drinking water well, and aldicarb would be classified as a restricted use pesticide due to ground water concerns. (Aldicarb is already classified as a restricted use pesticide due to its acute toxicity.) Additionally, the Agency is seeking public comment as to what, if any, additional measures should be considered regarding the use of aldicarb and site-conditional restrictions.

The Agency will also require monitoring in those areas classified as having a medium tendency to leach. The data generated will be used to determine whether further regulatory action is required in these areas.

Finally, for those areas where there is the greatest likelihood of ground water contamination, states will need to implement, either for the entire state or for a county(ies) within the state, State Pesticide Ground Water Management Plans (MPs). Briefly, MPs are comprehensive plans which describe the measures states will impose to prevent ground water contamination. The Agency believes that MPs provide the best method of protection ground water pesticide contamination.

The Agency is soliciting public comment on a number of issues regarding its preliminary determination for aldicarb. Included are questions regarding the components of an MP, which assessment (hydrogeologic region or county) should be used in identifying those areas where contamination is most likely to occur, how should a localized risk/benefit analysis be performed and who should conduct it, and who is responsible for the costs associated with cleaning up ground water contamination.

CONTACT PERSON

Bruce Kapner
Special Review Branch, Registration Division
Office of Pesticide Programs (TS-767C)
401 M Street, S.W.
Washington, D.C. 20460
(703) 557-1170

DISCLAIMER: The information presented in this Pesticide Fact Sheet is for informational purposes only and may not be used to fulfill data requirements for pesticide registration or reregistration.

ALLETHRIN STEREOISOMERS

Reason for Issuance: Registration Standard
Date Issued: March 24, 1988
Fact Sheet Number: 158

1. DESCRIPTION OF CHEMICALS

The following chemicals are all synthetic pyrethroid insecticides. That is, they are synthetic duplicates of a component of pyrethrum which is extracted from chrysanthemum flowers. Introduced in 1949, Allethrin was the first synthetic pyrethroid. Bioallethrin and S-bioallethrin were introduced in 1969 and 1972, respectively

A. Common Name: Allethrin
   Generic Name: (2-methyl-1-propenyl) - 2-methyl-4-oxo-3-
                (2 propenyl)-2-cyclo-penten-1-yl ester or mixture
                of cis and trans isomers.
   Trade Name: Pynamin
   EPA Shaughnessy code: 004001 and 004002 (allethrin coil)
   Chemical Abstracts Service (CAS) Number: 584-79-2
   Producers: McLaughlin Gormley King
              Sumitomo Chemical Company
              Fairfield American

B. Common Name: d-trans Allethrin, Bioallethrin
   Generic Name: d-trans-chrysanthemum monocarboxylic
                ester of d 1-2-allyl-4-hydroxy-3-
                methyl-2-cyclo-penten-1-one
   Trade Name: Bioallethrin
   EPA Shaughnessy Code: 004003
   Chemical Abstracts Service (CAS) Number: 584-79-2
   Producers: McLaughlin Gormley King
              Roussel Uclaf

22

C. Common Name: S-bioallethrin; Esbiol
    Generic Name: d-trans-chrysanthemum monocarboxylic
                   acid ester of d-2-allyl-4-hydroxy-3-
                   methyl-2-cyclopenten-1-one
    Trade Name: Esbiol
    EPA Shaughnessy Code: 004004
    Chemical Abstracts Service (CAS) Number: 28434-00-6
    Producers: McLaughlin Gormley King
                Roussel Uclaf

D. Common Name: D-cis/trans allethrin; Pynamin Forte
    Generic Name: dl-3-allyl-2-methyl-4-oxo-2-cyclopentenyl
                   d-cis/trans chrysanthemate
    Trade Name: Pynamin Forte
    EPA Shaughnessy Code: 004005
    Chemical Abstracts Service (CAS) Number: 42534-61-2
    Producers: Sumitomo Chemical Co., Ltd.

2. USE PATTERNS AND FORMULATIONS

Application Sites: Broad spectrum insecticides and acaricides registered
                     for use on terrestrial food crops (vegetables,
                     citrus fruits, and orchard crops); terrestrial
                     nonfood uses (ornamental plants, turf, recreational
                     areas, and forest trees); greenhouse food and
                     nonfood crops (ornamentals and vegetables);
                     indoor and outdoor domestic dwellings; postharvest
                     use on fruit, vegetables and grains, and stored
                     food; commercial and industrial uses (food handling
                     establishments).

Types of Formulations: Pressurized liquids, mosquito coils, dusts,
                        emusifiable concentrates, soluble concentrate
                        liquids, and ready-to-use liquids. Almost
                        always formulated with a synergist and one or
                        more additional active ingredients.

Predominant uses and
Methods of Application:
                     Primarily indoor and outdoor use around the home
                     as foggers, plant, carpet and general purpose
                     aerosols, and mosquito coils to control common
                     pests including, but not limited to, ants, bedbugs,
                     carpet beetles, cockroaches, fleas, ticks, moths,
                     wasps and bees. Applied to crops foliarly by aerial
                     or ground equipment. Postharvest applications made
                     as an emulsive dip.

3. SCIENCE FINDINGS

Summary Science Statement

The Agency has very little acceptable toxicity data for the allethrin stereoisomers. There are no data available to assess the environmental fate characteristics of these compounds, including their potential to contaminate ground water. There are ecological effects data which show that the stereoisomers are highly toxic to fish and aquatic invertebrates, and essentially non-toxic to avian species. There are no acceptable residue data available to assess the adequacy of the current tolerances for allethrin.

Chemical Characteristics of the Technical Material

Physical State: Viscous oil; liquid, clear oil

Color: Pale yellow, yellow-orange, slightly brownish

Odor: Mild to slightly aromatic

Molecular weight and empirical formula: 302   - $C_{19}H_{26}O_3$

Solubility: Insoluble in water; miscible with petroleum oils, and soluble in paraffinic and aromatic hydrocarbons

Toxicology Characteristics

Acute toxicity: The acute oral toxicity of bioallethrin and and s-bioallethrin is low to moderate. Adequate data to discern other acute effects of these compounds are not available.

Subchronic toxicity: In a 90-day feeding study on bioallethrin, rats were administered 0, 500, 1500, 5000, and 10,000 ppm bioallethrin in the diet. A no-observed-effect-level (NOEL) was established at 1500 ppm based upon a decrease in body weight gain and increased levels of serum liver enzymes in females and increased liver weights in both sexes. This study, however, is presently classified as only supplementary, but may be upgraded upon submission of additional information.

Chronic toxicity: In a 6-month oral feeding study using beagle dogs, the animals were administered 0, 200, 1000, and 5000 ppm bioallethrin in the diet.   The NOEL was determined to be 200 ppm based on effects on the liver.

One rodent chronic feeding/oncogenicity study is available for d-cis/trans allethrin. In this study, rats were fed 0, 125, 500, and 2000 ppm of the test substance in the diet for 2 years. No oncogenic effects were observed. For systemic toxicity, the NOEL was determined to be 125 ppm based on decreased body weight gain and the presence of liver effects.

Teratogenicity: One teratology study conducted with bioallethrin is available. In this study, rats were dosed with 50, 125, and 195 mg/kg/day bioallethrin in the diet. The test compound did not induce developmental effects at the dose levels tested.

Mutagenicity: Two mutagenicity studies (DNA damage and reverse mutation) conducted with bioallethrin are negative for genetic damage.

Environmental Fate Characteristics

No data on the allethrin stereoisomers are available to assess the environmental fate and transport, and the potential exposure of humans and nontarget organisms. The potential of these compounds to contaminate ground water is unknown. Because the allethrins are thought to degrade rapidly in the environment, environmental fate data are being required on a "tiered" basis. This approach will permit the Agency to make a preliminary assessment of the persistence of these compounds. The requirement for additional testing will be deferred until evaluation of all data submitted under Tier I.

Ecological Characteristics

Avian Acute Oral Toxicity:

Species	Stereoisomer	LD_{50} or LC_{50}
Mallard Duck	Technical allethrin	>2000 mg/kg
Mallard Duck	D-cis/trans allethrin	5620 ppm
Bobwhite Quail	Bioallethrin	2030 ppm
Bobwhite Quail	D-cis/trans allethrin	5620 ppm

These data show that the allethrins are practically nontoxic to birds on both an acute and subacute exposure basis.

Freshwater Fish Acute Toxicity: Twenty-seven toxicity tests conducted with coldwater and warmwater fish species indicate that the allethrins are highly toxic to fish. The LC_{50} values ranged from 2.6 ppb (coho salmon -- bioallethrin) to 80 ppb (fathead minnow — S-bioallethrin).

Toxicity to Aquatic Invertebrates: Data show that allethrin is highly toxic to aquatic invertebrates with LC_{50} values of 5.6 ppb for stoneflies and 56 ppb for blackflies.

Toxicity to Non-target Insects: Although technical allethrin is moderately toxic to honey bees, the outdoor application rates are so low that even a direct application to bees is not likely to result in significant mortality.

Tolerance Assessment

The available data reviewed are insufficient to evaluate the adequacy of the established tolerances (covering postharvest use) for residues of allethrin in or on food/feed items (40 CFR 180.113). Allethrin is the only stereoisomer with established tolerances.

Because of insufficient residue chemistry and toxicity data for all of the allethrin stereoisomers, the Agency is unable to calculate an acceptable daily intake under the Tolerance Assessment System.

There are no Canadian or Mexican tolerances or Codex Maximum Residue Limits for residues of the allethrins in or on any plant commodity. Therefore, no compatibility questions exist.

4. REQUIRED UNIQUE LABELING

The Registration Standard for the allethrins contains no unique labeling requirements. It requires only updated environmental precautionary and disposal statements and a statement for outdoor use products that the product is highly toxic to fish.

5. SUMMARY OF REGULATORY POSITIONS AND RATIONALES

°The Agency is not starting a special review the allethrins.

°Since EPA believes that the allethrins may degrade rapidly in the environment, the Agency is requiring environmental fate data on a tiered basis. Additional data may be required upon evaluation of the tier I studies.

°The Agency is permitting registrants to use the technical product Esbiothrin as a respresentative test material for chronic studies on Bioallethrin and S-bioallethrin since it is a mixture of the two compounds, and they are of similar toxicity. Separate chronic studies are being required for Allethrin and D-cis/trans allethrin.

°The Agency is not requiring any endangered species restrictions since there is no evidence that the allethrins pose a hazard to endangered species from domestic indoor/outdoor uses.

°The Agency is not requiring any ground water advisory labeling, or reentry, spray drift, or protective clothing restrictions at this time.

°The Agency is not imposing restricted use classification on the allethrins.

°While the required data are under development all currently registered products containing the allethrins may be sold, distributed, formulated and used, provided that they are in compliance with all other terms specified in the Registration Standard.

6. SUMMARY OF MAJOR DATA GAPS

Toxicology

Acute Toxicity:

Acute oral LD_{50} toxicity (Allethrin, Pynamin-forte)
Acute dermal LD_{50} toxicity (Allethrin, Pynamin-forte, S-bioallethrin, Bioallethrin)
Acute inhalation LC_{50} Toxicity (all allethrins)
Eye irritation (all allethrins except Esbiothrin)
Dermal irritation (all allethrins except Esbiothrin)
Dermal sensitization (all allethrins)

Subchronic Toxicity

90-day feeding
  Rodent (all allethrins except Pynamin-forte)
  Nonrodent (all allethrins except Bioallethrin)
21-day dermal (all allethrins)
90-day inhalation (reserved for all allethrins)

Chronic Toxicity

Rodent feeding (all allethrins except Pynamin-forte)
Nonrodent feeding (all allethrins except Bioallethrin)
Rat oncogenicity (all allethrins except Pynamin-forte)
Mouse oncogenicity (all allethrins)
Rat teratogenicity (all allethrins except Bioallethrin)
Rabbit teratogenicity (all allethrins)
Reproduction (all allethrins)

Mutagenicity

Gene mutation (Allethrin, Pynamin-forte, S-bioallethrin)
Chromosomal aberration (all allethrins)
Other mechanisms of mutagenicity (all allethrins except Bioallethrin)

Special Testing

Metabolism (all allethrins)

Ecological Effects

Avian reproduction
Field testing - mammals and birds (reserved pending
  reproduction data)
Freshwater fish LC_{50} (typical EP)
Freshwater aquatic invertebrate LC_{50} (typical EP)
Acute estuarine and marine LC_{50} (fish, shrimp, oyster)
Fish early life stage and invertebrate life cycle
  (freshwater, estuarine)
Fish life cycle
Field testing (aquatic organisms)

Environmental Fate

TIER I

  DEGRADATION STUDIES - LAB
    Hydrolysis
    Photodegradation - water, soil, and air

  METABOLISM STUDIES - LAB
    Aerobic metabolism (soil and aquatic)
    Anaerobic metabolism in soil

  MOBILITY STUDIES
    Leaching/aged leaching
    Volatility (lab)

TIER II

Anaerobic aquatic metabolism	- Reserved
Volatility (field)	- Reserved
Field dissipation (soil)	- Reserved
Field dissipation (aquatic, sediment)	- Reserved
Field dissipation (soil, long-term)	- Reserved
Accumulation studies on rotational crops (confined)	- Reserved
Accumulation studies on rotational crops (field)	- Reserved
Accumulation studies on irrigated crops	- Reserved
Accumulation studies in fish	- Reserved
Accumulation studies in aquatic nontarget organisms	- Reserved
Reentry	- Reserved
Spray drift	- Reserved
Exposure	- Reserved

Product Chemistry

  Product Identity and Composition
  Analysis and Certification of Product Ingredients
  Physical and Chemical Characteristics

Residue Chemistry

  Nature of the Residue (Metabolism) in Plant and
   Livestock
  Residue Analytical Methods (may be required if
   additional metabolites of toxicological concern are
   identified)
  Stability Data
  Magnitude of Residue
    Crop field trials
    Postharvest treatment of fruits and vegetables
    Stored commodities
    Processing studies
    Meat/milk/poultry/eggs
    Food handling

7. CONTACT PERSON AT EPA

  Phillip O. Hutton
  Product Manager 17
  Registration Division (TS-767C)
  Office of Pesticide Programs
  Environmental Protection Agency
  401 M Street, S. W.
  Washington, D. C.   20460
  (703) 557-2600

DISCLAIMER: The Information presented in this Chemical Information Fact
             Sheet is for informational purposes only and may not be used
             to fulfill data requirements for pesticide registration and
             reregistration.

ALUMINUM AND MAGNESIUM PHOSPHIDE

Reason for Issuance: Amendment to Registration Standard
Date Issued: February 20, 1987
Fact Sheet Number: 118

1. Description of Chemicals

   Generic Names: Aluminum Phosphide, Magnesium Phosphide
   Trade Names: Phostoxin, Gastoxin, Quick Phos, Detia, etc.
   EPA Shaughnessy Codes: 066501 (aluminum phosphide), 066504
      (magnesium phosphide)
   Chemical Abstracts Service (CAS) Numbers: 20859-73-8 and
      12057-74-8
   Year of Initial Registration: 1978 and 1979
   Pesticide Type: Solid
   Chemical Family: Inorganic Phosphides
   U. S. Registrants: Degesch America, Inc.; Research Pro-
      ducts Co.; Pestcon Systems, Inc.; Bernardo Chemicals
      Ltd., Inc.; Phos-Fume Chemical Co.; Woodbury Chemical
      Co. of Missouri; Alpha Chemical Co.

2. Use Patterns and Formulations

   Application Sites: Indoor fumigation of agricultural
   food commodities, animal feeds, processed food commodities and non-food commodities (tobacco). Outdoor fumigation for burrowing rodent and mole control.

   Application Rates: 30 tablets or 75 pellets per square
   foot for fumigation of mills and warehouses; 1-4 tablets
   or 5-20 pellets for rodent burrows.

   Formulations: Tablets and pellets; powders in bags,
   envelopes and other types of containers.

3. Science Findings

   Summary Science Statement: The Agency has determined that
   the registered uses of this chemical will not generally
   cause unreasonable adverse effects to humans or the
   environment if used in accordance with the approved use
   directions and revised precautionary statements prescribed
   by the registration standard.

30

Chemical Characteristics: Solid, dark gray material (granules or powder); molecular weight of aluminum phosphide is 57.96; molecular weight of magnesium phosphide is 134.779; material must be protected from moisture in the atmosphere in air-tight containers; contact of the solid material with moisture in the air, or with water, or acids release phosphine, a highly toxic gas.

Toxicology Characteristics: Requirements for acute toxicity data have been waived because of the well known extreme inhalation toxicity of phosphine gas which it generates. Accordingly, aluminum and magnesium phosphide have been placed in toxicity category I, the highest toxicity category.

No chronic toxicology studies are required with respect to dietary exposure because there is not potential for dietary exposure (tolerances are set at limit of detection).

Toxicology studies on phosphine gas are required to assess the margins of safety for exposed workers and applicators because the Agency does not have adequate data to determine whether phosphine may cause any long term adverse effects to humans. These studies include 90 day inhalation, teratogenicity and mutagenicity testing.

Environmental Characteristics: Aluminum and magnesium phosphide react with moisture or water to release phosphine gas, which eventually dissipates into the atmosphere. The resulting material from the reaction is aluminum or magnesium hydroxide, a relatively inert and innocuous material, which is a constituent of clay. Exposure (monitoring data) and related information are required to help assess the margins of safety for applicators and workers exposed to phosphine gas.

Ecological Characteristics: Phosphine is a highly toxic gas to a wide range of living organisms. Indoor uses pose no risk to non-target organisms outside of the site to be treated. Outdoor end use products (i.e., rodent and mole control) must bear special precautionary labeling to protect endangered species. Manufacturing use products must bear environmental hazard statements for wildlife.

Tolerance Assessment: Tolerances have been established for raw agricultural commodities at a level of 0.1 ppm (40 CFR 180.225 and 180.375); processed foods 0.01 ppm (21 CFR 193.20 and 193.225); and animal feeds 0.1 ppm (40 CFR 561.40 and 561.268). Finished food and feed must be held 48 hours prior to being offered to the consumer; tobacco fumigated in hogsheads must be aerated 72 hours.

4. Summary of Regulatory Position

    --Aluminum and magnesium phosphide are Restricted Use
    Pesticides due to the extreme acute toxicity of phosphine
    gas which is released from the pesticide when it is
    exposed to moisture in the air. They may by used only by
    a certified applicator or by persons trained in accordance
    with the product manual working under the direct supervision
    and in the physical presence of the certified applicator.
    Physical presence means that the certified applicator
    must be available on the site or on the premises.

    --Respiratory protection is not required if the fumigant
    is applied from outside of an enclosed indoor area.
    However, if the applicator enters an enclosed indoor area
    to apply the fumigant, a NIOSH/MSHA approved canister
    respirator is required at the site. Exposure during
    application may not exceed 0.3 ppm phosphine as an 8
    hour time weighted average (TWA).   Engineering controls
    such as forced air ventilation are recommended as the
    primary means of meeting the exposure standard. Otherwise,
    an approved respirator must be worn.

    --Monitoring must be conducted with a low level detector
    device to assure that the exposure standard is not exceeded.
    A sufficient number of samples should be taken in places
    where worker exposure is likely to occur. It is recommended
    that the applicator or employer document exposure readings
    in an operation log or manual for each fumigation site.
    Once exposures have been adequately characterized for a
    particular site, subsequent monitoring is not routinely
    required for each aplication. However, spot checks should
    be made, especially if conditions significantly change
    or if a garlic odor is detected.

    --If monitoring shows that exposure is less than the
    standard, no respirator is required. If more than 0.3
    ppm TWA is encountered, a full face NIOSH/MSHA approved
    canister respirator is required up to 15 ppm phosphine.
    This type of respirator must be available during fumigation
    within an enclosed indoor area. If more than 15 ppm or
    unknown levels of phosphine are present, a NIOSH/MSHA
    approved self contained breathing apparatus (SCBA) is
    required. SCBA must be available at the site or locally
    such as at a fire department or rescue squad.

    --After application, no person may be exposed to more
    than 0.3 ppm phosphine (maximum concentration). Examples
    are: if the fumigated site leaks into an adjacent indoor
    area, during transfer of treated commodity or during
    reentry into an incompletely aerated space.

--All entrances to a fumigated site must be placarded (except for railroad hopper cars which must be placarded on both sides near the ladders and on the top hatch where fumigant was applied). A placard may only be removed after the commodity is completely aerated. To determine whether aeration is complete, each fumigated site or vehicle must be monitored and shown to contain 0.3 ppm or less phosphine gas in the air space around and, when feasible, in the mass of the commodity. If more than 0.3 ppm is detected, the placard must be transferred with the treated commodity. Persons transferring or handling incompletely aerated commodities must be informed of the presence of phosphine and adequate measures taken to prevent exposure to more than 0.3 ppm.

--At least two trained persons must be present when the product is applied from within a space to be fumigated or during reentry into a fumigated or partially aerated space.

5. Summary of Data Gaps:

90 day inhalation study in rats
Teratogenicity study in one species
Mutagenicity battery
Exposure (monitoring data and related information for major sites)

6. Implementation of Labeling:

March 1, 1987 -- Registrants submit revised labeling.

March 15, 1987 -- EPA returns approved new product manuals and labels.

May 31, 1987 -- Cases of product entering commerce after this date must contain a new EPA approved product manual. A stamp or similar notice on the outside of the case will read:

"The use of aluminum (or magnesium) phosphide is required to be used in accordance with the procedures and safeguards described in the product manual dated _____. A copy of that manual is enclosed in this case. You are required to review this manual before using this pesticide. If you are purchasing less than a full case, your retailer or distributor must provide you, without charge, a copy of this manual. Notice to Retailers -- Each purchaser of this pesticide, whether by the case lot or less than a case lot, must receive a copy of the manual described in this notice."

The phrase "entering commerce" means that registrants
may not release for shipment, distribute, sell offer
for sale, hold for sale, ship, deliver for shipment,
or receive and (having so received) deliver or offer
to deliver, to any person products which do not meet
these requirements after May 31, 1987.

January 31, 1988 -- All pesticide product containers
must bear an EPA approved label which complies with the
Registration Standard. In addition, an approved
product manual must be in each case. After this date
registrants may not release for shipment or other
persons may not distribute, sell, offer for sale, hold
for sale, ship, deliver for shipment, or receive and
(having so received) deliver or offer to deliver, to
any person products which do not meet these requirements.

7. Contact Person at EPA

Jeff Kempter, PM 32
Disinfectants Branch
Registration Division (TS-767C)
401 M Street, S.W.
Washington, D.C. 20460
Telephone: (703) 557-7470

DISCLAIMER: The information presented in this Pesticide
Fact Sheet is for informational purposes only and may not
be used to fulfill data requirements for pesticide registration and reregistration.

AMITRAZ

Reason for Issuance: Registration Standard
Date Issued: October 1987
Fact Sheet Number: 147

1. DESCRIPTION OF CHEMICAL

    - Generic Name: N'-(2,4-dimethylphenyl)-N-[[(2,4-dimethyl-phenyl)imino]methyl]-N-methylmethanimidamide
    - Common Name: Amitraz
    - Trade Name: Baam, Taktic
    - EPA Shaughnessy Code: 106201
    - Chemical Abstract Service (CAS) Number: 33089-61-1
    - Year of Initial Registration: 1975
    - Pesticide Type: Insecticide/Acaricide
    - Chemical Family: Diamidides
    - U.S. Producer: Nor-Am Chemical Company

2. USE PATTERNS AND FORMULATIONS

    - Application sites: pears
    - Types and methods of applications: aerial and ground application as a spray
    - Application rates: 0.75 lb active ingredient (ai)/A to 1.5 lb ai/A
    - Usual Carriers: Confidental business information

3. SCIENCE FINDINGS

    - In summary, amitraz is a diamidide compound of moderate mammalian acute toxicity but has been shown to be carcinogenic in mice. The Agency has concluded that amitraz is a borderline C/D carcinogen. Other toxicology studies do not demonstrate other significant chronic effects. Amitraz's behavior in the environment is not well defined. Its acute toxicity to birds is low, although it may affect reproduction, it is moderately toxic to aquatic species. Additional studies are being required for environmental fate and ecological effects.

Chemical Characteristics

- Technical amitraz is a straw colored crystalline solid. Its melting point is 86-87°C.
- It is extremely soluble in xylene and acetone and is very insoluble in water.
- Amitraz is relatively stable to heat, however the stability of amitraz in water is poor. Additional data are being required in order to determine the sensitivity of amitraz to metal ions and metal.

Toxicology Characteristics

- Acute oral: 523 mg/kg, Toxicity Category III.
- Acute dermal: >200 mg/kg, Toxicity Category II.
- Primary Eye Irritation: Non-irritating, Toxicity Category IV
- Acute Inhalation: 2.40 mg/l, Toxicity Category III.
- Primary Skin Irritation: Nonirritant, Toxicity Category IV.
- Skin Sensitization: Not a sensitizer.
- Major Routes of Exposure: Human exposure from amitraz applications is greatest from mixing and loading of pesticide formulation and applying it. Exposure can be reduced by the use of goggles or face shield and gloves and other protective clothing.
- Neurotoxicity: Amitraz is not expected to be a delayed neurotoxin because it is neither an organophosphate or an analog of a neurotoxic compound.
- Oncogenicity: A 2-year rat feeding and oncogenicity study was negative at 200 ppm. An 80-week mouse oncogenicity study demonstrated an increase in lymphoreticular tumors in female mice but not in males at 400 ppm. A 2-year mouse oncogenicity study demonstrated an increase in hepatocellular tumors in female mice at the 400 ppm level. The 2-year mouse study was referred to the Agency's Cancer Assessment Group (CAG) for evaluation. CAG determined, based on the weight of the evidence, that amitraz is in the lower portion of the Group "C" range. The SAP concluded that amitraiz should be classified as a Group "D", (not classifiable as to human carcinogenicity) because they believed the weight of the evidence was inadequate to clearly categorize the oncogenic potential of amitraz. The Agency has reassessed its own position in light of the SAP position and has now concluded that amitraz is a borderline Class C/D carcinogen.
- Metabolism: Available data suggest that amitraz is not readily absorbed in tissues and is excreted in the urine.
- Teratology: A teratology study in the rat was negative at 12 mg/kg body weight (bwt). A teratology study in the rabbit was negative at 25 mg/kg bwt.
- Reproduction: A 3-generation rat reproduction study was negative at 15 ppm.

- Mutagenicity: Available data show amitraz to be negative
  in the gene mutation, host mediated and dominant lethal
  test systems. Additional negative mutagenic studies were
  conducted in the Mouse Lymphoma Mutation Assay, Ames
  Bacterial Mutagenicity Test and in Unscheduled DNA
  Synthesis in Human Embryonic Cells.

Physiological and Biochemical Characteristics

- Mechanism of action: Amitraz causes depression of
  hypothalamic function.
- Metabolism and persistance in plants and animals: The
  metabolism of amitraz in plants has not been adequately
  described. Additional data are required. Data pertaining
  to the metabolism of amitraz in food producing animals are
  not required because there are no finite tolerances established for amitraz residues on crops involving livestock
  feed items.

Environmental Characteristics

- Available data are insufficient to fully assess the
  environmental fate of amitraz. Data gaps exist for all
  required studies.
- The available data do suggest that amitraz can leach
  slowly in certain soils. Data are required to assess
  amitraz's environmental fate and ability to leach
  through soils.

Ecological Characteristics

- Avian acute oral (LD_{50}) toxicity: 788 mg/kg for bobwhite
  quail (slightly toxic).
- Avian dietary (LC_{50}) toxicity: 7000 ppm for mallard duck
  (slightly toxic).
- Avian reproduction: No observable effect level (NOEL) is
  <40 ppm. A new study is required to demonstrate a NOEL.
- Freshwater fish acute (LC_{50}) toxicity: cold water species
  (rainbow trout)--0.74 ppm for technical; warm water species
  (bluegill)--1.34 ppb for technical.
- Aquatic freshwater invertebrates toxicity: Daphnia--35.0 ppb.
- Additional data are required to fully characterize the
  ecological effects of amitraz.

Required Unique Labeling Summary

   All manufacturing-use and end-use amitraz products
must bear appropriate labeling as specified in 40 CFR 162.10.
In addition to the above, the following information must
appear on the labeling:

- Manufacturing-use products must state that they are intended
  for formulation into other manufacturing-use products or
  end-use products only for use on pears, cattle and hogs.

- The Agency is removing the Restricted Use Classification
  for amitraz since the weight of the evidence is inadequate
  to clearly catagorize the oncogenic potential of amitraz.
  will specify the reason (oncogenicity).
- A reentry interval of 24 hours will continue to be required
  for the pear use until data required by the standard are
  received and evaluated.
- Worker protection statements will be required for all end-use
  amitraz products.

Tolerance Assessment

- The Agency is unable to complete a full tolerance assessment
  because the metabolism of amitraz in plants has not been
  adequately described.
- The available data support the established tolerances for
  pears, cattle and hogs.
- Established tolerances are published in 40 CFR 180.287 and
  they are:

Commodity	Parts Per Million
Pears	3.0
Apples	0.0
Cattle, fat	0.1
Cattle, meat byproducts (mbyp)	0.3
Cattle, meat	0.05
Milk	0.03
Goats, fat	0.0
Goats, mbyp	0.0
Goats, meat	0.0
Hogs, fat	0.3
Hogs, mbyp	0.3
Hogs, meat	0.05
Hogs, kidney and liver	0.2
Horses, fat	0.0
Horses, mbyp	0.0
Horses, meat	0.0
Sheep, fat	0.0
Sheep, mbyp	0.0
Sheep, meat	0.0

- The Agency established the above zero tolerances to reflect,
  in part, its conclusion of the Rebuttable Presumption Against
  Registration (RPAR) (1979) not to permit the use of amitraz
  on apples and therefore not to permit residues in apples or
  meat producing animals which consume apple processing waste.

- The acceptable daily intake (ADI) for amitraz is 0.0025 mg/kg/day based on a 2-year dog study with a NOEL of 10 ppm and a safety factor of 100. The maximum permitted intake (MPI) (based on a 60 kg person) is 0.15 mg/day. The published tolerances provide a theoretical maximum residue contribution (TMRC) to the daily diet of 0.0723 mg/day (for an average 1.5 kg daily diet) which accounts for 48.20% of the ADI.

4. SUMMARY OF REGULATORY POSITION AND RATIONALE

- The Agency has determined that it should continue to allow the registration of amitraz for use on pears, cattle and hogs. Other uses will be considered on a case-by-case basis.
- The toxicology data base for amitraz is complete. The Agency's has determined that amitraz should be classified as a Group C/D carcinogen.
- Available data are insufficient to fully assess the environmental fate of and the ecological effects from amitraz. Data are required to determine if amitraz will contaminate ground water.

5. SUMMARY OF MAJOR DATA GAPS

- The full complement of environmental fate data requirements including studies on degradation (hydrolysis and photolysis), soil metabolism, mobility, dissipation, and accumulation
- An avian reproduction study
- A plant metabolism study

CONTACT PERSON AT EPA

Dennis H. Edwards, Jr.
Product Manager (12)
Insecticide-Rodenticide Branch
Registration Division (TS-767)
Office of Pesticide Programs
Environmental Protection Agency
401 M Street, S.W.
Washington, D.C.   20460

ASSERT

Reason for Issuance: New Chemical Registration
Date Issued: April 1988
Fact Sheet Number: 159

1. Description of Chemical

   Chemical Name: A mixture of methyl-2-(4-isopropyl-4-methyl-5-oxo-2-
                  imidazolin-2-yl)-p-toluate and methyl-6-(4-isopropyl-4-
                  methyl-5-oxo-2-imidazolin-2-yl)-m-toluate

   CAS Number:

   Trade Name: ASSERT Herbicide

   EPA Shaughnessy Codes: 128842 and 128843

   Year of Initial Registration: 1988

   Pesticide Type: Herbicide

   U.S. and Foreign Producers: American Cyanamid Company

2. Use Patterns and Formulations

   Application Sites: Sunflowers, Wheat and Barley

   Types and Formulations: Liquid Concentrate

   Types and Methods of Application: The product is applied as a
                                     postemergence spray by air or
                                     ground equipment.

   Applicate Rates: 0.18 to 0.46 lb ai/acre.

   Usual Carrier: Water

3. Science Findings:

Summary Science Statement: All data required for registration of
this chemical are acceptable to the Agency. However, an additional
field dissipation study is being required due to leaching potential.
The current data base does not suggest any major toxicological problems.

Acute toxicology data on the technical grade active ingredient indicate
that the chemical is extremely caustic and corrosive. The toxicology
category is I (Danger), based on primary eye and skin irritation
studies.

Chemical Characteristics:

Physical State: Solid

Color: Off white to pale yellow

Odor: Musty odor

Melting Point: 113 °C - 153 °C (Range due to isomer ratio and
effects of impurities)

Density: 0.3 g/ml at 20 °C

Solubility: 875 ppm water; p-isomer 875 ppm in water at 25 °C;
m-isomer 1370 ppm in water at 25 °C

Dissociation Constant (Pka) 2.9 at 23.5 °C

Octanol/Water Partition Coefficient: p-isomer 35

m-isomer 66

Stability: Technical material was stable for 2 years at
ambient temperature.

Toxicology Characteristics:

Acute Toxicity: (End-use products)

Acute Oral Toxicity (Rat):   Greater than (>) 5000 milligrams/
                             kilogram (mg/kg),Toxicity Category
                             IV 1/

Acute Dermal Toxicity (Rabbit): Greater than 2000 mg/kg,
                                 Toxicity Category III 1/

Primary Eye Irritation (Rabbit): Toxicity Category III 1/

Primary Dermal Irritation (Rabbit): Toxicity Category III 1/

Acute Inhalation Toxicity (Rat): Greater than 518 mg/l
                                  Toxicity Category IV 1/

Dermal Sensitization (Guinea Pig): Not a sensitizer

Major Routes of Exposure: The major routes of expsoure are via
                           dermal and eye contact.

Chronic Toxicity: 90-Day Feeding Study (Rat) resulted in a no-
                   observable-effect level (NOEL) of 1000 ppm.

1/ Labeling statements required for: Toxicity Category III for acute
   dermal - "Harmful if absorbed through skin. Avoid contact with skin,
   eyes, or clothing. Wash thoroughly with soap and water after handling."
   Toxicity Category III for primary eye irritation - "Causes (moderate)
   eye injury (irritation). Avoid contact with eyes or clothing. Wash
   thoroughly with soap and water after handling." Toxicity Category IV -
   No precautionary statements required.

Major Routes of Exposures:

The Major routes of exposures are via dermal and eye contact.

Chronic Toxicity:

90-Day Feeding Study (rat) resulted in a (NOEL) of 1000 ppm.
One-year Feeding Study (dog) resulted in a NOEL of 1.25 mg/kg/day.
18-month Chronic Feeding/Oncogenicity Study (mice) resulted in a
NOEL of 750 mg/kg/day and no oncogenic effects at 750 mg/kg/day
for the hightest dose tested (HDT).
3-Generation Reproduction Study (rat) resulted in a maternal
toxicity NOEL of 25 mg/kg/day, and a fetotoxic NOEL of 250 mg/kg/day.
Teratology Study (rat) resulted in a maternal toxicity
NOEL of less than 250 mg/kg/day and a developmental toxicity
LEL (lowest effect level) of 500 mg/kg/day.
Teratology Study (rabbit) resulted in a maternal toxicity of 750
mg/kg/day, and a developmental toxicity NOEL of 250 mg/kg/day.
Mutagenicity (in vivo bone marrow test) - Not mutagenic.
Mutagenicity (Ames test) - Not mutagenic.
Mutagenicity (unscheduled DNA synethsis activity) - Not mutagenic.

Physiological or Biochemical Behavioral Characteristics:

Foliar absorption: Rapid.

Translocation: Systemic after absorption through either the
roots or foliage.

Mechanism of pesticidal action: Inhibits plant cell division of
rapidly growing tips of roots and shoots by inhibition of amino
acid synthesis.

Metabolism in Plants: Tolerant species metabolize the compound
to nonherbicidal metabolites.

Environmental Characteristics:

Absorption in basic soils types: Assert was very weakly absorbed on the
two sandy loam soils and on the two silt loam soils.

Environmental fate and surface and ground water contamination
concerns: Assert has the potential to leach and contaminate groundwater
at very low concentrations. Therefore, the Agency is requesting
additional field dissipation studies.

Resultant Average Persistence: Assert has a half-life of 25 to 45 days in soil.

Ecological Characteristics:

Avian Acute Toxicity (Mallard Duck)	> 2150 mg/kg
Avian Dietary Toxicity (Bobwhite Quail)	> 5000 ppm
Avian Dietary Toxicity (Mallard Duck)	> 5000 ppm
96-Hour Fish Toxicity (Rainbow Trout)	> 100 mg/l
96-Hour Fish Toxicity (Bluegill Sunfish)	> 100 mg/l
48-Hour Invertebrate Toxicity (Daphnia magna	> 100 mg/l
48-Hour Acute Toxicity (Honey Bee)	> 100 ug/bee

Assert demonstrates negligible toxicity to birds on both an acute and dietary basis. It is nontoxic to fish and invertebrates. It is not expected to adversely affect endangered/threatened species because of low toxicity.

Tolerance Assessment:

The nature of the residue is adequately understood, and adequate analytical methods (gas-liquid chromatography with a nitrogen sensitive detector) are available for enforcement purposes. Tolerances are established for residues of the herbicide AC 222,293 [Assert, a mixture of methyl-2-(isopropyl-4-methyl-5-oxo-2-imidazolin-2-yl)-p-toluate and methyl-6-(4-isopropyl-4-methyl-5-oxo-2-imidazolin-2-yl)-m-toluate] in or on the raw agricultural commodities sunflower seed at 0.10 parts per million (ppm), wheat grain and barley grain at 0.10 ppm, and wheat straw and barley straw at 2.00 ppm. The acceptable daily intake (ADI) based on the 1-year dog feeding study (NOEL of 6.25 mg/kg/day) and using a hundredfold safety factor is calculated to be 0.0625 mg/kg/day. The theoretical maximum residue contribution for these tolerances for a diet is calculated to be 0.00147 mg/kg/day. The current action will use 2.35 percent of the ADI. There are no other published tolerances for this chemical.

4. Summary of Regulatory Position and Rationale

None of the risk criteria listed in 40 CFR 154.7 have been exceeded for Assert. The Agency will issue a registration for this product.

Rationale: A review of available data indicate that no risk criteria have been exceeded or met for Assert.

The Agency is not imposing a ground water advisory statement for Assert.

Rationale: As the study was carried out, no residues were detected below the 3-inch soil layer. If residues are found to leach within the soil the Agency will request more data.

Additional residue data are not required to support the use on wheat Straw.

Rationale: The Agency has determined that available residue data adequately support the established tolerance on wheat straw.

The agency is requiring additional field dissipation studies to be submitted by 1990.

Rationale: Assert has potential to leach and contaminate groundwater at very low concentrations.

4. Summary of Major Data Gaps

None.

6. Contact Person at EPA

>   Robert J. Taylor
>
>   Product Manager 25
>
>   Registration Division (TS-767C)
>
>   Office of Pesticide Programs
>
>   Environmental Protection Agency
>
>   401 M Street SW.
>
>   Washington, DC   20460
>
>   Telephone: (703) 557-1800

DISCLAIMER: The information presented in this pesticide fact sheet is for informational purposes only, and may not be used to fulfill data requirements for pesticide registration and reregistration.

ASULAM

Reason for Issuance: Registration Standard
Date Issued: January 4, 1988
Fact Sheet Number: 153

1. Description of Chemicals

   Generic Names: Methyl sulfanilylcarbamate and sodium salt of methyl
                  sulfanilylcarbamate
   Common Names: Asulam and sodium salt of asulam
   Trade Name: Asulox
   EPA Shaughnessy Code: 106901 (asulam)
                         106902 (sodium salt of asulam)
   Chemical Abstracts Service (CAS) Number: 3337-71-1 (asulam)
                                            2302-17-2 (sodium salt of asulam)
   Year of Initial Registration: 1975
   Pesticide Type: Herbicide
   Chemical Family: Carbamate
   Producer: Rhone-Poulenc, Inc.

2. Use Patterns and Formulations

   Application Sites: Sodium salt of asulam is used for the postemergent control
   of broadleaf weeds, perennial grasses, and nonflowering plants on sugarcane
   and noncrop areas (such as rights-of-way), forestry sites (including Christmas
   tree plantations, site preparation, and conifer release), ornamentals,
   established turf, and aquatic sites (ditchbanks).

   Types of Formulations: 86.4% asulam technical manufacturing-use product;
   36.2% end-use product formulated as a soluble concentrate/liquid
   of the sodium salt of Asulam (equivalent to 3.34 pounds of asulam
   per gallon).

   Types and Methods of Application: Surface and aerial spray and spot
   treatment.

   Application Rates: From 1.67 to 6.68 pounds active ingredient per acre
   depending upon the use pattern and target weed species.

   Usual Carrier: Water.

47

3. Science Findings

Summary Science Statement: Asulam poses a limited oncogenic risk from dietary and worker exposure. The pesticide is tentatively classified as a Category C carcinogen (limited evidence of carcinogenicity in animals) according to EPA's Guidelines for Carcinogen Risk Assessment on the basis of a rat oncogenicity study which showed a statistically significant increase in benign adrenal gland pheochromocytomas in males at high dose only, and also statistically significant increase in thyroid gland C-cell carcinomas in low- and mid-dose males. Asulam did not produce a compound-related statistically significant increase in tumors in a mouse oncogenicity study. Available mutagenicity data are negative.

Tumors were produced in only one strain and sex of rodent, and in only a single experiment. Because of this rather limited evidence, quantification of human risk is considered inappropriate.

Asulam did not induce teratogenic effects (birth defects) or impair reproductive ability.

Data are insufficient to fully assess the acute toxicity of asulam to humans. However, available data on dermal and eye irritation and sensitization indicate that asulam is not acutely toxic to humans. In addition, data indicate that asulam is not acutely toxic to avian, aquatic, or mammalian species. Available data are insufficient to completely assess the environmental fate of asulam. Residues may carry over to subsequent crops and asulam may possibly contaminate ground water.

Chemical Characteristics:

Color: Cream to buff (technical [T])
         White (Pure Grade Active Ingredient [PGAI])

Physical State: Powder (T)
                 Crystalline (PGAI)

Odor: Slight (T)
        None (PGAI)

Melting Point: Approximately 135 °C (T)
                    (decomposes)
                143 to 144 °C (PGAI)
                    (decomposes)

Solubility: Ethanol - 12.4 g/100 mL
             Methanol - 35.6 g/100 mL
             Acetone - 38.5 g/100 mL
             Methyl ethyl ketone - 34.5 g/100 mL
             Dimethyl formamide - > 170 g/100 mL
             Water - 0.425 g/100 mL

Octanol/Water
Partition
Coefficient:

       0.30 at 0.01 m
       0.76 at 0.001 m
       1.76 at 0.0001 m

Toxicity Characteristics:

Acute Effects[1]:

> Primary Eye Irritation (Rabbit): Conjunctival irritation cleared by
> day 7; no observed corneal or iris irritation. Toxicity
> Category III.

> Primary Skin Irritation (Rabbit): No irritation. Toxicity Category
> IV.

> Dermal Sensitization: Negative.

Subchronic Effects:

> In a supplemental 6-month study dogs were dosed at 60, 300, and
> 1500 mg/kg in the diet. Low-dose females and males and
> females at high dose showed increased thyroid weights.
> Relative thyroid/body ratio was increased in high-dose males
> and females and, when compared to controls, high-dose males
> and females showed lower mean body weight gain. The tentative
> no-observed-effect level (NOEL) is 60 mg/kg/day. The study
> may be upgraded if appropriate clinical measurement summary
> tables and other data are submitted.

Chronic Effects:

> A 2-year feeding study in Charles River CD rats at doses of 0,
> 1000, 5000, and 25,000 ppm showed statistically significant
> increases in benign adrenal gland pheochromocytomas in high-dose
> males and also significant increases in thyroid gland C-cell
> carcinomas in low- and mid-dose males. The systemic NOEL =
> 36 mg/kg/day (720 ppm) based on thyroid follicular hyperplasia
> in mid- and high-dose males.

> A 18 month oncogenicity study in Carworth CI-1 mice at doses of
> 0, 1500 and 5000 ppm showed a statistically significant positive
> trend for undifferentiated sarcoma of the skin and subcutis in males.
> However, these tumors were not considered compound-related because:
> 1) they occurred only at the high dose in low incidence (4/46),
> 2) the incidence was not statistically significant though showed

---

[1] See 40 CFR 162.10 for discussion of toxicity categories and companion
labeling requirements.

a positive trend and 3) a parasitic skin infection was present which could have been related to the occurrence of skin tumors. The Agency is requiring a repeat mouse oncogenicity study since 1) the maximum tolerated dose (MTD) was not achieved, 2) two doses rather than three specified in Guidelines were used, and 3) evidence that the test animals were in poor health demonstrated by chronic kidney inflammation and parastic infection.

Developmental Toxicity:

Rabbits were dosed with 0, 150, 300, and 750 mg/kg/day asulam in the diet. The NOEL was 750 mg/kg/day (highest dose tested). Rats were dosed with 0, 50, 1000, and 1500 mg/kg/day. The NOEL was 1500 mg/kg/day (highest dose tested). Not teratogenic.

Two-Generation Reproduction - Rat:

Charles River CD rats were administered 0, 1000, 5000, and 25,000 ppm asulam in the diet. At 5000 and 25,000 ppm there were fewer live births per litter and also slightly lower fertility index in F_1 parents (second generation). The NOEL for reproductive effects is 1000 ppm. Did not impair reproductive ability.

Mutagenicity:

A cell transformation assay was negative. A provisionally acceptable study was negative for dominant lethal effects.

Physiological and Biochemical Behavior Characteristics:

Foliar Absorption: Applied after emergence, asulam is readily absorbed.

Translocation: Asulam may be taken up either by roots or leaves and moves to other parts of the plant.

Mechanism of Pesticidal Action: In the meristematic regions of plants (the growing points are areas of rapidly dividing cells at the tip of the stem, root, or branch), asulam interferes with the process of cell division and expansion.

Environmental Characteristics:

Available data are insufficient to assess the environmental fate of asulam.

Persistence and Absorption: Asulam appears to be resistant to hydrolysis. It is degraded in soil under aerobic conditions with half-life apparently from one to several days with most of degraded material bound to soil. Under anaerobic (flooded) conditions, the amount of bound residues appears to decrease and degradation rates decrease.

Leaching: Asulam appears to be mobile to very mobile in sand,
loamy soil, loam and clay loam soil. In soil columns both
the parent compound and degradates leached and were found as
19 to 88 percent of the applied material in the column
leachates.

Crop Rotation: Available data indicate that residues may occur
in crops grown in areas treated in prior years with asulam.

Environmental Fate and Surface and Ground Water Contamination Concerns:
Ability of asulam residues to carry over to subsequent crops
will be fully evaluated when the required crop rotation data
are submitted. Asulam may possibly contaminate ground water.

Additional data are required before the Agency can fully
assess the potential for ground water contamination.

Ecological Characteristics:

Avian Acute Oral Toxicity (40% formulation): > 4000 mg/kg
(mallard duck, partridge, pheasant, and pigeon)
(> 1600 mg/kg when adjusted to 100% asulam).

Avian Subacute Dietary Toxicity (60% formulation): > 75,000 ppm
(pheasant and mallard duck)
(> 45,000 ppm when adjusted to 100% asulam).

Freshwater Fish Acute Toxicity:
Bluegill sunfish (technical): > 180 ppm
Rainbow trout (60% formulation): > 5000 ppm
Channel catfish (60% formulation): > 5000 ppm

Freshwater Invertebrate Acute Toxicity (Daphnia): 27 ppm.

Estuarine and Marine Invertebrate Acute Toxicity:

Fiddler crab: > 100 ppm
Grass shrimp: > 100 ppm

Beneficial Insects Acute Contact Study (Honey Bees): 1.28 percent
mortality at 36.26 micrograms per bee.

Available data suggest that asulam is practically nontoxic to birds,
warmwater fish, and estuarine/marine species, and relatively
nontoxic to honey bees. Asulam is not expected to pose a
hazard to nontarget organisms; however, additional data are
required to be certain.

4. Tolerance Assessment

    Tolerances have been established for residues of asulam in or on the raw
    agricultural commodity sugarcane at 0.1 ppm (40 CFR 180.360).

    No Mexican tolerance or Codex Alimentarius Commission Maximum Residue
    Limits have been established for residues of asulam. A negligible
    residue tolerance is in effect in Canada for asulam residues in or on
    flax.

    The Agency is unable to complete a full tolerance reassessment because
    the available asulam toxicology and residue data do not fully support
    the established tolerance. There are data gaps for metabolism in
    animals, residue analytical method, and magnitude of residue in sugarcane.

    The acceptable daily intake (ADI) is a way of expressing the amount of
    a substance that the Agency believes, on the basis of the results of data
    from animal studies and the application of "safety" or "uncertainty" factors,
    may safely be ingested by humans without risk of adverse health effects. The
    ADI is expressed in terms of milligrams (mg) of the substance per kilogram
    (kg) of body weight per day (mg/kg/day).

    The Agency has calculated a provisional acceptable daily intake (PADI) of
    0.05 mg/kg/day for asulam on the basis of missing data and on a 2-generation
    reproductive effects study with a reproductive NOEL of 50 mg/kg/day (1000 ppm)
    but no NOEL for systemic effects and a 1000-fold safety factor.

    The established tolerances for asulam have a theoretical maximum residue
    contribution (TMRC) of 0.0001837 mg/kg/day, which utilizes 0.3673 percent
    of the PADI.

5. Summary of Regulatory Position and Rationale

    Special Review Status: EPA will not place asulam into Special Review
    because none of the risk criteria for initiating Special Review has
    been met.

    Registration of New Uses of Asulam: The Agency will issue registrations
    for substantially similar asulam products. However, new uses will be
    approved only on a case-by-case basis after considering the effects on
    the TMRC, the MPI (Maximum Permissible Intake), and the oncogenic
    risks.

Unique Label Precautionary Statements:

End-Use Products

ENVIRONMENTAL HAZARDS

Do not apply directly to water or wetlands (swamps, bogs, lagoons, marshes, or potholes). Do not clean equipment or dispose of equipment washwater in a manner that will contaminate water resources.

CROP ROTATION STATEMENT

Do not rotate with any crop which is not registered for use with Asulam for one year following the last application of this chemical.

CROP USE STATEMENTS

Do not treat sugarcane within 90 days of harvest. Do not graze or feed sugarcane forage and fodder to livestock.

6. Summary of Major Data Gaps (Data Required and Due Date)

Product Chemistry	
(All)	August 1988 and February 1989
Residue Chemistry	
Animal Metabolism	August 1989
Residue Analytical Methods	August 1989
Magnitude of Residues	August 1989 and February 1990
Environmental Fate	
Hydrolysis	November 1988
Photodegradation (water, soil)	November 1988
Aerobic and Anaerobic Soil Metabolism	May 1990
Aerobic and Anaerobic Aquatic Metabolism	May 1990
Leaching and Adsorption/Desorption	February 1989
Small-Scale Prospective Field Leaching Study	May 1988**
Field Dissipation (aquatic sediment and forestry)	May 1990
Rotational Crops (confined and field)	May 1991 and April 1992
Irrigated Crops	May 1991
Toxicology	
Acute Toxicity	November 1988
Subchronic Toxicity (21-day dermal)	February 1989
Mutagenicity (Ames test and chromosomal aberration)	November 1988 and February 1989
General Metabolism	February 1990
Oncogenicity (mouse)	April 1992

**Date protocols are due. After acceptance, the Agency will provide timeframes for submission of the reports.

Ecological Effects
    Freshwater Fish LC_{50} (coldwater fish)   November 1988
    Acute LC_{50} Estuarine and Marine
      Organisms (oyster study)   November 1988
    Acute LC_{50} Freshwater Invertebrates   February 1989

7. Contact Person at EPA

Richard F. Mountfort
U.S. Environmental Protection Agency
TS-767C
401 M Street SW.
Washington, DC 20460
(703) 557-1830

DISCLAIMER: The information presented in this Pesticide Fact Sheet is for information purposes only and may not be used to fulfill data requirements for pesticide registration and reregistration.

AVERMECTIN B$_1$

Reason for Issuance: Update—First Food Use (Cotton)
Date Issued: May 1989
Fact Sheet Number: 89.1

1. Description of Chemical

   Generic Name: Avermectin B$_1$ [A mixture of avermectins
                      containing $\geq$ 80% avermectin B$_{1a}$ (5-0-
                      dimethyl avermectin A$_{1a}$) and $\leq$ 20% avermectin
                      B$_{1b}$ (5-0-demethyl-25-de(1-methylpropyl)-25-
                      (1-methylethyl) avermectin A$_{1a}$)]
   Common Name: None assigned
   Trade Names: Affirm®, Agrimec®, Avid®, MK-936, Zephyr®
   Other Name: Abamectin
   EPA Shaughnessy Code: 0122804
   Chemical Abstracts Service (CAS) Numbers: 65195-55-3 and
                                               65195-56-4
   Year of Initial Registration: 1986
   Pesticide Type: Insecticide/Miticide
   Chemical Family: Avermectins (macrocylic lactones isolated
                      from soil organism Streptomyces avermitilis).
   U.S. Producers: Merck and Co., Inc.

2. Use Patterns and Formulations

   Application Sites: Control of imported fire ants on turf,
      lawns, and other non-crop areas; control of mites and
      other insects on shadehouse, greenhouse, and field-grown
      flowers, foliage plants, and other ornamentals, and on
      cotton.

   Type of Formulations: 0.011% insecticide bait, 2.0% Spray
   (0.15% EC), 70% technical.

   Method of Application: Bait broadcast (ground or air application)
      and individual mound to mound treatment (fire ants). Foliar
      spray (ground application) mixed with water for use on field
      grown flowers, foliage plants and ornamentals. Foliar spray
      (ground or air application) mixed with water for use on cotton.

   Rates of Application: For Fire Ants - Use rate is 50mg active
      ingredient (ai) per acre (1 pound (lb) product/acre);
      For Field Grown Flowers ands Ornamentals - Use rate is
      0.005 to 0.01 lb ai/acre (4 to 8 oz product/acre);
      For Cotton - Use rate is 0.01 to 0.02 lb ai/acre (8 to 16 oz
      product/acre) and 3 applications per season.

Usual Carriers: Pregelled defatted corn grit carrier
and water.

3.   Science Findings

Summary Science Statement

Technical avermectin exhibits high mammalian acute
toxicity. It is not considered to be mutagenic and
does not sensitize skin. It is not readily absorbed
by mammals and the majority of the residue is excreted
in the feces within 2 days. The 24-month rat chronic
feeding /oncogenicity study and 94-week mouse chronic
toxicity oncogenicity study were negative for oncogenic
potential. The results of a series of developmental
toxicity studies (rat, rabbit, mouse) have been
evaluated and showed that avermectin B_1 produces
developmental toxicity (cleft palate) in the CF_1
mouse. Toxicology data were also evaluated for the
delta-8,9-isomer of avermectin B_1 which is a plant
photodegradate that can range between 5 and 20 percent
of the residue on/in cottonseed. This isomer possesses
avermectinlike toxicological activity. It was
concluded that the delta 8,9-isomer also produces
developmental toxicity (cleft palate) in mice, but
not in rats.

In addition to avermectin and its delta 8,9-isomer,
toxicology data were also evaluated for the "polar
degradates" of avermectin, which constitute a large
percentage (up to 70%) of the total residue on cottonseed. Review of the toxicology data indicated that
these polar degradates do not possess avermectin-like
toxicological activity and for this reason need not
be included in the tolerance expression for residues
in/on cottonseed.

Sufficient data are available to characterize
avermectin from an environmental fate and ecological
standpoint. Avermectin is extremely toxic to mammals
and aquatic invertebrates and highly toxic to fish
and bees. Avermectin is relatively non-toxic to birds.
Based upon terrestrial residue analysis, aquatic runoff modeling and cluster analysis it appears that
certain endangered species may be impacted by the
use of avermectin on cotton. EPA and the Fish and
Wildlife Service are in the process of updating the

cotton cluster. The results when available will be applicable to the registration of avermectin on cotton.

Avermectin undergoes rapid photolysis, is readily degraded by soil microorganisms and, due to its binding properites and low water solubility, is expected to exhibit little or no potential for leaching; however, a complete assessment cannot be made until additional leaching and soil dissipation data are submitted.

Chemical Characteristics (Technical Grade):

Physical State: Crystalline powder
Color: Yellowish-white
Odor: Odorless
Melting Point: 155 – 157 °C
Vapor Pressure: Being tested, expected to be
    extremely low
Density: 1.16 $\pm$ 0.05 at 21 °C
Solubility: Insoluble in water ($\leq$ 5 ug/mL),
    readily soluble in organic solvents
pH: NA. The avermectin molecule has neither acidic
    nor basic functional groups
Octanol/Water Partition Coefficient: 9.9 x 10^3

Toxicological Characteristics:

Technical Grade Avermectin B$_1$

o Dermal Sensitization: Negative for skin sensitization

o Acute Oral LD$_{50}$ – Rat: 10.6 mg/kg (males),
   11.3 mg/kg (females)

o 14-Week Oral Study – Rat: NOEL $\geq$ 0.4 mg/kg/day (HDT)

o 18-Week Oral Study – Dog: NOEL = 0.25 mg/kg/day

o Teratology Study – Rat: Negative for terata up to
   1.0 mg/kg/day.

o Teratology Study – Rabbit: Negative for
   terata up to 1.0 mg/kg/day.

o Teratology Studies – Mouse: Teratogenic LEL = 0.4 mg/
   kg/day (cleft palate); Teratogenic NOEL =

o   Maternotoxicity Studies - Mouse: LEL = 0.075 mg/kg/
    day (lethality); NOEL = 0.05 mg/kg/day

o   2-Generation Reproduction Study - Rat: NOEL = 0.12
    mg/kg/day; LEL = 0.40 mg/kg/day (increased retinal
    folds in weanlings, increased dead pups at birth,
    decreased viability indices, decreased lactation
    indices, decreased pup body weights)

o   1-Year Oral Study - Dog: NOEL = 0.25 mg/kg/day;
    LEL = 0.50 mg/kg/ day (mydriasis in males and
    females)

o   94-Week Chronic Toxicity/Oncogenicity Study - Mice:
    Oncogenic potential: Negative up to 8 mg/kg/day
    (HDT); Systemic NOEL = 4 mg/kg/day; Systemic LEL =
    8 mg/kg/ day (Dermatitis in males, extramedullary
    hematopoiesis in the spleen in males, increased
    mortality in males, tremors and body weight loss
    in females)

o   2-Year Chronic Toxicity/Oncogenicity Study - Rats:
    Oncogenic potential: Negative up to 2.0 mg/kg/day
    (HDT); Systemic NOEL = 1.5 mg/kg/day; Systemic
    LEL = 2.0 mg/kg/day tremors in both sexes)

o   Metabolism Study - Rat: The metabolic T
    1/2 in rats is 1.2 days

o   Ames Mutagenicity Assay: Negative

o   Mutagenicity Assay for Chromosomal Aberrations
    in vitro in Chinese Hamster Ovary Cells

o   Rat Hepatocyte Mutagenicity Study: Under conditions
    of the study, abamectin (0.3 and 0.6 mM) caused
    an induction of single strand DNA breaks in rat
    hepatocytes in vitro: no effect was observed when
    the assay was carried out on hepatocytes from rats
    dosed in vivo at the LD_{50} dose level (10.6 mg/kg)

o   In Vivo Bone Marrow Mutagenicity Cytogenic Study:
    Negative in male mice at doses of 1.2 and 12.0
    mg/kg

Toxicity Studies on the Delta-8,9-Isomer of Avermectin

o Acute Oral LD_{50} - Mouse: > 80 mg/kg (HDT)
   (males and females)

o Teratology Study - Rat: Negative for terata
   up to 1.0 mg/kg/day (HDT)

o Teratology Studies - Mouse: Teratogenic LEL =
   0.10 mg/kg/day (cleft palate); Teratogenic NOEL =
   0.06 mg/kg/day

o Maternotoxicity Studies - Mouse: LEL = 0.50 mg/kg/
   day (lethality); NOEL = 0.10 mg/kg/day

o 1-Generation Reproduction Study - Rat: NOEL =
   0.4 mg/kg/day (HDT)

o Ames Mutagenicity Assay: Negative

Toxicity Studies on the "Polar Degradates" of Avermectin

o Acute Oral LD_{50} - Mouse: > 5000 mg/kg (HDT)

o Teratology Study - Mouse: Negative for terata up
   to 1.0 mg/kg/day (HDT)

o Teratology Study (polar degradates derived from
   citrus-treated fruit) - Mouse: Negative for
   terata up to 1.0 mg/kg/day (HDT)

o Ames Mutagenicity Assay: Negative

Ecological Characteristics:

Avian Oral (Bobwhite quail): LD_{50} > 2000 mg/kg;
LC_{50} = 3102 ppm
Avian Dietary (Mallard duck): LC_{50} = 383 ppm
Freshwater Fish (Bluegill): LC_{50} = 9.6 ppb
Rainbow trout: LC_{50} = 3.2 ppb
Estuarine Fish (Fathead minnow): LC_{50} = 15 ppb
Oyster Embryo Larvae: LC_{50} = 430 ppb.
Acute Freshwater Invertebrate (Daphnia):
LC_{50} = 0.22 ppb
Acute Estuarine Invertebrate (Shrimp, mysid):
LC_{50} = 0.02 ppb

Environmental Characteristics:

Avermectin is stable to hydrolysis at ph 5, 7, and 9
and thus is not expected to hydrolyze in the environment.
It photodegrades rapidly in water and soil with half-
lives of less than 12 hours and 1 day respectively.
Soil metabolism studies conducted in darkness indicate
degradation does occur with a half-life of 2 weeks
to 2 months under aerobic conditions. Anaerobic
degradation is slower. It is not expected to accumulate
in fish. Avermectin's solubility in water is determined
to be 7.8 ppb. The field dissipation study indicates
that avermectin, when applied in the bait formulation
directly to the soil, dissipates with a half-life of
about a week but may persist longer if the bait is
shaded. Due to its binding properties and low water
solubility, Avermectin is expected to exhibit little
or no potential for leaching; however, a determination
cannot be made until the results of the analyses of
the three remaining soil core replicates are submitted
and evaluated. Also an absorption/desorption leaching
study must be conducted since avermectin has shown
conflicting results in soil thin-layer chromatographic
(TLC) (immobile) and soil column studies.

Tolerance Assessment

A Section 408 tolerance under the Federal Food, Drug,
and Cosmetic Act has been established for residues
of avermectin B_1 and its delta 8,9-isomer in/on
the following raw agricultural commodity (RAC) (40
CFR 180.___)

Commodity	ppm
Cottonseed	0.005

The acceptable daily intake (ADI), based on a NOEL of
0.12 mg/kg/day from a 2-generation rat reproduction
study and safety factor of 300, is 0.0004 mg/kg/body
weight (bwt) day.

Because of developmental effects seen in animal
studies the Agency used the rat reproduction study
with a 300 fold safety factor to assess chronic
dietary exposure and establish an ADI. The 300 fold
safety factor was employed to account for (1)
inter- and intra-species differences (2) and pup
death observed in the reproduction study. (3)

maternal toxicity (lethality) NOEL = 0.05 mg/kg/day,
and (4) cleft palate in the mouse teratology study
with the isomer, NOEL = 0.06 mg/kg/day. The theoretical maximum residue contribution (TMRC) from the
proposed tolerance as well as pending tolerances on
celery, pears and tomatoes and temporary tolerances
on citrus with secondary residues in meat and milk
is 0.000052 mg/kg/day. This is equivalent to about
13 percent of the ADI. This analysis used tolerance
level residues and 100 percent of crop treated.
The TMRC from cotton only is .000001 representing
less than 0.1 of the ADI.

Because of adverse developmental effects seen in
animal studies detailed acute dietary exposure analysis
for this tolerance and pending tolerances for this
chemical was also conducted using a NOEL of 0.06
mg/kg body weight for developmental effects. The
food uses evaluated were the same as those evaluated in
the chronic exposure analysis. The acute exposure
analysis estimated the distribution of single-day
exposures for the overall U.S. population and certain
population subgroups. The analysis evaluated the
individual food consumption, as reported by respondents
in the 1977-78 USDA Food Consumption Survey, and
accumulated exposure to avermectin for each food
consumed for which a tolerance is being evaluated.
Each analysis assumed that avermectin residues were
present at tolerance level in all foods consumed.
The toxicologic endpoint pertained to developmental
toxicity. The subgroup of interest in this analysis
was women aged 13 and above, which has the subgroup
most closely approximating women of child-bearing
age. Based upon this analysis the Margin of Safety
(MOS) for the average woman of child bearing age was
calculated to be 1579. None of the target population
is expected to have a MOS less than 250.

The nature of the residue in cottonseed is adequately
defined. The residue of concern is the parent and
its delta 8,9-isomer. Based on (1) no accumulation
of avermectin in tissues or milk (from a ruminant
metabolism study) (2) absence of measurable residues
(<2 ppb) in cottonseed treated at exaggerated
application rates and (3) feeding restrictions for

cotton foliage the Agency has concluded that there
is no reasonable expectation of finite residues in
milk, eggs, meat, or poultry and no processing
data or food/feed additive tolerances are needed
for the use on cotton.

There are no Canadian or Mexican tolerances and no
Codex Maximum Residue Limits (MRLS) have been
established for avermectin B_1 and its delta 8,9-isomer
in/on cotton.   Therefore, no compatibility problem
exists.

4.   Summary of Regulatory Position and Rationale

o The Agency has determined that it should allow the
conditional registration of abamectin for agricultural use
to control mites on cotton.   Adequate data are available
to assess the acute and chronic toxicological effects
of abamectin to humans.   However since long-term fish,
aquatic and mammalian data are lacking and additional
leaching and soil dissipation data are required, the
registration is being conditionally approved with an
expiration date of March 31, 1992.   Due to the conditional
status of the registration the Agency is also establishing
the tolerance for this pesticide on cottonseed with an
expiration date of March 31, 1993 to cover residues
expected to be present during and for one year after the
period of conditional registration.

o In view of the high toxicity of technical abamectin
to fish, aquatic invertebrates and mammals and the potential
hazard associated with exposure from the use on cotton the
risk criteria for restricted use classification is exceeded
and thus the Agency, is restricting use on cotton to certified
applicators.

o Additional data are required to more adequately define
the hazards to mammals, fish, and aquatic invertebrates.
According to EPA's Ecological Effects Standard Evaluation
Procedures presumption of unacceptable risk is triggered when
the estimated environmental concentration (EEC) exceeds
the bird or mammal LC_{50}, or 1/2 the aquatic LC_{50} or EC_{50}.
According to EPA's assessments, these criteria are exceeded
for mammals and aquatic invertebrates for the cotton use
(mammal 1-day LC_{50} = 2.5 ppm, EEC = 4.8 ppm, aquatic invertebrate (freshwater) EC_{50} = 0.22 ppb, EEC = 0.6 ppb; estuarine
invertebrate EC_{50} = 0.02 ppb, EEC = 0.6 ppb).

o Because of adverse developmental effects seen in animal studies non-dietary exposure analysis was also conducted with respect to exposure to mixer/loaders, applicators and harvesters. Based upon surrogate exposure data; persons wearing long pants, long-sleeved shirts, rubber gloves and dermal absorption data in the monkey the calculated MOS for cotton crop applicators and workers were found to exceed 100 in all instances. This MOS is sufficient to adequately protect these workers.

5. Summary of Data Gaps

Name of Study	Reference Number	Due Date
Fish Life Cycle Test	§72-5	October 1991
Mesocosm Aquatic Study	§72-7	October 1991
Simulated Mammal Field Test	§71-5	October 1991
Soil Absorption/Desorption Results of the Analysis of the Remaining Soil Core Samples for the Field Dissipation Study	§163-1 §164-1	June 1990 July 1989

6. Required Unique Labeling Summary

o All products registered for use on cotton must bear the following restricted use labeling statements:

RESTRICTED USE PESTICIDE

Toxic to Fish, Mammals and Aquatic Organisms

For Retail sale to and use only by Certified Applicators or Persons under their direct supervision and only for those uses covered by the Certified Applicator;s certification

o The following use limitiations must appear on products registered for use on cotton:

Do not apply more than 48 fl oz per acre per year.

Do not apply within 20 days of harvest.

Do not graze or feed cotton foliage.

Do not reenter treated areas until sprays have dried.

Do not apply when weather conditions favor drift from target areas.

Do not apply this product through any type of irrigation system.

o Personal protective equipment and work safety statements must appear on the label of products registered for use on cotton and shade house, greenhouse and field grown flowers, foliage plants and ornamentals.

6.   Contact Person at EPA

George T. LaRocca
Product Manager 15
Insecticide-Rodenticide Branch
Registration Division (H7505C)
Office of Pesticide Program
U.S. Environmental Protection Agency
401 M Street SW.
Washington, DC   20460

Office location and telephone number:

Rm. 204, CM #2
1921 Jefferson Davis Highway
Arlington, VA   22202
(703) 557-2400

DISCLAIMER: The information presented in this Chemical Information Fact Sheet is for informational purposes only and may not be used to fulfill data requirements for pesticide registration and reregistration.

BACILLUS THURINGIENSIS

Reason for Issuance: Registration Standard
Date Issued: December 1988
Fact Sheet Number: 93.0

1. DESCRIPTION OF THE MICROBIAL PESTICIDE

Generic Name: Bacillus thuringiensis

Common Name: Bt

Trade and Other Names

Trade names for Bacillus thuringiensis subspecies kurstaki
include: Dipel, Thuricide, Bactospeine, Leptox, Novobac,
Bug Time, Cekubacillina, Attack, Foray, and Javelin.

Trade names for Bacillus thuringiensis subspecies israelensis
include: Bactimos, Teknar, and Vectobac.

The trade name for Bacillus thuringiensis subspecies aizawai
is Certan.

The trade name for Bacillus thuringiensis subspecies san diego
is M-One.

The trade name for Bacillus thuringiensis subspecies tenebrionis
is Trident.

EPA Shaughnessy Codes
(OPP Chemical Codes)

Microbial Pesticide Name: Bacillus thuringiensis (all
OPP Chemical Code: 006401   subspecies)

Microbial Pesticide Name: Bacillus thuringiensis subsp.
OPP Chemical Code: 006401   israelensis

Microbial Pesticide Name: Bacillus thuringiensis subsp.
OPP Chemical Code: 006402 kurstaki

Microbial Pesticide Name: Bacillus thuringiensis subsp.
OPP Chemical Code: 006403 aizawai

Microbial Pesticide Name: Bacillus thuringiensis subsp.
OPP Chemical Code: 128946 san diego

Microbial Pesticide Name: Bacillus thuringiensis subsp.
OPP Chemical Code: 006405 tenebrionis

Year of Initial Registration: 1961

Pesticide Type: Insecticide

U.S. and Foreign Producers:

Bacillus thuringiensis subsp. israelensis

- Abbott Laboratories, Chicago, IL
- Duphar B.V., Weesp, Holland
- Novo Industri AS, Copenhagen, Denmark
- Zoecon Corporation, A Sandoz Company
  Dallas, Texas

Bacillus thuringiensis subsp. kurstaki

- Abbott Laboratories, Chicago, IL
- Duphar B.V., Weesp, Holland
- Novo Industri AS, Copenhagen, Denmark
- Sandoz Crop Protection Corp.,
  Des Plaines, IL
- Ecogen, Inc., Langhorne, PA

Bacillus thuringiensis subsp. aizawai

- Sandoz Crop Protection Corp.,
  Des Plaines, IL

Bacillus thuringiensis subsp. san diego

- Mycogen Corporation, San Diego, CA

Bacillus thuringiensis subsp. tenebrionis

- Sandoz Crop Protection Corp.,
  Des Plaines, IL

2. USE PATTERNS AND FORMULATIONS

Target Pests: <u>Bacillus thuringiensis</u> subsp. <u>israelensis</u>:
mosquito (larvae), fungus gnats (larvae), and black flies
(larvae);
<u>B.t.</u> subsp. <u>aizawai</u>: Greater wax moth (larvae);
<u>B.t.</u> subsp. <u>kurstaki</u>: Lepidopterous larvae
<u>B.t.</u> subsp. <u>san diego</u>: Colorado potato beetle
(larvae) and elm leaf beetle (larvae and adults)
<u>B.t.</u> subsp. <u>tenebrionis</u>: Colorado potato beetle

Registered Uses:

<u>B.t.</u> subsp. <u>israelensis</u>: Terrestrial food crop use on
pastures;

Aquatic food crop use on rice;

Aquatic nonfood crop use on brackish
water, mangrove swamps, salt marshes,
tidal water, drainage systems,
irrigation systems, flood water areas,
woodland pools, standing water,
polluted water, sewage waste
lagoons, ponds, lakes, and
streams;

Greenhouse nonfood crop use on ornamental plants;

Domestic outdoor use on standing water
around the dwellings.

<u>B.t.</u> subsp. <u>aizawai</u>: Indoor use on empty honeycombs.

<u>B.t.</u> subsp. <u>kurstaki</u>: Terrestrial food crop uses on
cotton, corn, soybeans, sorghum (grain crop),
small grains, hops,
fruits (banana,blueberry, caneberries,
cranberry, currant, citrus fruits, grapes, kiwi fruit,
pome fruits, stone fruits, small fruits, strawberry,
tropical fruits, vegetables
(artichoke, asparagus, avocado, beans, beets, carrots,
celery, cole crops, cucumber, dandelion, eggplant,
endive, lentils, lettuce, melons, okra, onions,
parsley, parsnip, peas, pepper, potato, pumpkin,
radish, rutabaga, safflower, spinach, squash,

sugar beets, sugar maple, sunflower,
sweet potato, Swiss chard, ti,
tomato, watercress, watermelon), nuts (nut crops, nut
trees, peanut, walnut),   flavoring and spice crops,
garlic, horseradish, mint, salisfy, and forage
crops, alfalfa, hayage, pastures and
rangelands;

Terrestrial nonfood crop uses on
tobacco, ornamental flowering and
herbaceous plants, ornamental and/or shade trees, and
ornamental turf;

Aquatic food crop use on rice and wild
rice;

Greenhouse food crop uses on beans,
beets, carrots, celery, cole crops,
cucumber, eggplant, endive, flavoring and spice crops,
garlic, lentils, lettuce, melons, onions, parsley,
peas, peppers, potato, radish,
spinach, squash, sweet potato,
strawberry, tomato.

Greenhouse nonfood crop uses on
agricultural research crops and
ornamental flowering, herbaceous
and woody plants;

Forestry uses on forest trees;

Indoor uses on stored birdseed; herbs,
spices, and condiments; grain crops;
peanuts; agricultural and oil seeds;
soybeans, sunflower, and tobacco
(including flue-cured).

B.t. subsp. san diego: Terrestrial food crop uses on eggplant, potato and tomato;
Terrestrial nonfood crop uses on elm
trees.

B.t. subsp. tenebrionis: Terrestrial food crop use on potatoes.

Methods of Application: Hand sprayer; water treatment by
aerial or ground equipment; soil
application by drip or overhead
irrigation systems; foliar application
by aerial, conventional ground or

hand-held equipment and
center-pivot irrigation systems;
sprayer or sprinkler cans.

Formulations: Technicals, formulation intermediate, dusts,
granular, pelleted/tablet, wettable powder, emulsifiable
concentrate, flowable concentrate, ready-to-use, and pressurized
liquid.

3.   SCIENCE FINDINGS

The Agency reviewed and evaluated available data, including both data
submitted to the Agency in support of registration of B. thuringiensis as an
active ingredient and data from the published literature. This information
served as the basis for issuance of a draft Registration Standard in 1986.
On October 10, 1986, the Agency informed the public regarding the availability of
the draft Registration Standard for comment in the FEDERAL REGISTER, 51 FR 37488.
In the comment period of two months, five commenters responded to the Agency.
All comments have been taken into consideration in the issuance of this final
Registration Standard. The specific comments, as well as the Agency's responses,
are present in the public docket assigned to this Registration Standard.

In the 1986 draft Registration Standard, the Agency concluded that adequate
data were available to assess the toxicological and other biological effects of
B. thuringiensis on mammals, that no data gaps existed in the toxicology data
base, and that there were no substantial human or environmental safety concerns
except for certain endangered lepidopteran insect species. Although substantial
gaps were found to exist in the ecological effects data base, there were no
substantive concerns regarding unreasonable adverse effects of B. thuringiensis
for the registered products. Therefore, the Agency concluded that the use of B.
thuringiensis products could be continued, and that products could be used as
registered, with only minor precautionary labeling changes and additional
nontarget organism data being required.

Since the issuance of the 1986 draft Registration Standard, the Agency has
revised the Pesticide Assessment Guidelines Subdivision M, reassessed hazard to
endangered species, and has reviewed additional data on B. thuringiensis. The
Agency has also more keenly focused on product identity; i.e., strain-to-strain
variability within subspecies designations. Current methods applicable to strain
identification have advanced considerably since the initial B. thuringiensis
registration (1961). As with conventional chemical products, it is essential to
know product identity so that the applicability of test results, i.e. toxicity
testing and host range, can be related to specific products. A major focus of
this Registration Standard is to obtain state-of-the-art identification data on
B. thuringiensis strains. Each registered strain must be placed in a recognized
culture collection and is subject to the data requirements of this Standard.

Although the data submitted to the Agency since 1986 show rodent and nontarget organism effects, these data do not change the assessment of the draft 1986 Registration Standard that there is no evidence of any substantial human or environmental safety concerns related to current uses of B. thuringiensis. Furthermore, there is no evidence that B. thuringiensis poses a health risk via the oral route of exposure.

The nontarget organism effects include acute toxicity in birds (178 ppm LD_{50} and 1 ppm NOEL) and suggest adverse effects in freshwater fish, plants, aquatic invertebrates, and rare beneficial insects. The contribution of inert ingredients to toxicity as well as the relevance to environmental risk of routes of exposure and dose levels used in testing have yet to be determined. Therefore, whether these results show hazard to the environment cannot be determined until additional data are submitted.

Endangered Species

Risk to federally listed endangered species cannot be fully determined at this time. However, the U.S. Fish and Wildlife Service (FWS) has determined (1987) that certain uses of the subspecies kurstaki jeopardize the continued existence of the Kern primrose sphinx moth, Lange's metalmark butterfly, Smith's blue butterfly, El Segundo blue butterfly, Oregon silverspot butterfly, San Bruno elfin butterfly, Lotis blue butterfly, and the Schaus swallowtail butterfly. An earlier consultation (1984) with FWS addressing rangeland/pastureland pesticide use found that B. thuringiensis (subspecies not specified) jeopardize the existence of the Kern Primrose Sphinx moth, Delta Green Ground Beetle, and the Valley Elderberry Longhorn Beetle. As some B. thuringiensis products have activity against dipterans, these products could effect any endangered dipterans that may be listed in the future.

Tolerances

The 40 CFR 180.1011 tolerance exemption for B. thuringiensis is currently being reevaluated by the Agency. This reevaluation will continue as data submitted in response to the Registration Standard are reviewed. Areas which the Agency wishes to address in this reevaluation include the scope of the exemption (e.g. whether asporogenic strains should be included), quality assurance measures, limits or restriction on the presence of beta-exotoxin and whether or not the mouse subcutaneous test should be replaced with a mouse intraperitoneal test.

The 40 CFR 180.1001(c) inert tolerance exemption for B. thuringiensis fermentation solids and/or solubles is also currently being reevaluated by the Agency as product analysis data are submitted in response to the Registration Standard.

4. SUMMARY OF THE REGULATORY POSITIONS

a. The Agency is requiring special tests designed to more accurately characterize strains of Bacillus thuringiensis. These data will be used to reclassify registered strains into groups of strains with similar characteristics.

b. Registrants and applicants must identify the number of B. thuringiensis strains present in their products; each strain is subject to all the data requirements applicable to the use pattern(s) of that strain.

c. Upon review of the data submitted in response to the Registration Standard, the Agency will reevaluate the current tolerance exemption for B. thuringiensis (40 CFR 180.1011) and the inert tolerance exemption for B. thuringiensis fermentation solids and/or solubles (40 CFR 180.1001 (c)).

d. Testing requirements set forth for food use products (CFR 40 180.1011) will be required for nonfood uses of B. thuringiensis as well. Analysis for production batch contaminating microorganisms and their metabolites is not currently required of nonfood use products, such as those used in forestry programs. In order to assure public and applicator safety, the Agency believes such testing requirements need to be imposed on all B. thuringiensis products.

e. The Agency is not requiring data on ground water.

f. The Agency is not requiring Endangered Species labeling at this time.

g. The Agency is not initiating a Special Review of B. thuringiensis at this time.

h. B. thuringiensis does not meet the criteria for Restricted Use at this time.

i. The Agency is not requiring a reentry interval for B. thuringiensis.

j. Protective clothing is not required for users of any B. thuringiensis products at this time.

k. Each registered strain is to be deposited in a recognized culture collection.

6. SUMMARY OF MAJOR DATA GAPS

Due to the current Agency grouping of B. thuringiensis active ingredients by subspecies and the variability of strains within a subspecies, data supporting one strain within a subspecies may not be adequate to support another strain in the same subspecies. Each distinct strain, including each strain within a subspecies (active ingredient), is subject to all the data requirements applicable to the use pattern(s) of that strain. In order to utilize data submitted to the Agency prior to the issuance of the Registration Standard, registrants must show that the strain used in testing is substantially similar to the strain present in the product currently. Data gaps exist for all

requirements listed in the data tables of the Registration Standard.
However, some of these gaps may be filled if the registrant shows that the strain
used in testing is substantially similar to the strain present in the product
currently.

7.   CONTACT PERSON AT EPA

Phillip O. Hutton
Product Manager (17)
Insecticide Rodenticide Branch
Registration Division (TS-767C)
Office of Pesticide Programs
Environmental Protection Agency
401 M Street, S. W.
Washington, D. C.   20460

Office location and telephone number:

Room 207, Crystal Mall #2
1921 Jefferson Davis Highway
Arlington, VA   22202
(703) 557-2690

DISCLAIMER: The information in this Pesticide Fact Sheet
is a summary only and is not to be used to satisfy data
requirements for pesticide registration and reregistration.
The complete Registration Standard for the pesticide may be
obtained from the National Technical Information Service.
Contact the Product Manager listed above for further
information.

BENDIOCARB

Reason for Issuance: Registration Standard
Date Issued: June 1987
Fact Sheet Number: 195

1. Description of chemical

    Common Name: Bendiocarb
    Chemical Name: 2,2-dimethyl-1,3-benzodioxol-4-yl methylcarbamate
    CAS Number: 22781-23-3
    EPA Shaughnessy Code: 105201
    Trade names: Dycarb, Ficam, Garvox, Multamat, NC6897
                 Niomil, Seedox, Tatoo, and Turcam
    Pesticide type: Insecticide
    Chemical family: Carbamate
    Basic Registrant: NOR-AM Corporation

2. Use Patterns and Formulations

    Pests Controlled: Household pests, mosquitoes, ornamental
                           pests, and imported fire ants.
    Application Sites: Household, commercial buildings, outdoor
                           mosquito control, greenhouse, ornamentals,
                           aquatic nonfood (sewers), rodent burrows,
                           wood and wood structure protection, fire
                           ant mounds and food handling areas.
    Formulation Types: Dust, Wettable Powder, Ready to Use
                           Solution, Granular, Soluble Concentrate
                           Liquid, Shelf Paper.

    Application Methods: Broadcast, Foliar Application,
                         Dusting, Crack and Crevice, Soil
                         Drench, and Premise Treatment.

Of the total amount of bendiocarb applied in the United States
in 1983-1935, ~71% was used in professional applications
to household, commercial, and structural sites, ~14% was
used on ornamentals (including turf), ~12% was used on
domestic indoor and outdoor sites, and ~3% was used for
mosquito control. Application rates range from 1.52-32 oz
ai/100 gal, 1-4 lb ai/A, 0.005-0.01 oz ai/burrow for rodent
burrows, 0.0067-0.0133 oz ai/mound for fire ant mounds,
0.0026-0.011 lb/ai/A for mosquito control, and 0.07-0.14 oz
ai/manhole for use in sewers. Bendiocarb may be formulated
with petroleum distillates, piperonyl butoxide, and pyrethrins.

Single active ingredieht formulations consist of 0.5 and
1.0% Dusts, 2.5 and 5.0% Granulars, 20 and 76% Wettable Powders,
1% Wetttable Powders/Dusts, 0.03175, 0.38 and 0.45% Impregnated
Shelf Paper, 2.64 lb/gal Flowable Concentrates, 1.67 and 2 lb/gal
Soluble Concentrates/Liquids, and 1.5 and 2 lb/gal Ready To Use,
Bendiocarb is generally applied by broadcast or directed
spray using ground equipment. Application of the 5% Granulars,
76% Wettable Powders, 2.64 lb/gal Flowable Concentrates, the
1.67 and 2 lb/gal Soluble Concentrates/Liquids, and the 1.5
and 2 lb/gal Ready To Use formulations is recommended for use
by Pest Control Operators only.

3. Science Findings

   Chemical Characteristics

Color: White (Crystalline)
Odor: Odorless to very slight odor
Melting point: 129-130 C
Bulk Density: 0.52 g/ml (loose), 0.62 g/ml (compacted),
Vapor Pressure: 5.0×10^{-6} mm Hg at 25°C
Stability: Chemically stable during 16 months storage at
           40°C in sealed containers. No thermal breakdown
           at temperatures of up to 100°C
Solubility: Very slightly soluble in water and kerosene,
             highly soluble in nonpolar organic solvents.

   Toxicology Characteristics

Acute oral (rats): 45-48 mg/kg (male) (Toxicity Category I)
                   34-40 mg/kg (female)
Acute dermal (rats): 566 mg/kg (Toxicity Category II)
Acute inhalation: Data not available
Primary eye irritation: Slight (Toxicity Category III)
Primary skin irritation: Mild (Toxicity Category III)
Skin sensitization: Not a skin sensitizer

Subchronic oral: No Observed Effect Level (NOEL) = 0.5 mg/kg/day
                  for cholinesterase inhibition
Subchronic dermal: NOEL = 50 mg/kg/day for 80W formulated product

Chronic effects: Rat: NOEL = 0.5 mg/kg/day; effects noted were
                  cholinesterase inhibition and increased lenti-
                  cular opacity
                  Dog: NOEL = 2.5 mg/kg/day; cholinesterase
                  inhibition was the only effect noted
Oncogenicity: No evidence of oncogenic effects
Teratogenicity: Not a teratogen
Reproductive effects: Does not exhibit reproductive effects
Mutagenicity: Data not available

Physiological and Biochemical Characteristics

Metabolism: Bendiocarb is rapidly absorbed from the intestine
               and excreted primarily in urine. Little excretion
               occurs via the feces. The principal metabolite
               is the hydrolysis product of bendiocarb (NC-7312).
               Bendiocarb was not stored in organ tissues.

Dermal absorption: 53-64% on occluded skin in humans
                     8-16% on non-occluded skin in humans

Environmental Characteristics

Data are generally not available to evaluate the environmental
fate of bendiocarb

Ecological Characteristics

Avian acute toxicity: 3.1 mg/kg (mallard duck)
                        19 mg/kg (bobwhite quail)
Avian subacute toxicity: 477 ppm in the diet (mallard duck)
                           1770 ppm in the diety (bobwhite quail)
Avian reproduction: Reproductive impairment may occur when
                      used on turf
Freshwater fish acute toxicity: 1.2-1.5 ppm (trout)
                                  0.47-1.67 ppm (bluegill sunfish)
Freshwater invertebrates: 29.2 ppm (Daphnia magna)
Estuarine and marine organisms: Data not available
Honey bees: Highly toxic

Tolerance Assessment

Bendiocarb is not currently used on agricultural commodities.
Therefore, no residue tolerances have been established for
food crops, processed foods, or animals and animal byproducts.

No numerical food or feed additive tolerances have been
established for residues of bendiocarb in food/feed resulting
from its use in food handling and animal feed handling establishments. However, the prescribed conditions for use of
bendiocarb in these sites are set out in 21 CFR 193.152 and
21 CFR 561.191.

No tolerances or exemptions from the requirement of tolerances
have been established for residues of bendiocarb in or on
agricultural commodities. Requests for tolerance are currently
pending for a number of crops.

No tolerances or exemptions from the requirement of tolerances
for bendiocarb residues in animal commodities exist.
Animal metabolism and residue data submitted in support of
proposed tolerances are currently under review; therefore,

no conclusions regarding the nature of the residue in animals
will be made at this time.

No Canadian or Mexican tolerance or Codex MRL (Maximum Residue
Levels) related to uses of bendiocarb in food handling establishments have been established.

4. Summary of Regulatory Positions and Rationales

   --Bendiocarb will not be put into Special Review at this
     time. Potential risks to avian and aquatic species must
     be evaluated based on actual residue monitoring and field
     studies.

   --Bendiocarb use on commercial turf is restricted to use
     by certified applicators due to avian and aquatic toxicity.

   --Reentry intervals are not being established for turf and
     ornamental uses of bendiocarb.

   --Endangered species labeling statements are required for
     the mosquito adulticide and turf uses

   --Food additive tolerances are required for processed food
     and feed resulting from application in food/feed processing
     areas, and for residues resulting in food from use in
     shelf paper.

5. Summary of Outstanding Data Requirements	Approximate Time Frame
Product Chemistry Data	6-15 months
Residue Chemistry Data	
171-4 - Residue analytical	48 months
171-4 - Storage stability data	48 months
171-4 - Magnitude of the Residue - Processed foods commercial food/feed handling estab.	12 months
Environmental Fate Studies	9-27 months
Reentry Data	
132-1 - Foliar Dissipation	27 months
Spray Drift Studies	12 months
Toxicology	
81-3 - Acute Inhalation - Rat	9 months
82-4 - 90-Day Inhalation - Rat	15 months
84-2 - Mutagenicity studies	9-12 months

Wildlife and Aquatic Organisms
  70-1 - Aquatic Residue Monitoring   48 months
  71-5 - Actual Field Testing for Mammals   48 months
       and Birds
  72-3 - Estuarine & Marine Organism Acute LC_{50} 12 months
  72-4 - Fish Early Life-Stage and
       Aquatic Invertebrate Life Cycle   15 months
  72-6 - Aquatic Organism Accumulation   12 months

Nontarget Insect
  141-2 - Honey bee - toxicity of residues
        on foliage   15 months

F. Contact Person

Dennis H. Edwards
Product Manager 12
Insecticide-Rodenticide Branch
Registration Division (TS-767C)
401 M St., S.W.
Washington, DC 20460

Tel: 703-557-2386

DISCLAIMER:

THE INFORMATION PRESENTED IN THIS CHEMICAL INFORMATION FACT
SHEET IS FOR INFORMATIONAL PURPOSES ONLY AND NOT TO BE USED
TO FULFILL DATA REQUIREMENTS FOR PESTICIDE REGISTRATION AND
REREGISTRATION.

BIFENTHRIN

Reason for Issuance: Amended Registration First Food Use
Date Issued: October 1988
Fact Sheet Number: 177

1. DESCRIPTION OF CHEMICAL

Generic Name: (2-methyl[1,1-biphenyl]-3-yl)-methyl-3-(2-chloro-3,3,3-
            trifluoro-1-propenyl)-2,2-dimethyl cyclopropanecarboxylate
Common Name: Bifenthrin
Other Proposed Names: Biphenate, biphenthrin
Trade Names: Brigade, Capture, Talstar
Code Number: FMC 54800
EPA Shaughnessy No.: 128825
Chemical Abstracts Service (CAS) Number: 82657-04-3
Year of Initial Registration: 1985
Pesticide Type: Insecticide/Miticide
Chemical Family: Pyrethroid
U.S. and Foreign Producers: FMC Corporation

2. USE PATTERNS AND FORMULATIONS

Application Sites: Greenhouse Ornamentals, Cotton
Types and Methods of Application: Foliar Spray
Application Rates: 0.004 to 0.02 lb ai per 10 gallons of spray
Types of Formulations: 90% technical, 10 WP, and 2EC
Limitations: RESTRICTED USE PESTICIDE. Toxic to fish and aquatic
   organisms. For retail sale to and use only by Certified Applicators,
   or persons under their direct supervision and only for the uses covered
   by the certified applicator's certification.

3. SCIENCE FINDINGS

Summary Science Statement: Bifenthrin, a synthetic pyrethroid
   insecticide/miticide, is very highly toxic to fish and aquatic
   invertebrates. It is highly toxic to mammals by the oral route,
   with rat oral LD_{50} values of 70.1 and 53.8 mg/kg for males and
   females, respectively. There was no evidence of any oncogenic
   effects in a 2-year dietary (0, 12, 50, 100 and 200 ppm) study

in rats.   However, a mouse feeding oncogenicity study (87 weeks for
males, 92 weeks for females) with dose levels of 0, 50, 200, 500, and
600 ppm showed a significantly elevated incidence of leiomyosarcoma
of the urinary bladder of males at 600 ppm.   On the basis of a statistically highly significant (p = 0.00053) dose-related trend of increased
tumor incidence (and the significantly increased incidence at 600 ppm)
of leiomyosarcoma of the urinary bladder in male mice, the Agency has
classified bifenthrin as a class C (possible human) oncogen.   A Q_1*
of 5.4×10^{-1} $(mg/kg/day)^{-1}$ in human equivalents has been calculated.
The "worst case" incremental dietary oncogenic risk associated with
complete treatment of the cotton crop and maximum allowable residues
is calculated to be 6.0×10^{-6}.   Actual risk would be less since no
more than 25 percent of the cotton crop is likely to be treated.   It
has not demonstrated any teratogenic effects at the highest levels
tested.

Chemical Characteristics:

Physical State: Solid
Color: Off-white to pale tan
Odor: Very weak, aromatic
Boiling Point: Not applicable
Melting Point: 68 to 70.6 degrees C
Vapor Pressure: 1.81×10^{-7} torr at 25 degrees C
Solubility: Water: < 0.1 ppb
Organic Solvents: Soluble in methylene chloride, chloroform, acetone,
                   ether, and toluene.
Dissociation Constant: Not applicable due to the extremely low water
                        solubility.
Octanol/Water Partition Coefficient: $> 1 \times 10^6$
pH: 5.4 to 6.0
Oxidizing or Reducing Action: Has not demonstrated potential for
                               acting as an oxidizing or reducing
                               agent under normal handling and use
                               conditions.
Storage Stability: Preliminary studies indicate excellent storage
                    stability.

Toxicology Characteristics:

Acute Oral LD_{50} - Rat: 53.8 mg/kg (females), 70.1 mg/kg (males)
   (Category III)
Acute Dermal LD_{50} - Rabbit: > 2000 mg/kg (Category III)
Teratology - Rat: Maternal toxicity NOEL = 1 mg/kg/day; Developmental
   toxicity NOEL = 1 mg/kg/day
Teratology - Rabbit: Maternal toxicity NOEL = 2.67 mg/kg/day;
   Developmental toxicity NOEL > 8 mg/kg/day
2-Generation - Rat: Maternal toxicity NOEL = 30 ppm
   Reproductive toxicity NOEL > 100 ppm
   Developmental toxicity NOEL > 100 ppm
90-Day Feeding - Rat: NOEL = 50 ppm (2.5 mg/kg/day)
13-Week Feeding - Dog: NOEL = 2.21 mg/kg/day

1-Year Feeding - Dog: NOEL = 0.75 mg/kg/day
2-Year Feeding/Oncogenicity Study - Rat:
   Systemic toxicity NOEL = 50 ppm (2.5 mg/kg/day)
   Oncogenic toxicity LEL > 200 ppm (10 mg/kg/day)
87-Week Oncogenicity - Mouse:
   Oncogenic LEL = 50 ppm (7.5 mg/kg/day)
Mutagenicity:
   - Positive assay in mouse lymphoma, forward mutation
    (gene mutation)
   - Negative in other assays including CHO cells, Ames test, UDS up
    to 2.5 uL/mL, and in vivo rat bone marrow cells (categories of
    gene mutation, genotoxicity, and chromosome aberration)

Physiological and Biochemical Behavioral Characteristics:

   Foliar Absorption: Not absorbed
   Translocation: Not translocated
   Mechanism of Pesticide Action: Neurotoxicity characteristic of
    pyrethroid insecticides

Environmental Characteristics:

   Absorption and Leaching in Soil: Immobile in soils with high
    exchange capacity containing large amounts of organic matter, clay
    and silt. Low mobility in soils with low exchange capacity such as
    sandy soil low in organic matter.

   Hydrolysis: Bifenthrin is expected to be stable to hydrolysis.
    Additional data would be needed to more fully characterize its
    activity.

   Aerobic Soil Metabolism: Bifenthrin will aerobically degrade in soil
    with a half-life of 3 to 8 months. Additional information would be
    needed to more fully characterize its activity.

   Environmental Fate and Surface and Ground Water Contamination
    Concerns: No concerns at this time.

   Exposure of Humans and Nontarget Organisms to Chemical or Degradates:
    Applicator exposure in greenhouse and field use (cotton) and, if
    improperly used, exposure to fish and invertebrate animals.

   Exposure During Reentry Operations: No special precautions needed in
    greenhouses once spray residues are dry; however, for cotton use
    the following precautions are required:

        Do not apply this product in such a manner as to
        directly or through drift expose workers or other
        persons. The area being treated must be vacated
        by unprotected persons. Do not enter treated
        areas without protective clothing until sprays
        have dried. Because certain States may require

more restrictive reentry intervals for crops
treated with this product, consult your State
Department of Agriculture for further information.

Ecological Characteristics:

Hazards to Fish and Wildlife - Birds: Mallard duck LD_{50} < 2150 mg/kg
Bobwhite quail LD_{50} 1800 mg/kg; low acute toxicity to birds but
concern about possible bioaccumulation.
Fish - Bluegill sunfish: 96-hr LD_{50} 0.18 parts per billion (ppb);
rainbow trout 96-hr LD_{50} 0.10 ppb; very high toxicity to fish plus
concern with persistence in water and on soil and organic material
found in water.
Aquatic Invertebrates - Daphnia magna: 48-hr EC_{50} 1.6 ppb; very high
toxicity to freshwater invertebrate animals. Mysidopsis bahia:
96-hr LC_{50} = 3.97 parts per trillion; very high toxicity to estuarine
invertebrate animals. Concern about persistence in water. Additional
data are needed to fully characterize bifenthrin toxicity to aquatic
organisms.

Tolerance Assessment:

Commodities	Parts Per Million (ppm)
Cottonseed	0.5
Meat, fat, and meat byproducts of cattle, goats, hogs, horses, and sheep	0.1
Milk	0.02

The acceptable daily intake, based on a NOEL of 1.5 mg/kg/day from a
1-year dog feeding study and a safety factor of 100 is 0.015 mg/kg
body weight/day. The theoretical maximum residue contribution from
the above tolerances is 0.000445 mg/kg body weight/day.

Reported Pesticide Incidents: None

4. SUMMARY OF REGULATORY POSITION AND RATIONALE

Bifenthrin, formulated as a wettable powder and an emulsifiable
concentrate, is registered for use on greenhouse ornamentals and
cotton. Because of toxicity to fish and aquatic invertebrates,
precautionary labeling is required to warn against contamination of
bodies of water, endangered species restrictions, special equipment
for closed loading systems, and protective clothing for applicators
and mixers/loaders when applying products containing bifenthrin.

5.   SUMMARY OF DATA GAPS

Name of Study	GRN	Due Date
Estuarine Mollusc Acute Test	72-3	Received, under review
Fish Life Cycle	72-5	August 1988
Simulated or Actual Aquatic Field Test	72-7	January 1989
Volatility (Lab)	163-2	August 1989
Aquatic Invertebrate Life Cycle (Fresh Water)	72-4	October 1989
Aquatic Invertebrate Life Cycle (Estuarine)	72-4	October 1989
Soil Dissipation	164-1	November 1990
Confined Crop Rotation	165-1	October 1991

6.   CONTACT PERSON AT EPA

George T. LaRocca
Product Manager (15)
Insecticide-Rodenticide Branch
Registration Division (TS-767C)
Office of Pesticide Programs
Environmental Protection Agency
401 M Street SW.
Washington, D.C.   20460
Office location and telephone number:
Room 211, Crystal Mall #2; (703) 557-4421
1921 Jefferson Davis Highway
Arlington, VA   22202

DISCLAIMER: The information presented in this Fact Sheet is for informational purposes only and may not be used to fulfill data requirements for pesticide registration and reregistration.

BROMINE CHLORIDE

Reason for Issuance: New Chemical
Date Issued: October 6, 1987
Fact Sheet Number: 146

1. Description of Chemical

   Generic and Common Name: bromine chloride

   Trade Names: Bromine Chloride, Bromine Chloride Disinfectant

   EPA/OPP Pesticide Chemical Code: 020504

   Chemical Abstracts Service (CAS) Number: 13863-41-7

   Year of Initial Registration: 1987

   Pesticide Type: Disinfectant, Algicide, Bactericide, Slimicide

   U.S. Producers: Ethyl Corporation, Great Lakes Chemical Corp.

2. Use Patterns and Formulations

   Application sites: Manufacturing Use Product for reformulation or repackaging
   or for use as a disinfectant, algicide, slimicide, bactericide for wastewater,
   commercial and industrial recirculating cooling water systems and industrial
   once through cooling water systems.

   Type of formulation: liquid under pressure

   Methods of application: injected via closed system into wastewater or
   cooling water.

   Application rates: up to 1.7 lb per 10,000 gallons of wastewater or recirculating cooling water; up to 0.85 lb per 10,000 gallons of once through
   cooling water.

   Usual carriers: none

3. Science Findings

   Summary Science Statement: Based on the reviews of the data submitted,
   when used in accordance with widespread and commonly recognized practice,
   bromine chloride will not generally cause unreasonable adverse effects on
   the environment.

83

Chemical Characteristics:

Physical state: liquid under pressure

Odor: halogen (bromine/chlorine)

Boiling Point: 5° C

Melting Point: -66° C

Flashpoint: Non-flammable, but may ignite combustibles

Unusual handling characteristics: stored under pressure in 150 lb. and larger containers. In case of spills or leaks, use a self-contained breathing apparatus at or above 5 ppm chlorine and a full face, canister or cartridge respirator approved by NIOSH for chlorine below 5 ppm. Use lime slurry or dry soda ash to neutralize spills. Vapor releases should be controlled with water fog. Cylinders of anhydrous ammonia should be available for decontamination of bromine choride fumes.

Toxicological Characteristics:

Acute toxicity studies have been waived because bromine chloride breaks down into bromine gas and chlorine gas upon exposure to air. Both bromine gas and chloride are Class I poisons and the toxicity of bromine chloride is assumed to be the same as for bromine and chlorine gas. Information which EPA already has for chlorine is as follows:

Acute oral toxicity: N/A since it is a gas at room temperature.

Acute inhalation toxicity: Toxicity Category I. Inhalation LC_{50} (rat) = 293 ppm for 1 hr. Inhalation LD_{LO} (human) = 430 ppm for 1 hr. Inhalation TC_{LO} (human) = 15 ppm for pulmonary problems.

Primary skin and eye irritation: Toxicity Category I. Capable of causing severe burns.

Major routes of exposure: inhalation and skin and eye contact.

Problems which are known to have occurred with this chemical: none for bromine chloride, although incidents/accidents have occurred with chlorine gas.

Chronic, subchronic, mutagenic and teratogenic data are not required for the registration of bromine chloride for manufacturing use and for the industrial uses described above since the product is applied through a closed system.

Physiological/Biochemical Behavioral Characteristics: N/A
Environmental Characteristics: Upon application to water, bromine chloride immediately dissociates into hypobromous acid (HOBr) and hypochlorous acid (HOCl).

Ecological Characteristics:

EPA has reviewed approximately ten studies which compare the toxicity of
HOBr to HOCl and concluded that HOCl appears to be more toxic than HOBr.
Since there are currently registered products which produce HOCl (e.g.,
chlorine and sodium hypochlorite), products which produce HOBr are there-
fore considered to be as toxic or less toxic than those which produce
HOCl.

Acute avian, fish, aquatic invertebrate, estuarine and marine organism
toxicity studies for technical grade bromine chloride have been waived
because organisms are not exposed to this chemical. However, the following studies are required to be conducted on HOBr, the pesticidal chemical
to which organisms are exposed, as a condition of registration:

 a. 96 hour LC_{50} for one coldwater species (rainbow trout) and one
 warmwater species (bluegill sunfish).

 b. Acute 48 hour LC_{50} for an aquatic invertebrate (Daphnia magna).

 c. Residue monitoring study for freshwater and estuarine/marine use
 patterns, using once-through cooling systems.

 d. 96 hour LC_{50} study for sheepshead minnow.

 e. 96 hour LC_{50} study for shrimp.

 f. 96 hour embryo-larvae or the 48 hour shell deposition study for
 American oysters.

Efficacy review results: Although EPA does not require efficacy data for
non-health related label claims, EPA reviewed several available studies
which demonstrate that bromine chloride is efficacious for the uses
described above.

Tolerance Assessment: N/A

4. Summary of Regulatory Position and Rationale

The applications for registration of this chemical meet the following criteria
for conditional registration set forth in 51 FR 7628 dated March 5, 1986:

 a. "Since the data requirement was imposed, there has been insufficient
 time for the data to have been generated." EPA reached a final decision
 concerning the data requirements for this chemical in August 1987, which
 did not give the registrants adequate time to generate the data prior to
 submitting an application. Normally, studies are required to be conducted
 on the technical grade of the active ingredient. However, in this case,
 EPA decided to require testing only on a breakdown product (hypobromous
 acid), which the registrants could not have foreseen.

b. "During the period of the conditional registration, use of the pesticide will not cause unreasonable adverse effects." EPA has concluded that the toxicity to aquatic organisms is the same or less than the toxicity of currently registered pesticides for use in industrial cooling water. Therefore, this criterion has been met.

c. "Use of the pesticide is in the public interest." EPA has reviewed studies which demonstrate that use of bromine chloride will be cost effective, that the product is efficacious and that the current NPDES discharge limit of 0.2 mg/l of total residual oxidant will be met with bromine chloride. In addition, the risk associated with bromine chloride will be no greater than and possibly less than the risk associated with currently registered products containing chlorine. Therefore, it is in the public interest to conditionally register bromine chloride.

As long as the data which are required are submitted within the time frames specified below, products containing bromine chloride will be conditionally registered. If data are not submitted on time, these products will be subject to cancellation.

5. Summary of Major Data Gaps	Deadline for Submission
96 hr. LC_{50} for rainbow trout or bluegill	6/30/88
48 hr. LC_{50} for aquatic invertebrate	6/30/88
96 hr. LC_{50} for sheepshead minnow and shrimp	6/30/88
96 hr. embryo larvae or 48 hr. shell deposition for American oyster	6/30/88
Residue monitoring for freshwater and estuarine environments during once through cooling water use	3/30/89

6. Contact Person at EPA

Jeff Kempter, PM 32
Disinfectants Branch
Registration Division (TS-767C)
401 M St., S.W.
Washington, D.C. 20460
Telephone: 703-557-3964

DISCLAIMER: The information presented in this Chemical Information Fact Sheet is for information purposes only and may not be used to fulfill data requirements for pesticide registration and reregistration.

CADMIUM SALTS

Reason for Issuance: Final Determination of Special Review
Date Issued: August 5, 1987
Fact Sheet Number: 103.1

1. Description of chemicals

   Chemical names: cadmium carbonate
                   cadmium chloride
                   cadmium sebacate
                   cadmium succinate
                   anilinocadmium dilactate

   Common names: Same as above

   Trade names: None

   EPA Shaughnessy codes: 012901, 012902, 012903,
                          012904, 064601

   Chemical abstracts service (CAS) numbers: 134A, 135, 136A,
                                             136B, 051D

   Years of Initial Registration: 1949-1952

   Pesticide type: fungicides

   Chemical family: cadmium salts

   U.S. producers: W. A. Cleary and Mallinckrodt

2. Use Patterns and Formulations

   Application sites: Golf course and home lawn turf.

   Types of formulations: Wettable powders, dusts, and granulars.

   Types and methods of application: Ground application by
                                     hand-held sprayers and boom
                                     sprayers.

3.   Science Findings

     Physical and Chemical Characteristics-

          Physical state: solid
          Boiling point: 765° C
          Melting point: 321° C

     Human Toxicology Characteristics-

          °Acute toxicity:

               Moderate to moderately high (Toxicity Categories III
               and II); specific values are unavailable for each
               compound since there are no technical registrations
               and there are data gaps on formulated products.

               Acute effects to kidneys are formation of fatty bodies
               in the kidneys and degeneration of renal tubules.

          °Chronic toxicity:

               Oncogenic as demonstrated in laboratory animal and
               human epidemiological studies:

                    Rat chronic inhalation study -- LOEL 12.5 ug
                         Cd chloride/m^3 (lowest dose tested) for lung
                         tumors.

                    Rat chronic injection study -- 3.6% Cd chloride
                         (lowest concentration tested) caused testicular
                         and pancreatic islet tumors.

                    Epidemiological studies of factory workers --
                         chronic exposure to cadmium oxide and dust
                         has shown statistically significant increases
                         in the incidence of lung tumors.

               Kidney effects of proteinuria, glucosuria, excretion
               of amino acids and decreased renal function:

                    Rat drinking water study (24 wks) -- NOEL 10 mg/L
                         (lowest dose tested) for proteinuria.

                    Epidemiological study of factory workers exposure to
                         cadmium oxide dust (50 yrs) -- LOEL 2 ug/m^3 for
                         renal tubular proteinuria.

Mutagenic effects from 36 studies on various cadmium
compounds are equivocal; depending on protocol and
end point examined, results vary.

Teratogenic, fetotoxic and reproductive effects
have been shown in laboratory animal studies
however, the data are inadequate to support that
cadmium would produce these types of effects in
humans. Further, the data suggest that these
effects are dependent on routes of administration
which may not be analagous to human exposures
from the pesticidal use.

4.   Summary of regulatory position and rationale

   On October 10, 1986, a FEDERAL REGISTER Notice (51 FR 36524)
was published concerning the Special Review of all pesticide
products containing cadmium compounds registered for use on golf
course and home lawn turf. The Notice announced EPA's Preliminary
Determination to cancel all registrations and deny applications
of all pesticide products that contain cadmium compounds (salts
of chloride, sebacate, succinate, carbonate and anilino cadmium
dilactate) as active ingredients that are registered for use on
turf of golf courses and home lawns based on oncogenic and kidney
risks. The Support Document included an assessment of the hazard
to applicators, the benefits of use on golf courses and home
lawns, and a discussion of measures to reduce exposure to applicators.

   In evaluating the hazards to applicators, the Agency concluded that there is up to a 10^{-4} lifetime risk of oncogenicity
to golf course applicators applying cadmium to greens and tee
areas (with hand-held sprayers) and low kidney Effect Ratios
(0.1 to 81) for both golf course and home lawn applicators.
Based on available data, the Agency determined that some of
the alternatives do not appear to pose a greater health hazard
than cadmium.

   The Agency also reviewed the benefits of cadmium on golf
courses and home lawns. Estimates indicate that approximately
30,000 lbs of cadmium are used as turf fungicides annually.
Nearly all of the use is on golf courses. The impact of cancellation would result in a cost increase of approximately $500
annually for each of the affected golf courses and a negligible
cost increase to homeowners.

   In weighing the risks and benefits, the Agency reviewed
a number of options to reduce the risk to applicators. It
considered requiring protective clothing (i.e., elbow length

gloves and half-face respirators) and prohibiting the use
of hand-held sprayers as alternatives to cancellation. It
concluded that for golf course applicators, elbow length
gloves would not sufficiently reduce kidney risks. Requiring
half-face respirators could reduce oncogenic risks one order
of magnitude but the Agency does not believe that this equipment would be rigorously utilized. Prohibiting the use of
hand-held spray equipment could reduce both oncogenic and kidney
risks. For home lawn applicators, kidney risks would not be
sufficiently reduced by requiring elbow length gloves, as
the Agency believes this equipment would not be rigorously
utilized for the home lawn use pattern. Prohibiting the use
of hand-held spray equipment would be tantamount to cancellation
due to a lack of suitable alternative application equipment.

   In consideration of the toxicological effects of cadmium
compounds, the estimated potential risks of these effects to
applicators, the lack of effective measures to mitigate these
unacceptable risks, the availability of effective alternatives
and an estimated minor economic impact to users, the Agency
concluded that the risks of continued use of cadmium pesticide
products outweighed the benefits. Therefore, the Agency proposed
cancellation of all pesticidal uses of cadmium compounds.

   During the comment period that followed publication of the
Preliminary Determination, the Agency received three comments
and additional use information. All comments have been considered and are addressed in the Final Determination. The
new use information indicated that most golf course applicators
use power spray equipment (i.e., mini-boom sprayers and walking
boom sprayers) rather than hand-held sprayers to apply cadmium to
greens and tee areas. The Agency reviewed its existing data base
and determined that it lacked adequate data on this application
method. In order to assess the risk of oncogenic and kidney
effects from the use of this equipment, the Agency has required
applicator exposure data through a Data Call In Notice. These
data are required to be submitted to the Agency by July 1988
at which time the Agency will reassess the applicator risks
from the use of power spray equipment and determine if further
regulatory action is necessary.

   The weight of evidence continues to lead the Agency to
conclude that the oncogenic and kidney risks associated with the
use of cadmium on golf course greens and tee areas (with hand-held
sprayers) and the kidney risks associated with the use of cadmium
on golf course fairways and home lawns outweigh the low benefits
and that cancellation of these uses is the only appropriate
action. As the Agency cannot currently assess the risks to golf
course applicators applying cadmium to greens and tee areas by
mini-boom sprayers, it will allow continued use of cadmium on
golf course greens and tee areas with modifications in the terms
and conditions of registration such as "Restricted Use," application
to greens and tee areas only, application by mini-boom sprayers
only, and protective clothing (chemical resistant gloves, long
sleeve shirts and long legged pants during application and

chemical resistant aprons during mixing and loading) requirements
until applicator exposure data is received and reviewed.

5. Contact person at EPA: Valerie Meredith Bael
                             Environmental Protection Agency
                             Office of Pesticide Programs
                             Registration Division (TS-767C)
                             401 M Street, S.W.
                             Washington, D.C. 20460

CARBOFURAN

Reason for Issuance: Special Review—Preliminary Determination
Date Issued: January 5, 1989
Fact Sheet Number: 197

1. DESCRIPTION OF CHEMICAL

Common Name: Carbofuran

Chemical Name: 2,3-dihydro-2,2-dimethyl-7-benzofuranyl
methylcarbamate

Chemical Family: Carbamate

Trade Name: Furadan®

EPA Shaughnessy Code: 090601

Chemical Abstracts Service (CAS) Number: 1563-66-2

Year of Initial Registration: 1969

Pesticide Type: Insecticide, nematicide

U.S. Producer: FMC Corporation

2. USE PATTERNS AND FORMULATIONS

Carbofuran is currently registered on a variety of fruit
and field crops, vegetables, tobacco, ornamentals, and
forest tree seedlings. Approximately 7 to 10 million
pounds of active ingredient (lb ai) are applied to these
sites per year. From 6 to 9 million lb ai of the annual
usage is accounted for by the granular formulation. The
carbofuran granular formulation was placed in Special
Review in 1985 based on the avian hazard.

Types and Methods of Application: Aerial and ground.

Application Rates: 0.1 to 19.9 lb ai/acre (granular)
0.05 to 10.1 lb ai/acre (flowable)

Types of Formulations: Granular, flowable, and wettable
powder formulations and a spike
product.

92

3.   SCIENCE FINDINGS

Chemical Characteristics:

Physically, technical carbofuran is a white crystalline
solid that has a melting point of 153 to 154 °C (pure material).
Carbofuran is stable under natural or acidic conditions and is
unstable under alkaline conditions.

Toxicological Characteristics:

The Agency evaluated information concerning the hazard
to humans from carbofuran and its major alternatives. Based
on the available data, carbofuran does not appear to pose
a chronic health hazard because it has not shown positive
oncogenic, teratogenic, or reproductive effects. The data
base is complete and is considered acceptable. The data
bases for carbofuran's alternatives do not suggest adverse
health effects however the data bases are not complete so a
full conclusion cannot be drawn. The Agency has required
that these data be submitted to complete the data bases.

Based on data on acute health effects, the acute oral
hazard of carbofuran is the same order of magnitude as
fonophos, phorate, and terbufos, but is less than aldicarb,
and greater than the other major alternatives.

Environmental Characteristics:

The Agency also evaluated the potential for ground
water contamination from carbofuran. The environmental
fate data indicate that carbofuran is highly mobile and
has a potential to leach. Simulation modeling supports
this hypothesis. The environmental fate data indicate
that under conditions of low pH and low temperature,
residues of carbofuran could persist after leaching into
ground water. Since these conditions are not widespread
in the United States, most leaching of carbofuran will
probably not result in significant concentrations at the
wellhead. Monitoring information for Long Island, New
York; Maryland; and Massachusetts show the highest and
most frequently found residues in ground water. Concentrations above 36 parts per billion, the draft lifetime Health
Advisory Level, will probably only occur in localized,
worst-case situations. The Agency will be requiring the
registrants to revise the product labels' ground water
advisory statement.

Ecological Effects:

To evaluate the avian hazard from the granular formulation, the Agency evaluated the risk to birds based on
(1) acute avian toxicity, (2) exposure, (3) field studies,
(4) bird kill incidents, and (5) population effects.

Based on laboratory data, the Agency concluded that
granular carbofuran is acutely toxic to birds, and that
a single granule may kill a small bird. Birds are expected
to be present at the time of carbofuran application.
Dietary exposure occurs from direct ingestion of granules
and exposure from ingestion of contaminated soil invertebrates
such as earthworms. Predatory birds may be secondarily
exposed to carbofuran by feeding on contaminated vertebrates
such as small birds.

There were 6 field studies conducted at 11 locations
that investigated the loss of birds from label-directed,
soil-incorporated uses of 10G and 15G applied as band
and in-furrow applications and 10G using specialized
equipment. All studies consistently resulted in bird
mortality, regardless of application rate or methods
which employed commonly practiced techniques for soil
incorporation of granules. Both direct and secondary
poisoning occurred.

Bird kill incidents from direct poisoning from carbofuran
granules have occurred in several crops in various areas
of the country and Canada. The types of birds varied and
included both migratory and nonmigratory birds. Bird
mortality was frequently associated with at-planting
application, but has occurred with other uses throughout
the year. Direct poisoning of birds has caused over 40
reported bird kill incidents.

Secondary poisoning incidents have also occurred and
involved bald eagles, red-tailed hawks, red-shouldered hawks,
northern harriers, and others.

The direct and secondary bird kill incidents that have
been reported underestimate the number of incidents actually
taking place because of the problems associated with the
reporting of bird kill incidents and with carcass removal
by predators.

Populations of declining or endangered species may
be present in areas where granular carbofuran is applied.
The Agency cited documented population declines of the
red-shouldered hawk, loggerhead shrike, field sparrow,
Henslow's sparrow, and others. Statistically significant
declines have been measured for several species.

While the Agency does not consider granular carbofuran to be the sole causative factor in the decline of the bird species discussed, carbofuran is one of the most highly toxic pesticides to which these birds are exposed. Given its widespread use in agriculture, carbofuran is likely to be responsible for bird deaths in these species. The Agency concluded that granular carbofuran can, therefore, be an important additive factor in the declines.

The Fish and Wildlife Service's Division of Endangered Species and Habitat Conservation (DESHC) indicated in its Biological Opinion for carbofuran that the Aplomado falcon, Attwater's greater prairie chicken, and Aleutian Canada goose were the bird species jeopardized by the use of carbofuran and indicated that the use be eliminated in certain areas. DESHC also indicated that the bald eagle, whooping crane, and Mississippi sandhill crane may be adversely affected. DESHC recommended prohibiting the use of carbofuran in certain areas to avoid impact on these species.

The Agency has examined other statutes that are intended to protect birds and that compliment FIFRA. The Migratory Bird Treaty Act prohibits the taking "by any means or in any manner" individual birds of migratory species that are listed in the Act's regulations. Birds of more than 20 such species have been reported killed by carbofuran. Likewise, the Bald and Golden Eagle Protection Act prohibits takings of the bald and golden eagles and the Endangered Species Act prohibits taking of threatened or endangered species. A number of bald eagles killed by carbofuran have been reported and the Fish and Wildlife Service has determined that carbofuran use threatens the continued existance of several endangered species.

The Agency has concluded that in general carbofuran poses the greatest risk to birds as compared with other granular pesticides, including its alternatives. This conclusion was based on estimations of the numbers of LD50s per square foot of treated ground according to labeled use rates and methods. The field studies and reported bird kill incidents for carbofuran confirm the Agency's conclusion that carbofuran poses a high risk. This approach for comparative risk analysis can be used by the Agency to identify other high risk pesticides for which regulatory action would be appropriate.

4. BENEFITS ANALYSIS

The Agency analyzed the benefits of carbofuran use on 10 sites. The percentage of granular carbofuran use on these sites is as follows: 68 percent for corn, 14 percent for sorghum, 5 percent for soybeans, 2 percent for rice, 5 percent for peanuts, and 2 percent for tobacco. Also, less than 1 percent is used on each of the following sites: cotton, cranberries, sunflowers, and pineseed orchards. These uses encompass over 95 percent of the granular carbofuran usage and about 85 percent of all carbofuran formulation usage.

If carbofuran is not available for treatment of the 10 sites, the Agency estimated an annual grower impact that ranged from approximately $22.8 to $33.0 million. The largest economic impact from cancellation of granular carbofuran will be for rice since no registered alternatives are available for control of the rice water weevil. The Agency estimates a grower impact to be $12.2 million annually; a $6.1 million decrease in Federal deficiency payments to rice growers would indicate a loss to society of $6.1 million.

Corn is the major use site for carbofuran, and cost-effective, efficacious alternatives are available. No changes in costs of production, yields, or revenues are expected. The corn insecticide market is highly competitive, and viable alternatives with similar pesticide performance are available at comparable cost per acre.

The carbofuran market for corn has been declining since 1978, and current usage is approximately one-third the level it was in 1978. By 1986, the market share held by carbofuran dropped to less than 15 percent where, in terms of acre treatments, it ranked fourth out of the five major corn insecticides. The reasons for the decline are not clear, but could include loss in efficacy, spectrum of control, and others.

Carbofuran is applied to nonflooded cranberries in Washington and Oregon to control the black vine weevil. Carbofuran is the only pesticide registered for black vine weevil larvae control. Acephate is an efficacious insecticide for control of the adults. The impact on cranberries, without considering acephate's use, is expected to occur over a 7-year period due to the perennial nature of the crop. Overall impacts could range from $7 million to $7.7 million over this period.

For the remaining crops, the Agency does not anti-cipate major impacts. The overall economic impact from cancellation is not expected to result in significant changes in either production costs or outputs.

The Agency also evaluated aspects of carbofuran use that are not easily quantifiable. For example, only one carbamate (trimethacarb) would be available for corn growers who rotate organophosphate and carbamate insecticides to delay development of resistance in soil pests, although the Agency recognizes that some cross-resistance with organophosphates could occur. Also, carbofuran has residual and systemic properties and a broad spectrum of control. However, repeated use of carbofuran may lead to an apparent increase in soil microbial populations that are capable of reducing its effectiveness.

5. SUMMARY OF REGULATORY POSITION AND RATIONALE

In weighing the risks and benefits, the Agency reviewed a number of options other than cancellation to reduce the risk to birds. Among these measures were (1) additional precautionary labeling regarding the hazard to birds, (2) limiting carbofuran use to certain months of the year, (3) limiting application geographically, and (4) implementing a risk reduction program. The Agency evaluated these measures and determined that they would not adequately mitigate the risk.

As a result, the Agency is proposing to cancel granular carbofuran use on all sites. The decision to cancel granular carbofuran use is based on the conclusion that the risk to birds outweighs the benefits of use. Because of the substantial risks and substantial benefits associated with the use of carbofuran on rice to control the rice water weevil, the Agency has requested specific additional information pertaining to the associated risks, benefits, usage, and additional means of control.

6. CONTACT PERSON

Jay Ellenberger
Special Review Branch
Special Review and Reregistration Division
Office of Pesticide Programs (TS-767C)
401 M Street, S.W.
Washington, D.C. 20460
(703) 557-7400

DISCLAIMER: The information presented in this Pesticide Fact Sheet is for informational purpose only and may not be used to fulfill data requirements for pesticide registration or reregistration.

CARBON TETRACHLORIDE (and others)

Reason for Issuance: Special Review Action Document Issued
Date Issued: August 1987
Fact Sheet Number: 102.1

1. DESCRIPTION OF CHEMICALS

Chemical names: Carbon tetrachloride - CCl_4
Carbon disulfide - CS_2
Ethylene dichloride - $C_2H_2Cl_2$
Chloroform - $CHCl_3$

Other names: Carbon disulfide - carbon bisulfide
Ethylene dichloride - EDC

Trade names: Carbon tetrachloride - Benzinoform, Carbona,
Dowfume 75, ENT 4705, Flukoide, Halon 104

EPA Shaughnessy Code:
Carbon disulfide	016401
Ethylene dichloride	042003
Chloroform	020701
Carbon tetrachloride	016501
CCl_4 with EDC	016502
CCl_4 with EDB and EDC	016503

Chemical Abstracts Service (CAS) number:
CS_2	-	75-15-0
EDC	-	107-06-2
$CHCl_3$	-	67-66-3
CCl_4	-	56-23-5

Pesticide Type: Fumigant

Year of Initial Registration: 1956

Chemical family: Chlorinated hydrocarbons

2. USE PATTERNS AND FORMULATIONS

   Application sites: Harvested grains throughout storage,
                        transfer, milling, distribution, and
                        processing phases.

   Type of formulations: Gas

   Types and methods of application: Fumigation

3. SUMMARY OF REGULATORY POSITION AND RATIONALE

   The Environmental Protection Agency (EPA) is proposing to
revoke the exemption from the requirement of a tolerance and
the food additive regulations for carbon tetrachloride, carbon
disulfide, ethylene dichloride, and chloroform. The tolerance
exemptions and food additive regulations are for the use of these
chemicals as grain fumigants.

   This action is being taken because all products containing
CCl_4, CD, EDC, or chloroform as an active ingredient for use as a
grain fumigant have been cancelled either voluntarily by the
registrants or by administrative action.

   The data presently available to the Agency indicate that
carbon disulfide, ethylene dichloride, and chloroform are not
persistent and there is no anticipation of significant residues
resulting from the last legal application of these chemicals, on or
before June 30, 1986. There is no need to establish action
levels to replace the existing tolerance exemptions upon their
revocation.

   These limited data also indicate the presence of CCl_4
residues in or on raw grain, intermediate grain products, and
ready-to-eat grain products. Although these data suggest that
residues of carbon tetrachloride are decreasing with time, it
would be difficult to utilize these data to establish tolerances
or action levels to replace the current tolerance exemption for
CCl_4 upon its revocation. It is doubtful that the presence of
low levels of residues of this pesticide for this short-term period
would pose a risk to the public health.

   EPA has discussed the appropriateness of action levels with
the Food and Drug Administration (FDA). They agree that action
levels should not be established and that FDA will not take enforcement action if residues are detected in grain or grain-based
consumer products after the exemptions are removed if it appears
such residues resulted from the legal use of the fumigants on or
before June 30, 1986. FDA has notified the Agency in writing
of this position.

4. CONTACT PERSON

    Mark T. Boodee,
    Special Review Branch,
    Registration Division,
    Office of Pesticide Programs (TS-767C),
    401 M Street, S.W.,
    Washington, D.C. 20460,
    (703) 557-7402.

    DISCLAIMER: The information presented in this Pesticide Fact
    Sheet is for informational purposes only and may not be used
    to fulfill data requirements for pesticide registration or
    reregistration.

CHLORIMURON ETHYL (CLASSIC)

Reason for Issuance: New Chemical Reg.
Date Issued: July 30, 1987
Fact Sheet Number: 82.1

1. Description of Chemical

   Generic Name: Ethyl 2-[[[[(4-chloro-6-methoxyprimidin-
                   2-yl)amino]carbonyl]amino]sulfonyl]benzoate

   Common Name: Chlorimuron ethyl
   Trade Name: DuPont Classic Herbicide; DPX-F6025
   EPA Shaughnessy Code: 128901
   Chemical Abstracts Service (CAS) Number: 90982-32-4
   Year of Initial Registration: 1986
   Pesticide Type: Herbicide
   Chemical Family: Sulfonylurea
   U.S. and Foreign Producers: E.I. du Pont de Nemours
                                & Company

2. Use Patterns and Formulations

   Application Site: Soybeans
   Types and Methods of Application: Postemergence and preemergen
                                      foliar by ground equipment.

   Application Rates: 1/2 to 1 1/2 ounces active ingredient
                       per acre (oz ai/A) preemergence.

                       1/8 to 3/16 (oz ai/A) postemergence.

   Types of Formulation: 25% dispersible granule
   Usual Carrier: Water

3. Science Findings

   Summary Science Statement:

       All data required for registration of this chemical
       are acceptable to the Agency.

       Chlorimuron ethyl has low acute toxicity (Category
       III for acute dermal and primary eye irritation and
       Category IV for all other forms of acute toxicity.)

101

It was not oncogenic to mice or rats, not teratogenic
to rabbits or rats, and not mutagenic. The pesticide
is foliarly absorbed and translocated within the plant.
It works by inhibition of cell division in shoots and
roots. The major degradation pathway is hydrolysis.

The pesticide will leach in some soils and has the to
potential contaminate ground water at very low concentrations. Chlorimuron ethyl demonstrates negligible
toxicity nontoxic to birds and slightly toxic to fish
and invertebrates. The nature of residues in plants
is adequately understood and adequate methodology is
available for enforcement of a tolerance of 0.05 part
per million (ppm) on soybeans.

Chemical Characteristics:

Physical State: Solid
Color: Off-white to pale yellow
Odor: None
Melting Point: 181 °C
Density: 1.51 gram/cubic centimeter (g/cc)
Solubility in various organic solvents at 25 °C

	g/100 mL
Acetone	7.05
Acetonitrile	3.10
Benzene	0.815
Ethyl acetate	2.36
Ethyl alcohol	0.392
n-hexane	0.006
Methyl alcohol	0.740
Methylene chloride	15.3
Xylenes	0.283

Solubility in Water at Controlled pH:

pH	Solubility grams/100cc
1.3	.015
1.9	.015
2.5	.015
4.2	.041
5.0	.090
5.8	.990
6.5	4.500
7.0	12.000

Handling Characteristics: Store product in original container only.

Toxicology Characteristics:

Acute Toxicity:

Acute Oral Toxicity (Rat):   Greater than (>) 5000 milligrams/
                             kilogram (mg/kg)
                             Toxicity Category IV[1]/

Acute Dermal Toxicity (Rabbits):   > 2000 mg/kg
                                   Toxicity Category III[1]/

Primary Eye Irritation (Rabbit):   Draize score = 1
                                   Toxicity Category III[1]/

Primary Dermal Irritation (Rabbit):   Primary irritation
        score from .13 to .63   Toxicity Category IV[1]/

Primary Skin Irritation (Guinea Pig):   Not irritating
                                        Toxicity Category IV[1]/

Dermal Sensitization (Guinea Pig):   Not a sensitizer
                                     Toxicity Category IV[1]/

----

[1]/ Labeling statements required for: Toxicity Category
     III for acute dermal – "Harmful if absorbed
     through skin. Avoid contact with skin, eyes,
     or clothing. Wash thoroughly with soap and
     water after handling."
     Toxicity Category III for primary eye irritation –
     "Causes (moderate) eye injury (irritation). Avoid
     contact with eyes or clothing. Wash thoroughly
     with soap and water after handling."
     Toxicity Category IV – no precautionary statements
     required.

Major Routes of Exposure:

The major routes of exposure are via dermal and eye contact.

Chronic Toxicity:

90-Day Feeding Study (Mouse) resulted in a no observable-effect
level (NOEL) of 18.75 mg/kg/day.

90-Day Feeding Study (Dog) resulted in a NOEL of
2.5 mg/kg/day.

One-Year Feeding Study (Dog) resulted in a NOEL of
6.25 mg/kg/day.

18-Month Chronic Feeding/Oncogenicity Study (Mouse) resulted in a NOEL of 3.75 mg/kg/day and no oncogenic effects at 187.5 mg/kg/day for the highest dose tested (HDT).

Chronic Feeding/Oncogenic Study (Rat) resulted in a NOEL of 12.5 mg/kg/day and no oncogenic effects at 125 mg/kg/day (HDT).

2-Generation Reproduction Study (Rat) resulted in a maternal toxicity NOEL of 12.5 mg/kg/day and a fetotoxic NOEL of 1.25 mg/kg/day.

Teratology Study (Rat) resulted in a maternal toxicity NOEL of 30 mg/kg/day, and a developmental toxicity NOEL of 30 mg/kg/day.

Teratology Study (Rabbit) resulted in a maternal toxicity of 60 mg/kg/day and a developmental toxicity NOEL of 15 mg/kg/day.

Mutagenicity (in vivo Bone Marrow Test) - not mutagenic.

Mutagenicity (Ames test) - not mutagenic.

Mutagenicity (Unscheduled DNA Synethsis Activity) - not mutagenic.

Physiological or Biochemical Behavioral Characteristics:

Foliar absorption: Rapid.

Translocation: Systemic after absorption through either the foliage or the roots.

Mechanism of pesticidal action: Inhibits plant cell division of rapidly growing tips of roots and shoots by inhibition of amino acid synthesis.

Metabolism in plants: Tolerant species metabolize the compound to nonherbicidal metabolites.

Persistence in plants: Does not persist in plants.

Environmental Characteristics:

Adsorption and leaching in basic soil types: Chlorimuron ethyl was very weakly absorbed on the two sandy loam soils and only weakly absorbed on the two silt loam soils. Absorbed radioactivity was readily desorbed from the sandy loam soils, but was more tightly retained on silt loams. Chloroimuron ethyl had low mobility on Keyport silt loam, intermediate mobility on Flanagan silt loam and Cecil sandy loam, and high mobility on Woodstown sandy loam.

Microbial breakdown: Initial deactivation of the molecule is through
hydrolysis followed by complete metabolism to low molecular
weight compounds through normal soil microbial degradation.

Loss from photodecomposition/volatilization:
Photodegradation of chlorimuron ethyl is not a major degradation
pathway, but did proceed at twice the rate in exposed samples
compared to the nonexposed samples.

Bioaccumulation:
The octanol/water partition coefficient (K_{ow}) of 1.3 and
available information show the hydrolysis products have a
lower K_{ow}. Since the correlation between octanol/water partitioning
and fish accumulation is only accurate within a factor of 100,
chlorimuron ethyl and its degradation products have potential
to accumulate in fish to levels 130 times higher than levels
in water.

Resultant average persistence:
Chlorimuron ethyl has a half-life of 7.5 weeks in soil.

Environmental fate and surface and ground water contamination
concerns:
Chlorimuron methyl has the potential to leach and contaminate
ground water at very low concentrations.

Exposure of humans and nontarget organisms to pesticide or
degradates:
Human risk from exposure is minimal because of low acute
toxicity (Category III and IV). Nontarget organism risk from
exposure is minimal because maximum expected residues on soil
and water do not approach the toxicity values for the organisms
tested.

Ecological Characteristics:

Avian Acute Toxicity (Mallard Duck):	> 2510 mg/kg
Avian Dietary Toxicity (Bobwhite Quail):	> 5620 ppm
Avian Dietary Toxicity (Mallard Duck):	> 5620 ppm
96-Hour Fish Toxicity (Rainbow Trout):	> 12 mg/L
96-Hour Fish Toxicity (Bluegill Sunfish):	> 10 mg/L
48-Hour Invertebrate Toxicity (Daphnia magna)	> 10 mg/L
48-Hour Acute Toxicity (Honey Bees):	12.5 ug/bee

Chlorimuron ethyl demonstrates negligible toxicity to birds on
both an acute and dietary basis. It is slightly toxic to fish
and invertebrates. It is not expected to adversely affect
endangered/threatened species because of low toxicity and low
application rate.

Tolerance Assessment:

The nature of the residue in plants is adequately understood. The residue of concern is chlorimuron ethyl. Adequate methodology (high pressure liquid chromatography [HPLC] using a photoconductivity detector) is available for enforcement.

Tolerances are established for residues of the herbicide chlorimuron ethyl (ethyl 2-[[[[(4-chloro-6-methoxypyrimidin-2-yl)amino]carbonyl]amino]sulfonyl] benzoate in or on soybeans at 0.05 ppm.

The acceptable daily intake (ADI) based on the 1-year dog feeding study (NOEL of 6.25 mg/kg/day) and using a hundredfold safety factor is calculated to be .0625 mg/kg/day. The maximum permissible intake (MPI) for a 60 kg human is calculated to be 3.70 mg/day. The theoretical maximum residue contribution (TMRC) for this tolerance for a 1.5 kg diet is calculated to be .000012 mg/day. The current action will use .0184 percent of the ADI. There are no other published tolerances for this chemical.

4. Summary of Regulatory Position and Rationale

a. None of the risk criteria listed in 40 CFR 154.7 have been exceeded for chlorimuron ethyl. The Agency will issue registrations for substantially similar products and will issue significant new uses.

Rationale: A review of available data indicate that no risk criteria have been exceeded or met for chlorimuron ethyl.

b. The Agency is not imposing a ground water advisory statement for chlorimuron ethyl.

Rationale: Even though chlorimuron ethyl has the potential to leach in some soils and contaminate ground water at very low levels, the available data do not indicate a concern for ground water contamination. This was determined by consideration of the low application rates (ounces per acre) and the very low levels expected to reach ground water.

c. The Agency is not requiring reentry data or protective clothing for chlorimuron ethyl.

Rationale: Human risk from exposure is minimal because of the low acute toxicity.

d. The Agency is not imposing endangered species labeling for the use on soybeans.

Rationale: Based on the low toxicity and low application rate, chlorimuron ethyl is not expected to adversely affect endangered animal species. Refer to sections on ecological characteristics and endangered species in Chapter III for detailed information on available data.

e. Additional residue data are not required to support the use on soybeans.

Rationale: The Agency has determined that available residue data adequately support the established tolerance on soybeans.

F. Additional data are being required on droplet size spectrum, drift field evaluation, seed germination/seedling emergence, vegetative vigor, and aquatic plant growth.

Rationale: The agency has determined that the potential for off target movement and carryover of this chemical are not adequately understood. This lack of definition, coupled with the low volume application of this herbicide, make these studies necessary.

5. Summary of Major Data Gaps

None.

6. Contact Person at EPA

Robert J. Taylor
Product Manager 25
Registration Division (TS-767C)
Office of Pesticide Programs
Environmental Protection Agency
401 M Street SW.
Washington, DC 20460
Telephone: (703) 557-1800

Disclaimer: The information presented in this Pesticide Fact Sheet is for informational purposes only, and may not be used to fulfill data requirements for pesticide registration and reregistration.

CHLORINATED ISOCYANURATES

Reason for Issuance: Registration Standard
Date Issued: March 8, 1988
Fact Sheet Number: 165

1. Description of Chemicals

    a. Chemical Name: 1,3-Dichloro-s-triazine-2,4,6 (1H,3H,5H) trione.

        Common Names: Dichloro-s-triazinetrione, dichloroisocyanuric acid.

        Trade Name: ACL 70.

        EPA/OPP Pesticide Chemical Code: 081401.

        CAS Registry No: 2782-57-2.

        Empirical Formula: $C_3HN_3O_3Cl_2$.

    b. Chemical Name: 1,3-Dichloro-s-triazine-2,4,6 (1H,3H,5H) trione potassium salt.

        Common Names: Potassium dichloro-s-triazinetrione; Potassium dichloroisocyanurate.

        Trade Names: ACL 59; P.D.I.C.

        EPA/OPP Pesticide Chemical Code: 081403.

        CAS Registry No: 2244-21-5.

        Empirical Formula: $C_3N_3O_3Cl_2K$.

c. Chemical Name: 1,3-Dichloro-s-triazine-2,4,6 (1H,3H,5H) trione
    sodium salt.

    Common Names: Sodium dichloro-s-triazinetrione; Sodium
    dichloroisocyanurate.

    Trade Names: ACL 60; CDB 60; CDB 63; S.D.I.C.

    EPA/OPP Pesticide Chemical Code: 081404.

    CAS Registry No: 2893-78-9.

    Empirical Formula: $C_3N_3O_3Cl_2Na$.

d. Chemical Name: 1,3,5-trichloro-s-triazine-2,4,6 (1H,3H,5H)
    trione.

    Common Names: Trichloro-s-triazinetrione; trichloroisocyanuric
    acid; trichloroisocyanurate.

    Trade Names: ACL 90 Plus; CDB 90; T.I.C.A.; TCCA.

    EPA/OPP Pesticide Chemical Code: 081405.

    CAS Registry No: 87-90-1.

    Empirical Formula: $C_3N_3O_3Cl_3$.

e. Chemical Name: [Mono-(1,3,5-trichloro)tetra
    (1-potassium 3,5-dichloro)] penta-s-triazinetrione.

    Common Name: Penta-s-triazinetrione.

    Trade Names: ACL 66; DS.

    EPA/OPP Pesticide Chemical Code: 081406.

    CAS Registry No: 30622-37-8.

    Empirical Formula: $(C_3N_3O_3Cl_2K)_4 \cdot C_3N_3O_3CL_3$.

f. Chemical Name: 1,3-Dichloro-s-triazine-2,4,6 (1H,3H,5H) trione
    sodium salt dihydrate.

    Common Names: Sodium dichoro-s-triazinetrione dihydrate; sodium
    dichloroisocyanurate dihydrate.

    Trade Names: ACL 56; CDB Clearon; DICD.

    EPA/OPP Pesticide Chemical Code: 081407.

    CAS Registry No: 51580-86-0.

    Empirical Formula: $C_3N_3O_3Cl_2N_2 \cdot 2H_2O$.

g. For All Six Chemicals:

Year of Initial Registration: 1958

Pesticide Type: Disinfectant; sanitizer; algaecide; fungicide

Chemical Family: Halogenated triazines

U.S. and Foreign Producers: Olin Corporation; Monsanto Company; Fallek Chemical Corporation; ICI America, Inc.; ICD Chemicals, Inc.; Nissan Chemical Ind., Ltd.; Shikoku Chemicals Company; 3-V Chemical Corporation.

2.   Use Patterns and Formulations

Application Sites: Major use is swimming pools. Minor uses include industrial water cooling systems; oil recovery systems; sewage systems; industrial preservatives; food processing/service industries; laundry sanitizer uses; household areas; mold and mildew; egg sanitizing and poultry drinking water sanitization.

Types of Formulations: Solids; aqueous dilutions.

Types and Methods of Application: Added as solids to swimming pools or industrial cooling water systems; aqueous solutions applied by mopping, scrubbing, flooding, sponging, spraying and immersing to food and non-food contact surfaces.

Application Rates: Sanitizer/disinfectant rates range from 47-4784 ppm available chlorine depending on the degree and circumstance of use, i.e., sanitizing, disinfecting, mold and mildew control; precleaned, non-precleaned, porous, non-porous, food contact surfaces. Swimming pools require a final residual level of 1-3 ppm; start up or super chlorination 21.3 ppm; periodic maintenance of 0.2-4.7 ppm. Industrial water cooling systems require an initial dose of 2-600 ppm and subsequence doses ranging from 0.5-107 ppm.

Usual Carriers: Water

3.   Science Findings

There are no significant human toxicological or environmental fate data gaps for the chlorinated isocyanurates. Uncertainty exists in the areas of chronic effects to aquatic organisms.

Chemical Characteristics:

Color: White.

Physical State: Crystalline Solid.

Odor: Slight odor of chlorine.

Melting Point: 225-250° C.

Bulk Density: Powder, 0.50-0.65 g/ml; regular, 0.82-1.0 g/ml; granular, 0.85-0.96 g/ml; extra granular, 0.92-0.95 g/ml.

Solubility: Dichloro-s-triazinetrione, 0.8 g/100 g water at 25°C; potassium dichloro-s-triazinetrione, 9 g/100 g water at 25°C; sodium dichloro-striazinetrione, 24.8 g/100 g water at 26.8°C; trichloro-s-triazinetrione, 1.2 g/100 g water at 25°C; penta-s-triazinetrione, 2% in water at 30°C; sodium dichloro-s-triazinetrione dihydrate, 26.2 g/100 ml water at 25°C.

Vapor Pressure: Very small, impossible to measure.

Stability: Stable and relatively inert when dry.

Unusual Handling Characteristics: Strong oxidizing agents which should be kept away from heat and flames.

Toxicology Characteristics:

Acute Oral Toxicity: Toxicity category III or IV

Acute Inhalation Toxicity: Vaporization and respirable dust are not expected from the registered patterns of use.

Primary Eye Irritation: Toxicity category I

Primary Skin Irritation: Toxicity category III or IV

Problems which are known to have occurred with the use of the chlorinated isocyanurates, i.e., Pesticide Information Monitoring Service: None

Subchronic and Chronic Toxicity:

Isocyanurate related effects observed in the subchronic studies included hyperplasia of the urinary blader and calculi in male mice and hyperplasia of the lining of the urinary bladder in male rats. In the chronic study in rats, toxicity secondary to formation of calculi in the kidney and urinary bladder was observed in the male rats during the first 12 months of dosing at 5375 ppm. No effects were observed at 2400 ppm. No oncogenic effects were observed in the lifetime studies in mice and rats. The chlorinated isoyanurates do not induce reproductive effects in rats or teratogenic effects in rats or rabbits; they are not mutagenic. The chlorinated isocyanurates have low oral and dermal toxicity and high primary eye irritation toxicity.

Physiological and Biochemical Behavioral Characteristics: N/A

Environmental Characteristics:

Based on the available data, there are no apparant human exposure, groundwater, reentry, or drift issues associated with the registered patterns of use.

Ecological Characteristics:

The chlorinated isocyanurates are slightly toxic to birds, but are acutely toxic to fish and invertebrates. The chronic effects from these chemicals to aquatic organisms are not fully known, however.

Based on the registered patterns of use, no exposure to nontarget insect species or endangered species is expected.

Tolerance Assessment:

Uses of the chlorinated isocyanurates in food/feed handling establishments as sanitizing solutions on food processing equipment and utensils, and on other food contact articles are regulated by the Food and Drug Administration according to 21 CFR 178.1010(b)(2) and (c)(2).

The provisions of the Federal Food, Drug, and Cosmetic Act (Section 402) require the establishment of or an exemption from a tolerance for the use of chlorinated isocyanurate formulations as poultry drinking water sanitizers.

Agricultural uses of the chlorinated isocyanurates for the commercial treatment of eggs for human consumption are regulated under the United States Department of Agriculture (USDA) egg grading and egg products inspection programs. USDA requires that to be accepted under these programs, all formulations must be in compliance with 21 CFR 178.1010 as sanitizing solutions for food contact surfaces.

4.   Summary of Regulatory Position and Rationale

The chlorinated isocyanurates are not being placed into Special Review. Available data indicate that the chlorinated isocyanurates do not pose a risk of serious acute injury to humans, avian species and aquatic organisms. Except for the mechanical damage resulting from calculi in the urinary bladder of male mice and rats dosed with saturated solutions of sodium isocyanurate in the drinking water, these chemicals produce no chronic toxicity effects.

The Agency has determined that the present precautionary statements for persons handling or applying the chlorinated isocyanurates are sufficient for the labels for manufacturing-use products. Currently registered manufacturing-use labels contain the signal word "Danger" and the associated precautionary labeling statements required for products with Category I acute eye irritation effects.

The Agency will not require nontarget insect studies because no honey bee exposure is expected from the registered patterns of use.

The Agency will not impose a special label advisory statement for endangered species at this time because the chlorine released from the chlorinated isocyanurates is so volatile there is little liklihood of exposure based on the registered patterns of use or from accidental spills.

Because the Agency has data in its files which adequately demonstrates the efficacy of hypochlorous acid, no additional microbiological efficacy data will be required for manufacturing-use products which are to be reformulated into antimicrobials.

The Agency is requiring the establishment of or exemption from a tolerance for the use of chlorinated isocyanurate formulations as poultry drinking water sanitizers and as egg sanitizers.

While data gaps are being filled, currently registered manufacturing-use products containing chlorinated isocyanurate may be sold, distributed, formulated, and used, subject to the requirements of the chlorinated isocyanurate registration standard. The Agency does not normally cancel or withhold registration for previously registered use patterns simply because data are missing or inadequate. Data required under that standard will be reviewed and evaluated, after which the Agency will determine if additional regulatory changes are necessary.

Unique Warning Statements: None

5.   Summary of Data Gaps

Wildlife and Aquatic Organisms:

Actual Field Testing - Aquatic Organisms: Due May 1990
                                 Protocol: Due September 1988

6.   Contact Person at EPA

Walter C. Francis, PM Team (32)
Disinfectants Branch
Registration Division (TS-767C)
401 M. Street, S.W.
Washington, DC 20460
Telephone: (703) 557-3964

DISCLAIMER: The information present in this Chemical Information Fact Sheet is for informational purposes only and may not be used to fulfill data requirements for pesticide registration and reregistration.

CHLORPROPHAM

Reason for Issuance: Registration Standard
Date Issued: December 23, 1987
Fact Sheet Number: 150

1. DESCRIPTION OF CHEMICAL

   Generic Name: Isopropyl N-(3-chlorophenyl) carbamate

   Common Name: Chlorpropham

   Trade Names: Beet-Kleen, Furloe, Sprout Nip, Spud-Nic, Tater-
                 pex, Triherbicide-CIPC, Unicrop CIPC, Chloro IPC

   EPA Shaughnessy Code: 018301

   Chemical Abstracts Service (CAS) Number: 101-21-3

   Year of Initial Registration: 1962

   Pesticide Type: Herbicide and plant growth regulator

   Pests Controlled: Suckers on tobacco plants, sprouting in
                      stored potatoes, broadleaf weeds and grasses.

   Chemical Family: Carbamate

   U. S. and Foreign Producers: Pennwalt Holland B.V.
                                 (Netherlands), PPG Industries, Inc.,
                                 Chemical Div.-U.S., Universal Crop
                                 Protection Ltd.

114

2. USE PATTERNS AND FORMULATIONS

Application Sites: Terrestrial food and nonfood crop and
ornamentals.

Types and Methods of Application: Chlorpropham is a selective
preplant incorporated, preemergence, and postemergence
herbicide and plant growth regulator. Chlorpropham may be
applied by ground or by air.

Application Rates: Alfalfa 1-6 pounds active ingredient per acre
(lb ai/A); beans (lima and snap) 4 lb ai/A;
perennial grasses (seed crop) no rate given; flowers
(annual, biennial, perennial (bulbs)) 4-6 lb ai/A;
garlic 2-4 lb ai/A; spinach 1-2 lb ai/A; clovers
2-4 lb ai/A; onions 4-8 lb ai/A; ornamentals 4-8 lb
ai/A; safflower 3-6 lb ai/A; blackberries, raspberries
6 lb ai/A; blueberries 8-12 lb ai/A; cranberries 10-20
lb ai/A; southern peas 4-6 lb ai/A; soybeans 2-4 lb ai/A.
sugarbeet (seed crop) 3-4 lb ai/A; and tomatoes 4 lb ai/A.

Types of Formulations: 98% Technical Grade Active Ingredient (TGAI);
5%, 10.3% and 20% active ingredient (ai) granule (G);
11.9% ai, 15% ai, 22.2% ai, 25% ai, 36% ai and
47% ai emulsifiable concentrate (EC); 46% ai, 46.5%
ai, 49.65% ai, 78.4% ai and 78.5% ai liquid ready
to use (RTU).

Usual Carrier: Water

3. SCIENCE FINDINGS

Summary Science Statement: The current data base for
chlorpropham is insufficient with extensive data gaps in all
areas..

Insufficient data are available to permit a reliable prediction
of the leaching potential of chlorpropham. Taking into account
chlorpropham's high solubility and relative stability in
water, in addition to the known mobility of a related chemical,
propham, chlorpropham can be expected to leach and might enter
ground water.

Chemical Characteristics: T = Technical
                            P = Pure Active Ingredient

  Physical state - (T) fused solid

  Color - (T) off white to light brown

  Density, bulk density, or -   (T) ca. 1.2 gram/milliliter
  Specific gravity   (P) 1.180 at 30°C

  Solubility - (P) 102.5 parts per million (ppm) in water,
                 24°C

  Melting point - (T) 37-40°C
                    (P) 39° C

Toxicology Characteristics

  Acute Toxicity - No acceptable data are available on the
    acute toxicity, primary eye irritation or dermal
    irritation.

  Chronic and Subchronic Toxicity - No available data are
    available on the subchronic toxicity, oncogenicity, or
    metabolism of chlorpropham. The available data on terato-
    genicity and reproduction are acceptable.

    Teratogencity:

    Rat:
      Maternal Toxic No Observed Effect Level (NOEL) = 100 mg/kg/day
      This is the dose level that produces no observable effects in
      pregnant rats.
      Developmental Toxic NOEL = 350 mg/kg/day

    Rabbit:
      Maternal Toxic NOEL = 250 mg/kg/day
      This is the dose level that produces no observable effects in
      the embryo or fetuses of rabbits.
      Developmental Toxic NOEL = 125 mg/kg/day

    Reproduction:

    Rat:
      Reproductive NOEL $\geq$ 10000 ppm (highest dose tested (HDT))
      Systemic NOEL = 1000 ppm (lowest dose tested (LDT))

    Mutagenicity - The single acceptable mutagenicity
      study (gene mutation) was negative.

Physiological and Behavioral Characteristics

Translocation - Chlorpropham may be translocated from the
roots into the shoots.

Mechanism of Pesticide Action - Chlorpropham suppresses
plant transpiraton and respiration, and inhibits root and
epicotyl growth.

Metabolism and Persistence in Plants and Animals - The
metabolism of chlorpropham in growing plants has been
adequately described. The herbicide is translocated
from roots into shoots and residues include chlorpropham,
isopropyl 3-chloro-6-hydroxycarbanilate, isopropyl 3-chloro-
4-hydroxycarbanilate, 1-hydroxy-2-propyl-3-chlorocarbanilate
(isopropyl-OH-CIPC), isopropyl 3-chloro-2-hydroxycarbanilate,
and 3-chloroaniline. Additional data are required regarding
the metabolism of chlorpropham in stored potato tubers
treated postharvest and in livestock (ruminants and poultry).

Environmental Characteristics

Available data are insufficient to fully assess the
environmental fate of chlorpropham. The data requirement for
a hydrolysis study has been satisfied.

A hydrolysis study showed that chlorpropham is relatively
stable in sterile water in the dark. After 32 days in aqueous
buffered solutions at pH 4, 7, and 9 held in the dark at
40°C, about 90% of the applied chlorpropham remained undegraded.

The remaining environmental fate studies are inadequate,
but supplementary data indicate that chlorpropham (parent compound)
dissipates with a half-life of <14 days in the upper 3 inches
of silty clay loam and silt loam soils regardless of site or
application procedure (incorporated or surface-applied).

Fish accumulation data indicate that chlorpropham
bioaccumulated in the skinless fillet of a bluegill sunfish
to 100 times the levels in water.

Supplementary data indicate that chlorpropham accumulated
in rotational crops planted 12 months after treatment.

Reentry data are not required because available
toxicological data do not indicate a need for reentry data.

The following studies are required: photodegradation in water and on soil, aerobic and anaerobic soil metabolism, leaching and absorption/desorption, volatility (lab), field dissipation, irrigated crops and fish accumulation. Additional rotational crop studies (confined and field) are also required.

The Agency is concerned about pesticide residues reaching ground water. The potential for chlorpropham to reach ground water cannot be assessed since no leaching data are available. Taking into consideration chlorpropham's high solubility and its relative stability in water and the mobility of a related chemical, propham, chlorpropham can be expected to leach and thus might enter ground water.

Chlorpropham is the subject of a ground water DCI notification and additional data are needed to fully characterize the potential for it to enter ground water.

Ecological Characteristics

   Hazards to Fish and Wildlife - A supplementary study
      indicates that chlorpropham is practically nontoxic to water
      fowl (mallard median lethal dose (LD_{50}) is greater than
      2000 milligrams per kilogram (mg/kg)).

   Core studies indicate that chlorpropham is moderately
      toxic to coldwater and warmwater freshwater fishes (bluegill
      sunfish median lethal concentration (LC_{50})= 6.3-6.8
      parts per million (ppm); rainbow trout LC_{50} = 3.02-5.7
      ppm).

Tolerance Assessment

   Tolerances have been established for residues of
chlorpropham in or on a variety of raw agricultural plant
commodities, meat, milk, and eggs (40 CFR 180.181 and 40 CFR
180.319).

   Results of Tolerance Assessment - Due to the lack of
      acceptable plant and animal (livestock) metabolism data,
      storage stability data, and residue data, a conclusive
      tolerance reassessment cannot be conducted.

   Based on chronic effects observed in a two-generation rat
      reproduction study (slow weight gain; microscopic lesions
      in kidneys, spleen, liver and marrow; gross spleenic
      lesions; and organ weight changes in the liver and spleen)
      a Provisional Acceptable Daily Intake (PADI) has been

established at 0.2 mg/kg/day based on a NOEL of 50 mg/kg/day
and an uncertainty factor of 300. [An uncertainty factor
of 100 was used to account for the inter -and intraspecies
difference and a factor of 3 was used to account for the
inadequate data base for chronic toxicity].

The Theoretical Maximum Residue Contribution (TMRC) to
the human diet was based upon published tolerances. The TMRC
for 22 subgroups of the U.S. population ranged from 0.0182-
0.1154 mg/kg/day which occupies 9-58% of the PADI. Upon
receipt of the requested residue chemistry and toxicology
data, the chlorpropham tolerances will be reassessed.

Reported Pesticide Incidents

    There are no Pesticide Incident Monitoring System (PIMS)
reports or accident reports concerning chlorpropham.

4. SUMMARY OF REGULATORY POSITION AND RATIONALE

Warning Statements Required on Labels:

Manufacturing-Use Products

    Do not discharge effluent containing this product into
lakes, streams, ponds, estuaries, oceans, or public waters
unless this product is specifically identified and addressed
in an NPDES permit. Do not discharge effluent containing
this product into sewer systems without previously notifying
the sewage treatment plant authority. For guidance, contact
your State Water Board or Regional Office of the EPA.

End-Use Products (Terrestrial Food and Non-food Crop)

    Do not apply directly to water or wetlands (swamps,
bogs, marshes, potholes).. Do not apply where runoff is likely
to occur. Do not contaminate water by cleaning of equipment
or disposal of wastes.

5. SUMMARY OF MAJOR DATA GAPS:

DATA	DUE DATE
PRODUCT CHEMISTRY	6 to 15 Months

RESIDUE CHEMISTRY
```
  Nature of Residue (metabolism)              18 Months
  Residue Analytical Method                   15 Months
  Storage Stability                           18 Months
  Magnitude of the Residue for         18 to 24 Months
    Each Food Use
  Magnitude of the Residue in                 15 Months
    Drinking and Irrigation water
```

TOXICOLOGY
```
  Acute Oral Toxicity (rat)                    9 Months
  Acute Dermal Toxicity (rabbit)               9 Months
  Acute Inhalation Toxicity (rat)              9 Months
  Primary Eye Irritation                       9 Months
  Primary Dermal Irritation                    9 Months
  Dermal Sensitization                         9 Months
  90 Day Feeding (rodent)                      15 Months
                 (non-rodent)                 18 Months
  21 Day Dermal (rabbit)                       12 Months
  Smoke Inhalation                             15 Months
  Chronic Toxicity (rodent and non-rodent)    50 Months
  Oncogenicity (rat and mouse)                50 Months
  Structural Chromosomal Aberration           12 Months
  Other Mechanisms Mutagenicity               12 months
  General Metabolism                          24 months
```

ENVIRONMENTAL FATE
```
  Photodegradation (water and soil)            9 Months
  Metabolism (aerobic and anaerobic soil)     27 Months
  Leaching and Adsorption/Desorption           7 Months
  Dissipation (soil)                          27 Months
              (soil, long-term)             Reserved
  Accumulation
    Rotational Crops (confined)               39 Months
    Rotational Crops (field)                Reserved
    Irrigated Crops                           39 Months
    In Fish                                   12 Months
```

WILDLIFE AND AQUATIC ORGANISMS
```
  Residue Monitoring                           9 Months
  Acute Avian Oral Toxicity                    9 Months
  Avian Subacute Dietary                       9 Months
  Freshwater Fish Toxicity                     9 Months
  Acute Toxicity to Freshwater                 9 Months
    Invertebrate
```

Acute Toxicity to Estuarine and Marine Organisms	12 Months
Fish Early Life Stage and Aquatic Invertebrate Life Cycle	15 Months
Fish Life Cycle	27 Months

6.   CONTACT PERSON AT EPA:

Robert J. Taylor, Product Manager (25)
401 M. Street, S.W.
Washington, D.C.   20460
(703) 557-1800

DISCLAIMER: The information presented in this Pesticide Fact Sheet is for informational purposes only and may not be used to fulfill data requirements for pesticide registration and reregistration.

COAL TAR/CREOSOTE

Reason for Issuance: Registration Standard
Date Issued: April 1988
Fact Sheet Number: 161

1. DESCRIPTION OF CHEMICALS

Chemical names: Creosote is a complex, heterogeneous mixture
                   of chemicals derived from the fractional dis-
                   tillation of coal or wood tar. Most coal tar/
                   creosote formulations for wood preservation
                   should conform to the eight standards of the
                   American Wood Preservers Association's(AWPA).
                   The standards consist of physical properties
                   and percentages of specified distillation
                   ranges.

Common names: Coal tar/creosote

Pesticide type: Wood preservative, fungicide, bacteriocide, in-
                 sect repellent for wood boring insect

Chemical family: Organic compounds

Major U.S. Producers: Koppers, Aristech Chemical, Allied Chemical,
                       Reilly Tar and Chemical Company

2. USE PATTERN AND FORMULATIONS

Application Sites: Wood and wood products, such as

        °Pressure treated wood for construction
        °Groundline treat of utility poles
        °Home and farm use
        °Pole framing
        °Pilings
        °Railroad tie repair

Types of Formulations: End use products vary from liquid to vis-
                        cous semi-solid tars and pitches

Method of Application: Pressure treatment and non-pressure treat-
                        ments; brush-on, spray-on and dipping
                        (soaking)

122

3. SCIENCE FINDINGS

Chemical characteristics

Physical state - viscous to oily liquid
Color - Black to yellowish-green
Odor - Naphthalene-like (petroleum)
Boiling point - Variable
Specific gravity - 1.08 and above at 20°C
Solubility - Soluble in organic solvents, e.g., benzene,
   ether, alcohol, acetone; slightly soluble in water

Toxicological Characteristics:

Acute toxicty - Because of the complex nature of coal tar/creosote and its constituents, acute toxicity of differing formulations may vary considerably. Eye and skin burns, conjunctivitis and skin sensitization have been reported in conjunction with worker exposures.

Oral - Moderately toxic

Dermal - Moderately toxic

Inhalation - Undetermined

Eye Irritation - Corrosive or highly irritating to eyes

Skin Irritation - Undetermined

Dermal Sensitization - Undetermined

Oncogenicity: On a qualitative basis, there is substantial evidence of oncogenicity associated with coal tar/creosote products. There are no epidemiological studies of workers using coal tar/ creosote products in wood treatment plants. However, epidemiological studies of coke oven workers, who are exposed to coal tar compounds from coke oven emissions, reveal increased incidences of lung, bladder, prostate, pancreas, and intestinal cancer. Coal tar and many of its constituents have been well characterized as oncogens in animal studies. However, because of the complexity of coal tar/creosotes, available studies are not adequate for quantitative risk assessment. An epidemiological study would provide the best basis for assessing the human oncogenic risk of coal tar/ creosote.

The Agency is requiring oncogenicity studies in mice by the dermal route, and is requiring that a preliminary epidemiological study be conducted to collect data to be used in designing a comprehensive epidemiological study.

In addition, the Agency is requiring exposure studies among treatment plant workers exposed to coal tar/creosote. These studies, together with mouse dermal oncogenicity studies, will permit the Agency to conduct a quantitative risk assessment.

Mutagenicity: Coal tar/creosote formulations have been shown to elicit mutagenic responses in laboratory test species including microbial and mammalian test systems. Creosote showed a positive response in the Ames Test for mutagenicity. The Agency is requiring additional mutagenicity studies.

Teratogenicity: Data are not available to assess the teratogenic or fetotoxic effects of coal tar/creosote products. These data are required.

Environmental characteristics: Data are not available on the environmental fate characteristics of coal tar/creosote or its individual constituents. These data are required. Testing procedures and test materials will be selected based on evaluation of product chemistry and other required data.

Ecological Characteristics: The Agency has very limited data regarding availability and toxicity of coal tar/creosote constituents to fish and wildlife species.

Laboratory bioassays indicate the coal tar products are generally moderately toxic to aquatic organisms. Because the amount and type of exposure in the field is unknown, it is not possible to estimate risks to aquatic organisms. Other than laboratory data, there is no evidence to date of environmental hazards to marine organisms (including endangered species) from its use in treating wood used in aquatic sites.

Aquatic toxicity studies are required, and if significant toxicity is found, studies will be required using treated wood to simulate actual environmental exposures to freshwater and marine species.

Tolerance Assessment: There are no direct food or feed uses registered for coal tar/ creosote. Moreover, label restrictions prohibit the use of coal tar or creosote products registered for farm and home use to treat wood intended to be used in a manner in which the preservative may become a component of food or feed. Therefore the Agency has not established tolerances or exemptions from tolerance in raw agricultural commodities, processed food and feed, or animals under the Federal Food, Drug and Cosmetic Act (FFDCA). No clearances are required under the FFDCA to support the registered uses of coal tar/creosote.

Because there are no food or feed uses, certain subchronic and chronic studies, such as rat metabolism, chronic feeding studies in rats and dogs, and reproduction studies are not required.

4.   SUMMARY OF REGULATORY POSITIONS AND RATIONALES

---The Agency is continuing requirements imposed under the
   Special Review of coal tar/creosote products. These
   include restricted use classification and protective
   clothing and equipment requirements intended to reduce
   exposure of applicators and treatment plant workers.

---The Agency is requiring an evaluation of the permeability
   of various protective clothing materials to determine
   which offer the greatest protection from coal tar/creosote
   products.

---The Agency is requiring that registrants develop and maintain composite test materials(CTM) representing each of
   the eight standard products defined by the American Wood
   Preserver's Association(AWPA). The physical/chemical
   specifications of each product should be submitted along
   with the procedure for developing the composite. The
   Agency requires that each composite be analyzed for the
   identity and quantity of individual constituents.

---The Agency is requiring toxicology testing using composite
   materials that are representative of typical AWPA standard
   products of coal tar/creosote. This is necessary because
   coal tar/creosote products consist of a large number of individual constituents, testing of all of which is impractical.

---The Agency is requiring ecological effects data on AWPA
   standard products of coal tar/creosote typically used for
   the treatment of wood for the freshwater and saltwater
   environments.

---All environmental fate data requirements for terrestial and
   aquatic nonfood uses (40 CFR 158.130) are required. Test
   materials for these studies will be determined by the
   Agency after it has received and reviewed the product chemistry and other required data for the eight composites of
   AWPA standard products.

---The Agency is requiring registrants to conduct a preliminary
   epidemiological study prior to requiring a comprehensive epidemiological study in order to develop baseline data on the
   nature and degree of risks to wood treatment plant workers.

---Exposure and work activity data are required to assess the
   quantity and nature of coal tar/creosote products to which
   treatment plant workers are exposed. Such studies will be
   useful in designing an appropriate epidemiological protocol
   or in quantitative risk assessment.

5. SUMMARY OF LABELING STATEMENTS

    Labeling requirements for coal tar/creosote products are out-
lined in Part IV.D of the registration standard. These requirements
parallel those presented in the Federal Register Notice, 51 FR 1334,
dated January 10, 1986. Specific labeling language is outlined for
for wood preservative products for each of the following five use
categories:

    1. Pole Framing, Piling applications and Railroad Tie Repair
       uses
    2. Pressure Treatment uses
    3. Groundline Treatment of Utility Poles
    4. Home and Farm Use
    5. Non-Pressure Treatment

With the exception of the Pole Framing, Piling Applications and
Railroad Tie Repair use category, product labels are required to
carry "Restricted Use Pesticide" classification and associated
language. The following types of label statements are required to
appear on the labels of all coal tar/creosote products.

    --- Precautionary statements, including signal word DANGER based
        on potential eye effects

    --- Statements of practical treatment, including a statement not
        to induce vomiting if swallowed

    --- Protective clothing statements, including gloves, boots and
        coveralls impervious to coal tar/creosote products; require-
        ments for the decontamination and cleaning of such clothing.

    --- Requirements for cartridge or canister respirators for work-
        ers in treatment plants and for use around farms and homes
        where inhalation exposure cannot be avoided.

    --- A prohibition against indoor application and prohibition
        against use of treated wood products indoors.

    --- A prohibition against the use of home and farm products in
        such a way that residues might transfer to crops, food,
        feed or animals, such as crop storage buildings or beehives,
        or where animals may be directly exposed to the pesticide,
        such as brooding pens.

    --- A prohibition against use where irrigation or drinking water
        contamination may result.

    --- Environmental hazard statements concerning toxicity to fish.

    --- Identification as a toxic hazardous waste if disposed of.

6.   SUMMARY OF MAJOR DATA GAPS

Product Chemistry
  --Product identity
  --Analysis and certification
    of product ingredients
  --Physical/chemical characteristics

Environmental Fate
  --Hydrolysis
  --Photodegradation in water
  --Aerobic soil metabolism
  --Anarobic aquatic metabolism
  --Aerobic aquatic metabolism
  --Leaching and adsorption/desorption
  --Soil dissipation
  --Aquatic dissipation

Toxicology studies
  --Acute toxicity
  --Subchronic dermal and inhalation
  --Oncogenicity (dermal)
  --Teratogenicity
  --Mutagenicity

Ecological effects studies
  --Avian acute toxicty
  --Acute freshwater fish/invertebrates
    and estuarine organisms

Special studies
  --Protective clothing permeability
    and durability studies
  --Exposure studies (Protocol)
  --Worker activities studies
  --Epidemiology feasibility study

7.   CONTACT PERSON AT EPA

      Lois A. Rossi (PM-21)
      Registration Division (TS-767C)
      U.S. Environmental Protection Agency
      401 M St., SW
      Washington, D.C. 20460
      Telephone: Area Code (703) 557-1900

DISCLAIMER: The information presented in this Pesticide Fact
Sheet is for informational purposes only and may not be used to
fulfill data requirements for pesticide registration and reregistration.

COUMAPHOS

Reason for Issuance: Registration Standard
Date Issued: September 27, 1989
Fact Sheet Number: 207

1. <u>DESCRIPTION OF CHEMICAL</u>

Common Name: Coumaphos
Chemical Family: Organophosphate
Pesticide Type: Insecticide/acaricide
Chemical Name: 0,0-diethyl 0-(3-chloro-4-methyl-2-oxo-
2H-1-benzopyran-7-yl) phosphorothioate
Trade Names: Bay 21/199, Asuntol, Muscatox, Resitox,
Baymix, Meldane, Co-Ral and Negashunt
Other Chemical
Nomenclature: 0-3-chloro-4-methylcoumarin-7-yl
0,0-diethyl phosphorothioate; 3-chloro-7-
diethoxyphosphino-thioyloxy-4-
methylcoumarin; 0-(3-chloro-4-methyl-2-
oxo-2H-1-benzopyran-7-yl) 0,0-diethyl
phosphorothioate (Chemical Abstracts,
9th Collective Index); 3,chloro-7-hydroxy-
4-methylcoumarin 0-ester with 0,0-diethyl
phosphorothioate (8th Collective Index);
0,3-chloro-4-methyl-2-oxo-2H-chromen-7-yl
0-0-diethyl phosphorothioate;
[0-(3-chloro-4-methyl-7-coumarinyl)]
0,0-diethyl phosphorothioate; 0,0-diethyl
0-(3-chloro-4-methyl-7-courmarinyl)
phosphorothioate; phosphorothioic acid 0-
(3-chloro-4-methyl-2-oxo-2H-1-benzopyran-
7-yl) 0,0-diethyl ester; 3-chloro-4-
methylumbelliferone, 0-ester with 0,0-
dietyl phosphorothioate; 0,0-diethyl
0-(3-chloro-4-methylumbelliferone
thiophosphate
Year of Initial Registration: 1958
CAS Registry Number: 56-72-4
EPA Pesticide Chemical Code (Shaughnessy Number): 036501
U.S. Manufacturer: Bayvet, a division of Cutter Laboratories

USE PATTERNS AND FORMULATIONS

Coumaphos is applied as a direct animal treatment to control arthropod pests of beef cattle, dairy cattle, sheep, goats, horses and swine. It is used to treat swine bedding. Registered control claims are for face flies, horn flies, fly larvae, cattle grubs, ticks (including ear tick), lice, mites, screwworms, sheep keds and fleeceworms. Methods of application consist of dusts, sprays, dips, pour-ons, dust bags and backrubber oilers. Annual usage is 264,000 to 525,600 lbs (1986 estimate). The predominate use is on beef cattle (98%). A relatively small amount is used on dairy cattle (<2%) and swine (<1%).

Current Status and Summary Science Statement

Toxicity data requirements for registration of products containing coumaphos (including acute toxicity testing on end-use product formulations) have been met, except for a 21-day dermal toxicity study, a non-rodent chronic toxicity study, a reproduction study, and a structural chromosome aberration study. Technical coumaphos is highly acutely toxic by the oral and inhalation routes of exposure (Toxicity Category I and II, respectively) and moderately acutely toxic by the dermal route of exposure (Toxicity Category III) based on studies using rats, rabbits and guinea pigs. Technical coumaphos causes only mild eye and dermal irritation (Toxicity Category III and IV, respectively), and is nonsensitizing. End-use product formulations fall in a range of Toxicity Categories from I to III. Coumaphos does not produce organophosphate-type delayed neurotoxicity, based on acute neurotoxicity testing in hens. The oncogenic potential of coumaphos is satisfactorily defined. In vitro microbial studies for gene mutation and DNA damage coumaphos did not cause a mutagenic response, and when tested in the rat and mouse, there were no carcinogenic effects noted. Coumaphos is not a developmental toxicant, or teratogen based on findings in studies utilizing rats and rabbits. Results of a chronic feeding study using rats show that cholinesterase (plasma and erythrocyte) is the primary target of coumaphos. Decreased body weight gain is a secondary effect. In a rat metabolism study, coumaphos was rapidly excreted. There were no dose-related changes in metabolism or evidence of activation/bioaccumulation.

The coumaphos data base for ecological effects testing is complete, with the exception of two special studies. Based on the results of laboratory studies using birds, fish, and aquatic invertebrates, technical coumaphos is moderately acutely toxic to fish and very highly acutely toxic to birds and aquatic invertebrates. Coumaphos is moderately toxic to birds on a subacute (dietary) basis. Aquatic invertebrates may be potentially exposed to hazardous levels of coumaphos resulting from washing-off of the material from the backs of newly treated

cattle which have entered a body of water. Aquatic residue
monitoring is required to assess the potential hazards. Due to
the potential for avian exposure resulting from birds feeding in
cattle lots and on the backs of cattle, Tier I avian field
testing is required to assess possible effects to birds resulting
from the direct treatment to livestock.

The environmental fate profile for coumaphos is adequately
delineated for the registered use pattern, except for a
groundwater assessment. Coumaphos is relatively immobile in aged
sandy loam soil, based on findings in a column leaching study.
There are no immediate concerns for groundwater contamination
from non-point source application of coumaphos. However, the
potential does exist for localized, point source contamination in
animal treatment areas (particularly where animals are dipped),
and as a result of associated disposal practices. Due to
increased Agency sensitivity in the area of pesticides and
groundwater contamination, environmental fate studies are
required so that the Agency can assess coumaphos's potential for
point source contamination.

Most of the residue chemistry conclusions drawn in the 1981
Standard have been reversed in the current Standard. Residue
chemistry data requirements were not imposed in the 1981
Standard. Since issuance of that Standard, the Agency has
published residue chemistry guidelines (Pesticide Assessment
Guidelines, Subdivision O, 1982, EPA-540/9-82-023) and other
Federal Register (FR) Notices which provide a more stringent
interpretation of the existing regulations. As a result of these
new guidelines, data are now required in the area of animal
metabolism, storage stability and method validation. No changes
to coumaphos tolerances are indicated at this time.

The Agency is unable to totally assess the safety of current
tolerances and establish an acceptable daily intake (ADI) value
for coumaphos because of the absence of chronic toxicity studies
(reproduction and dog chronic toxicity), and outstanding residue
chemistry data. However, a preliminary dietary exposure
analysis has been performed for coumaphos. Based on the results
of this analysis, current coumaphos tolerances are considered to
be adequate to protect the public health. When the remaining
data requirements have been fulfilled, the Agency will perform a
final reassessment of coumaphos tolerances.

Chemical/Physical Characteristics of the Technical Material

Empirical Formula: $C_{14}H_{16}ClO_5PS$
Molecular Weight: 362.8
Color: grey to tan
Physical state: powder to granules
Odor: characteristic sulfur
Melting Point: 90 to 95 $^\circ$C

Boiling Point: 20 $^\circ$C at 10^{-7} mmHg
Solubility: (at 20 $^\circ$C):   g/100 mL at 20 C

acetone		23.82
methylene chloride		6.39
denatured alchohol		0.90
xylenes		0.90
hexanes		0.07
water	insoluble at	0.002
octanol		0.13
odorless mineral spirits		0.09
diethyl phthalate		21.50

Vapor Pressure: 1 X 10^{-7} mmHg
Density, Bulk Density, or
Specific Gravity: granules: 30.06 lb/cu ft, loose; 30.85
                    lb/cu ft, packed. mhammermilled: 24.35
                    lb/cu ft, loose; 30.51 lb/cu ft, packed
pH: 7.23 at 1 g/100 mL
Stability: hydrolyses slowly under alkaline conditions;
             stable under normal storage conditions and
             use; incompatible with piperonyl butoxide
Storage Stability: Stable (<6% loss) in glass vials up
                     to 8 weeks at -12 to 50 C, dry and
                     at pH 4-10, at 83% moisture, exposed
                     to aluminum, and stainless steel;
                     stable exposed to sunlight for 4
                     days.

Toxicology Characteristics

Acute Oral: Toxicity Category I (LD_{50} of greater than 240 mg/kg
              in males rats and 17 mg/kg in female rats)
Acute dermal: Toxicity Category III (LD_{50} of greater than 2400
                mg/kg in rabbits
Acute inhalation: Toxicity Category II (LC_{50} dose for a 1-hour
                    is 341 mg/m^3 in female rats and
                    greater than 1080 mg/m^3 in male rats)
Primary eye irritation: Toxicity Category III, mild
                          eye irritation reported
Primary dermal irritation: Toxicity Category IV, very minor
                             dermal irritation reported
Skin sensitization: No observable evidence of dermal
                      sensitization
Delayed Neurotoxicity: Did not induce delayed neurotoxicity in
                         an acceptable study in hens.
Subchronic non-rodent/
     rodent studies: None available. Not required
                       since chronic data supercede
                       need for subchronic testing.
21-day dermal toxicity: Required Study
Chronic toxicity: Dog study is required. Rat study NOEL is 0.07
                    mg/kg for decreased cholinesterase activity.
Oncogenicity: The mouse and rat chronic toxicity/
                oncogenicity studies did not reveal any evidence
                that coumaphos is oncogenic.

Mutagenicity: Negative in all areas of mutagenicity tested.   A
                structural chromosomal abberation study is
                required.
Teratogenicity: Rat teratology study NOEL and LEL were 5 and 25
                mg/kg (based on the observation of cholinergic
                effects), respectively.   The developmental NOEL
                was greater than 25 mg/kg (HDT).   Rabbit
                teratology study maternal NOEL and LEL were
                2.0 and 18.0 mg/kg, respectively;
                developmental NOEL was greater than 18.0 mg/kg
                (HDT).
Reproduction: Required study
Metabolism: In a rat metabolism study, coumaphos was rapidly
                excreted.   No dose-related changes in metabolism or
                evidence of activation/bioaccumulation were noted
                in this study.

Environmental Characteristics

Based on the results of a column leaching study, coumaphos
can be characterized as persistent, but immobile in sandy loam
soils.   There are no immediate concerns for groundwater
contamination from non-point source application of coumaphos.
However, the potential does exist for localized, non-point source
contamination in animal treatment areas (particularly where
animals are dipped), and as a result of associated disposal
practices.   In order to evaluate the potential for point source
contamination, special studies are required: a photodegradation
study in soil, a photodegradation study in water, an
adsorption/desorption study, a hydrolysis study and a
retrospective field dissipation study.

Ecological Characteristics

Based on the results of acceptable laboratory data,
technical coumaphos is characterized as highly to very highly
toxic to birds, moderately toxic to fish and highly toxic to
aquatic invertebrates:

  - Acute LD_{50} (mallard): 29.4 mg/kg
  - Acute LD_{50} (pheasant): 7.94 mg/kg
  - Dietary LC_{50}:
      401 ppm (mallard)
      82 ppm (bobwhite)
  - Freshwater invertebrates toxicity (96-hr LC_{50}) for
    amphipods: 0.15 ppb
  - Fish acute toxicity (96-hr LC_{50}) for rainbow trout: 5900
    ppb
  - Fish acute toxicity (96-hr LC_{50}) for bluegill sunfish:
    5000 ppb

Results of laboratory testing, in conjunction with
theoretical monitoring, indicate that aquatic invertebrates may
be potentially exposed to hazardous levels of coumaphos resulting

from washing-off of the material from the backs of newly treated cattle which have entered a body of water, such as a pond or stream.   To evaluate the potential risk, a residue monitoring study is required.   There is a potential for avian exposure resulting from birds feeding in cattle feedlots and on the backs of cattle.   Tier I avian field testing is required to assess possible effects to birds resulting from direct treatment to livestock.

Tolerance Assessment

U.S. tolerances are established for residues of the insecticide coumaphos, O,O-diethyl O-(3-chloro-4-methyl-2-oxo-2H-1-benzopyran-7-yl) phosphorothioate, and its oxygen analog, O,O-diethyl O-3-chloro-4-methyl-2-oxo-2H-1-benzopyran-7-yl-phosphate, in or on raw agricultural products as follows (40 CFR 180.189):

   o   1 ppm in or on meat, fat, and meat byproducts of cattle, goats, hogs, horses, poultry[1], and sheep
   o   0.5 ppm in milk fat (reflecting negligible residues in milk)
   o   0.1 ppm in eggs

Most of the residue chemistry conclusions drawn in the 1981 Standard have been reversed.   No residue chemistry data requirements were imposed in the 1981 Standard.   Since issuance of that Standard, the Agency has published residue chemistry guidelines (Pesticide Assessment Guidelines, Subdivision O, 1982, EPA-540/9-82-023) and other FR Notices which provide a more stringent interpretation of the existing regulations.   As a result of these new guidelines, data are now needed in the area of animal metabolism, storage stability and method validation.

The Provisional Acceptable Daily Intake (PADI) for coumaphos is 0.0007 mg/kg/day and is based on the 2-year rat feeding/oncogenicity study NOEL of 0.07 mg/kg/day (based on plasma cholinesterase inhibition in females) and uncertainty factor of 100.   The Anticipated Residue Contribution (ARC) for the United States population is 0.000127 mg/kg/day, occupying 18.2% of the PADI.   The two highest calculated exposures for the population subgroups are children 1 to 6 years of age [ARC occupies 33.6% of the PADI] and children 7 to 12 years of age [ARC occupies 25.6% of the PADI].   Based on these calculations, coumaphos applied at the currently registered application rates would not be expected to exceed established tolerances.

The Agency is unable to totally assess the safety of current tolerances and establish an acceptable daily intake (ADI) value for coumaphos because of the absence of chronic toxicity studies (reproduction and dog chronic toxicity), and outstanding data in

---

[1]   There are no longer any federally registered uses for poultry/poultry houses.   Therefore, the Agency intends to revoke the tolerances for poultry and eggs.

the area of animal metabolism, method validation and storage
stability. When the required data have been submitted and
evaluated, the Agency will perform a final reassessment of
coumaphos tolerances.

4. SUMMARY OF REGULATORY POSITIONS AND RATIONALES

   - The Agency is not initiating a Special Review for coumaphos.
No Special Review concerns were identified for this chemical by
the Agency during its review of the current data base.

   - The Agency is classifying coumaphos 11.6% EC and 42%
flowable concentrate end-use products as restricted use due to
acute oral hazards.

   - The Agency will approve new food/feed tolerances for
coumaphos on a case-by-case basis.

   - Environmental fate testing is required to evaluate the
potential for coumaphos to impact groundwater or surface water
resulting from point source application.

   - A special aquatic residue monitoring study is required.

   - Special Tier I avian field testing is required.

   - The Agency will revoke the poultry and egg tolerances,
since coumaphos is no longer federally registered for use on
poultry or in poultry houses.

   - Unique labeling statements are required:

      o Restricted-use classification is required for coumaphos
11.6% EC and 42% flowable concentrate formulations.

      o Special disposal instructions are required for products
bearing directions for use a livestock dip treatment.

      o Labels bearing directions for use on goats and sheep must
be amended to specify a preslaughter interval (PSI) of 3 days.

      o Product labels must bear revised and updated fish and
wildlife statments.

      o Worker safety and protective clothing statements are
required for products falling in Toxicity Category I or II.

      o Each end-use product label must be revised to reflect the
appropriat signal word and precautionary statements assigned to
it based on the results of acceptable acute toxicity testing.

      o Revised labeling must be submitted for those products
which do not contain directions for use specifying a maximum
single application rate expressed in terms of: (1) amount of

active ingredient per animal; (2) a maximum seasonal application rate or number of applications permitted per season; and (3) a minimum interval between applications, revised labeling must be submitted.

SUMMARY OF OUTSTANDING DATA REQUIREMENTS

Toxicology

21-Day Dermal Toxicity	1 Year
Dog Chronic Toxicity	4 Years
Reproduction Study	4 Years
Chromosome Aberration	1 Year

Environmental Fate/Exposure

Photodegradation in Water and Soil	1 Year
Adsorption/Desorption	2 Years
Special Retrospective Field Dissipation Study	2 "
Hydrolysis Study	1 "

Fish and Wildlife

Monitoring for Aquatic Invertebrate Mortality and Residues in Water	3 Years
Tier I avian field testing	3 Years

Residue Chemistry

Metabolism data – Animals	1 Years
Residue Analytical Methods	1 "
Storage Stability Data	1 "

Product Chemistry

Remaining Data Gaps	1 -2 Years

6.   Contact Person at EPA

     George LaRocca.
     Product Manager (15)
     Insecticide-Rodenticide Branch
     Registration Division (H7505C)
     Environmental Protection Agency
     Washington, DC 20460
     Tel. No. (703) 557-24006

DISCLAIMER: The information presented in this Chemical Information Fact Sheet is for informational purposes only and may not be used to fulfill data requirements for pesticide registration and reregistration.

CRYOLITE

Reason for Issuance: Registration Standard
Date Issued: April 1988
Fact Sheet Number: 2.1

1. DESCRIPTION OF CHEMICAL

   Generic name: Sodium aluminofluoride or
      (Chemical)   sodium fluoaluminate

   Common name: Cryolite

   Trade Names: Kryocide and Prokil

   Other Chemical
   Nomenclature: trisodium hexafluoroaluminate, sodium
                  hexafluoroaluminate, and sodium aluminum
                  fluoride

   EPA Pesticide Chemical (Shaughnessy) Number: 075101

   Chemical Abstracts Service (CAS) number: 15096-52-3
                                             (Merck Index)

   Year of initial registration: 1959

   Pesticide type: Insecticide

   Chemical family: Inorganic fluorine compound

   U.S. Registrants: Agchem Division of Pennwalt Corp., Amvac
                      Chemical, Gowan Co. and Moyer Products,
                      Inc.

2. USE PATTERNS AND FORMULATIONS

   Application sites: Terrestrial food crop use on apples,
                      beans, beets, broccoli, Brussels
                      sprouts, cabbage, cantaloupes, carrots,
                      cauliflower, collards, cranberries,
                      cucumbers, eggplant, grapefruit,

136

grapes, kale, kohlrabi, kumquats,
lemons, lettuce, limes, melons, mustard
greens, oranges, pears, peppers,
pumpkins, radishes, squash,
strawberries, tangerines, tomatoes,
turnips, and watermelons.

Terrestrial non-food crop use on
ornamental trees and shrubs (including
nursery stock).

Formulations: Wettable powders and dusts

Methods of application: Aerial or ground, foliar application as a spray or dust

Application rates: rates from 5 to 78.98 lb ai/A

3. SCIENCE FINDINGS

Summary Science Statement

     Technical cryolite is mildly toxic on an acute oral, dermal and
inhalation basis (Toxicity Category IV, III and III, respectively.)
Cryolite does not demonstrate a teratogenic, fetotoxic, or mutagenic
potential. The Agency has concerns about the potential adverse
health effects of fluoride to bones, as well as the potential adverse
cosmetic effects of fluoride to the teeth of children, resulting from
the pesticidal application of cryolite. These concerns are based
upon findings of crippling skeletal fluorosis in adults who were
chronically exposed to relatively high levels of fluoride and
extensive epidemiological studies with large populations of children
carried out over the last 40 years[1]. The Agency cannot set a
reference dose (RfD) for cryolite until additional data are
submitted, specifically, a rat metabolism (pharmacodynamic) study to
quantitate the bioavailability of fluoride from cryolite.
Information derived from this study should allow the Agency to
establish a RfD for cryolite. None of the tolerances for cryolite
are adequately supported, and residue data are required to support
all crop uses.

_____

     In 1986, the Agency's Reference Dose Work Group
established a reference dose (RfD) of 0.12 mg/kg/day for
fluoride (designed to prevent the development of crippling
skeletal fluorosis), and also established an additional value of
0.06 mg/kg/day for fluoride. This additional value is designed
to prevent the development of objectionable dental fluorosis in
children.

There are no groundwater concerns identified for this chemical. A confined rotational crop study is required because preliminary residue chemistry data show tolerance-exceeding levels of fluorine and/or cryolite in primary crops following direct treatment with cryolite, allowing for a possibility that residues will also occur in rotational crops.

Cryolite is not expected to pose an acute hazard to fish and wildlife based on acceptable acute and subacute studies which show that technical cryolite is practically nontoxic to birds and bees, and only slightly toxic to fish and aquatic invertebrates. However, aerial application of cryolite at ≥ 30 lb ai/A may pose a chronic hazard for aquatic invertebrates. Therefore, an aquatic invertebrate life cycle study is required to support these uses.

Chemical/Physical Characteristics of the Technical Material

Chemical/Physical
Characteristics: Cryolite is a naturally occurring mineral (large deposits occurring in Greenland and the Urals), or it may be synthetically produced by the reaction of aluminum oxide, sodium chloride, and hydrogen fluoride.

Color: white (natural); white to yellow-brownish white (synthetic)
Physical state: powder (natural); glassy powder at 20° C (synthetic)
Odor: none (natural and synthetic)
Melting Point: 1000° C (Merck Index)

Toxicology Characteristics

Acute Oral: Toxicity Category IV (LD_{50} > 5 g/kg in the rat)

Acute dermal: Toxicity Category III (LD_{50} > 2.1 g/kg in the rabbit)

Acute inhalation: Toxicity Category III (LD_{50} >2.06 mg/L and < 5.03 mg/L in the rat

Primary dermal irritation: Toxicity Category IV (rabbit: P.I. score = 0.0; not an irritant)

Primary eye irritation: Toxicity Category III (in rabbit: moderate conjunctival irritation that disappeared within 7 days)

Skin sensitization: Non-sensitizing

Delayed Neurotoxicity: Cryolite is not an organophosphate and therefore a study is not required.

Subchronic dermal toxicity: Data gap. A 21-day dermal toxicity
                               study is required.

Subchronic feeding studies: Studies reviewed were classified
                             supplementary because a NOEL for
                             fluoride accumulation in the bone
                             could not be established.

Mutagenicity: Cryolite was negative in the Ames assay for
               mutagenic activity; in the DNA repair test using
               Escherichia coli for genotoxic effects; and in a
               rat in vivo cytogenetics assay for structural
               chromosome aberrations.

Teratogenicity: No teratogenic or fetotoxic effects noted at
                 the highest dose tested in a rat study (NOEL >
                 3000 mg/kg/day).

Metabolism: Data gap. A rat metabolism (pharmacodynamic) study is
             required.

Chronic Toxicity: Data gap. Chronic toxicity studies (rat and
                   dog) are required.

Oncogenicity: Data gap. Oncogenicity studies (rat and mouse)
               are required.

Reproduction: Data gap. A 2-generation rat reproduction study
               is required.

Magnitude of the Residue in Plants: Data gaps. Field crop
                                     trials and processing studies
                                     are required.

Environmental Characteristics

Cryolite is only slightly mobile in soil. It does not
hydrolyze, but rather dissociates, yielding free fluorine at pH 5, 7,
and 9. Cryolite is not expected to contaminate groundwater, since it
is only slightly mobile in soil. Submitted residue chemistry data
show that tolerance-exceeding levels of fluoride and/or cryolite will
occur in food crops following direct treatment with cryolite and
hence the possibility of residues occurring in rotational crops. A
confined rotational crop study is required. The requirement for a
field rotational crop study is reserved, pending receipt and
evaluation of the confined study.

Ecological Characteristics

Based on acceptable acute data, technical cryolite has been determined to be practically non-toxic to birds and bees, and slightly toxic to freshwater fish and aquatic invertebrates.

- Acute LD_{50} (bobwhite):
  >2,150 mg/kg (practically nontoxic)
- Dietary LC_{50} (mallard duck and bobwhite):
  >10,000 ppm (practically nontoxic)
- Freshwater invertebrate toxicity (96-hr LC_{50}) for
  Daphnia pulex: 10 ppm; for Daphnia magna: >100 ppm
  (slightly toxic)
- Fish acute toxicity (96-hr LC_{50}) for rainbow trout: 47
  ppm (slightly toxic)
- Fish acute toxicity (96-hr LC_{50}) for bluegill sunfish:
  >400 ppm (practically nontoxic)
- Acute toxicity to honeybees: 1.45% mortality at 217 mg
  per bee

The aquatic estimated environmental concentration (EEC) resulting from indirect application (i.e., runoff and spray drift) o cryolite exceeds the criteria for requiring the aquatic invertebrate life cycle test (i.e., >1% of the acute toxicity value for the most sensitive species) when cryolite is applied aerially at $\geq$30 lb ai/acre. Therefore, the test is required for all products allowing for application under these conditions.

Tolerance Assessment

Tolerances for residues of cryolite in or on food commodities ar published in 40 CFR 180.145. These tolerances are expressed in term of combined fluorine for residues of the insecticidal fluorine compounds cryolite and synthetic cryolite. They are set at 7 ppm fo all listed commodities.

Data are not available regarding the metabolism of cryolite in plants. Although the Agency recognizes that traditional plant metabolism studies using radiolabeled materials may not be useful or practical for cryolite (an inorganic compound), studies showing the form of fluoride (cryolite per se vs. free fluoride ion) in or on raw agricultural plant commodities could provide useful information regarding the nature of the residue as consumed by humans and livestock. Development of an analytical method for distinguishing between cryolite per se and free fluoride ion is required. This method must be used in selected residue field trials so that a residue profile (cryolite per se vs. total fluoride) can be developed.   The metabolism of cryolite in animals is not adequately understood.   No data have been submitted pertaining to the metabolism of cryolite in ruminants or poultry.   This requirement is deferred until the required rat metabolism (pharmacodynamic) study has been submitted and reviewed.

An adequate method is available for enforcement of tolerances and data collection for residues of cryolite in or on plant commodities. The limit of detection for fluorine is 0.1 ppm.   Cryolite is not expected to degrade during storage because it is a naturally occurring mineral.

Field trial studies are required for all crops for which there are cryolite tolerances.   Processing studies are also required. Tolerances (or feeding/grazing restrictions) must be proposed and appropriate supporting residue data submitted for the raw agricultural commodities and feed items - bean vines and hay. Separate tolerances must be proposed for succulent and dry beans. Food/feed additive tolerances must be proposed for the combined fluorine residues of cryolite and synthetic cryolite in grape pomace, raisins, and raisin waste, and in paste, puree and catsup of tomatoes, since residues have been observed to concentrate in these commodities.

4. SUMMARY OF REGULATORY POSITIONS AND RATIONALES

 - Cryolite does not meet the criteria for restricted use classification.

 - Cryolite is not a candidate for Special Review.

 - There are no groundwater concerns identified for this pesticide.

 - The Agency will not approve any significant new food uses until the Agency has received residue chemistry data and toxicology data ((rat metabolism (pharmacodynamic) study) sufficient to allow the Agency to perform a tolerance reassessment.

SUMMARY OF REQUIRED LABEL MODIFICATION

 - An updated Environmental Hazard Statement is required.

SUMMARY OF OUTSTANDING DATA REQUIREMENTS

Due Date[2]

Toxicology

_____

[2] Due date is measured from the date of receipt of the Standard by the registrant unless otherwise specified.

21-day dermal toxicity	9 months
Metabolism (pharmacodynamic)	24 "

Due Date

Chronic Toxicity (rodent and non-rodent) 50 Months	
Oncogenicity (rat and mouse)	50 "
Reproduction (rat)	39 "

Environmental Fate/Exposure

Confined rotational crop study	39 "

Fish and Wildlife

Aquatic invertebrate life-cycle study	15 "

Residue Chemistry

Residue data - raw agricultural commodities	18 "
Processing studies	24 "
Plant metabolism	18 "

Product Chemistry

Majority of Data	9 -15 "

6.   Contact Person at EPA

William H. Miller
Product Manager (16)
Insecticide-Rodenticide Branch
Registration Division (TS-767)
Environmental Protection Agency
Washington, DC 20460

Tel. No. (703) 557-2600

DISCLAIMER: The information presented in this Chemical Information Fact Sheet is for informational purposes only and may not be used to fulfill data requirements for pesticide registration and reregistration.

CYFLUTHRIN

Reason for Issuance: Unconditional Registration
Date Issued: December 30, 1987
Fact Sheet Number: 164

1. DESCRIPTION OF CHEMICAL

   Generic Name: Cyfluthrin (Cyano (4-fluoro-3-phenoxyphenyl)
                  methyl 3-(2,2-dichloroethenyl)-2,2-dimethyl
                  cyclopropanecarboxylate
   Common Name: Cyfluthrin
   Trade Name: Laser
   EPA Shaughnessy Code: 128831-5
   Chemical Abstracts Service (CAS) Numbers: 68359-37-5
   Year of Initial Registration: 1987
   Pesticide Type: Insecticide
   Chemical Family: Synthetic pyrethroids
   U.S. Producers: Mobay Chemical Company;
                    Miles Laboratories, Inc.

2. USE PATTERNS AND FORMULATIONS

   Application Sites: General indoor and domestic outdoor
                       (ornamental areas)
   Method of Application: localized spot treatment,
                           crack and crevice and space sprays

   Formulation Types: a) Laser Ant and Roach Killer
                          0.1% Cyfluthrin
                          1.0% Piperonyl Butoxide
                          0.05% Pyrethrins

                       b) Laser Flying Insect Killer
                          0.04% Cyfluthrin
                          0.72% Tetramethrin
                          1.00% Piperonyl Butoxide

                       c) Laser Flea Killer Spray
                          0.01% Cyfluthrin
                          0.8% Chlorpyrifos
                          2.5% Pyrethrins
                          1.0% Piperonyl Butoxide

143

          d) Laser Ant and Roach Killer II
            0.10% Cyfluthrin
            1.00% Propoxur
            0.25% Pyrethrins
            1.00% Piperonyl Butoxide

          e) Laser House and Garden Multipurpose
            Insect Killer
            0.04% Cyfluthrin
            0.35% Tetramethrin
            1.00% Piperonyl Butoxide

Application Rates: 0.04 and 0.1% sprays

Usual Carriers: organic solvents; water

Limitations: For Domestic Indoor and Outdoor
              Use Only

3. SCIENCE FINDINGS

Summary Science Statement:

    Technical cyfluthrin does not exhibit high mammalian
toxicity. It is not considered to be mutagenic or
teratogenic and is not a skin sensitizer. The results
of the acute toxicity on the domestic use formulations
indicates that the chemical is of low toxicity. Cyfluthrin
has been tested in several studies (rat, mouse, chicken,
dog) for possible delayed type neurotoxicity and has been
shown to be neurotoxic when administered at relatively
high dosage levels under certain conditions. However
the neurotoxic dose levels were considerably higher
than those which elicited other signs of toxicity
indicating that cyfluthrin should not exhibit delayed
type neurotoxicity under conditions of use. Chronic
feeding and oncogenic studies indicate that cyfluthrin
is not an oncogen.

    Sufficient data are available to characterize cyfluthrin
from an environmental fate and ecological effects standpoint.
Cyfluthrin is extremely toxic to fish and aquatic organisms
but is practically non-toxic to upland game birds and
waterfowl. An acute contact LD_{50} study indicated that
cyfluthrin is toxic to the honey bee with an LD_{50} = 0.037
mg/bee. Cyfluthrin undergoes rapid photolysis, is readily
degraded by soil and is relatively insoluble in water.
There is little or no potential for leaching.

A tolerance assessment is not needed because the registered use pattern is for non-crop/non-food use. There are no data gaps.

Chemical/Physical Characteristics of the Technical Material

Physical State: liquid
Color: dark amber
Odor: aromatic solvent odor at room temparature
Melting Point: not applicable
Vapor Pressure: 2.1 x 10-8 mbars
Density: 0.830 + .005 g/ml
Solubility: 0.002 mg/ml at 20°C in water
pH: 5.6
Octanol/Water Partition Coefficient: 4.2×10^4

Toxicological Characteristics:

Technical Cyfluthrin

Acute Oral (Mouse):   males: 291 mg/kg   Toxicity Category II
   83.6%   females: 609 mg/kg

Acute Dermal:
LD_{50} > 5000 mg/kg   Toxicity Category III
21-day Dermal (rat):
LD50 > 5000 mg/kg   Toxicity Category III
Primary Dermal Irritation (rabbit):
none observed   Toxicity Category IV
Skin Sensitization (guinea pig):
not a sensitizer
Acute Inhalation: LC_{50} (4 hour)| males - >0.735 mg/l
                                     | females - 0.200 - 0.735 mg/l

Teratology: (rat) Maternal NOEL = 3 mg/kg/day
                    Maternal LEL = 10 mg/kg/day
                    Fetotoxic NOEL = 30 mg/kg/day
                    Teratogenic NOEL = 30 mg/kg/day

2-Year Feeding/Oncogenicity: (rat) Oncogenic NOEL = 22.5 mg/kg/day
                                   Systemic NOEL = 2.5 mg/kg/day
                                   Systemic NOEL = 7.5 mg/kg/day

Gene Mutation: negative
Structural Chromosome Aberration: negative
Unscheduled DNA Synthesis: negative

Laser Formulations

Acute Oral Toxicity in rats: Toxicity Category IV
LD_{50} = >5000mg/kg
Acute Dermal Toxicity in rats: Toxicity Category III
LD_{50} = >2000mg/kg
Acute Inhalation: Toxicity Category IV
LC_{50} = >5mg/l
Primary Eye Irritation: Toxicity Category IV
Primary Dermal Irritation: Toxicity Category IV
PIS = 0.9
Dermal Sensitiztion: not a sensitizer

C. Physiological and Biological Characteristics

    The mode of action in biological systems is stomach and
    contact exhibiting neuropathological characteristics typical
    of pyrethroid insecticides. Slight repellant effect.
    Foliar absorption: N/A
    Translocation: N/A

D. Environmental Characteristics

    Adequate data are sufficient to define the fate of cyfluthrin
    in the environment. Cyfluthrin is stable to hydrolysis
    at environmental pH and temperature and to photolysis.
    It photodegrades rapidly with a half-life of <2days at ph
    9. The major degradate is 4-fluoro-3-phenxyl-benzaldehyde
    Under the conditions of the soil TLC test using various
    soils, (aged and unaged) cyfluthrin residues are considered
    immobile in soils with a half-life of 56 to 63 days in
    German loam and sandy loam, respectively. Anaerobic
    conditions did not alter either degradation rate or
    products. Cyfluthrin's solubility in water is determined
    to be 2ppb (20°C). It has a bioaccumulation factor in
    fish of 858X. Residues are depurated radidly in untreated
    water. Accumulated residues are found in non-edible
    tissue. Cyfluthrin and its degradates do not leach into
    the soil. There are no concerns at this time in regard
    to ground water.

E.   Ecological Characteristics

Avian Acute Oral: Bobwhite quail - $LD_{50} > 2000$ mg/kg
Avian Subacute Dietary: Mallard duck - $LC_{50} > 5000$ ppm
                         Bobwhite quail - $LC_{50} > 5000$ ppm
Freshwater Fish: Bluegill - $LC_{50} = 1.5$ ppb
                  Rainbow Trout - $LC_{50} = 0.68$ ppb
Freshwater Invertebrate: Daphnia magna $LC_{50} = 0.14$ pptr
Marine/Estuarine Invertebrate: Mysid shrimp - $LC_{50} = 2.42$ pptr
                                Eastern Oyster $EC_{50} = 3.2$ pptr
Marine/Estuarine Fish: Sheepshead minnow - $LC_{50} = 4.05$ ppb

4.   Summary of Regulatory Position and Rationale

The Agency has determined that it should allow the
registration of cyfluthrin for domestic indoor and outdoor
use for control of household pests, such as fleas,
cockroaches, garden insects. Adequate data are available
to assess the acute toxicological effects of cyfluthrin to
humans. Since no crop, food or feed uses are proposed, a tolerance assessment is not necessary. As is typical of synthetic
pyrethroids, technical cyfluthrin is highly toxic to aquatic
organisms (invertebrates and fish). However, the proposed use
patterns should not pose any environmental hazard. None of the
criteria for unreasonable adverse effects listed in section
162.11(a) of Title 40 of the U.S. Code of Federal Regulations
have been met or exceeded for this use.

5.   Summary of Major Data Gaps

There are no data gaps for the domestic indoor/outdoor use.

6.   Contact Person at EPA

George T. LaRocca
Product Manager (15)
Insecticide-Rodenticide Branch
Registration Division (TS-767C)
Office of Pesticide Programs
Environmental Protection Agency
401 M Street, S. W.
Washington, D. C.   20460

Office location and telephone number:
Room 211, Crystal Mall #2
1921 Jefferson Davis Highway
Arlington, VA   22202
(703) 557-2400

CYPERMETHRIN

Reason for Issuance: Registration Update
Date Issued: January 3, 1989
Fact Sheet Number: 199

1. Description of Chemical

   Generic Name: (+/-)alpha-cyano-(3-phenoxyphenyl)methyl(+)-cis,trans-
                    3-(2,2-dichloroethenyl)-2,2-dimethylcyclopropanecarboxylate*
                    [containing 14% 1S-cis-S, 1R-cis-R, 1S-trans-S and 1R-trans-R
                    and 11% 1R-cis-S, 1S-cis-R, 1R-trans-S and 1S-trans-R). . . .]
   Common Name: Cypermethrin
   Trade Names: Ammo™; Cymbush®; Demon®; Cynoff™
   Other Names: Barricade; CCN52;
                    Cymperator; Cyperkill; Folcord;
                    Kafil; Super; NRDC 149;
                    Siperin; Ripcord
   EPA Shaughnessy Code: 109702
   Chemical Abstracts Service (CAS) Number: 66841-24-5
   Year of Initial Registration: 1984
   Pesticide Type: Pyrethroid-like; Insecticide/Miticide
   Chemical Family: Pyrethroid
   Manufacturers: FMC Corporation; ICI Americas, Inc.;
                    Shell International Chemical Company, Ltd.
                    (London)

2. Use Patterns and Formulations

   Application Rates: Applied to cotton, lettuce (head) and pecans at a
      rate up to 0.1 pounds active ingredient per acre. It may also be
      applied into overhead sprinkler irrigation water. Cypermethrin is
      also applied by Pest Control Operators as a crack, crevice, and spot
      spray treatment in and around areas including, but not limited to,
      stores, warehouses, industrial buildings, houses, apartment buildings,
      greenhouses, laboratories, and on vessels, railcars, buses, trucks,
      trailers, and aircraft.

   Also it may be used in nonfood areas of schools, nursing homes, hospitals,
      restaurants, and hotels; and food manufacturing, processing, and
      servicing establishments; as barrier treatments; and as an insect
      repellent for horses and ponies.

   Usual Carrier: Water and oil

148

Type of Formulations: 30.6%, 25.3%, and 36.6% emulsifiable concentrate and 88% technical

Limitations:

o   Registrations are being extended with an expiration date of June 15, 1989. Tolerances expire December 31, 1989.

o   RESTRICTED USE PESTICIDE - Extremely toxic to fish. For retail sale to and use only by Certified Applicators, or persons under their direct supervision, and only for those uses covered by the Certified Applicator's Certification.

o   REENTRY STATEMENTS - Do not treat areas while unprotected humans or domestic animals are present in the treatment areas.

Do not allow entry into treated areas without protective clothing until sprays have dried.

o   Do not apply within 21 days of harvest.

o   Do not graze livestock in treated orchards or cut treated cover crops for feed.

o   CROP ROTATION RESTRICTION - Do not plant rotational crops within 30 days of last application.

3.   Science Findings

Cypermethrin, a pyrethroid, is extremely toxic to fish. It is highly toxic to bees exposed to direct treatment on blooming crops or weeds. Cypermethrin has a low toxicity to mammals. EPA's review of the cypermethrin pond study indicated that the data, as presented, were not adequate or scientifically complete to allow the Agency to evaluate the actual impact that the use of cypermethrin would have an aquatic life forms. Therefore, ICI/FMC must repeat this study. The EPA Peer Review Committee completed its evaluation of cypermethrin with respect to its oncogenic potential and concluded that the data available for cypermethrin provide limited evidence of oncogenicity for the chemical in female mice. According to EPA Guidelines for Carcinogen Risk Assessment *(Federal Register* September 24, 1986), the Committee classified cypermethrin as a weak Category C oncogen (possible human carcinogen with limited evidence of carcinogenicity in animals). That is, cypermethrin produced benign lung adenomas at the highest dose level in only one sex and species of animal (female mice) and was not considered strong enough to warrant a "quantitative estimation of human risk."

Chemical/Physical Characteristics of the Technical Grade:

Physical State: Solid

Color: Colorless crystals
Odor: Odorless
Molecular Weight: 416.3
Molecular Formula: $C_{22}H_{19}O_3NCl_2$
Melting Point: 60 to 80 °C
Boiling Point: 170 to 195 °C
Density: 1.249 g CM^3 at 20 °C
Vapor Pressure: 8×10^{-4} at 80 °C; 1×10^{-7} at 20 °C
Solubility in Various Solvents:

Water	4 ppb
Base	as water
Acid	as water
Propylene glycol	insoluble (less than 0.5%)
Methanol	very soluble (approx. 75%)
Acetone	completely miscible
Cyclohexanone	completely miscible
Hexane	slightly soluble (7%)
Xylene	completely miscible
Methylene dichloride	completely miscible

Stability - as neat material: No detectable decomposition at normal ambient temperatures and for at least 3 months at 50 °C to date

- as dilute aqueous solution: Slowly hydrolyzes at pH 7 and below. Hydrolyzes more rapidly at pH 9.

Slow photodegradation in sterile solution in sunlight (< 10% in 32 days).

Toxicology Characteristics of the Technical Grade:

o Acute Oral LD_{50} - Rat: LD_{50} = 247 (187-326) mg/kg (males)
LD_{50} = 309 (150-500) mg/kg (females)

o Acute Dermal LD_{50} - Rabbit: LD_{50} > 2460 mg/kg

o Primary Dermal Irritation - Rabbit: PIS = 0.71 (not irritating)

o Primary Eye Irritation - Rabbit: Mild irritation

o Skin Sensitization - Guinea Pig: May cause allergic skin reactions

o Subchronic Oral - Rat: NOEL of 75 ppm for pharmacological effects. NOEL of 150 ppm for toxic effects.

o Chronic Toxicity - Rat: NOEL = 150 ppm
                             LEL = 1500 ppm (HDT)

o Oncogenicity - 24-Month Mouse: Positive neoplastic
   response in lung tissue. Increased incidence of benign
   adenomas in females (only), statistically significant
   at 1600 ppm (HDT). No evidence of oncogenicity in
   the rat at up to 1500 ppm (HDT).

o Teratogenicity - Rabbit: Not teratogenic at 30 mg/kg/day.
                             (HDT).
                 - Rat: Not teratogenic at 70 mg/kg/day (HDT).

o Reproduction - 3-Generation Rat: NOEL for adverse reproductive effects = 750 ppm (HDT), NOEL for systemic
   effects = 50 ppm, LEL = 150 ppm (decreased body weight
   gain in maturing pups).

o Mutagenicity - Ames Test: Not mutagenic.

o Mutagenesis - Host-Mediated Assay: Not mutagenic at
   50 mg/kg.

o Mutagenesis - Dominant Lethal: Not mutagenic at 25 mg/kg.

Physiological and Biochemical Characteristics:

The mode of action in biological systems is stomach and contact
   exhibiting neurotoxicological characteristics typical of pyrethroid
   insecticides. Slight repellent effect.

Foliar Absorption - N/A.

Translocation - N/A.

Environmental Characteristics:

Adequate data are sufficent to define the fate of cypermethrin in
   the environment. Cypermethrin is stable to hydrolysis, with an
   estimated T 1/2 exceeding 50 days at environmentally expected
   temperatures and pH values. Cypermethrin is extremely stable to
   photolysis in water with an estimated T 1/2 exceeding 100 days at
   environmentally expected temperatures and pH values. Photoproducts
   produced included DCVA, 3-phenoxybenzaldehyde, and 3-phenoxybenzoic
   acid. Cypermethrin photodegrades rapidly on soil surfaces (T 1/2:
   8 to 16 days) to many photoproducts, the major ones identified as
   3-phenoxybenzoic acid and compound XIV. Cypermethrin degrades in
   soil under laboratory conditions. The rate is more rapid on
   sandy clay and sandy loam soils than on clay soils and more rapid
   on soils lower in organic matter content and cation exchange
   capacity under all aerobic conditions. The T 1/2 in aerobic

soils ranged from 2 to 8 weeks. In sterile aerobic soils,
cypermethrin degraded with a T 1/2 of 20 to 25 weeks indicating
that microbes play a signicant role in soil degradation.
Cypermethrin degraded more slowly under anaerobic or waterlogged
conditions with the major metabolite 3-phenoxybenzoic acid produced.
Under aerobic conditions the major metabolites produced were DCVA
and 3-phenoxybenzoic acid. Cypermethrin does not leach significantly in soil. It has a low solubility in water (0.2 ppm) and,
consequently, high adsorption characteristics. The leaching
potential for its degradates may be higher. Under field conditions,
runoff of cypermethrin has been shown to occur to some degree and
was probably due to physical transport of the solid particles via
erosion. Cypermethrin itself degrades rapidly in the field with
a T 1/2 of 4 to 12 days. The persistence of the major aerobic
soil metabolites is not known. Both accumulation and depuration
of cypermethrin residues will occur in trout and catfish. Bioconcentration factors of approximately 1200X were calculated in
rainbow trout in a flowthrough study., The data requirement for
accumulation of cypermethrin in rotational crops has not been
satisfactorily completed.

Ecological Effects Characteristics:

o Avian acute oral LD_{50} - Mallard Duck: > 4640 mg/kg

o Avian dietary LC_{50} - Mallard Duck and Bobwhite Quail:
   LC_{50} > 20,000 ppm

o Avian reproduction - Mallard Duck and Bobwhite Quail:
   NOEL > 50 ppm (HDT)

o Fish acute 96-hour LC_{50} - Rainbow Trout = 0.82 ppb

o Fish acute 96-hour LC_{50} - Bluegill Sunfish = 1.78 ppb

o Aquatic invertebrate acute LC_{50} - Daphnia magna = 0.26 ppb

Tolerance Assessments:

Section 408 tolerances under the Federal Food, Drug, and Cosmetic Act
are established until December 31, 1989 for residues of the
insecticide cypermethrin [(+/-)alpha-cyano-(3-phenoxyphenyl)
methyl (+-cis,trans-3-(2,2-dichloroethenyl)-2 panecarboxylate]
and its metabolites* 3-PB Acid and DCVA in or on the following
raw agricultural commodities:

Commodity	ppm
Cattle, fat	0.05
Cattle, meat	0.05
Cattle, meat byproducts	0.05

Commodity	ppm
Cottonseed	0.5
Goats, fat	0.05
Goats, meat	0.05
Goats, meat byproducts	0.05
Hogs, fat	0.05
Hogs, meat	0.05
Hogs, meat byproducts	0.05
Horses, fat	0.05
Horses, meat	0.05
Horses, meat byproducts	0.05
Lettuce (head)	10.00
Milk	0.05
Pecans*	0.05
Sheep, fat	0.05
Sheep, meat	0.05
Sheep, meat byproducts	0.05

The acceptable daily intake (ADI) is calculated to be 0.01 mg/kg/day based on a dog study with a NOEL of 1.0 mg/kg/day and using a safety factor of 100. The maximum permissible intake (MPI) is calculated to be 0.60 mg/kg/day for a 60-kg person. Published tolerances result in a theoretical maximum residue contribution (TMRC) of 0.002773 mg/kg bwt/day. The existing TMRC is equivalent to 27.7 percent of the ADI. No additional data are required to support the current crop tolerances listed in 40 CFR 180.418.

4. Summary of Regulatory Position and Rationale

Adequate data are available to assess the acute and chronic toxicological effects of cypermethrin to humans. The Agency's review of the field study (§72-7) found that the data, as presented, were not scientifically adequate nor sufficiently complete to allow the Agency to evaluate the actual impact that the use of cypermethrin would have on aquatic life forms. ICI/FMC disagreed with the Agency's conclusion and will submit additional information necessary in order for the Agency to complete the risk assessment.

On the basis of this information, EPA seriously considered whether to issue new conditional registrations for cypermethrin. Based on the information submitted, on January 3, 1989 the Agency issued new conditional registrations of cypermethrin which will expire on June 15, 1989.

The Agency issued these new conditional registrations for this short period of time to ICI and FMC in light of their agreement to:

a. Submit all data generated in the course of the cypermethrin Alabama pond study pertaining to runoff and residues in water and sediment to the Agency by January 1989.

b. Submit all other data now in existence and not previously
    submitted to the Agency, as well as all other data generated
    in the future in the course of the cypermethrin Alabama pond
    study, as soon as is practical.

c. Conduct an aquatic mesocosm study (a simulated 2-year field
    study) for which EPA will develop and provide the protocol no
    later than April 15, 1989, in case ICI/FMC do not persuade
    EPA that the current cypermethrin pond study is acceptable.

d. Within 30 days of receipt of an EPA protocol for an aquatic
    mesocosm study, provide the Agency with written unconditional
    acceptance of the protocol and an unconditional commitment to
    conduct the study through completion. For any modification
    of the protocol to be valid, it must be agreed to by EPA
    within the above-mentioned 30-day period.

EPA has concluded that the continual use of cypermethrin for this short
   period of time will not cause a significant increase in the risk of
   adverse effects to the environment.

The Delaney Clause in section 409 of the Federal Food, Drug, and Cosmetic
   Act bars the establishment of food additive regulations for substances
   which induce cancer in man or test animals. Since cypermethrin has
   been found to produce an oncogenic response in test animals, no 409
   tolerances will be granted. Cypermethrin as parent does not leach
   significantly in soil. At this time, there are no concerns for
   ground water contamination.

5. Summary of Major Data Gaps

    Simulated and/or actual field study (§72-7)

6. Contact Person at EPA

    George T. LaRocca
    Product Manager (15)
    Insecticide-Rodenticide Branch
    Registration Division (H7504C)
    Office of Pesticide Programs
    Environmental Protection Agency
    401 M Street SW.
    Washington, DC 20460
    Office location and telephone number:
    Room 211, Crystal Mall #2
    1921 Jefferson Davis Highway
    Arlington, VA 22202
    Phone: (703) 557-2400

DISCLAIMER: The information presented in this Pesticide Fact Sheet is for
informational purposes only and may not be used to fulfill data requirements
for pesticide registration and reregistration.

DALAPON

Reason for Issuance: Registration Standard
Date Issued: July 31, 1987
Fact Sheet Number: 141

DESCRIPTION OF CHEMICAL

Dalapon (free acid)
    Chemical Name: 2,2-dichloropropionic acid
    CAS Registry No.:   75-99-0
    Shaughnessy No.:   28901

Dalapon Sodium Salt
    Chemical Name: sodium 2,2-dichloropropionate
    CAS Registry No.:   127-20-8
    Shaughnessy No.:   28902

Dalapon Magnesium Salt
    Chemical Name: magnesium 2,2-dichloropropionate
    CAS Registry No.:   N/A
    Shaughnessy No.:   28903

Year of Initial Registration: 1969

Trade Names: 2,2-dichloropropanoic acid, sodium and magnesium
    2,2-dichloropropanoate, Basfapon, Basfapon B, Basfapon N,
    Dalapon 85, dalapon magnesium, Dalapon-Na, dalapon sodium,
    Ded-Weed, Devipon, Dowpon M, DM Dalapon, DPA (JMAF), Gramevin,
    magnesium dalapon, magnesium salt of dalapon, proprop,
    Radapon, Revenge, sodium dalapon, and sodium salt of dalapon.

Pesticide Type: Herbicide

U.S. and Foreign Producers: Dow Chemical Company; BASF
    Aktiengesellschaft; Crystal Chemical Inter-America; SDS
    Biotech Corp.; Sintesul; and Vertec Chemical Corp.

USE PATTERNS AND FORMULATIONS

Application Sites: Terrestrial, food crops (root and tuber
    and legume vegetables; citrus, pome, stone and small fruits;
    tree nuts; cereal grains, coffee beans; sugarcane; birds-
    foot trefoil); terrestrial, nonfood crops (fallowland;
    pastureland; rangeland; noncrop areas; ornamental grasses,
    shrubs and trees); aquatic, food crop (taro paddy banks);

aquatic nonfood sites (drainage ditches; industrial waste
disposal systems, irrigation ditchbanks); forestry (forest
seedlings).

Predominant Uses: Primarily an industrial/noncrop herbicide
(70 percent); agricultural use (30 percent) primarily
on sugarcane and potatoes.

Types and Methods of Application: Conventional aerial or ground
equipment as fall or spring preplant, preemergence, or postemergence application, depending on the crop or site situation.

Application Rates: 0.075 lb active ingredient (ai) per 5,000
square feet to 25.5 lb ai per acre (A) for industrial/non-
crop use; 0.12-17 lb ai/A for food uses.

Types of Formulations: Wettable powder, soluble concentrate/
solid, impregnated materials, granular and technical.

Usual Carriers: Water

SCIENCE FINDINGS

Summary Science Statement: Available data are insufficient to
assess the toxicological effects of dalapon. Only one acceptable oncogenicity study is available and complete evaluation
of this study is not possible without historical control data.
This study did show an increased incidence of benign lung tumors.

There are no acceptable environmental fate studies. Preliminary data, however, do indicate there may be a potential for
groundwater contamination.

There are insufficient data available to fully assess ecological effects from the use of dalapon. However, there is a
threat to endangered plants based on the use pattern of dalapon.

Chemical Characteristics: Product chemistry data are unacceptable,
except dissociation constant as noted below, and are required.

Dissociation constant
(magnesium salt) = pka = 1.84 at 0.05M

Toxicology Characteristics: While published registrant data indicate low toxicity, the Agency has only limited toxicological
data on dalapon as discussed below:

In an acceptable two-year oncogenicity study, mice were fed
doses of 0, 2, 60 and 200 mg/kg/day. All doses (2, 60 and 200
mg/kg/day) showed a doubling of the incidence of benign lung
tumors in the male mice compared with controls. Historical
control data are needed to allow final evaluation of the study.

A supplementary study is available on the teratogenic
potential of dalapon in the rat. No teratogenic response
was reported. However, a NOEL for fetal and maternal
toxicity was not determined.

Physiological and Behavioral Characteristics:

Foliar Absorption: Dalapon is absorbed by roots and
leaves but is easily washed off the foliage.

Translocation: Dalapon translocates readily throughout
the plant and accumulates in young tissue; it is not
degraded in plants.

Mechanism of Pesticide Action: The mode of action of dalapon
involves interference of meristematic activity in root
tips and apical meristems, reduction of wax formation
on leaf surfaces, and alteration of cell membranes.

Environmental Characteristics: There are no acceptable studies
to assess the environmental fate of dalapon. Supplementary
data do indicate that there may be a potential for groundwater
contamination because dalapon appears to be highly soluble
in water, to leach readily and to be moderately persistent.
Dalapon has been designated as one of the pesticides for
analysis in the National Pesticides in Well Water Survey.

Ecological Characteristics:

Avian Dietary Toxicity - Japanese quail   = > 5000 ppm
                         Ring-necked
                           pheasant   = > 5000 ppm
                         Mallard duck   = > 5000 ppm

Freshwater Fish Toxicity - Bluegill   = 105 ppm

Aquatic Invertebrates Toxicity - Daphnia pulex = 11 ppm

Potential Problems Related to Endangered Species: The Office
of Endangered Species has issued a jeopardy opinion which
indicates that several plant species could be in jeopardy
if herbicides are used within or adjacent to their habitat.
This opinion applies to dalapon's registered forest uses.

Tolerance Assessment: Tolerances have been established for
residues of dalapon in a variety of raw agricultural commodities and meat byproducts (40 CFR 180.150(a)(b)) and in
processed food (21 CFR 193.105) and feed (21 CFR 561.110).

The toxicology data for dalapon are insufficient to determine an Acceptable Daily Intake (ADI). Prior to reevaluation of dalapon, an ADI of 0.08 mg/kg/day was considered
appropriate based on a NOEL of 8 mg/kg/day from a 2-year
rat study and an uncertainty factor of 100. This study

has since been determined to be unacceptable. However,
until such time as data are available and an ADI can be
established, the rat study will be used to determine a
provisional ADI (PADI). Because of the study's deficiencies, an uncertainly factor of 1000 was used to arrive at
a PADI of 0.008 mg/kg/day. Comparing the percent of crop
treated to the PADI results in a Theoretical Maximum Residue Contribution (TMRC) for the U.S. population of 0.00036
mg/kg/day. The TMRC occupies 4.5 percent of the PADI.

Pesticide Incident Monitoring System (PIMS): PIMS files contain
22 incident reports involving dalapon, nine of which involved
dalapon in combination with other ingredients. Twelve were
related to agricultural use. Equipment failure occurred in
four of the incidents. Fifteen of the incidents involved
exposure to humans; while medical treatment was administered
in 14 of these incidents, none required hospitalization.

SUMMARY OF REGULATORY POSITION AND RATIONALE

Summary of Agency Position: The Agency is requiring registrants of dalapon to submit additional data as identified
in the Registration Standard and summarized in the following section. The Agency will not establish any new food
use or register any significant new uses until adequate
data are available to fully assess dalapon.

Unique Warning Statements Required on Labels: Unique labeling
is not imposed in the Registration Standard. Endangered
species labeling, however, is required for forest uses.
This labeling is addressed in Pesticide Registration
Notice (PR) 87-4 dated May 1, 1987.

SUMMARY OF MAJOR DATA GAPS

Data	Due Date (From Issuance of Standard)
Product Chemistry	6-12 months
Residue Chemistry	15-18 months
Environmental Fate	9-39 months
Toxicology	9-50 months
Ecological Effects	9-12 months

CONTACT PERSON AT EPA: Robert J. Taylor
Office of Pesticide Programs, EPA
Registration Division (TS-767C)
401 M Street SW.
Washington, DC 20460
Phone: (703) 557-1800

DISCLAIMER: The information presented in this Fact Sheet is
for information purposes only and may not be used to fulfill
data requirements for pesticide registration and reregistration.

2,4-DB

Reason for Issuance: Registration Standard
Date Issued: October 3, 1988
Fact Sheet Number: 179

1. Description of Chemical(s)

   Generic Name(s): 4-(2,4-dichlorophenoxy)butyric acid (2,4-DB); 2,4-DB
   sodium salt; 2,4-DB, dimethylamine salt; 2,4-DB, butyl
   ester; 2,4-DB, 2-butoxyethyl ester; and 2,4-DB, isooctyl
   ester.
   Common Name(s): 2,4-DB, and its sodium salt, amine and esters.
   Trade Name(s): 2,4-DB is available under the trade names, Butoxone and
   Butyrac, formulated as an amine salt or ester.
   EPA Shaughnessy Codes: 030801 (acid)
                      030804 (sodium salt)
                      030819 (dimethylamine salt)
                      030853 (2-butoxyethyl ester)
                      030856 (butyl ester)
                      030863 (isooctyl ester)
   Chemical Abstracts Service (CAS) Number(s): 94-82-6 (acid)
                                                10433-59-7 (sodium salt)
                                                2755-42-1 (dimethylamine salt)
                                              32357-46-3 (2-butoxyethyl ester)
                                              6753-24-8 (butyl ester)
                                            1320-15-6 (isooctyl ester)
   Year of Initial Registration: 1958
   Pesticide Type: Herbicide; plant growth regulator
   Chemical Family: Chlorinated phenoxys
   U.S. and Foreign Producers: 2,4-DB technical products are manufactured
                                         by both U.S. and foreign companies.

2. Use Patterns and Formulations

   Registered Uses: Terrestrial Food

   Predominant Uses: Agricultural crops, alfalfa (16-67% of total usage),
      soybeans (27-68%), peanuts (5-13%), and clover (less than 2%).

   Pests Controlled: Broadleaf weeds

   Formulation Types Registered: Liquid
      (emusifable concentrate, soluble concentrate)

159

Method of Application: Ground equipment and aircraft.

3. Science Findings

Existing data are inadequate to assess the carcinogenic potential of
    2,4-DB. In a rat study, 2,4-DB was not oncogenic. In a mouse
    oncogenic study there was a weak but possible dose-relationship
    involving hepatocellular carcinomas in males. A new study or
    historical control data in certain tumor incidences in the strain
    of mice at the testing facility are required before the significance
    can be determined. Additional data are needed to determine the
    teratogenic potential of 2,4-DB.

2,4-DB is generally formulated as a sodium salt, an amine salt or an
    ester. The amine and ester forms may differ in biological activity
    and environmental fate from the parent compound. Data are needed on
    each amine and ester to enable a complete assessment.

Concern about possible groundwater contamination exists for the family
    of 2,4-D compounds (2,4-D, 2,4-DB, and 2,4-DP). Additional data and
    a label warning statement are required.

Chemical Characteristics: (2,4-DB acid technical)

Physical State: Crystalline solid
Color: White to light brown
Odor: Slight phenolic
Melting Point: 116 to 119 °C
Solubility: Highly soluble in acetone, benzene, carbon tetrachloride,
    diesel oil and kerosene; slightly soluble in water.
Vapor Pressure: Negligible at 25 °C
Stability: Stable

Toxicological Characteristics: (Note: 2,4-DP acid is test material.)

Acute Oral Toxicity - Rats: 2.33 g/kg (males), 1.54 g/kg (females),
    1.96 g/kg (males), 1.47 g/kg (females); Toxicity Category III

Acute Dermal Toxicity - Rabbits: > 2 g/kg; Toxicity Category III

Primary Eye Irritation - Rabbits: All irritation cleared at 7 days;
    Toxicity Category III

Primary Skin Irritation - Rabbit: No irritation at 27 and 72 hours;
    Toxicity Category IV

Subchronic Toxicity - Dog:
    Lowest-observed-effect level (LOEL) = 25 mg/kg/day
    No-observed-effect level (NOEL) = 8 mg/kg/day
    Mortality, body weight depression, histolopathological findings
    at LOEL and highest dose level (80 mg/kg/day)

Chronic Toxicity:

Chronic Feeding/Oncogenicity - Rats:
Systemic LOEL = 30 mg/kg/day for non-oncogenic effects
Systemic NOEL = 3 mg/kg/day for non-oncogenic effects
Decreased mean body weight gains, change in blood chemistry and
hematology parameters and significantly lower heart weights than
control. Not oncogenic under the conditions of the study.

Oncogenicity Study - Mice:
Weak, but possible dose-relationship involving hepatocellular
carcinomas in males. A new study is required or this study may
be upgraded if historical control data on certain tumor incidences
in the strain of mice at testing facility are submitted and justification that female mice of this strain were tested at or close to
the Maximum Tolerated Dose (MTD).

Teratology Study: No acceptable data available.

Reproduction Study:
NOEL = 15 mg/kg/day
At highest dose (75 mg/kg/day) ovarian weight was significantly less
in dams mean body weights were lower, fewer pups were born per litter,
and extremely high pup mortality occurred during lactation period.

Mutagenicity Study:
Ames study negative with and without metabolic activation.
Chinese Hamster Ovary assay with activation suggests a weak
mutagen immediately below doses causing high levels of cytotoxicity.
In Chinese Hamster Ovary assay a significant increase in chromosomal
aberration with 17.25-hour exposure, but no increase with 2-hour
exposure. Unscheduled DNA synthesis, no evidence of induction.

Physiological and Behavioral Characteristics:

Mechanism of Pesticide Action: Phenoxy herbicides (including 2,4-DB)
are hormone weed killers affecting the activity of enzymes, respiration,
and cell division.

Environmental Characteristics: No data are available.

Ecological Characteristics: (Note: all figures are LC_{50} values)

Avian Toxicity: No acute oral studies available. Dietary studies
indicate 2,4-DB acid is practically nontoxic (> 5000 ppm, respectively)
to waterfowl and upland game birds.

Aquatic Organism Toxicity: 2,4-DB acid is moderately to slightly toxic to freshwater fish (18 ppm fathead minnow, 7.5 to 17 ppm bluegill sunfish, 2.0 to 14 ppm rainbow trout). No studies are available for freshwater invertebrates.

Nontarget Insect Toxicity: 2,4-DB dimethylamine salt has a low toxicity to bees.

Nontarget Plants/Endangered Species: Since 2,4-DB is a broadleaf herbicide, a potential hazard exists for nontarget plants. Hazard assessments for endangered species cannot be completed until additional data are received.

Tolerance Assessment:

Tolerance Established: Tolerances of 0.2 ppm have been established for 2,4-DB and its metabolite 2,4-D in or on alfalfa, clover, mint hay, peanuts, soybeans, soybean hay, and birdsfoot trefoil (40 CFR 180.331).

Results of Tolerance Assessment: A provisional acceptable daily intake of 0.01 mg/kg/day for 2,4-DB acid has been established based on a rat chronic feeding study with NOEL of 3 mg/kg/day and utilizing a 300X safety factor. A Theoretical Maximum Residue Contribution (TMRC) for the U.S. population was calculated to be 0.000083 mg/kg/day which utilizes 0.83 % of the PADI

Reported Pesticide Incidents: Based on the Pesticide Incident Monitoring System files covering the period of 1966 to 1979, reports were received concerning off-target movement for unspecified 2,4-D (family) compounds. The incidents involved drift from aerial (173 reports) and ground (104 reports) applications, as well as volatilization (35 reports) and resulted in damage to nontarget crops and other desirable plants.

4. Summary of Regulatory Position and Rationale

Summary of Agency Position: The Agency has concluded that existing data are inadequate to assess the carcinogenic potential of 2,4-DB. Under conditions of the study 2,4-DB was not oncogenic in the rat. In a mouse oncogenic study with 2,4-DB acid there was no conclusive evidence of oncogenicity. However, there was a weak but possible dose relationship involving hepatocellular carcinomas in males. This mouse oncogenicity was deficient because historical control data on certain tumor incidences in the strain of mice at the testing facility were not submitted. Unless the data are submitted, the study must be repeated.

Data are being required on the salt, ester and amine formulations of 2,4-DB as well as on the acid. The Agency will not establish any significant new food use tolerances or register any significant new uses at this time.

Unique Label Warning Statements: Groundwater Advisory - (end-use
products [EPs]): "This product can reach ground water as a result
of mixing and loading. To minimize groundwater contamination from
spills during mixing and loading and cleaning of equipment, take the
following steps:

Mixing and Loading: When mixing, loading or applying this product,
wear chemical resistant gloves. Wash nondisposable gloves
thoroughly with soap and water before removing.

The mixing and loading of spray mixtures into the spray equipment
must be carried out on an impervious pad (i.e., concrete slab,
plastic sheeting) large enough to catch any spilled material. If
spills occur, contain the spill by using absorbent material (e.g,
sand, earth or synthetic absorbent). Dispose of the contaminated
absorbent material by placing in a plastic bag and following
disposal instructions on this label.

Triple rinse empty containers and add the rinsate to the mixing tank.

Cleaning of Equipment: When cleaning equipment, do not pour the
washwater on the ground; spray or drain over a large area away
from wells and other water sources."

5. Summary of Major Data Gaps

The following data are required for 2,4-DB acid. The Agency is also
requiring data on each individual salt, ester and amine of 2,4-DB.
Specific requirements are detailed in the Data Tables, Appendix I
of the Registration Standard, which can be obtained from the
Product Manager listed below.

Study	Due Date - From Date of Standard
Product Chemistry	6 - 15 months

Residue Chemistry:
   Plant and animal metabolism
   Analytical methods
   Residue studies
	18 - 24 months

Toxicology: Acutes, subchronic
   oral, 21-day dermal,
   second species oncogenicity
   or historical data from mouse
   oncogenicity study, chronic
   feeding (nonrodent)
   teratogenicity, mutagenicity
	6 - 50 months

Ecological Effects: 9 - 12 months
  Avian oral/dietary,
  aquatic organisms
  (freshwater fish/
  invertebrates),
  nontarget insect and
  plant studies

Environmental Fate: 9 - 27 months
  All studies

6. Contact person at EPA

  Mr. Richard F. Mountfort
  Product Manager (23)
  Fungicide-Herbicide Branch
  Registration Division (TS-767C)
  Office of Pesticide Programs
  Environmental Protection Agency
  401 M Street SW.
  Washington, DC 20460
  Telephone: (703) 557-1830

DISCLAIMER: The information presented in this Pesticide Fact Sheet is for
informational purposes only and may not be used to fulfill data requirements
for pesticide registration and reregistration.

DCPA

Reason for Issuance: Registration Standard
Date Issued: June 6, 1988
Fact Sheet Number: 166

1. DESCRIPTION OF CHEMICAL

Generic Name:	Dimethyl tetrachloro-terephthalate
Common Names:	No common name has been assigned; DCPA is commonly used; other names in use are chlorothal and chlorothal-dimethyl
Trade Names:	Dacthal●
EPA Shaughnessy Code:	078701
Chemical Abstracts Service (CAS) Number:	1861-32-1
Year of Initial Registration:	1958
Pesticide Type:	Herbicide
Chemical Family:	Chlorinated benzoic acids
U.S. Producer:	Fermenta Plant Protection Company

2. USE PATTERNS AND FORMULATIONS

Application sites: Terrestrial food crops (Agricultural Crops), Terrestrial Nonfood Crops (Agricultural Crops), Ornamental Plants and Forest Trees, Domestic Outdoor (Ornamental Plants and Forest Trees).

Types of formulations: Formulation intermediates containing
50, 75 and 90 percent DCPA; Wettable Powders containing
25, 50, 60 75, and 90 percent DCPA and granulars containing 1.15 to 24.0 percent DCPA.

Types and methods of application:

Applied with ground or aerial equipment to soil
pre-emergence to weed seed germination, broadcast or
in bands, either post-plant, post-transplant or at
layby. Applied to ornamental turf and lawns with
either spray or granular equipment before weed seed
germination.

Usual carrier: Water and fertilizers.

3.   SCIENCE FINDINGS

Summary Science Statement: DCPA and its metabolites
appear to have low acute and chronic toxicity based on
the limited studies that are available. However, these
products contain 2,3,7,8-tetrachlorodibenzo-p-dioxin
(2,3,7,8-TCDD) and hexachlorobenzene (HCB) as impurities
from the manufacturing of technical DCPA. These impurities
have chronic toxicological properties (including oncogenic,
teratogenic, fetotoxic, mutagenic or adverse effects on
immune response in mammals) that are of particular concern
in the reregistration of DCPA pesticide products. The
Agency has classified both these impurities as Probable
Human Carcinogens (Group B2). The Agency has performed
preliminary assessments of the risks posed by these impurities and is requiring environmental fate and residue
chemistry studies on the impurities in order to refine
these risk assessments.

Chemical characteristics:

Physical state:	Crystalline solid
Color	White
Odor	Odorless
Melting point	156°C
Solubility	Insoluble in water, high in aromatic hydrocarbon solvents

<u>Toxicological Characteristics of DCPA</u>:

- Acute toxicity: Data gap.

- Dermal sensitizaton: Not a sensitizer in guinea pig.

- Subchronic toxicity: Data gap.

- Chronic toxicity: In an acceptable chronic oral
  toxicity of DCPA in dogs, four male and four female
  beagle dogs were dosed with DCPA at 100, 1,000, and
  10,000 ppm in their diet. No compound-related effects
  were observed. The NOEL was 10,000 ppm.

  In a chronic toxicity study in rats, 35 male and 35
  female rats were dosed with DCPA at 100, 1,000 and
  10,000 ppm in their diets for 2 years. At termination
  of the study, the kidney weights in males and the
  adrenal weights of females were significantly higher than
  controls. The NOEL was 1,000 ppm (50 mg/kg/day). The
  LEL was 10,000 ppm (500 mg/kg/day).

- Oncogenicity: Data gap.

- Reproductive toxicity: Data gap.

- Teratology: Twenty-five bred female Charles River
  rats were fed 500, 1,000, and 2,000 mg/kg/day. No
  compound related effects were observed in the dams
  or the fetuses at the highest dose tested.

- Mutagenicity: DCPA did not induce toxic effects in in
  an <u>in vivo</u> dominant lethal study when administered to
  male rats in a single treatment of 3.16, 31.6 and 316
  mg/kg prior to mating.

- Metabolism: Data gap.

<u>Physiological and Biochemical Characteristics</u>:

- Mechanism of pesticidal action: DCPA appears to inhibit
  the normal cell division of root tips of a wide spectrum
  of plants. The precise mechanism of this effect is
  not understood.

- Metabolism and persistence in plants and animals:
  <u>Plants</u>: Radio-labeled DCPA was not translocated from
  treated leaves of cotton plants. If applied to soil
  or nutrient solution roots of cotton plants appeared
  to absorb it and translocate it to the stems and foliage
  of the plants. Monomethyl tetrachloroterephthalate
  (MTP) and tetrachloroterephthalic acid (TPA) and DCPA
  have been detected in residue studies from numerous
  agricultural commodities.

<u>Animals</u>: Non-radiolabeled studies have detected DCPA, MTP and TPA in ruminants and poultry. ^{14}C-DCPA studies are needed for both metabolism of DCPA in animals and plants

<u>Environmental Characteristics</u>:

- Environmental fate data include a hydrolysis study which indicates that DCPA is stable to hydrolysis for up to 36 days.

- Data gaps exist for all other environmental fate requirements.

- DCPA and its degradates were detected in ground water samples in four states. The highest level detected was 1139 parts per billion (ppb). The Agency has proposed a drinking water health advisory level of 3500 ppb. A ground water monitoring study is required to assess the extent of ground water contamination.

<u>Ecological Characteristics</u>:

- Acute avian oral toxicity: data gap.

- Avian dietary toxicity: data gap.

- Avian reproductive toxicty: data gap.

- Freshwater fish acute toxicity: LC_{50} = >100 to >320 ppm (practically nontoxic) for bluegill sun fish and LC_{50} = 30 ppm (slightly toxic) for rainbow trout.

- Freshwater invertebrate acute toxicity: EC_{50} ranged from 27 ppm to 135 ppm (slightly to practically nontoxic) for <u>Daphnia magna</u>. LC_{50} >100 ppm (practically nontoxic) for <u>Chironomus plumosus</u>. LC_{50} >6.2 ppm (moderately toxic) for <u>Gammarus pseudolimnacus</u>.

- Estuarine and marine organisms acute toxicity: LC_{50} > 1,000 ug/L (practically nontoxic) for Brown shrimp (<u>Penaeus aztecus</u>). LC_{50} = 620 ug/L (practically non-toxic) fpr Eastern oyster (<u>Crassostrea virainica</u>). LC_{50} = >1,000 ug/L (practically nontoxic) for Sheepshead Minnow (<u>Cyprinodon variegatus</u>).

- Available information suggests there is no acute hazard to endangered aquatic species. No terrestrial endangered species have been associated with the use patterns of DCPA products. No data are available to evaluate the hazard to endangered avian species or endangered aquatic plant species.

Tolerance assessment:

- Tolerances have been established for residues of DCPA on
  raw agricultural commodities. (See table on the following
  page for a listing of tolerances.)

- Using a Provisional Acceptable Daily Intake (PADI)
  of 0.5 mg/kg/day based on the NOEL of 50 mg/kg/day
  observed in a chronic toxicity rat study, the Maximum
  Permissible Intake (MPI) for a 60 kg person is 30.0
  mg/kg/day. Using this value, the Agency calculates
  that existing tolerances occupy 1.3 % of the ADI.

4. SUMMARY OF REGULATORY POSITIONS

A. The Agency will not initiate a Special Review on DCPA
    at this time. There are presently no chronic toxico-
    logical concerns for exposures to DCPA; however
    concerns for the chronic toxicological effects of
    the two manufacturing impurities 2,3,7,8-TCDD and
    HCB have not been resolved by the available
    information.

    At the present time the Agency does not consider the
    risks due to 2,3,7,8-TCDD from the use of DCPA be
    unreasonable. The highest risk estimated was 10^{-6}
    for agricultural applicators and PCOs.

    The Agency is concerned about the risks due to HCB
    from the use of DCPA which are 10^{-6} for dietary
    exposure and 10^{-4} for agricultural applicators,
    PCOs and a child exposed while playing on a treated
    lawn. The applicator exposures can be reduced by
    using protective clothing. The uncertainties in
    the exposure estimates used to assess the risk to
    children playing on a DCPA-treated lawn are so great
    that the Agency believes that the risk assessment
    cannot be used to determine whether criteria for
    inititing a Special Review have been exceeded.
    These exposure esitimates will be refined when data
    on foliar and soil exposure to HCB become available.
    The Agency believes the dietary risk to HCB from
    DCPA uses is acceptable while required metabolism
    and residue data are being developed.

B. The Agency is not classifying any DCPA uses as
    being for restricted use. As discussed above, the
    uncertainties in the exposure estimates are very
    great. Accordingly, the Agency is unable to conclude
    that the risk posed by DCPA warrants its classifica-
    tion as a restricted use pesticide. The Agency will
    reconsider this position when the data it is requiring
    become available.

Table 1.   Tolerances in parts per million (ppm) for residues of DCPA and metabolites MTP and TPA.

Raw Agricultural Commodity	Residues of DCPA Parts per Million			
	U.S.	Canada	Mexico	Codex
Beans, field dry	2	–	–	–
Beans, mung, dry	2	–	–	–
Beans, snap, succulent	2	–	–	–
Broccoli	1	–	–	–
Brussels sprouts	1	–	–	–
Cabbage	1	–	–	–
Cantaloupes	1	–	–	–
Cauliflower	1	–	–	–
Collards	2	–	–	–
corn, field, fodder	0.4	–	–	–
corn, field, forage	0.4	–	–	–
Corn, grain (including field and pop)	0.05	–	–	–
corn, pop, fodder	0.4	–	–	–
Corn, pop, forage	0.4	–	–	–
Corn, sweet (K + CWHR)	0.05	–	–	–
Corn, sweet, fodder	0.4	–	–	–
Corn, sweet, forage	0.4	–	–	–
Cottonseed	0.02	–	–	–
Cress, upland	5	–	–	–
Cucumbers	1	–	–	–
Eggplant	1	–	–	–
Garlic	1	–	–	–
Honeydew melons	1	–	–	–
Horseradish	2	–	–	–
Kale	2	–	–	–
Lettuce	2	–	–	–
Mustard, greens	5	–	–	–
Onions	1	–	–	–
Peas, southern, black-eyed	2	–	–	–
Peppers	2	–	–	–
Pimentos	2	–	–	–
Potatoes	2	–	–	–
Radish, roots	2.0	–	–	–
Radish, tops	15.0	–	–	–
Rutabagas	2	–	–	–
Soybeans	2	–	–	–
Squash, summer	1	–	–	–
Squash, winter	1	–	–	–
Strawberries	2	–	–	–
Sweet potatoes	2	–	–	–
Tomatoes	1	–	–	–
Turnips	2	–	–	–
Turnips, greens	5	–	–	–
Watermelons	1	–	–	–
Yams	2	–	–	–

C.   The Agency is requiring registrants to certify
     that the levels of 2,3,7,8-TCDD and HCB in DCPA used
     to formulate their products do not exceed 0.1 ppb
     and 0.3 percent, respectively. This measure will
     assure that the levels of these impurities in commer-
     cially available DCPA products do not exceed the
     reported maximum levels upon which the Agency based
     its risk assessment. Registrants must also analyze
     their products for other species of dioxins and
     establish certified limits for these impurities as
     well.

D.   The Agency will not register any significant new uses
     of DCPA until product chemistry, toxicology and
     residue chemistry data gaps have been filled.

E.   The Agency is requiring ground water monitoring
     studies to assses the extent of ground water
     contamination. Because of the low toxicity of DCPA,
     the low levels observed in ground water to date, and
     the limited number of observations of ground water
     contamination, the Agency finds that additional
     regulatory action is not warranted.

F.   The Agency is requiring dietary exposure information
     on DCPA impurities in order to determine the nature
     and magnitude of 2,3,7,8-TCDD and HCB residues. If
     these data show that these impurities or their
     metabolites accumulate in DCPA-treated agricultural
     commodities to levels that raise concerns about
     dietary risk, the Agency may find that additional
     regulatory action is warranted.

G.   The Agency is requiring dislodgeable residue and
     foliar dissipation data on HCB in order to estimate
     dermal exposure to this impurity. These data are
     needed to refine the estimates of HCB exposure to
     farm workers and to users of DCPA-treated lawns.

H.   The Agency is requiring the use of protective clothing
     and equipment for all uses of DCPA. End-use products
     registered for agricultural use or for professional
     use on ornamental turf must be labeled as follows:

          USE ONLY WHEN WEARING THE FOLLOWING PROTECTIVE
          CLOTHING AND EQUIPMENT DURING MIXING/LOADING AND
          APPLYING, REPAIR AND CLEANING OF MIXING, LOADING,
          AND APPLICATION EQUIPMENT, DISPOSAL OF THE
          PESTICIDE, AND EARLY REENTRY INTO TREATED AREAS:
          Wear a longsleeved shirt and long-legged pants
          and chemical resistant gloves.

> IMPORTANT! If pesticide comes in contact with
> skin, wash off with soap and water. Always
> wash hands, face, and arms with soap and water
> before smoking, eating, drinking, or toileting.
> AFTER WORK: Before removing gloves, wash them
> with soap and water. Take off all work clothes
> and shoes. Shower using soap and water. Do not
> reuse contaminated clothing. Clothing worn
> during work must be laundered separately from
> household articles. Clothing that becomes heavily
> contaminated or drenched must be destroyed
> according to State and local regulations.
> HEAVILY CONTAMINATED OR DRENCHED CLOTHING CANNOT
> BE ADEQUATELY DECONTAMINATED.

End use products registered for homeowner uses must be
labled as follows:

> USE ONLY WHEN WEARING THE FOLLOWING PROTECTIVE
> CLOTHING: Wear long-sleeved shirt and long-legged
> pants and chemical resistant gloves.

> IMPORTANT! If the pesticide comes in contact
> with skin wash off with soap and water. Always
> wash hands, face, and arms with soap and water
> before smoking, eating, drinking, or toileting.

> AFTER USE: Before removing gloves, wash them
> with soap and water. Take off all work clothes
> and shoes. Shower using soap and water. Do not
> reuse contaminated clothing. Clothing worn
> during use must be laundered separately from
> household articles.

I. The Agency is imposing an interim 24-hr reentry
interval for agricultural crop uses of DCPA pesticide
products until required reentry data have been found
to support a different reentry interval. End-use
products registered for agricultural crop use must
be labled as follows:

> Do not enter treated areas for 24 hours after ap-
> plication unless wearing long sleeved shirt and
> long pants.

J. The Agency is imposing a required precautionary
statement, environmental hazard statements, and
statement of practical treatment, as follows:

All products must bear the following precautionary
statement:

Causes moderate eye irritation. Avoid contact
with eyes or clothing. Wash thoroughly with soap
and water after handling.

All manufacturing use products must bear the following
environmental hazard statement:

Do not discharge effluent containing this product
into lakes, streams, ponds, estuaries, oceans, or
public waters unless this product is specifically
identified and addressed in an NPDES permit. Do
not discharge effluent containing this product
to sewer systems without previously notifying
the sewage treatment plant authority.
For guidance, contact your State Water Board or
Regional Office of the EPA.

All nongranular end-use products must bear the following
environmental hazard statements:

Do not apply directly to water or wetlands (swamps,
bogs, marshes, and potholes). Do not contaminate
water when disposing of equipment wastewater.

All granular end-use products must bear the following
environmental hazard statements:

Collect and incorporate granules spilled on the
soil surface. Do not apply directly to water or
wetlands (swamps, bogs, marshes, and potholes).
Do not contaminate water by cleaning of equipment
or disposal of wastes.

All products must bear the following statements of
practical treatment:

If in Eyes: Flush with plenty of water. Call a
physican if irritation persists.

K. The Agency has determined that the following studies
will receive priority review:

158.120 Product Chemistry

61-2   Description of Beginning Materials and
       Manufacturing Process
61-3   Discussion of Formation of Impurities
62-1   Preliminary Analysis

<u>158.125 Residue Chemistry - DCPA</u>

171-4   Nature of Residue (Metabolism)
          -Plants
          -Livestock
171-4   Residue Analytical Method
          -Plant Residues
          -Animal Residues
171-4   Storage Stability Data

<u>158.125 Residue Chemistry - Impurities</u>

171-4   Nature of Residue (Metabolism)
          -Plants
          -Livestock
171-4   Residue Analytical Method
          -Plant Residues
          -Animal Residues
171-4   Storage Stability Data

<u>158.130 Environmental Fate</u>

163-1   Leaching and Adsorption/Desorption
132-1   Foliar Dissipation
133-3   Dermal Exposure (Conditional, at option of Registrant)
133-4   Inhalation Exposure (Conditinal, at option of
          Registrant)
  --   Ground Water Monitoring

5.   <u>SUMMARY OF MAJOR DATA GAPS</u>

Product Chemistry

Toxicology:
    Acute testing
    Subchronic testing
    Chronic testing
    Special testing

Environmental Fate:
    Photodegradation
    Metabolism Studies - Laboratory
    Mobility Studies
    Dissipation Studies - Field
    Accumulation Studies
    Sub-division K, Reentry Studies
    Ground water Monitoring

Ecological Effects:
    Avian Testing
    Aquatic Crganism Testing
    Nontarget Area Phytotoxicity

Residue Chemistry for DCPA Products:
  Metabolism in Plants and Livestock
  Residue Analytical Methods for Plants and Animals
  Storage Stability Data for Raw Agricultural Commodities
  Magnitude of Residues in Food and Feed Commodities

Residue Chemistry for Impurities in DCPA Products:
  Metabolism in Plants
  Residue Analytical Methods for Plant Residues
  Magnitude of Residues in Raw Agricultural Commodities

6. CONTACT PERSON AT EPA

Richard F. Mountfort
Product Manager (23)
Fungicide-Herbicide Branch
Registration Division (TS-767C)
Office of Pesticide Programs
Environmental Protection Agency
401 M Street, S.W.
Washington, DC 20460

Office Location and Telephone Number:
Room 237, Crystal Mall Building #2
1921 Jefferson Davis Highway
Arlington, VA 22202
703-557-1830

DISCLAIMER: The information presented in this Pesticide Fact
Sheet is for informational purposes only and may not be used
to fulfill data requirements for pesticide registration and
reregistration. The complete Registration Standard for DCPA
may be obtained from the National Technical Information Service
(NTIS), Port Royal Road, Springfield, VA 22152 (Telephone
No. 703-487-4650). Price for paper copies vary depending on
the length of the document, microfiche copies are $5.95 each.
When ordering you must furnish the NTIS with the stock number.
The stock number may be obtained from the contact person
identified above.

DIAZINON

Reason for Issuance: Registration Standard
Date Issued: December 1988
Fact Sheet Number: 96.1

1. DESCRIPTION OF CHEMICAL

    Common Name: Diazinon

    Trade and Other Names: Spectracide, D.Z.N., Knox-Out

    EPA Shaughnessy Codes: 057801

    Chemical Abstracts Service (CAS) Number: 333-41-5

    ENT Registry Number: 19507

    Year of Initial Registration: 1956

    Pesticide Type: Insecticide

    Chemical Family: Organophosphate

    U.S. and Foreign Producers: Ciba-Geigy, Trans Chemic Industries Inc.,
                    and Makhteshim Agan (America) Inc.

2. USE PATTERNS AND FORMULATIONS

    Application: Aerosols, sprays, pet collars, ear tags, dips, ground
            blast, aerial, and soil incorporation.

    Annual Usage: 10 million pounds active ingredient (1985 data).

    Predominant Use(s): Agricultural, Home and Garden uses.

    Types of Formulations: Dusts, emulsifiable concentrates, granules,
            impregnated materials, liquids, microencapsulated, pressurized
            sprays, soluble concentrates, wettable powders.

3.   SCIENCE FINDINGS

Summary Science Statement

Diazinon is not oncogenic in the Fisher F344 rat or in the B6C3F1 mouse. Diazinon does not induce developmental toxicity in rats or rabbits at dose levels up to and including 100 mg/kg/day (highest dose tested).

Based on acceptable laboratory data, technical diazinon is characterized as very highly toxic to waterfowl on an acute oral basis, with an LD_{50} of 6.38 mg/kg for mallard ducks. Avian dietary studies characterized diazinon as highly toxic to upland game birds with a dietary LC_{50} of 245 ppm for bobwhite quail. Supplemental data characterize diazinon as very highly toxic to waterfowl with a dietary LC_{50} of less than 47 ppm for mallard ducks. End-use formulations of diazinon are characterized as very highly toxic to waterfowl, upland game birds and songbirds on acute oral and dietary basis. Technical diazinon and its end-use formulations are characterized as very highly toxic to aquatic organisms. It is considered highly toxic non-target insects.

Diazinon degrades rapidly under aerobic, anaerobic, aquatic anaerobic and sterile soil conditions. Microbial degradation appears to be the major pathway for the degradation of diazinon. The most probable mechanism responsible for degradation under sterile and anaerobic soil conditions appears to be chemical hydrolysis in acidic soils. Supplemental hydrolysis data indicate that diazinon is stable with respect to hydrolysis at pH 7 and 9 but hydrolyzes in non-sterile water with a pH of 5.

The major soil degradate is oxypyrimidine. Oxypyrimidine is more persistent than diazinon under aerobic and sterile, anaerobic and anaerobic aquatic soil conditions.

Diazinon with a 4 day EC50 of 4.14 mg/L and a 7 day EC50 of 3.7 mg/L is characterized as being moderately toxic to freshwater green alga. Diazinon caused greater than 25% detrimental effect in plant vigor in tomatoes, cucumbers, onions, and carrots. Diazinon caused a greater than 25% detrimental effect in seed germination in oats, tomatoes, and carrots. No detrimental effect was seen for seedling emergence.

Chemical Characteristics of the Technical Material

Physical State: Liquid

Color: Amber and brown

Odor: Mild, sweet, and aromatic

Molecular Weight and Formula: (304.3) $C_{12}H_{21}N_2O_3PS$

Boiling Point: 83-84°C at 0.002 mm Hg

Vapor Pressure: 1.4×10^{-4} mm at 20°C

Density: 1.12 g/ml at 20°C

Solubility in various solvents: In petroleum oils, 60 ppm in water at
                                 25°C and 40 ppm in water at 20°C.

Toxicology Characteristics

Acute Oral: LD50= 618 mg/kg[*]

Acute Dermal: LD50 > 2000 mg/kg[*]

Primary Dermal Irritation: Non-irritating[*]

Primary Eye Irritation: Non-irritating[*]

Dermal Sensitization: Positive response in 10% of the human volunteers
                       tested[*]

Acute Inhalation: LC50=3.5 mg/L[*]

Acute Delayed neurotoxicity: Data gap

Subchronic toxicity: Data gap

Oncogenicity: (mouse and rat) Not oncogenic

Chronic Feeding: Data gap

Metabolism: Data gap

Teratogenicity: (Rabbit and Rat) No developmental toxicity at doses up to
                 and including 100 mg/kg/day.

---

[*] This information may not be applicable to all currently registered
manufacturing use products (MUPs) of diazinon. The Agency is requiring
additional toxicity studies to determine the toxicological similarities
or dissimilarities of the registered MUPs.

Reproduction: Data gap

Mutagenicity: Data gap

Major routes of exposure

Dermal, oral and inhalation

Physiological and Biochemical Characteristics

Mechanism of Pesticidal Action: Cholinesterase Inhibition

Metabolism and Persistence in Plants and Animals: Data gap

Environmental Characteristics

Diazinon degrades rapidly under aerobic, anaerobic, aquatic anaerobic and sterile soil conditions. The major soil degradate is oxypyrimidine. Oxypyrimidine is more persistent than diazinon under aerobic and sterile, anaerobic and anaerobic aquatic soil conditions. Diazinon's potential to contaminate groundwater is unknown.

Ecological Characteristics

Avian acute toxicity:
    3.2 mg/kg (Red-winged blackbird)-Very highly toxic
    6.3 mg/kg (Mallard duck)   -Very highly toxic
    10 mg/kg (Bobwhite quail)   -Very highly toxic

Avian dietary toxicity:
    < 47 ppm (Mallard duck)   -Highly toxic
    245 ppm (Bobwhite quail)   -Highly toxic

Freshwater fish acute toxicity:
    Warmwater LC50 = 136 ug/L (Bluegill sunfish) -Highly toxic
    Coldwater LC50 = 90 ug/L (Rainbow Trout)   -Very highly toxic

Marine fish acute toxicity:
    LC50 = 1400 ug/L (Sheepshead minnow)-Moderately toxic

Freshwater invertebrate toxicity:
    LC50 = 0.2 mg/L (Gammarus fasciatus)-Very highly toxic

Marine invertebrate toxicity: Data gap

Non-Target Insects:
    LD50 = 0.22 ug/bee (contact: Honey Bees)-Highly toxic
    LD50 = 0.2 ug/bee (oral: Honey Bees)   -Highly toxic

TOLERANCE ASSESSMENT

Tolerances have been established for residues of diazinon in a variety of
raw agricultural commodities, in meat, fat and meat byproducts (40 CFR
180.153), food additives (40 CFR 185.1750) and in food handling establishments (40 CFR 185.1750), and feed handling/processing establishments
(40 CFR 186.1750). Tolerances for residues of diazinon are currently
expressed as residues of diazinon per se.

Codex MRL's, Canadian, and Mexican tolerances have been established for
many of diazinon's registered uses. Compatibility of these tolerances to
that of U.S. tolerances cannot be determined until all additional metabolism
and residue studies are available.

Based on inhibition of plasma cholinesterase observed in a 90 day rat
feeding study a NOEL of 0.009 mg/kg/day was established. A provisional
acceptable daily intake (PADI) has been established at 0.00009 mg/kg/day
utilizing an uncertainty factor of 100. The PADI is provisional because the
existing data base on diazinon is lacking chronic feeding studies, and a
multi-generation reproduction study.

4.   Required Unique Labeling

     o Restricted use Statement for all commercial outdoor
        uses (e.g., turf, and agricultural).

     o Restricted use Statement for residential products in toxicity category
        I or II (danger or warning).

     o Homeowner Protection Statements for indoor and outdoor
        application.

     o Institutional use Protection Statements for hospitals and schools.

     o Feed and Food Handling Establishments statements.

     o Worker Protection Statements for Toxicity Categories I, II and III end-
        use formulations.

     o 24-Hour Interim Reentry Interval (commercial and greenhouse use).

5.   Summary Of Regulatory Positions

o The Agency is deferring a decision at this time on whether to place diazinon into Special Review for its potential hazard to avian species resulting from its use on agricultural crops, on turf and other grassy sites (e.g., athletic fields, recreational parks, home lawns).

o The Agency is classifying all commercial outdoor uses (agricultural crops, ornamentals, and turf) of diazinon for restricted use, based upon its known toxicity to birds and aquatic species.

o All diazinon end-use products that are in Toxicity Category I or II (DANGER or WARNING) and bear product labeling that directly recommends residential use or reasonably can be interpreted to permit residential use are classified for restricted use. Such products may be used only by certified applicators or persons under their direct supervision. In the past, the Agency has allowed these types of products to be labeled , "For Agricultural Use Only" or "For PCO Use Only" in an attempt to limit use to commercial or trained applicators. However, these statements are unenforceable.

o   The Agency will is requiring the following testing of a series of typical end use products: acute oral, acute dermal, primary dermal irritation, primary eye irritation, dermal sensitization, and acute inhalation if appropriate. The Agency will reserve alternative product formulations testing, pending submission and review of toxicity testing on the stabilized technical diazinon products (manufacturing use products).

o The U.S. Fish and Wildlife Service, Division of Endangered Species and Habitat Conservation (DESCH) has determined that certain uses of diazinon, including uses on corn and sorghum may jeopardize the continued existence of endangered species. Based on this determination, DESHC specified reasonable and prudent alternatives to avoid jeopardizing the continued existence of the identified species by these uses. EPA is developing a program to reduce or eliminate exposure to these species to a point where use does not result in jeopardy. PR Notices 87-4 and 87-5, which specified labeling requirements designed to reduce or eliminate exposure to endangered species, have been withdrawn. The Agency will issue a notice of any necessary regulatory actions when the program is developed.

o The Agency will require each registrant of a manufacturing use product to submit the following toxicity studies on their current formulations: acute oral, acute dermal, acute inhalation, primary dermal irritation, primary eye irritation, dermal sensitization, and a 6-week rat feeding study. The Agency may require additional toxicity testing based upon its evaluation of these studies.

o The Agency will impose a 24-hour interim reentry interval for all commercial uses of diazinon (greenhouses, agricultural).

o The Agency is revising worker safety and protective equipment statements for end use products containing diazinon.

o The Agency is not imposing a ground water contamination advisory statement for diazinon products at this time. The Agency will assess the potential of diazinon for groundwater contamination after receipt and review of environmental fate data and will determine whether regulatory action is necessary.

o The Agency is imposing additional statements for all end-use products intended for use in and around the home.

o The Agency has determined that diazinon products must bear revised and updated labeling for hazards to nontarget species.

o The Agency will propose tolerance revocation for rutabagas, red chicory tops, and dandelions (40 CFR 180.153).

o Residue data must be submitted and tolerances must be proposed for corn fodder and forage, and either sorghum forage and fodder, or wheat forage, hay and straw, and soybean straw and hay.

o For the following crops; sorghum fodder and forage, soybean straw and hay, and sugarcane forage, the registrant is given the choice of developing and submitting data in support of tolerances, or of adding label restrictions aganist the feeding and grazing of treated crops to livestock. Each registrant must inform the Agency by 90 days of receipt of this Registration Standard which option he chooses. If he selects the label restrictions, labeling submitted at the 9 month deadline must include the grazing/feeding restrictions.

o The Agency will not grant any significant new tolerances or any sigificant new food uses for diazinon until the required residue chemistry and toxicology studies have been submitted and reviewed.

o The Agency is not requiring additional residue data to support the established tolerances for diazinon in or on guar beans and coffee beans.

o The Agency will revise commodity definitions for certain raw agricultural commodities listed in 40 CFR 180.153.

   a) Tolerance listing "peas with pods (determined on peas after removing any shell present when marketed)" will be revised to read, "peas, succulent".

   b) Tolerance listing "bean forage" will be revised to read, "bean vines".

   c) the tolerance listing for "wheat forage and straw" was omitted from listing and will read "wheat forage and straw 0.05 ppm".

o The common name "diazinon" will appear before the chemical name on the pesticide label. Labels must be revised to reflect this.

o Petroleum distillates and xylene based solvents must be declared as inert ingredients.

o The Agency has identified certain data that will receive immediate review when submitted;

      158.240 Residue Chemistry
- Plant and Animal Metabolism
- Special Storage Stability (EUP)

      158.290 Environmental Fate
- Hydrolysis
- Photolysis

      158.340 Toxicology
- Acute Toxicity Studies (MUP)
- 6 week Feeding study (MUP)
- Neurotoxicity study (MUP)
- Acute Toxicity Studies (TEP)

      158.490 Ecological Effects (all)

6.   Summary of Major Data Gaps

	Timeframe Ranges
Toxicology	12-50 Months
Environmental Fate/Exposure	9-39 Months
Ecological effects	9-30 Months
Residue Chemistry	18-24 Months
Product Chemistry	9-12 Months

7.   <u>CONTACT PERSON AT EPA</u>

George LaRocca
Product Manager (15)
Insecticide Rodenticide Branch
Registration Division (TS-767C)
Office of Pesticide Programs
Environmental Protection Agency
401 M Street, S. W.
Washington, D. C. 20460

Office location and telephone number:

Room 204, Crystal Mall #2
1921 Jefferson Davis Highway
Arlington, VA 22202
(703) 557-2400

DISCLAIMER: The information in this Pesticide Fact Sheet
is a summary only and is not to be used to satisfy data
requirements for pesticide registration and reregistration.
The complete Registration Standard for the pesticide may be
obtained from the National Technical Information Service.
Contact the Product Manager listed above for further
information.

2,4-DICHLOROPHENOXYACETIC ACID

Reason for Issuance: Registration Standard
Date Issued: September 1988
Fact Sheet Number: 94.2

DESCRIPTION OF CHEMICAL

Generic Name: 2,4-Dichlorophenoxyacetic Acid

Common Name: 2,4-D (includes parent acid as well as salt, amine and ester derivatives)

Trade Name: 2,4-D is available under a large selection of trade names, most often formulated as an inorganic salt, amine or ester.

EPA Chemical Code: 030001 (Acid)

Chemical Abstracts Service (CAS) Number: 94-75-7 (Acid)

Year of Initial Registration: 1948

Pesticide Type: Herbicide; Plant Growth Regulator.

Chemical Family: Chlorinated phenoxy

U.S. and Foreign Producers: 2,4-D technical products are manufactured by a large number of companies, both U.S. and foreign.

USE PATTERNS AND FORMULATIONS

Registered Uses: Terrestrial, food and nonfood; aquatic, food and nonfood; domestic; and forestry.

Predominant Uses: Postemergent weed control in agricultural crops (approximately 57 percent of total usage; over 45 percent of total usage is on wheat and corn; 20 percent of total usage on pastures and rangelands; other major crops are sorghum, other small grains, rice and sugarcane); the remainder is used on noncrop areas, with a small amount used as a plant growth regulator (in filberts, citrus and potatoes).

185

Formulation Types Registered: Granular; amine and ester
    liquids; aerosol spray (foam).

Methods of Application: Aerial and ground equipment,
    knapsack sprayers, pressure and hose-end applicators,
    and lawn spreaders.

SCIENCE FINDINGS

Summary Science Statement: The Agency's Office of Pesticide
    Programs (OPP) has classified 2,4-D as a Group D oncogen
    (not classifiable as to human carcinogenicity) because
    existing data are inadequate to assess the carcinogenic
    potential of 2,4-D.   Accidental human poisoning with
    2,4-D, which resulted in severe neurotoxicity, has been
    reported; adequate neurotoxicity studies are not
    available.   While published data indicate that 2,4-D may
    be teratogenic, an acceptable rat teratology study is
    negative; a study in rabbits is needed.

2,4-D is often formulated as various esters and amines.
    These formulations may affect the physical
    characteristics, biological activity and environmental
    fate of the parent compound.   Data are needed on each
    ester and amine before the Agency can completely assess
    2,4-D.

Although laboratory data demonstrate that 2,4-D is
    mobile in soils, its potential to contaminate
    groundwater is limited by its rapid rate of degradation
    and uptake by target plants.   However, residues of 2,4-D
    have been detected in groundwater, mostly from point
    sources, such as mixing, loading and disposal.

Certain formulations of 2,4-D are highly toxic to fish
    and/or aquatic invertebrates.   Other formulations, for
    which the Agency has data, are in the range of
    moderately toxic to practically nontoxic to nontarget
    organisms.   The Office of Endangered Species has issued
    biological opinions indicating that certain endangered
    species may be in jeopardy from the use of 2,4-D.

Chemical Characteristics (Acid):

    Physical state - Flakes, powder, and crystalline powder
        and solid
    Color - White to light tan
    Odor - Phenolic to odorless
    Melting point - 135-142°C
    Boiling point - 160°C at 0.4 mm Hg
    Solubility - Soluble in acetone, ethanol, aqueous
        alkali, alcohols, diethyl ether, ethyl ether,

isopropanol, methyl isobutyl ketone, most organic
solvents; insoluble in benzene, petroleum oils
Vapor pressure - 0.4 mm Hg at 160°C
Stability - Stable to melting point

Toxicology Characteristics (Acid, except as noted):

Acute Toxicity:

2,4-D Acid -

Oral (rat):   639 mg/kg (males); 764 mg/kg (females);
                Toxicity Category III[1]

Inhalation (rat):   1.79 mg/L; Toxicity Category III

Dermal Sensitization (guinea pig):   Not a sensitizer.

2,4-D Sodium Salt -

Oral (rat):   876 mg/kg (males); 975 mg/kg (females);
                Toxicity Category III

Dermal (rat):   >2000 mg/kg; Toxicity Category III

Diethanolamine Salt (Manufacturing-Use Product) -

Oral (rat):   >2000 mg/kg (males); 1605 mg/kg
                (females); Toxicity Category III

Dermal (rabbit):   >2000 mg/kg (males and females);
                     Toxicity Category III

Inhalation (rat):   >3.8 mg/L; Toxicity Category III

Primary Eye (rabbit):   Severe irritation and corneal
          ulcer not resolved 21 days post-treatment;
          Toxicity Category I

Primary Dermal (rabbit):   No signs of dermal
          irritation; Toxicity Category IV

Dermal Sensitization (guinea pig):   Not a dermal
          sensitizer.

Butoxyethyl Ester -

Oral (rat):   866 mg/kg; Toxicity Category III

---

[1]For a description of Toxicity Categories, see 40 CFR 156.10.

Dermal (rabbit):   >2000 mg/kg (females: 1829 (mg/kg
                      (males); Toxicity Category III

Inhalation (rat):   >4.6 mg/L; Toxicity Category III

Primary Eye (rabbit): Very mild eye irritation resolved
       in 72 hours; Toxicity Category III

Primary Dermal (rabbit): Very slight erythema cleared
       in 72 hours; Toxicity Category III.

Dermal Sensitization (guinea pig): Was a sensitizer in
       two tests and not a sensitizer in a third
       test.

Isooctyl Ester –

Oral (rat):   982 mg/kg (males); >720 <864 mg/kg
                 (females); Toxicity Category III

Dermal (rabbit):   >2000 mg/kg; Toxicity Category III

Isobutyl Ester –

Oral (rat):   700 mg/kg (males); 553 mg/kg (females);
                 Toxicity Category III

Dermal (rabbit):   >2000 mg/kg; Toxicity Category III

Isopropyl Ester –

Oral (rat):   640 mg/kg (males); 440 mg/kg (females);
                 Toxicity Category II

Dermal (rabbit):   >2000 mg/kg; Toxicity Category III

Inhalation (rat):   >4.97 mg/L; Toxicity Category III

Primary eye (rabbit): All irritation cleared at 4
       days; Toxicity Category III

Primary dermal (rabbit): No irritation at 72 hours;
                          Toxicity Category IV

Dermal Sensitization (guinea pig):   Nonsensitizer

Subchronic Toxicity: No acceptable data are available on
    2,4-D. The requirement for subchronic oral studies on
    the acid is waived because chronic studies are required;
    a subchronic dermal study is required.   Subchronic
    studies are required for the esters and amines.

Chronic Toxicity:

  Oncogenicity (rats): No observed effects level (NOEL)
    for systemic effects - 1 mg/kg/day; lowest observed
    effects level (LOEL) for systemic effects - 5
    mg/kg/day; further evaluation needed to determine
    if maximum tolerated dose was reached.

  Oncogenicity (mice): NOEL for systemic effects - 1
    mg/kg/day; LOEL for systemic effects - 15
    mg/kg/day; further evaluation needed to determine
    if maximum tolerated dose was reached.

  Teratology (rats): Fetotoxicity (delayed ossification)
    LOEL 75 mg/kg/day and NOEL 25 mg/kg/day; Maternal
    toxicity NOEL 75 mg/kg/day (highest dose tested).

  Reproduction (rats): NOEL 5 mg/kg/day.

Major Routes of Exposure: The major route of exposure is
  dermal; respiratory exposure is negligible.

Physiological and Behavioral Characteristics:

  Foliar Absorption: 2,4-D is absorbed through the roots
    and/or leaves depending upon the type of
    formulation. A rain-free period of 4 to 6 hours
    usually is adequate for uptake.

  Translocation: Following foliar absorption, 2,4-D
    translocates within the phloem, probably moving
    with the food material. Following root absorption,
    it may move upward in the transpiration stream.
    Translocation rate is influenced by the growth rate
    of the plant. Accumulation occurs principally at
    the rapid growth regions of shoots and roots.

  Mechanism of Pesticide Action: 2,4-D acid stimulates
    nucleic acid and protein synthesis affecting the
    activity of enzymes, respiration and cell division.
    Broadleaf plants exhibit malformed leaves, stems
    and roots.

Environmental Characteristics:

  Absorption and Leaching: 2,4-D is mobile to highly
    mobile in five soil types. Based on available
    data, aged 2,4-D residues are only slightly mobile.

  Microbial Breakdown: 2,4-D degrades rapidly in aerobic
    silty clay and loam soil systems.

Bioaccumulation: Available data indicate a low
    potential for 2,4-D to accumulate in fish.

Resultant Average Persistence: In aerobic silty clay
    and loam soils, 1.9-2.2 percent of applied 2,4-D
    remained at 51 days post-treatment; in four other
    soils, only 0.7-2.5 percent remained at 150 days
    post-treatment.

Environmental Fate and Surface and Groundwater
    Contamination Concerns: Although laboratory data
    demonstrate that 2,4-D is mobile in soils, its
    potential to contaminate groundwater is limited by
    its rapid rate of degradation and uptake by target
    plants. However, residues of 2,4-D have been
    detected in groundwater, mostly from point
    sources, such as mixing, loading and disposal.

Exposure of Humans and Nontarget Organisms: Accidental
    human poisoning with 2,4-D, which resulted in
    severe neurotoxicity, has been reported. Reports
    have been received concerning off-target movement
    of 2,4-D resulting in damage to crops or other
    desirable plants.

Exposure during Reentry Operations: Based on available
    data, 2,4-D products are of low toxicity (Toxicity
    Categories III and IV). Because of these low levels
    of toxicity, reentry is not a concern.

Ecological Characteristics (detailed information can be
    obtained from the Registration Standard):

Avian Toxicity: Acceptable data indicate that 2,4-D
    acid can be characterized as moderately toxic to
    practically nontoxic to avian species on an acute
    basis. Butyl ester can be characterized as
    practically nontoxic on an acute and chronic basis.

Fish Toxicity: Acceptable data indicate that 2,4-D acid
    and certain of its salts, esters and amines can be
    characterized in the range of moderately toxic to
    practically nontoxic to fish. However, the
    compounds N-oleyl-1,3-propylenediamine salt, N,N-
    dimethyloleyl-linoleylamine, butyl ester,
    butoxyethanol ester and propylene glycol butyl
    ether ester can be characterized as highly toxic to
    fish, based on the following toxicity values:

    N-oleyl-1,3-propylene-
        diamine salt   0.3 ppm (bluegill sunfish)
                        0.8 ppm (channel catfish)

```
N,N-dimethyloleyl-
    linoleylamine   0.64 ppm (rainbow trout)
Butyl ester         0.49-2.82 ppm (cutthroat trout)
                    0.5-2.8 ppm (lake trout)
                    0.4-0.96 ppm (rainbow trout)
                    0.29-0.3 ppm (bluegill sunfish)
Butoxyethanol ester 0.65 ppm (rainbow trout)
                    0.76-1.2 ppm (bluegill sunfish)
                    3.3 ppm (fathead minnow)
                    0.78-1.35 ppm (channel catfish)
Butoxypropyl ester  5.4 ppm (rainbow trout)
Propylene glycol
   butyl ether ester 0.33-2.8 ppm (cutthroat trout)
                    0.39-2.93 ppm (lake trout)
                    0.95-1.44 ppm (rainbow trout)
                    0.56-0.67 ppm (bluegill
                    sunfish)
```

Freshwater Invertebrates Toxicity: Of those compounds
    for which the Agency has data, reported toxicity
    values indicate that the compounds can be
    characterized as slightly toxic to practically
    nontoxic, excepted as noted below. The compounds
    set forth below have toxicity values which
    characterize them as highly toxic to aquatic
    invertebrates.

```
Dimethylamine       0.15 ppm (grass shrimp)
Isooctyl ester      0.5 ppm (waterflea)
Butoxyethanol ester 1.7-6.4 ppm (waterflea)
                    2.2 ppm (seed shrimp)
                    2.6 ppm (sow bug)
                    0.44-6.1 ppm (side swimmer)
                    0.39-0.79 ppm (midge)
Propylene glycol
   butyl ether ester 0.1-14 ppm (waterflea)
                    0.42 ppm (seed shrimp)
```

Estuarine and Marine Organisms Toxicity: Acceptable
    data are available only for the butoxyethanol ester
    which report toxicity values of 5.0 mg/L (longnose
    killifish), 2.6 mg/L (Eastern oyster) and 5.6 mg/L
    (brown shrimp), which indicate that the material is
    moderately toxic to estuarine and marine organisms.

Effects on Plants: Limited plant protection studies are
    available. In a spray drift study, two application
    methods were compared as to quantity and pattern of
    deposition. No difference was found between the
    amine derivatives (diethanolamine and
    dimethylamine). With these amines, drift was
    observed beyond 225 feet from the site of

application. No residues, attributable to drift,
were found when applied postemergent to wheat or
corn.

The toxicity of butoxyethanol ester was tested on
four species of algae. Toxicity values ranged from
75 mg/L to 150 mg/L.

Nontarget Insects: There is sufficient information to
characterize 2,4-D as relatively nontoxic to honey
bees, when bees are exposed to direct treatment.

Potential Problems Related to Endangered Species: The
Office of Endangered Species has determined that
certain uses of 2,4-D may jeopardize the continued
existence of endangered species or critical habitat
of certain endangered species.

Tolerance Assessment:

Tolerances Established: Tolerances and food and feed
additive regulations have been established for
residues of 2,4-D in a variety of raw agricultural
commodities and meat byproducts (40 CFR 180.142),
and in processed food (40 CFR 185.1450) and feed
(40 CFR 186.1450).

Results of Tolerance Assessment: A provisional
acceptable daily intake (PADI) of 0.003 mg/kg/day
for 2,4-D acid has been established based on a two-
year rat feeding study. Compound-related effects
were observed in the kidneys of both male and
female rats. The LOEL was 5 mg/kg/day and the NOEL
was 1 mg/kg/day. An uncertainty factor of 100 was
used to account for the inter- and intraspecies
differences. An additional uncertainty factor of 3
was used since there is no dog study available and
no information available that indicates the dog is
less sensitive than the rat.

Reported Pesticide Incidents: Based on the Pesticide
Incident Monitoring System files, covering the period
1966 to 1979, reports were received concerning the off-
target movement of 2,4-D in unspecified formulations,
esters and amines. The incidents involved drift from
aerial (173 reports) and ground (104 reports)
applications, as well as volatilization and drift (35
reports) and resulted in damage to off-target crops or
other desirable.plants.

SUMMARY OF REGULATORY POSITION AND RATIONALE

Summary of Agency Position: OPP has classified 2,4-D as a
    Group D oncogen (not classifiable as to human
    carcinogenicity).   EPA is, however, requiring additional
    data, including additional information on oncogenicity
    and teratogenicity and neurotoxicity studies, for
    further evaluation of 2,4-D.   Data are being required on
    the ester and amine formulations of 2,4-D as well as on
    the acid.   EPA will not establish any significant new
    food use tolerances or register any significant new uses
    at this time.

Additional data are needed to thoroughly evaluate the
ecological effects of 2,4-D and its potential to
contaminate groundwater.

EPA is developing a program to reduce or eliminate
exposure to endangered species from the use of 2,4-D to
a point where use does not result in jeopardy, and will
issue notice of any labeling revisions when the program
is developed.   Endangered species labeling is not
required at this time.

Unique Warning Statements Required on Labels:

Manufacturing-Use Products: "Do not discharge effluent
    containing this product into lakes, streams, ponds,
    estuaries, oceans, or public waters unless this
    product is specifically identified and addressed in
    an NPDES permit.   Do not discharge effluent
    containing this product to sewer systems without
    previously notifying the sewage treatment plant
    authority.   For guidance, contact your State Water
    Board or Regional Office of the EPA."

End-Use Products:

Aquatic Uses.   "Drift or runoff may adversely affect
    nontarget plants.   Do not apply directly to water
    except as specified on this label.   Do not
    contaminate water when disposing of equipment
    washwaters."

Nonaquatic Uses.   "Drift or runoff may adversely affect
    nontarget plants.   Do not apply directly to water
    or wetlands (swamps, bogs, marshes, and potholes).
    Do not contaminate water when disposing of
    equipment washwaters."

End-Use Products - Certain Formulations: End-use
    products containing the following formulations must

contain the above environmental precautions modified to indicate that the product is toxic either to fish or aquatic invertebrates:

Toxic to Fish - N-Oleyl-1,3-Propylenediamine salt
                N,N-Dimethyloleyl-Linoleylamine
                Butyl ester
                Butoxylethanol ester
                Propylene glycol butyl ether ester

Toxic to Aquatic Invertebrates - Dimethylamine
                                 Isooctyl ester

All End-Use Products: The following statements are required in the use directions for all end-use products:

Liquid Formulations: "This product can reach groundwater from mixing and loading. To minimize groundwater contamination from spills during mixing, loading and cleaning of equipment, take the following steps:

"Mixing and Loading: When mixing, loading or applying this product, wear chemical resistant gloves. Wash nondisposable gloves thoroughly with soap and water before removing.

"The mixing and loading of spray mixtures into the spray equipment must be carried out on an impervious pad (i.e., concrete slab, plastic sheeting) large enough to catch any spilled material. If spills occur, contain the spill by using an absorbent material (e.g, sand, earth or synthetic absorbent). Dispose of the contaminated absorbent material by placing in a plastic bag and following disposal instructions on this label.

"Triple rinse empty containers and add the rinsate to the mixing tank.

"Cleaning of Equipment: When cleaning equipment, do not pour the washwater on the ground; spray or drain over a large area away from wells and other water sources."

Granular Formulations: "This product can reach ground-water from improper handling. To minimize ground-water contamination from spills during loading and cleaning of equipment, take the following steps:

"Handling: When handling this product, wear chemical resistant gloves. Wash nondisposable

gloves thoroughly with soap and water before
removing. If spills occur, collect the material
and dispose of by following disposal instructions
on this label.

"Cleaning of Equipment: When cleaning equipment,
do not pour the washwater on the ground; spray or
drain over a large area away from wells and other
water sources."

End-Use Products - Certain Food/Feed Uses. Labels for
products registered for certain food/feed uses must
contain revised use directions pertaining to
appropriate preharvest, pregrazing and preslaughter
intervals; allowable range of diluent; and/or
maximum seasonal application rate and/or number of
applications.

SUMMARY OF MAJOR DATA GAPS. The following data are required
for 2,4-D acid. The Agency is also requiring data on each
individual ester and amine of 2,4-D. Specific requirements
are detailed in the Data Tables, Appendix I of the
Registration Standard, which can be obtained from the Product
Manager listed below.

Study	Due Date - From Date of Standard
Product Chemistry	6-15 months

Residue Chemistry:
  Plant and animal metabolism
  Analytical methods
  Residue studies
	18-24 months

Toxicology:
  Primary Eye and Dermal Irritation
  21-Day Dermal
  Chronic Toxicity (nonrodent)
  Teratogenicity (rabbit)
  Mutagenicity
  Metabolism
  Special Dermal (Neurotoxicity)
  Reserved: Oncogenicity (two species)
	9-50 months

Ecological Effects:
  Avian Dietary
  Aquatic Organism (freshwater fish
    and invertebrates; estuarine
    and marine organisms; accumulation)
  Phytotoxicity (Tier II)
	9-18 months

Study   <u>Due Date – From</u>
                                        <u>Date of Standard</u>

Environmental Fate: 9–50 months
  Hydrolysis
  Photodegradation (water, soil, air)
  Metabolism (anaerobic soil; aerobic
    and anaerobic aquatic)
  Leaching and Adsorption/Desorption
  Volatility (lab and field)
  Dissipation (soil, aquatic and forestry)
  Accumulation (confined rotational
    crops; irrigated crops; fish
    and aquatic nontarget organisms)
  Spray drift

CONTACT PERSON AT EPA: Mr. Richard Mountfort
                          Product Manager (Team 23)
                          Fungicide-Herbicide Branch
                          Registration Division (TS-767C)
                          Office of Pesticide Programs, EPA
                          Washington, DC 20460

                          Telephone: (703) 557-1830

DISCLAIMER: The information in this Pesticide Fact Sheet is
a summary only and may not be used to fulfill data
requirements for pesticide registration and reregistration.

2-(2,4-DICHLOROPHENOXY) PROPIONIC ACID*

Reason for Issuance: Registration Standard
Date Issued: September 1988
Fact Sheet Number: 180

DESCRIPTION OF CHEMICAL(S)

Generic Name(s): 2-(2,4-dichlorophenoxy)propionic acid
                2,4,DP, diethanolamine salt
                2,4-DP, dimethylamine salt
                2,4-DP, butoxyethyl ester
                2,4-DP, isooctyl ester

Common Names(s): 2,4-DP, or dichlorprop, and its amines and esters.

Trade Name(s): 2,4-DP is available under a number of trade names, most often formulated as an amine salt or ester.

EPA Shaughnessy Codes: 031401 (acid)
                        031416 (diethanolamine salt)
                        031419 (dimethylamine salt)
                        031453 (butoxyethyl ester)
                        031463 (isooctyl ester)

Chemical Abstracts Service (CAS) Number: 120-36-5 (acid)
                                          84731-66-8 (diethanolamine salt)
                                          53404-32-3 (dimethylamine salt)
                                          53404-31-2 (butoxyethyl ester)
                                          28631-35-8 (isooctyl ester)

Year of Initial Registration: 1968 (acid).

Pesticide Type: Herbicide; plant growth regulator.

Chemical Family: Chlorinated phenoxy.

U.S. and Foreign Producers: 2,4-DP technical products are manufactured by a number of companies, both U.S. and foreign.

*and amine salts and esters

197

USE PATTERNS AND FORMULATIONS

Registered Uses: Terrestrial Nonfood (golf courses, noncrop and
    recreational areas); Aquatic Nonfood (drainage ditchbanks);
    Domestic Outdoor (home lawns); Forestry (conifer/pine release
    forest plantation site preparation).

Predominant Uses: Ornamental turf, e.g., golf courses and home lawns
    (approximately 57% of total usage); rights-of-way/roadways
    (approximately 42% of total usage).

Formulation Types Registered: Granular; liquid (emulsifiable concentrate,
    soluble concentrate, ready-to-use); aerosol spray.

Methods of Application: Ground equipment and aircraft.

SCIENCE FINDINGS

Existing data are inadequate to assess environmental and health
effects for 2,4-DP. Available data indicate that 2,4-DP is not
oncogenic in the rat, but there is a positive trend for mutagenic
effects. There is also some potential for toxicity to aquatic
organisms. Additional chronic toxicological testing is not required
since there are no registered food uses for 2,4-DP.

2,4-DP is generally formulated as an amine salt or ester. The
amine and ester forms may differ in biological activity and
environmental fate from the parent compound. Data are needed on
each amine and ester, as well as the acid, to enable a complete
assessment.

Concern about possible groundwater contamination exists for the
family of 2,4-D compounds (2,4-D, 2,4-DB and 2,4-DP). Additional
data and a label warning statement are required.

Chemical Characteristics. (2,4-DP acid technical)

Physical State - Crystalline solid; flakes.
Color - White to tan.
Odor - Slight phenolic.
Molecular weight - 235.1.
Empirical formula - $C_9H_8Cl_2O_3$.
Melting point - 116 - 118°C.
Solubility - Highly soluble in acetone, benzene, carbon
    tetrachloride, diesel oil and kerosene; slightly soluble
    in water.
Vapor pressure - Negligible at 20° C.
Specific gravity - 1.186 at 20° C.
Stability - Stable to melting point.

Toxicological Characteristics. (NOTE: 2,4-DP acid is test material
unless specified otherwise).

Acute Toxicity - Oral LD_{50} (mouse): 500 mg/kg (males)
                                 620 mg/kg (females)
                          (rat): 700 mg/kg (males)
                                 500 mg/kg (females)

                          Toxicity Category III

Subchronic Toxicity - Oral studies in rats and dogs were inconclusive.
    Some kidney and liver effects were noted.

Chronic Toxicity -

    Chronic Feeding/
    Oncogenicity (rat): LOEL for systemic effects 11 mg/kg (male),
                        13 mg/kg (female); NOEL 4 mg/kg.
                        No significant increase in tumor
                        incidence in any organ. Kidney and
                        liver toxicity and decreased body
                        weight/food efficiency at highest dose
                        level. No oncogenic effects under
                        conditions of the study.

    Oncogenicity (mouse): LOEL for systemic effects 25 mg/kg;
                          NOEL not established.
                          No oncogenic effects under conditions
                          of the study, but study inadequate
                          due to numerous reporting deficiencies.

    Teratology: No acceptable data.

    Reproduction: LOEL for maternal and developmental toxicity
       (rat)   100 mg/kg; NOEL 50 mg/kg.
                  No reproduction or fertility effects under
                  conditions of the study, but study inadequate
                  due to numerous reporting deficiencies.

    Mutagenicity: There is a positive trend for mutagenic
                  effects for both 2,4-DP acid and 2,4-DP
                  butoxyethyl ester. Additional data are
                  required for each 2,4-DP compound.

    Metabolism: No acceptable data.

Physiological and Behavioral Characteristics.

    Mechanism of Pesticide Action: Phenoxy herbicides (including
       2,4-DP) are hormone weed killers affecting the activity
       of enzymes, respiration and cell division.

Environmental Characteristics.

No data are available.

Ecological Characteristics.

Avian Toxicity: No acute oral studies available.
Dietary studies indicate 2,4-DP butoxyethyl
ester is practically nontoxic (LC_{50} >10,000 ppm)
to waterfowl and upland game birds.

Aquatic Organism Toxicity: Acceptable data indicate 2,4-DP
butoxyethyl ester is highly toxic
to fish and freshwater invertebrates.

Non-Target Insect Toxicity: No data are available.

Non-Target Plants/
Endangered Species: Since 2,4-DP is a broadleaf herbicide a
potential hazard exists for non-target plants.
Hazard assessments for endangered species
cannot be completed until additional data
are received.

Tolerance Assessment. There are no registered food or feed uses for
2,4-DP acid or its amines or esters that require the establishment
of tolerances.

Reported Pesticide Incidents. Based on the Pesticide Incident
Monitoring System files covering the period 1966 to 1979, reports
were received concerning off-target movement for unspecified
2,4-D (family) compounds. The incidents involved drift from
aerial (173 reports) and ground (104 reports) application, as
well as volatilization (35 reports) and resulted in damage to
non-target crops and other desirable plants.

SUMMARY OF REGULATORY POSITION AND RATIONALE

Summary of Agency Position. The Agency is requiring additional
toxicological data for 2,4-DP and will further evaluate the
mutagenic and teratogenic potential of 2,4-DP when additional
data are received. Environmental effects will also be evaluated
when data gaps for fish/wildlife and environmental fate are
filled. The Agency will not establish any significant new uses
until Registration Standard data are reviewed.

Unique Label Warning Statements.

1. Environmental Hazards - (butoxyethyl ester).

   Manufacturing-use: "This product is toxic to fish."

   End-use: "This product is toxic to fish and aquatic invertebrates."

2. Groundwater/Protective Clothing - (end-use products).

   "This product can reach groundwater as a result of mixing/loading
   or from improper handling. To minimize groundwater contamination
   from spills during mixing/loading/cleaning of equipment, take
   the following steps:"

   (Liquid only) - "Mixing and loading: When mixing, loading or
   applying this product, wear chemical resistant gloves. Wash
   nondisposable gloves thoroughly with soap and water before
   removing. The mixing and loading of spray mixtures into the
   spray equipment must be carried out on an impervious pad (i.e.,
   concrete slab, plastic sheeting) large enough to catch any
   spilled material. If spills occur, contain the spill by using
   an absorbent material (e.g., sand, earth or synthetic absorbent).
   Dispose of the contaminated absorbent material by placing in a
   plastic bag and following disposal instructions on this label.
   Triple rinse empty containers and add the rinsate to the mixing
   tank."

   (Granular only) - "Handling: When handling this product, wear
   chemical resistant gloves. Wash nondisposable gloves thoroughly
   with soap and water before removing. If spills occur, collect
   the material and dispose of by following disposal instructions
   on this label."

   (Liquid and granular) - "Cleaning of equipment: When cleaning
   equipment, do not pour the washwater on the ground; spray or
   drain over a large area away from wells and other water sources."

SUMMARY OF MAJOR DATA GAPS: The following data are required for 2,4-DP acid. The Agency is also requiring data on each individual amine and ester of 2,4-DP. Specific requirements are detailed in the Data Tables, Appendix I of the Registration Standard.

	Due Date (From Issuance of Standard)
Product Chemistry	6 – 15 months
Toxicology:	9 – 18 months
Acutes	
Subchronic oral	
and dermal	
Teratogenicity	
Mutagenicity	
Ecological Effects:	9 – 12 months
Avian oral/dietary	
Aquatic organisms	
(freshwater fish/	
invertebrates)	
Non-target insect	
and plant studies	
Environmental Fate:	9 – 27 months
Hydrolysis	
Metabolism	
Mobility	
Dissipation	
Fish Accumulation	

CONTACT PERSON AT EPA

    Mr. Richard Mountfort
    Product Manager 23
    Fungicide-Herbicide Branch
    Registration Division (TS-767C)
    Office of Pesticide Programs, EPA
    Washington, DC 20460

    Telephone: (703) 557-1830

DISCLAIMER: The information presented in this Pesticide Fact Sheet is for informational purposes only and may not be used to fulfill data requirements for pesticide registration and reregistration.

DICHLORVOS

Reason for Issuance: Registration Standard
Date Issued: October 1987
Fact Sheet Number: 134

1. DESCRIPTION OF CHEMICAL

    Generic Name: 2,2-dichlorovinyl dimethyl phosphate
    (Chemical)   hexahydro-4,7-methanoindene

    Common Name: Dichlorvos

    Trade and
    Other Names: 2,2-dichloroethenyl dimethyl phosphate;
                  DDVP (USA); DDVF (USSR), Nogos; Vapona;
                  Dedevap; Mafu; Oko; Atgard; No-pest;
                  Herkol; Ciovap; Ravap; Vaponite; Canogard;
                  Equigard; Task and Riton.

    EPA Shaughnessy Code: 084001

    Chemical Abstracts Service (CAS) Number: 62-73-7

    Year of Initial Registration: 1948

    Pesticide Type: Insecticide

    Chemical Family: Organophosphate

    U.S. and Foreign Producers: Amvac Chemical Corp.; SDS Biotech
                                 Corp.; Prentiss Drug and Chemical
                                 Co.; Dow Chemical Corp; Kaw Valley;
                                 MGK Co.; Denka Chemia B.V.; Kenco
                                 Chemical and Mfg. Corp.; Wesley
                                 Industries. Inc; Fermenta Animal
                                 Health; and E.I. du Pont de Nemours
                                 and Company

2.   USE PATTERNS AND FORMULATIONS

Application Sites: Dichlorvos is used in areas where flies,
mosquitoes, gnats, cockroaches and other
nuisance insect pests occur. Use sites
include the following:

in and around domestic dwellings (contract,
crack and crevice, bait, resin strip, and
space spray); direct application to
domestic food animals (spray, pour-on,
dip, back rubber, face oiler, and paint
and impregnated material); direct application
to domestic animals (spray, flea collar,
and tag); in and around premises housing
animals (food and nonfood) (space spray,
bait, resin strip, and contact); ornamental
and food crop greenhouse use (contact and
space spray); mushroom houses (contact
and space spray); tobacco warehouses
(contact and space spray); postharvest
tomato treatment (dust applied to fruit
in containers); sewage plants/sewage
systems (resin strips and direct application to sewage); aircraft, buses and other
commercial transportation vehicles (space
spray); nonperishable bulk-stored raw
agricultural commodities (RACs) (including
animal feed, beans [dried type], cocoa
beans, coffee beans, grain crops such as
corn, nut crops, peanuts, peas [field],
soybeans, and tobacco) (resin strip and
space spray); nonperishable packaged or
bagged RACs (including beans [dried
type], cocoa beans, coffee beans, grain
crops such as corn, nut crops, peanuts,
peas, and soybeans) (space spray); nonperishable packaged or bagged processed
agricultural commodities (including
cereals, crackers, flour, packaged cookies,
and sugar); lawns (foliar application);
ornamental turf (foliar application);
ornamental plants (foliar and bark
application); noncrop aquatic areas
(ground or air equipment and resin strip);

eating establishments, and food
processing and storage areas (space
spray, contact, bait, and resin strip);
other commercial, institutional and
industrial areas (space spray, contact
bait, and resin strip); enclosed outdoor
utility equipment (space treatment);
gypsy moth traps and malaise (tent)
traps (resin strip); figs (EPA SLN
No. CA-830045) (ground and air equipment); urban and rural outdoor areas
(food and nonfood); including screwworm
adult suppression (EPA SLN Nos. TX-780056,
AZ-8003100, NM-790003, CA-810040) and
recreational areas (spray, bait, and
resin strip); and miscellaneous sites
including epcot display crops (EPA SLN
Nos. FL-820051, FL-820054, FL-820055),
rat receptacle bags, refuse and solid
waste containers and sites (spray, bait,
and resin strip).

Formulation Types: Emulsifiable concentrates, soluble
concentrate liquids, granulars, pressurized
liquids and dusts, impregnated materials,
pellets/tablets, liquids (ready to use),
wettable powders and dusts.

3. SCIENCE FINDINGS

Summary Science Statement

Dichlorvos is an organophosphate insecticide with moderate
to high acute toxicity. Dichlorvos is classified as a B_2
"Probable Human Carcinogen", is a potent cholinesterase
inhibitor, and may cause liver effects from subchronic and
chronic exposure. Dichlorvos is potentially highly to very
highly toxic to fish and other wildlife. It may have the
capacity under various use patterns, i.e., fig and mosquito
larvicide use, to cause adverse effects to aquatic invertebrates,
fish, and birds. Dichlorvos degrades fairly rapidly in soil
and water and is therefore not expected to reach groundwater.
However, the environmental fate is largely uncharacterized,
and additional data are needed before definitive conclusions
are reached. A tolerance reassessment of dichlorvos is not
possible because of gaps in the toxicology data base and
because many of the tolerances are not adequately supported.

Chemical/Physical Characteristics of the Technical Material

Physical State: Liquid
Color: Colorless to light amber
Molecular Weight and Formula: 221.0 - $C_4H_7Cl_2O_4P$
Boiling Point: 117° C at 10 mm Hg
Specific Gravity: 1.42 at 25° C
Density: 1.65-1.67 g/ml
Vapor Pressure: 0.032 mm Hg at 32 °C
Solubility in Various Solvents: Slightly soluble in water
                                 and kerosene; readily solu-
                                 ble in most organic solvents
Stability: Stable in the presence of hydrocarbon solvents;
            undergoes hydrolysis in the presence of water
            and is readily decomposed by strong acids and bases
Miscibility: Miscible with most organic solvents and aerosol
              propellants
Corrosion Characteristics: Corrosive to steel; noncorrosive
                            to stainless steel, aluminum,
                            nickel, Hostelloz 13, Teflon
Chemical Relationships: Trichlorfon and naled are chemically
                         related to dichlorvos. Naled degrades
                         to dichlorvos in plants, animals, and
                         soil. In cattle and poultry, naled is
                         debrominated to dichlorvos which further
                         degrades to dichlorodesmethyl-dichlorvos
                         (major pathway) or desmethyl-dichlorvos
                         (minor pathway). Trichlorfon degrades
                         to dichlorvos in soil and alkaline pond
                         water, and possibly in plants and animals.

Toxicology Characteristics

Acute Oral: Toxicity Category II (56 and 80 mg/kg in female
             and male Sherman rats, respectively).

Acute Dermal: Toxicity Category I (75 and 107 mg/kg in
               female and male Sherman rats, respectively).

Primary Dermal Irritation: Toxicity Category IV based on
                            mild dermal irritation reported
                            in a rabbit study.

Primary Eye Irritation: Toxicity Category III based on
                         a lack of corneal injury and only
                         mild redness and chemosis at 24
                         hours postapplication of 1.67 mg/kg
                         technical dichlorvos in rabbits'
                         eyes.

Skin Sensitization: Data Gap

Acute Inhalation: Toxicity Category I based on a toxicity value of >193 mg/m^3 in rats.

Delayed Neurotoxicity: Data Gap

Subchronic Oral (rodent) Testing: Data Gap

Oncogenicity: Classified as a B$_2$ "probable human carcinogen" This determination for oncogenicity is based on a draft report of a review of two rodent studies sponsored by the National Toxicology Program in which there were significantly increased incidences of forestomach squamous cell carcinoma/papillomas in female mice and pancreatic acinar adenomas, mononuclear cell leukemia and lung adenomas in male rats and mammary gland fibroadenomas in female rats. The potency or Q$_1$* is 2.9 x 10^{-1} (mg/kg/day)$^{-1}$.

Chronic Feeding: Based on a dog chronic feeding study with increased relative liver weights in males at 32 ppm and above, and enlargement of liver cells in both sexes at 32 ppm and above, the no-observable-effect level (NOEL) is 0.08 mg/kg/day (3.2 ppm). A rodent feeding study is required.

Metabolism: Data Gap

Teratogenicity: Rodent teratology study is a data gap; based on an acceptable rabbit teratology study by inhalation, the NOEL for embyro/fetotoxicity is 2 ug/L based on decreased fetal weights.

Reproduction: Data Gap

Mutagenicity: Dichlorvos is a direct-acting (gene) mutagen, in bacteria, fungi, and mammalian cells in vitro. Dichlorvos was shown to be negative in micronucleus and sister-chromatid exchange assays conducted in mice, and was also negative in repeated dominant lethal assays.

Major Routes of Exposure: Dermal and respiratory exposure to mixers, loaders, applicators, reentry workers and occupants of treated buildings.

Environmental Characteristics

Numerous data gaps exist for environmental fate. Data reviewed by the Agency indicate that dichlorvos degrades fairly rapidly with half-lives of 2 to 8 hours in soils ranging in texture from sand to silt. The mobility of dichlorvos is inversely correlated with soil organic matter content. Preliminary data also suggest that dichlorvos is intermediately to very mobile in a variety of soils ranging in texture from sandy loam to clay. Groundwater contamination may not be a problem because of dichlorvos's rapid degradation. However, acceptable data are lacking, and are needed before the Agency can assess the potential for dichlorvos to contaminate groundwater.

Ecological Characteristics (technical grade)

Avian oral toxicity: Data Gap

Avian dietary toxicity: Acute toxicity values greater
(8-day)   than 1317 and 5000 ppm for
                          mallard ducks at 5 and 16 days
                          of age, respectively; for phea-
                          sants the acute toxicity value
                          is 568 ppm.

Freshwater fish acute toxicity: 0.1 ppm for rainbow
(96-hr. LC$_{50}$)   trout;
                                 0.23 ppm for striped
                                 mullet.

Estuarine/marine fish acute toxicity: 0.23 for striped
                                        mullet

Freshwater invertebrate toxicity: 0.00007 ppm for Daphnia
(48-hr. or 96 hr. EC$_{50}$)   pulex; 0.004 ppm for
                                    sand shrimp.

Estuarine/marine invertebrate toxicity: 0.0004 ppm for
                                          sand shrimp

Tolerance Reassessment

Analytical methodology for determining the levels of residues of dichlorvos in plants and animals is adequate for data collection of residues in plant commodities and in animal tissues and milk. Storage stability data demonstrate that residues of dichlorvos in or on frozen plant samples are stable up to 90 days after application. Residues in or on frozen animal tissues are stable up to 8 weeks after application. Additional storage data (length of time and conditions under which samples were stored) are required

in order to evaluate the adequacy of the dichlorvos tolerances.
If previously submitted samples or newly generated samples
were/are stored frozen for greater than 90 days (plants)
or 8 weeks (animals), additional storage stability data
will also be needed. Data on the magnitude and levels of
residues of dichlorvos in radishes, lettuce, tomatoes,
cucumbers, milk, eggs, and the meat, fat, and meat byproducts
of cattle, goats, hogs, horses, sheep, and poultry are
insufficient to determine the adequacy of the established
tolerances. Tolerances for residues of dichlorvos in or on
figs, dried figs, and mushrooms are adequately supported.
Data are required to support use of dichlorvos products in
food areas of food handling establishments. Processing and
cooking studies are required. Data reflecting the use of
dichlorvos on stored, unfinished tobacco are required. The
established tolerances for residues of dichlorvos in or on
bulk-stored nonperishable RACs are not adequately supported,
and data are required. The established tolerances are adequate
to cover residues of dichlorvos in or on packaged or bagged
nonperishable RACs and processed food resulting from the
application of dichlorvos vapors. However, the data indicate
that residues following use of aerosol treatments will result
in toleranceexceeding residues. Therefore, residue data and
an accompanying tolerance revision must be submitted. Also,
insufficient data are available regarding residues of dichlorvos
in the meat, fat, and meat byproducts of cattle, goats,
horses, sheep, and poultry and milk and eggs. This constitutes
a data gap.

Based on a 2-year dog feeding study with a NOEL of
0.08 mg/kg/day (3.2 ppm) for changes in liver weight and
enlarged liver cells, a provisional acceptable daily intake
(PADI) of 0.0008 mg/kg/day has been calculated using a one
hundred-fold uncertainty factor. The calculated estimate is
provisional because the existing data base for dichlorvos is
lacking a chronic rat feeding study, a reproduction study and
a rat teratology study. The Anticipated Residue Contribution
(ARC) for dichlorvos for the U.S. population average is
0.000416 mg/kg/day based on current tolerance levels, cooking
data for small grains, and an estimate of percent crop treated.
The ARC occupies 52% of the PADI. For children 1 to 6 years
of age, the ARC occupies 122% of the PADI.

4. SUMMARY OF REGULATORY POSITIONS AND RATIONALES

   °   The Agency is considering further regulatory action for
all registered uses of dichlorvos because of concerns about the
risks of oncogenicity, cholinesterase inhibition, and liver
effects. The risks of concern are for the public from consumption

risks of oncogenicity, cholinesterase inhibition, and liver
effects. The risks of concern are for the public from consumption
of foods containing residues of dichlorvos, for those involved in
the application of dichlorvos, for workers reentering treated
areas, for residents/occupants of treated areas (including areas
treated with resin pest strips or other dichlorvos products), for
people exposed to pets treated with dichlorvos, and for pets
treated with dichlorvos.

o   In order to meet the statutory standard for continued
registration, the Agency has determined that all dichlorvos
products (excluding domestic uses [household sprays containing
0.5% or less active ingredient], resin strips and pet uses)
must be restricted for retail sale to and use by Certified
Applicators or persons under their direct supervision.   In
view of the potential health hazards associated with exposure
to dichlorvos, the Agency is concerned about exposure to
dichlorvos which may result from improper application or
use of dichlorvos products, and so is restricting use of
the pesticide.

o   In order to meet the statutory standard for continued
registration, the Agency has determined that all products
must contain a cancer hazard warning. The Agency believes
that incorporation of this statement affords the product
user with important information about the potential
oncogenic hazard associated with the use of the product.

o   The Agency has determined that in order to meet the statutory
standard for continued registration, all dichlorvos product
labels must bear protective clothing statments.

o   An interim 48-hour reentry interval is imposed for the
agricultural and commercial uses of dichlorvos until such
time as appropriate reentry data have been submitted and
evaluated. Exposure data are required to support
registration of total release foggers and aerosol products
intended for use in domestic dwellings.

o   Pesticide spray drift data requirements are imposed for the
fig use and outdoor aerosol and fogging mosquito use. Data
required include droplet size spectrum studies and field
evaluation of pesticide spray drift.

o   The Agency is requiring a special environmental monitoring
study in order to evaluate whether or not the fig and mosquito
control uses of dichlorvos may result in exposure of certain
terrestrial and aquatic organisms to potentially hazardous
levels of the pesticide.

o   The Agency has determined that endangered species label

o   In order to remain in compliance with FIFRA, the Agency has
    determined that endangered species label restrictions
    are necessary to protect endangered threatened species in
    aquatic areas. Labeling which prohibits the mosquito larvicide
    use of dichlorvos in the range of endangered and threatened
    species is required through PR Notice 87-4, which was issued
    in May, 1987. Additional endangered species labeling require-
    ments are reserved pending the results of a consultation with
    the Office of Endangered Species regarding use of dichlorvos
    on figs.

•   The Agency is deferring decisions concerning dichlorvos
    and groundwater contamination until such time as the
    information required by the Standard have been submitted
    and reviewed.

•   No new tolerances for raw agricultural commodities or
    new food uses will be granted until data gaps for
    residue chemistry and chronic toxicology have been
    tilled.

•   All products must bear appropriate labeling as specified
    in 40 CFR 162.10, PR Notices 83-2, 83-3, and 87-4,
    precautions and warnings listed in the Dichlorvos Use
    Index, and as indicated in the Registration Standard.

6.   SUMMARY OF OUTSTANDING DATA REQUIREMENTS

	Time Frame
Toxicology	
Dermal sensitization	9 Months
Dermal absorption	6 "
21-day dermal toxicity	9 "
Neurotoxicity	9 "
Subchronic oral toxicity - rodent species	12 "
Chronic feeding - rodent species	50 "
Teratogenicity - species other than the rabbit	15 "
Reproductive effects - rats (2-generation)	39 "
Metabolism	24 "
Environmental Fate/Exposure	
Hydrolysis study	9 Months
Aerobic soil metabolism study	27 "
Aerobic and anaerobic aquatic metabolism study	27 "
Leaching and adsorption/desorption study	12 "
Terrestrial field dissipation study	27 "

Aquatic (sediment) - field study	27	"
Photodegradation in water, soil and air	9	"
Volatility (lab) studies	12	"
Accumulation - irrigated crops	39	"
Reentry data	18	"
Dermal and inhalation Exposure	18	"
Spray drift data	18	"

Residue Chemistry

Livestock metabolism (direct animal treatment)	18	Months
Storage stability data	18	"
Residue analytical Method	15	"
Residue data (except	18	"
for figs, dried figs, and mushrooms)	18	
Processing and cooking studies	18	"
Residue data on stored, unfinished tobacco	18	"
Residue data -food handling establishements	12	"

Product Chemistry   9-15 Months

Fish and Wildlife

Acute avian toxicity	9	Months
Acute toxicity to fish	9	"
Acute toxicity to freshwater invertebrates	9	"
Acute toxicity to estuarine and marine organisms	9	
Fish early life stage and aquatic invertebrate life-cycle	15	"
Special environmental residue monitoring study	24	"
Avian reproduction	24	"
Fish life cycle study	27	"
Honeybee - toxicity of residues on foliage	15	"

7.   CONTACT PERSON AT EPA

George LaRocca
Product Manager (15)
Insecticide-Rodenticide Branch
Registration Division (TS-767C)
Office of Pesticide Programs
Environmental Protection Agency
401 M Street SW.
Washington, DC 20460

Office location and telephone number:
Room 204, Crystal Mall #2
1921 Jefferson Davis Highway
Arlington, VA 22202
(703) 557-2400

DISCLAIMER: The information presented in this Chemical Information
Fact Sheet is for informational purposes only and may not be used
to fulfill data requirements for pesticide registration and
reregistration.

DIFENZOQUAT

Reason for Issuance: Registration Standard
Date Issued: December 1988
Fact Sheet Number: 194

1. DESCRIPTION OF CHEMICAL

Generic Name: 1,2-dimethyl-3,5-diphenyl-1H-pyrazolium

Common Name: Difenzoquat

Trade and Other Names: AVENGE

EPA Shaughnessy Codes: 106401 (Salt)

Chemical Abstracts Service (CAS) Number: 43222-48-6 (Salt)

Year of Initial Registration: 1974

Pesticide Type: Herbicide

Chemical Family: Pyrazolium

U.S. and Foreign Producers: American Cyanamid Company

2.   USE PATTERNS AND FORMULATIONS

   Application: A postemergence herbicide to control wild oats in alfalfa
                  (seed crop in CA only), wheat and barley.

   Types and Methods of Application: Applied as a postemergence broadcast
                                        treatment by aerial and ground equipment.

   Application Rates: 0.6 - 1.0 lb cation/A

   Types of Formulations:

                        Single active ingredient formulations:

                        31.2% SC/L, 31.8% SC/L, and 62.5% SC/S

                        Difenzoquat can be tank mixed with bromoxymil,
                        metsulfuron methyl, chlorsulfuron, 2,4-D (2,4-
                        dichlorophenoxyacetic acid) and MCPA (2-methyl-
                        4-chlorophenoxyacetic acid).

3.   SCIENCE FINDINGS

   Summary Science Statement

   Chemical Characteristics of the Technical Material

   Physical State: Solid

   Color: White to off-white

   Odor: Odorless

   Molecular Weight and Formula: 360.4 - $C_{18}H_{20}N_2O_4S$ (Salt)

   Melting Point: 155 - 157°C   Difenzoquat decomposes above 160°C.

   Vapor Pressure: negligible

   Density: 41 lb/cu ft

   Solubility in various solvents: 76.5% in water at 23°C. Poorly soluble
                                     in most organic solvents.

Toxicology Characteristics

Difenzoquat is moderately toxic by the oral route (LD_{50} for male rats -
270 mg/kg) and slightly toxic via the dermal route of exposure (LD_{50} for male
rabbits - 3540 mg/kg). Difenzoquat was found to be slightly irritating to the
eyes and moderately irritating to abraded skin and non-irritating to intact
skin. The inhalation route of exposure does not indicate any toxicity problems
[LC_{50} for male rats is greater than 298.2 mg/l (1 hr. exposure)].

Major routes of exposure: Dermal and inhalation.

Subchronic toxicity: No compound related effects were observed in 90-Day
                       Feeding Study. Systemic NOEL was 2500 ppm (62.5
                       mg/kg:HDT). (HDT = Highest Dose Tested).

Oncogenicity: Difenzoquat was negative for oncogenicity at the 5000 ppm
                (250 mg/kg:HDT) level in rats.

Teratogenicity: Difenzoquat was negative for teratogenicity, fetotoxicity
                and maternal toxicity in rats at the 2500 ppm (125
                mg/kg:HDT) level.

Reproduction: A 3-generation rat reproduction study found the parental NOEL
                was equal to or greater than 2500 ppm (125 mg/kg:HDT) and
                reproductive/developmental NOEL was 500 ppm (25 mg/kg). The
                only maternal effect observed in all three generations was a
                decreased body weight gain in the high dose group during the
                premating period.

Mutagenicity: Insufficient data available to evaluate the mutagenicity
                potential.

Environmental Characteristics

Difenzoquat is tightly bound to soil particles and does not readily leach.

Ecological Characteristics

Avian acute toxicity: Difenzoquat is slightly toxic to birds. Mallard ducks
                        (LC_{50} - 10,388 ppm) and bobwhite quail (LC_{50} - 4640
                        ppm).

Freshwater fish acute toxicity: Difenzoquat is slightly toxic to fish:
                        Bluegill (LC_{50} - 90.4 ppm) and Rainbow
                        Trout (LC_{50} - 76-99 ppm).

Freshwater invertebrate toxicity: Difenzoquat is moderately toxic to
                        freshwater invertebrates. Daphnia (LC_{50} -
                        2.6 ppm).

TOLERANCE ASSESSMENT

Tolerances have been established for residues of difenzoquat in a variety of raw agricultural commodities (40 CFR 180.369).

The current U.S. tolerances for difenzoquat range from 0.05 ppm in fat, meat, meat byproducts and wheat grain, 0.2 ppm for barley grain, and 20 ppm for wheat and barley straw. A Canadian maximum residue level of 0.1 ppm (negligible residue) has been established for wheat and barley grain. There are no Mexican Tolerances or CODEX MRLs for difenzoquat.

4.   Summary of Regulatory Positions

Based on available information, the Agency has made the following determinations:

· Difenzoquat does not exceed any risk criteria for Special Review.

· Difenzoquat does not meet the criteria for restricted use.

· The Agency is not imposing any special labeling for endangered or threatened species because of its use patterns of single application postemergence to wheat and barley, and because of low toxicity to fish and birds.

· Based on available toxicology data the Agency has no concerns with human exposure which would require special protective clothing.

· No groundwater advisory labeling is required because difenzoquat does not readily leach.

· No tolerances for feed items treated with difenzoquat will be issued until data gaps for animal metabolism and magnitude of residue have been fulfilled.

5.   Summary of Major Data Gaps   Timeframe
                                                            for
                                                          Submission[1]/

Toxicology

Acute Oral Toxicity in one sex (females)   9 months
Acute Dermal Toxicity in one sex (females)   9 months
Acute Inhalation Toxicity in one sex (females)   9 months
Dermal Sensitization   9 months
Subchronic Dermal (21-day)   9 months
Chronic Toxicity in one species (nonrodent)   50 months
Teratogenicity in one species (rat)   15 months
Oncogenicity in one species (mouse)   50 months
Mutagenicity
    (Gene mutation, chromosomal aberration,
    and direct DNA damage studies).   12 months
Metabolism   24 months

Environmental Fate

Hydrolysis   9 months
Photodegradation in water and on soil   9 months
Aerobic and anaerobic soil metabolism   27 months
Leaching and Adsorption/Desorption   12 months
Volatility (Lab)   12 months
Terrestrial field dissipation (soil)   27 months
Accumulation rotational crops (confined)   39 months
Fish accumulation   12 months

Product Chemistry

Product Identity and Disclosure of Ingredients   9 months
Description of Manufacturing Process   9 months
Discussion of Formulation and Impurities   9 months
Preliminary Analysis of Product Samples   12 months
Certification of Ingredient Limits   12 months
Analytical Methods to Verify Certified Limits   12 months
Physical and Chemical Characteristics   9 months

Residue Chemistry: Tolerance Reassessment

Nature of the Residue (Metabolism) in Livestock   18 months
Residue Analytical Methods   15 months
Storage Stability Data   18 months
Magnitude of Residue in Plants, and
    Fat/Meat/Meat of Animal Byproducts   18 months

1. Timeframe for submission is number of months from date of issuance of the
registration standard, December 23, 1988.

6.   CONTACT PERSON AT EPA

   Richard F. Mountfort
   Product Manager 23
   Fungicide/Herbicide Branch
   Registration Division (TS-767C)
   Office of Pesticide Programs
   Environmental Protection Agency
   401 M Street, S. W.
   Washington, D. C. 20460

   Office location and telephone number:

   Room 237, Crystal Mall #2
   1921 Jefferson Davis Highway
   Arlington, VA 22202
   (703) 557-1830

DISCLAIMER: The information in this Pesticide Fact Sheet
is a summary only and is not to be used to satisfy data
requirements for pesticide registration and reregistration.
The complete Registration Standard for the pesticide may be
obtained from the National Technical Information Service.
Contact the Product Manager listed above for further
information.

ETHEPHON

Reason for Issuance: Registration Standard
Date Issued: September 29, 1988
Fact Sheet Number: 176

1. Description of Chemical

   Generic Name: Ethephon
   Common Name: Ethephon
   Trade Names: Bromeflor, Cerone, Chlorethephon (New Zealand), Ethrel,
                   Florel, Prep and Flordimex
   EPA Shaughnessy Code: 099801
   Chemical Abstracts Service (CAS) Number: 16672-87-0
   Year of Initial Registration: 1973
   Pesticide Type: Growth Regulator
   U.S. and Foreign Producers: AmChem/Union Carbide
                                Rhone-Poulenc

2. Use Patterns and Formulations

   Application Sites: To enhance fruit ripening, flower initiation,
      fruit and leaf abscission, and breaking of apical dominance in
      apples, barley, wheat, blackberries, boysenberry, blueberry,
      cantaloupe, cherry (tart and sweet), cotton, cucumber, squash
      and pumpkin (hybrid seed production), figs, cucumber (pickling),
      filberts, grapes, lemon, tangerine, tangelo, pepper, pimento,
      tomato, walnut, tobacco (flue cured), ornamentals, guava,
      grapes for table and raisin production, pineapples, sugarcane,
      coffee beans, cottonseed, macadamia nuts, trees (forest and
      shelterbelt) and cranberries.

   Mechanism of Action: Generation of ethylene after application (ethylene
      is a naturally-produced plant hormone).

   Types of Formulations: Emulsifiable concentrate, soluble
      concentrates/liquids, ready-to-use liquids.

   Types and Methods of Application: Broadcast by ground or aerial
      equipment.

219

Application Rates: Vary from 0.08 to 2.0 pounds of active ingredient per acre depending upon the use site and desired effects.

3. Science Findings

Chemical Characteristics:

Empirical Formula: $C_2H_6ClO_3P$
Molecular Weight: 144.5
Physical State: Waxy, solid
Color: White
Melting Point: 74 to 75 °C
Solubility: Very soluble in water, alcohol, propylene glycol; very slightly soluble in aromatic solvents.

Toxicology Characteristics:

Acute Effects

Acute Oral Toxicity: 1.6 g/kg rats (Toxicity Category III)

Acute Dermal Toxicity: > 5 g/kg rabbits (Toxicity Category III)

Acute Inhalation: Not available

Primary Eye Irritation: Not available, but assumed to be corrosive based on the primary dermal results

Primary Skin Irritation: Score 6.75 (Corrosive) rabbits (Toxicity Category I)

Delayed Neurotoxicity Hen: No signs of neurotoxicity

Sub-Chronic Effects

No adequate data are available on subchronic oral toxicity in rats or dogs. However, the requirement for subchronic studies in the rodent and non-rodent will be waived if the required chronic studies are acceptable.

Chronic Effects

Sprague-Dawley Rats - 18 Months: NOEL for cholinesterase activity is 15 mg/kg/day (supplementary data).

Beagle Dogs - 2-Year: NOEL for source A* ethephon was 7.5 mg/kg/day (supplementary data).

Swiss Albino Mouse - 18-Month oncogenicity study: NOEL for RBC cholinesterase activity is < 4.5mg/kg/day which was the lowest dose tested (LDT) (supplementary data).

Oncogenicity

The available data are not adequate to assess the oncogenic potential of ethephon. In the rat and mouse studies submitted, only chronic effects were observed and these effects are discussed under the chronic toxicity section.

Teratology

Rat: NOEL for maternal and embryo/fetal toxicity is 600 mg/kg/day. NOEL for teratogenic effects is 600 mg/kg/day.

Rabbit: NOEL is 50 mg/kg/day based on fetal resorptions at higher dose levels tested.

Reproduction: No adequate data available.

Mutagenicity: Salmonella typhimurium indicate no mutagenic effect up to 1000 ug/100 uL without activation. No acceptable data are available for chromosomal aberrations, primary DNA damage, or other genotoxic effects.

Metabolism: A rat metabolism study is required.

Possible Presence of Contaminant of Toxicological Concern

The Agency finds that residues of monochloroacetic acid may be found in ethephon-treated commodities. Monochloroacetic acid is a potential degradation product of an impurity in ethephon, monochloroethyl ester of (2-chloroethyl)-phosphonic acid. Monochloroacetic acid is an extremely toxic metabolic inhibitor and has been prohibited from addition to food under 21 CFR 189.155. Analysis of certain food and feed crops for residues of this contaminant are required.

Major Routes of Exposure

Dermal, Inhalation

Physiological and Biochemical Behavioral Characteristics:

Metabolism: Acceptable metabolism data are available for plants and ruminants. The residue of concern in plant commodities, ruminant tissues and milk is ethephon per se. Additional data are required regarding metabolism in poultry tissues and eggs.

* In the study there were two, apparently different, sources of the chemical tested and reported as source A and source B.

Possible Presence of Impurities: Available data indicate that technical
ethephon products may contain 2-chloroethanol as an impurity.
2-Chloroethanol is extremely toxic via the inhalation route and
has caused human deaths. Because of its volatility, 2-choroethanol
is unlikely to be present in ethephon end use products in amounts
high enough to pose an inhalation hazard. However, the impurity
could pose a hazard when technical or manufacuring use products
are stored or used in poorly ventilated spaces. Under these
conditions, 2-chloroethanol vapors could accumulate to levels
which may be hazardous to workers in the area. The Agency is
requiring submission of product chemistry data to assess the
extent of hazard posed by 2-choroethanol.

Environmental Characteristics:

Preliminary Adsorption and Leaching Characteristics: Data are required
on leaching, volatility, and hydrolysis of ethephon to characterize
the potential to reach ground water.

Ecological Characteristics:

Data are sufficient to characterize ethephon as slightly toxic to
birds (Lc50 804 mg/kg quail, 3750 ppm ducks). Laboratory and
field studies indicate that ethephon is slightly toxic to fish
(Lc50 180 mg/L bluegill, 170 mg/L trout (average)).

Tolerance Assessment:

List of crops and Tolerances: (CFR 40 180.300)

Commodity	Tolerance (ppm)
Apples	5
Barley, grain	2.0
Barley, straw	10.0
Blackberries	30
Blueberries	20
Cantaloupes	2
Cattle, meat byproducts (mbyp)	0.1
Cattle meat	0.1
Cattle fat	0.1
Cherries	10
Coffee beans	0.1 (N)
Cottonseed	2.0
Cranberries	5
Cucumbers	0.1
Figs	5
Filberts	0.5
Goats, fat	0.1
Goats, mbyp	0.1
Goats, meat	0.1

Commodity	Tolerance (ppm)
Grapes	2.0
Hogs, fat	0.1
Hogs, mbyp	0.1
Hogs, meat	0.1
Horses, fat	0.1
Horses, mbyp	0.1
Horses, meat	0.1
Lemons	2
Macadamia nuts	0.5
Milk	0.1
Peppers	30
Pineapples	2
Pineapple fodder	3
Pineapple forage	3
Pumpkins	0.1
Sheep, fat	0.1
Sheep, mbyp	0.1
Sheep, meat	0.1
Tangerines	0.5
Tangerine hybrids	0.5
Tomatoes	2
Walnuts	0.5
Wheat, grain	2.0
Wheat, straw	10.0
Guavas	0.1
Sugarcane Hawaii only	0.1

40 CFR 185.2700 and 186.2700 (food and feed additive)

Barley milling fractions except flour	5.0
Wheat milling fractions except flour	5.0
Raisins	12.0
Raisin waste	65.0
Sugarcane molasses	1.5
Barley milling fractions except flour	5.0
Wheat milling fractions except flour	5.0

Results of Tolerance Assessment: The PADI for ethephon has recently been revised by the Agency ADI Committee and is now established at 0.005 mg/kg/day (0.5 mg/kg/day for a LEL, and an uncertainty factor of 100). This value is based on an LEL of 0.5 mg/kg/day for a decrease in plasma cholinesterase activity in a 16-day study in humans. An uncertainty factor of 100 is derived from a factor of 10 for the variation in the susceptibility of humans, and a factor of 10 for the use of an LEL instead of a NOEL. The PADI will be reevaluated when the required toxicity and residue chemistry data are submitted and evaluated.

4. Label Warning Statements:

All Manufacturing-Use Products

Do not discharge effluent containing this product into lakes, streams, ponds, estuaries, oceans, or public water unless this product is specifically identified and addressed in an NPDES permit. Do not discharge effluent containing this product to sewer systems without previously notifying the sewage treatment plant authority. For guidance contact your State Water Board or Regional Office of the EPA.

Technical grade ethephon must be stored and used in ventilated areas only.

End-Use Products

Do not apply directly to water or wetlands (swamps, bogs, marshes, and potholes). Do not contaminate water when disposing of equipment washwaters.

Mixers, loaders and applicators must wear a full face shield, long trousers, long sleeved shirt, gloves, and boots to avoid as much skin and eye contact as possible.

Do not enter treated fields within 24-hours after application.

The following interim pre-harvest intervals (PHI) must be included on end use product labels for the raw agricultural commodities listed. These interim PHI's may be revised after the required field residue data have been submitted and evaluated.

Apples minimum 7 days, barley 40 days, blackberries and
blueberries 42 days, cherries minimum 7 days, coffee
beans 14 days, cotton 14-21 days, cranberries 17-21 days,
cucumbers 17-21 days, figs 14 days, filberts 7 days,
grapes 14 days, guavas 7 months, melons 2 days, peppers
14 days, pineapples 2 days, tangerines 5-10 days,
tomatoes 14-20 days for processing and 3-6 days fresh
market California only, walnuts 5-10 days, wheat 40 days.

6.   Summary of Regulatory Positions

1.   Ethephon does not meet any of the criteria specified
in 40 CFR 154.7; therefore a Special Review is not
being initiated at this time.

2.   The Agency will not require restricted use classification
for ethephon end-use products.

3.   The Agency is deferring decisions concerning ethephon's
potential for contaminating ground water until information
on its environmental characteristics and fate have been
submitted and and reviewed.

4.   The Agency has identified certain data that will
receive immediate review when submitted.

5.   The Agency has determined that foliar and soil dislogeable
residue data are required to establish reentry intervals
for all crops. An interim reentry interval of 24 hours
is being imposed for all crops until final reentry reentry
intervals are established.

6.   Pre-harvest intervals are required on product labeling
for a variety of currently registered use sites.

7.   The Agency is requiring data on animal metabolism
as well as storage stability studies and residue studies for
poultry and eggs. In order to remain in compliance with FIFRA,
registrants must do one of the following:

a.   Submit data which demonstrate that no residues
remain in eggs and poultry as a result of feeding treated
commodities;

b.   Propose tolerances and provide appropriate
supporting data for residues in poultry tissues and eggs.

8. Additional residue data, including processing data, must be submitted for the following raw agricultural commodities: peppers, tomatoes, cucumbers, melons, lemons, tangerines, apples, cherries, blackberries, boysenberries, blueberries, cranberries, grapes, filberts, walnuts, barley (wheat data may substitute), wheat, (and wheat straw), coffee beans, cotton seed, figs, guavas, pineapples, sugarcane, and tobacco. For tobacco, pyrolysis data must be submitted. If residues concentrate in any of the processed products, the appropriate food additive tolerance(s) must be proposed. Specific data requirements may be found in the data tables.

9. The Agency is requiring the proposal of either a tolerance for sugarcane forage or a grazing prohbition for sugarcane forage.

10. The Agency has determined that the following revisions in the tolerances listed in 40 CFR 180.300 and 21 CFR 193.186 are necessary.

o The designation "N" (negligible) must be deleted from all tolerances entries.

o The commodity "pineapple fodder" must be deleted from 40 CFR 180.300.

o The tolerance for guava must be added to 40 CFR 180.300.

o The tolerance for raisins must be added to 21 CFR 193.186.

11. Product chemistry and residue data are required depicting residues of monochloroacetic acid in or on food and feed commidities following registered applications of ethephon. Monochloroacetic acid may be a toxic metabolic residue derived from an impurity in technical ethephon.

12. The Agency has identified 2-chloroethanol as a contaminant of toxicological concern. The Agency is requiring data to assess the extent of contamination with this substance. Additionally, the Agency is requiring that manufacturing use products bear a label statement advising users to store and use the product in well-ventilated areas.

13. While the data gaps are being filled, currently
registered manufacturing-use products and end-use products
containing ethephon as the sole active ingredient may be sold,
distributed, formulated, and used in the United States, subject
to the terms and conditions specified in this Standard.
Registrants must provide or agree to develop additional data,
required in the Registration Standard.

The Agency will issue registrations for substantially
similar products and new uses will be issued after considering
the effects on the theoretical maximum residue contribution
(TMRC) and the maximum permissable intake (MPI).

7. Summary of Major Data Gaps

Data Required	Due Date (After Issuance of the Standard)
Product Chemistry	6 - 12 Months
Hydrolysis	9 Months
Photodegradation	9 Months
Soil Metabolism	27 Months
Fish Accumulation	12 Months
Rotational Crops	39 Months
Leaching and Adsorption/ Desorption	12 Months
Terrestrial Field Dissipation	27 Months
Acute Estuarine and Marine Organisms	12 Months
Acute Inhalation	9 Months
Reproduction	39 Months
Mutagenicity	12 Months
Animal Metabolism	18 Months
Magnitude of Plant Residues	6 - 24 Months
Analytical Methods	15 Months
Storage Stability	18 Months
Chronic Toxicity Rodent	9 Months*
Oncogenicity data	9 Months*

*These studies may be upgraded or they will have to be repeated.

6. Contact Person at EPA

>   Robert J. Taylor
>   Product Manager 25
>   Fungicide–Herbicide Branch
>   Registration Division (TS-767C)
>   Office of Pesticide Programs
>   Environmental Protection Agency
>   401 M Street SW.
>   Washington, DC 20460
>   (703) 557-1800

DISCLAIMER: The information presented in this Pesticide Fact Sheet is for informational purposes only and may not be used to fulfill data requirements for pesticide registration and reregistration.

ETHION

Reason for Issuance: Registration Standard (Second Round Review)
Date Issued: September 30, 1989
Fact Sheet Number: 209

1. DESCRIPTION OF CHEMICAL

    Generic (Chemical) Name: 0,0,0',0'-Tetraethyl S,S'-
                    methylene bisphosphorodithioate

    Common Name: Ethion

    Trade and Other Names: Ethanox, Ethiol, Hylemox, Rhodiacide,
                    Rhodocide, Vegfru Fosmite, RP-Thion,
                    Tafethion.

    EPA Pesticide Chemical Code (Shaughnessy No.): 0584401

    Chemical Abstracts Service (CAS) Number: 563-12-2

    Year of Initial Registration: 1965

    Pesticide Type: Insecticide (non-systemic)

    Chemical Family: Organothiophosphate member of the
                    Organophosphate Family

    U.S. and Foreign Producers:
                    Cheminova (Denmark)
                    FMC Agricultural Chemical Group (USA)
                    M/S Pesticides India (India)
                    Rallis India Ltd.
                    Rhone-Poulenc Agrochemic (France)
                    Sintesul S.A. (Brazil)
                    Volrho Ltd. (India)

2. Use Patterns and Formulations

Application: A non-systemic insecticide for control of leaf-feeding insects, mites, and scale.

Registered Uses:

> Terrestrial Non-food Crops: Bermudagrass, junipers, ornamental evergreens, pine trees, lawns, ornamental turf, and ornamental plants.

> Greenhouse Non-food Crops: Ornamental plants.

> Domestic Outdoor Uses: Domestic dwellings, and lawns.

> Terrestrial Food Crop Use: Alfalfa (seed crop), almonds, apples, apricots, beans, cherries, chestnuts, corn (field), cotton, cucumbers, eggplants, filberts, grapefruit, grapes, lemons, limes, melons, nectarines, onions (dry & green), oranges, peaches, peanuts, pears, pecans, peppers, pimentos, plums, prunes, sorghum (grain & forage), sorghum (seed crop), squash (summer), strawberries, tangelos, tangerines, tomatoes, and walnuts.

Mode of Insecticidal Activity: Toxic action is exerted by inhibiting enzymes of the nervous system through inhibition of cholinesterase.

Method of Application: Ground and aerial foliar applications, furrow treatments by ground equipment, and seed treatment.

Annual Usage: 1.2 to 1.5 millon pounds of active ingredient are used in the United States.

Predominant Usage: Citrus accounts for 86% to 89% of the total pounds of ethion used in the United States. The remaining 11% to 14% is applied to cotton and a variety of fruit trees, nut trees, and vegetables. Approximately 55% to 70% of all domestically produced citrus fruits are treated with ethion. Less than 2% of the domestic cotton acreage and fewer than 10% of the fruit (other than citrus), vegetable, and nut acreage is treated with ethion.

Formulations: Emulsifiable solution 500 g/l
Wettable Powder 25%
Dusts 2%,3%,and 4%
Emulsifiable concentrates 4 and 8 lbs/gal
Granules 5% and 10%

3. Science Findings

Chemical Characteristics of Technical Material

Color: colorless to light brown or pale yellow.

Physical state: liquid at room temperature.

Odor: mild dithiophosphate

Melting point: -12°F to -15°F

Boiling point: $164-165^{o}$C at 0.3 mm Hg.

Specific gravity: $1.215-1.230$ at 20°C

Solubility: Practically insoluble in water; soluble in most
organic solvents.

Vapor pressure: 1.5×10^{-6}mm Hg at 25°C

Dissociation constant: Not applicable; TGAI does not disperse in
water.

Octanol/water partition coefficient: 10,000 [GC with FPD
detector].

pH: 5.1 in an equal volume of distilled water.

Storage stability: 95% of ethion remained unchanged during one
year of storage at 25°C. There was <2%
deterioration in three years of storage under
the same conditions. Recommended shelf life
is two years. Do not store below 20°F to
avoid crystal formation.

Corrosion: Ethion does not corrode stainless steel or aluminum.

Summary Science Statement

Acute oral toxicity studies show that technical ethion is highly toxic to mammals, particularly to females; [Toxicity Category I (females) and Toxicity Category II (males)]. Ethion has moderately high acute dermal and inhalation toxicities; [Toxicity Category II for dermal toxicity and Toxicity Category II (females) and Toxicity Category III (males) for acute inhalation]. Ethion caused slight redness of the conjunctiva of the eye and slight erythema on the skin. The effects cleared within 48 hours, thus classifying ethion in Toxicity Category IV. Ethion was not found to be a dermal sensitizer and does not cause acute delayed neurotoxicity.

Subchronic toxicity: No compound-related histomorphologic changes were found in a 90-Day Dog Feeding Study. Based upon the inhibition of plasma cholinesterase activity observed, the LEL is 2.5 ppm and the NOEL is 0.5 ppm.

Chronic toxicity: In a chronic toxicity study conducted with rats, a decrease in serum cholinesterase was observed in high dose males and females. No other effects were observed. Based on the cholinesterase inhibition, the systemic NOEL is 4 ppm and the LEL is 40 ppm.

Oncogenicity: Ethion was not found to be carcinogenic in rats and mice.

Teratogenicity: Studies conducted with rabbits and rats did not indicate that there were any structural or functional abnormalities in test animals. However, increases in the incidence of hyperactivity of dams and delayed ossification of pubes in fetuses were observed in the rat study. In the rabbit teratology study, there were increases in the incidence of orange-colored urine, decreases in body weight and food consumption, along with an increased incidence of fused sternal centra in fetuses of treated females. The three generation reproduction study did not show any compound related reproductive effects, a decrease in serum cholinesterase activity.

Mutagenicity: Assays on gene mutation, structural chromosomal aberration, and unscheduled DNA Synthesis indicate ethion is not mutagenic.

Reproduction and Fertility Effects: Ethion has a reproductive NOEL of 25 ppm (HDT) and a systemic NOEL of 25 ppm for males and 4.0 ppm for females. These findings were based on tests in which ethion was administered to groups of F_0, F_1, F_2, male and female rats (15 males/dose and 30 females/dose) at dietary concentrations of 0, 2, 4, and 25 ppm. A decrease in serum cholinesterase activity was observed in the F_1, and F_2 high-dose females (25 ppm).

Gene mutation test: An Ames assay showed that ethion at the concentration range of 0.625 to 10.0 nl/ml does not produce mutagenic effects in five strains of Salmonella typhimurium (TA-98, TA-100, TA-1535, TA-1537, and TA-1538).

Structural chromosome aberration test: An in vivo cytogenetic assay indicated that ethion at dose ranges of 4.7 to 47 mg/kg does not induce chromosome aberrations in rats.

Test for other genotoxic effects: An unscheduled DNA synthesis (USD) in rat primary hepatocytes showed no evidence of induction of UDS at dose levels of 100, 500, 2500, 5,000, and 10,000 ug/plate.

Information on Human Effects:

A group of adult male volunteers (10) was randomly divided into a control group (3 males) and a treatment group (6 males). The treatment group received ethion serially and orally (gelatin capsule) at dose levels of 0.05, 0.075, 0.1, and 0.15 mg/kg. Significant reduction of plasma cholinesterase activity was seen at doses of 0.075 mg/kg and above. Based upon this data a NOEL of 0.05 mg/kg and a LED of 0.075 mg/kg/day were established.

Environmental Characteristics

Environmental Fate: The Agency has determined that ethion appears to be resistant to hydrolysis (except at very alkaline pH), photolyzes in water and on soil (half-life of 58 and 51 days respectively) and its major metabolite is CO_2.

Groundwater: The Agency presently believes that ethion is not expected to leech. Available data indicate that it is immobile and only moderately persistent.

Ecological Characteristics

Aquatic Toxicity: Technical ethion is very highly toxic to freshwater and marine fish. Acute toxicity ranges from an LC_{50} of 49 ppb for Bluegill to an LC_{50} of 720 ppb for Cutthroat trout and flathead minnows.

Ethion appears to be a heavy bioaccumulator (1400 for whole fish on day 42).

Ethion is very highly toxic to freshwater invertebrates. The acute toxicity ranged from 0.056 to 7.7 ppb. The toxicity of ethion to marine/estuarine invertebrates is also very high, ranging from 5.6 ppb to 49 ppb.

Honeybee Acute Toxicity: Ethion was found to be practically non-toxic to honeybees (LD_{50} 20.55 ug/bee).

Endangered Species: Because of the demonstrated toxicity of ethion to nontarget fish and aquatic invertebrates, ethion has been identified by the Office of Endangered Species (OES), U.S. Fish and Wildlife Service (FWS) as being likely to negatively impact on endangered aquatic organisms when applied to certain crops (i.e. citrus, corn, sorghum and cotton). The Agency is developing a program to reduce or eliminate exposure to these vulnerable organisms, and will issue notice of any product labeling, other than those identified in this registration standard, when the program has been developed and implemented.

Worker Protection

Based on the acute hazard to exposed persons, the following interim reentry intervals will be imposed until appropriate exposure data has been evaluated by the Agency:

Crop	Reentry interval from last application
Citrus	30 Days
Peaches	14 Days
Nectarines	14 Days
Grapes	14 Days
All other crops	2 Days

The precautionary and protective clothing label statements listed below are required on all products containing ethion with the signal words DANGER or WARNING that are applied to agricultural sites, structural pest control sites, or greenhouses.

"Do not rub eyes or mouth with hands. If you feel sick in any way STOP work and get help right away. See Practical Treatment Section of this label."

"Do not apply this product in a way that will contact unprotected workers, either directly or though drift. Only protected handlers may be in the area during application."

"USE ONLY WHEN WEARING THE FOLLOWING PROTECTIVE CLOTHING AND EQUIPMENT DURING MIXING/LOADING, APPLICATION, REPAIR AND CLEANING OF MIXING, LOADING, AND APPLICATION EQUIPMENT, DISPOSAL OF THE PESTICIDE, AND EARLY REENTRY INTO TREATED AREAS: Protective suit of one or two pieces covering all parts of the body except head, hands, and feet; chemical resistant gloves; chemical resistant shoes (or chemical resistant shoe covers or chemical resistant boots); and a NIOSH or MSA approved respirator. In addition, mixer/loaders must wear a chemical resistant apron and face shield or goggles.

During equipment repair and cleaning, the respirator need not be worn. During early reentry after sprays have dried or dusts have settled and vapors have dispersed, the respirator need not be worn.

IF MIXING/LOADING IS PERFORMED USING A CLOSED SYSTEM, THE FOLLOWING PROTECTIVE CLOTHING AND EQUIPMENT MAY BE WORN AS AN ALTERNATIVE: long-sleeved shirt and longlegged pants; shoes and socks. Chemical resistant gloves must be available in the cab or cockpit and must be worn when exiting. This clothing is inadequate protection during equipment repair or cleaning, reentry, or pesticide disposal work.

IMPORTANT! If pesticide comes in contact with skin, immediately wash off with soap and water. Always wash hands, face, and arms with soap and water before smoking, eating, drinking, or when using the toilet.

AFTER WORK: Wash gloves with soap and water before removing them. Take off all work cloths and shoes. After removing clothing shower using soap and water, then put on clean clothes. Do not reuse contaminated clothing. Personal clothing worn during work must be laundered separately from household items. Store protective clothing separate from personal clothing. Clean or launder protective clothing after each use. Respirators must be cleaned and filters replaced according to instructions included with the respirator. Protective clothing and equipment that becomes heavily contaminated or drenched must be destroyed according to state and local regulations. HEAVILY CONTAMINATED OR DRENCHED CLOTHING CANNOT BE ADEQUATELY DECONTAMINATED.

DURING AERIAL APPLICATION, HUMAN FLAGGERS MUST BE IN A TOTALLY ENCLOSED VEHICLE."

"Do not enter or allow entry into treated areas until (sprays have dried/dusts have settled/vapors have dispersed, as applicable) to perform hand labor tasks. A person may enter the area to perform other tasks only if the person is wearing the personal protective clothing listed on the label."

"After (sprays have dried/dusts have settled/vapors dispersed, as applicable) do not enter or allow entry into the treated area until the reentry interval has expired, unless the person entering the treated area is wearing the personal protective equipment listed on the label for early reentry."

Tolerance Assessment

   Tolerances for ethion in or on raw agricultural commodities
and animal products are published in 40 CFR 180.173. Food
additive tolerances for ethion have been established for dried
tea and raisins and are published in 40 CFR 185.2750 (formerly 21
CFR 193.190). A feed additive tolerance for ethion has been
established for dehydrated citrus pulp and is published in 40 CFR
186.2750 (formerly 21 CFR 561.230). All tolerances are expressed
in terms of ethion and its oxygen analog (s-
[(diethoxyphosphinothioyl)thio)methyl]0,0-
diethylphosphorothioate].

Summary of Tolerances Issued for Ethion

Commodities	Tolerances (PPM) US	Canadian	MRL International Mexican	Codex
Almonds	0.1	0.1	0.1	0.1^1
Almond, hulls	5.0			
Apples	2.0	2	2	2^1
Apricots	0.1	none	none	0.1^1
Beans	2.0	1	2	$none^2$
Cattle, fat	2.5	2.5	none	
Cattle, meat	2.5	2.5	none	2.5^1
Cattle, mbyp	1.0	2.5	none	1^1
Cherries	0.1	none	none	0.1^1
Chestnuts	0.1	none	0.1	0.1^1
Citrus	2.0	2	2	2^1
Citrus, pulp	10			
Corn, fodder	14	none	none	$none^2$
Corn, forage	14	none	14	$none^2$
Corn, grain	0.1	none	0.1	0.05^3
Cottonseed	0.5	none	0.5	0.5^1
Cucumbers	0.5	0.1	0.5	0.5^1
Eggs	0.2	none	none	0.2^1
Eggplants	1.0	0.1	1	1^1
Filberts	0.1	none	0.1	0.1^1
Goats, fat	0.2	none	none	0.2^1
Goats, meat	0.2	none	none	0.2^1
Goats, mbyp	0.2	none	none	0.2^1
Grapes	2.0	2	2	2^1
Raisins	4.0			
Hogs, fat	0.2	none	none	0.2^1
Hogs, meat	0.2	none	none	0.2^1
Hogs, mbyp	0.2	none	none	0.2^1
Horses, fat	0.2	none	none	0.2^1
Horses, meat	0.2	none	none	0.2^1
Horses, mbyp	0.2	none	none	0.2^1
Melons	2.0	0.1	2	2^1
Milk, fat	0.5	none	none	0.2^4
Nectarines	1.0	none	none	1.0^1
Onions	1.0	0.1	1	1^1
Peaches	1.0	1.0	1.0	1.0^1
Pears	2.0	2	1	2^1
Pecans	0.1	none	0.1	0.1^1
Peppers	1.0	0.1	1	1^1
Pimentos	1.0	0.1	none	$none^2$
Plums (fresh prunes)	2.0	1.0		2.0^1
Poultry, fat	0.2	none	none	0.2^1
Poultry, meat	0.2	none	none	0.2^1

Summary of Tolerances Issued for Ethion

Commodities	Tolerances (PPM) US	Canadian	MRL International Mexican	Codex
Poultry, mbyp	0.2	none	none	0.2^1
Sheep, fat	0.2	none	none	0.2^1
Sheep, meat	0.2	none	none	0.2^1
Sheep, mbyp	0.2	none	none	0.2^1
Sorghum, forage	2.0	none	2.0	$none^2$
Sorghum, grain	2.0	none	2.0	$none^2$
Squash, summer	0.5	0.1	0.5	0.5^1
Strawberries	2.0	1.0	2.0	2.0^1
Tea (dried)	10.0	none	none	5.0^5
Tomatoes	2.0	0.5	2	2^1
Walnuts	0.1	none	0.1	0.1^1

1. Established Codex MRL is numerically identical to U.S. Tolerance.

2. No Codex MRL has been established, therefore, no questions of compatibility exist with respect to Codex MRL.

3. A Codex MRL (CXL) of 0.055 ppm exist for residues of ethion per se in or on maize. This level is lower than that of the U.S..

4. U.S. Tolerance is higher.

5. (Green, black tea): A decision regarding the potential for compatibility between the permanent Codex MRL and the U.S. tolerance will not be made until the adequacy of the U.S. tolerance has been ascertained.

Summary of Dietary Exposure Analysis

The Agency has concluded that the use of ethion will not result in chronic health effects. There does not appear to be a health risk from short term exposure to ethion residue on grapefruit, stone fruits other than peaches and plums, eggs, grains, meat nuts, and poultry. However, based on the limited data/information available, short term exposure to ethion from residues on peaches, vegetables, tea, and oranges may result in acute cholinesterase inhibition for selected TAS populations, infants and children. Also, short term exposure to ethion residues on apples and pears, grapes, melons, tomatoes dry beans, succulent beans, strawberries (pooled with citrus other than oranges and grapefruit), and plums or prunes may result in acute cholinesterase inhibition in all TAS population groups.

The Agency is initiating a Data Call-In to exaimine acute
dietary exposure; anticipated residue data are being called in
under 40 CFR 158.240 (Reduction of Residue).

SUMMARY OF REGULATORY POSITION

The Agency is considering further regulatory action based on
dietary exposure concerns. The Agency has determined that
certain current tolerances may not provide an adequate margin
of safety in humans. Short term exposure to ethion residues on
apples and pears, grapes, melons, tomatoes, dry beans,
succulent beans, strawberries (pooled with citrus other than
oranges and grapefruit), and plums and prunes may result in
acute cholinesterase inhibition for all TAS population groups.
The Agency also has determined that short term exposure to
ethion residues on peaches, vegetables, tea, and oranges may
result in acute cholinesterase inhibition in selected TAS
populations, infants and children.

The Agency is not imposing the Restricted Use Classification on
all pesticide products containing ethion. At present three
ethion products are classified Restricted Use because of their
high dermal and inhalation toxicity. Two of these products
contain 81.9% active ingredient and the other 81% active
ingredient. The Agency will make a decision regarding the
Restricted Use Classification on a product-by-product basis for
the remaining products after evaluating the product specific
toxicity data submitted in response to the Registration
Standard issued September 30, 1989.

No groundwater advisory labeling is required because ethion is
not expected to leach.

Based on the high acute toxicity and worker poisoning
incidents, the Agency is requiring label language stipulating
the use of protective clothing and the following reentry
intervals.

Citrus.................. 30 Days
Nectarines.............. 14 Days
Peaches................. 14 Days
Grapes.................. 14 Days
All other crops and uses. 2 Days

The Agency is imposing these reentry intervals as an interim
measure until reentry data can be generated.

No significant new food uses or increases of tolerances for
food or feed items treated with ethion will be considered
until the outstanding residue chemistry studies and residue
reduction information have been submitted and reviewed by the
Agency.

CONTACT PERSON AT EPA:

Product Specific Inquiries:
William Miller
Product Manager (Team 16)
Insecticide-Rodenticide Branch
Registration Division (H7505C)
Office of Pesticide Programs, EPA
Environmental Protection Agency
401 M Street, S. W.
Washington, DC 20460

Office location and telephone number:
Room 211, Crystal Mall #2
1921 Jefferson Davis Highway
Arlington, VA 22202
(703) 557-2600

Reregistration Document Inquiries:
Richard W. King
Review Manager
Reregistration Branch
Special Review and Reregistration
Division (H-7508C)
Environmental Protection Agency
401 M Street, S.W.
Washington. D.C. 20460

Office location and telephone number:
Room 1120 Crystal Mall #2
1921 Jefferson Davis Highway
Arlington, VA 22202
(703) 557-0304

DISCLAIMER: The information in this Pesticide Fact Sheet is a
summary only and is not to be used to satisfy data
requirements for pesticide registration and
reregistration. The complete Reregistration
Document for the pesticide may be obtained from
the National Technical Information Service.
Contact the Review Manager listed above for
further information.

ETHOPROP

Reason for Issuance: Registration Standard
Date Issued: June 30, 1988
Fact Sheet Number: 3.2

1. DESCRIPTION OF CHEMICAL

   Generic Name: O-ethyl S,S - dipropyl phosphorodithioate

   Common Name: Ethoprop

   Empirical Formula: $C_8H_{19} O_2PS_2$

   Trade Names: Mocap®, VC 9-104

   EPA Pesticide Chemical
     (Shaughnessy) Number: 041101

   Chemical Abstract Service Number (CAS): 13194-48-4

   Pesticide Type: Insecticide, Nematocide, and Fungicide

   Chemical Class: Organophosphate

   U.S. Registrant: Rhone - Poulenc, Inc.

2. USE PATTERNS AND FORMULATIONS

   Application Sites:

        Food crop application of ethoprop: bananas,
        broccoli (EUP), cabbage, califlower (EUP),
        corn grain, corn fodder and forage, cucumbers,
        mushrooms, okra, fresh corn including sweet
        corn (kernals plus cob with husk removed),
        lima beans, lima bean foage, snap beans, snap
        bean forage, peanuts, peanut hay, pineapples,
        pineapple fodder and forage, potatoes, soybeans,
        soybean forage and hay, sugarcane, sugarcene
        fodder and forage, and sweet potatoes (40 CFR
        §180.31 and §180.262). There are no internat-
        ional tolerances or Codex Maximum Residue Limits
        for residues of ethoprop. The ethoprop toler-
        ance (40 CFR §180.262) should be revised to read,
        "pineapples, pineapple fodder and forage,"
        instead of "Pineapple fodder and forage."

3.   SCIENCE FINDINGS

Summary Science Statement

Ethoprop is an organophosphate insecticide whose primary mechanism of toxicity is cholinesterase inhibition. Animal studies have shown that ethoprop inhibits plasma, erythrocyte, and brain cholinesterase activity. Technical ethoprop is highly toxic (Toxicity Category I to mammals when acutely administered orally, dermally, or via inhalation). Application of 0.1 ml of undiluted ethoprop (technical) into rabbit eyes for the eye irritation test produced 100% mortality within one hour post exposure (Toxicity Categroy I). No delayed neurotoxicity was observed in hens administered approximately 70% ethoprop.

Treatment-related inhibition of brain, erythrocyte, and plasma cholinesterase activities were observed in a rat sub-chronic feeding study (90 days). The pattern of cholinesterase inhibition suggested that the no observable effect level (NOEL) for cholinesterase depression was approximately 0.3 ppm (0.015 mg/kg/day). Cholinesterase inhibition (plasma and erythrocyte) was also judged to have occurred at the lowest dose tested (1.0 ppm).

The potential chronic toxicity, oncogenicity, teratology, and reproductive toxicity of ethoprop could not be fully eval-uated with the available data (see pages 4,5,6 for the details) Additional information is required.

An acceptable battery of mutagenicty tests (gene mutation, chromosomal abberation, and DNA damage) was submitted. In the presence of a metabolic activation system, ethoprop was determined to be a genotoxic agent in vitro in two types of chromosomal aberration studies. Ethoprop was not determined to be a mutagen in two gene mutations studies. For DNA damage, an assay determined ethoprop to be inactive (negative).

Four special studies are required to address certain questions which arose in the assessment of the ethoprop data-base. The first is designed to determine a NOEL for cholin-esterase inhibition in the rat, the second to resolve issues in the dog chronic feeding study (including the determination of a NOEL for cholinesterase inhibition in the dog), the third to confirm in vitro genotoxic findings with an in vivo cyto-genic assay, and the fourth to resolve the issue of eye lesions noted in the mouse oncogenicity study.

The Agency is unable to provide a quantitative estimate of the ethoprop potential for groundwater contamination.
Ethoprop was found to be very mobile in columns of loam soil
types. Additional studies are required to assess the potential for groundwater contamination including hydrolysis,
photolysis in water and soils, leaching of soil degradates,
and field dissipation.

Chemical/Physical Characteristics Of Ethoprop (Technical)

°Clear yellow tinted liquid with a strong mercaptan
 odor.

°Boiling point 86-91°C at 0.2 mm Hg.

°Solubility in water to 843 ppm at 21°C and soluble
 in most organic solvents.

°No corrosion observed on SAE type 1020 steel or
 aluminum foil of the type used to line a bag of
 granular formulation 332 hours at 21°C.

Toxicological Characteristics

Acute Oral Toxicity: Toxicity Category I [LD50 (mg/kg;
rats): males, 61.03 (49.19-75.01)];females, 32.8 (25.41-
42.44).

Acute Dermal Toxicity: Toxicity Category I [LD50 (mg/kg;
rabbits): 25.7 (14.44-45.83) mg/kg].

Acute Inhalation: Toxicity Category I [LC50 (mg/l;
rats): 0.12 (0.08-0.17)].

Primary Dermal Irritation: Toxicity Category I. All
rabbits died within 8 hours post exposure. No primary irritation index was obtainable.

Primary Eye Irritation: Toxicity Categroy I. All rabbits died within one hour post exposure.

Skin Sensitization: Data was not submitted, and not
required because of the early onset of death observed in
the other acute tests.

Delayed Neurotoxicity: No acute delayed neurotoxicity
was observed in the hen. The oral LD50 (hens) of Mocap EC
is 9.9 (8.8-12.1) mg/kg

Subchronic Dermal Toxicity: The study reviewed was considered supplementary because large variations exhisted in plasma and erythrocyte cholinesterase levels observed among the animals on test. An acceptable study is required.

Subchronic Feeding Studies: (Non-rodent) - An acceptable 13-week study was conducted using the beagle dog. Ethoprop (technical) was tested using dietary dose levels of 1.0, 3.0, and 100 ppm. Cholinesterase inhibition was the only treatment related effect noted. No toxicological significant inhibition of plasma or erythrocyte cholinesterase was judged to have occurred at the lowest dose tested (1.0 ppm; 0.075 mg/kg/day). The lowest effect level for cholinesterase depression was 3.0 ppm (0.225 mg/kg/day).

(Rodent) - The subchronic feeding study (rats; Charles River) is considered supplemental, and has been used to establish a provisional acceptable daily intake (PADI). Three groups of 25 albino rats per sex were fed ethoprop technical in the daily diet at doses of 0, 0.3, 1.0 or 100 ppm for 90 days. The pattern of cholinesterase inhibition suggests that the NOEL is approximately 0.3 ppm (0.015 mg/kg/day). A new subchronic rat study is not required provided an acceptable rat chronic feeding study is conducted. The data obtained from the chronic study will suffice in this case for the information needed for a subchronic oral toxicity study.

Chronic Toxicity

Due to various deficiences, the available studies for nonrodent and rodent chronic toxicity do not fulfill current requirements. Adequate data are needed. Beagle dogs were fed ethoprop in their diets at 0, 0.025, 1.0 and 10 mg/kg/day for 52 weeks. Plasma cholinesterase (ChE) was inhibited in females at all dose levels. Whereas, the mid- and high dose levels are effect levels for erythrocyte ChE inhibition, and the high-dose is an effect level for brain ChE inhibition. Also, all male dogs at all dose levels exhibited less weight gain than the control group. No NOEL could be determined from the study.
In order to determine a NOEL for plasma erythrocyte and and the issue of decreased male body weight gain in all test groups, the Agency is requiring an abbreviated study of 17 weeks.
In the chronic feeding study, Fisher 344 rats were fed 0, 4.5, 9.0 and 18 ppm of ethroprop for 12 weeks, and then placed on diets of 0, 49, 98 or 196 ppms for the remaining 52 weeks. Cholinesterase inhibition was observed at all the dose levels and a NOEl could not be determined. The MTD was considered to be the highest dose tested. The study is considered supplemental, and additional information is required.
The one year dog chronic feeding study is evaluated to be inadequate, but does not need to be repeated. The Agency

is requiring specified special studies to resolve the deficiencies noted in the one-year dog chronic feeding study.   The
Agency review for chronic and oncogenic potential in rats
(MRID 00138636) indicated that for the chronic feeding portion
of this study the classification was core - supplemental.   It
cannot be upgraded since no NOEL for cholinesterase inhibition
was observed.   At all doses of ethoprop tested, cholinesterase
inhibition of >20% was observed.   Additional information must
be submitted.   In addition, the Agency is requiring summary
incidence tables for clinical observations.

Oncogenicity

The requirements for oncognicity testing (two species)
are not fulfilled, and data must be submitted.   B6C3Fl mice
were fed 0, 15, 30, and 60 ppm of ethoprop in their diet
for 78 weeks.   The study did not demonstrate oncoginic effects
under the conditions of this study.   However, the highest dose
tested (60 ppm) is considered to be at least two times under
the maximum tolerated dose.   This study is considered supplemental.

Fisher 344 rats were fed 0. 4.5, 9.0, and 18 ppm of
ethoprop for 12 weeks, and then placed on diets containing 0,
49, 98, or 196 ppm of ethoprop for the remaining 52 weeks.
Under the conditions of this study, there was an increase in
the number of C-cell adenomas of the thyroid in males receiving the high-dose when compared to controls, and there
was a dose-related increase in the number of endometrial
polyps in females.   However, the total number of individual
tissues examined histologically per group was not presented.
Consequently, the incidence of these lesions cannot be determined or analyzed statistically, and an evaluation of the
oncogenic potential of ethoprop cannot be performed from
the reported study.   This study is considered by the Agency
to be supplementary.   Additional data are required.

Teratology

Ethoprop technical was adminstered by oral intubation
to groups of Sprague-Dawley rats at doses of 0. 0.16, 1.6,
and 16.0 mg/kg/day (MRID 00104532).   However, based on the
data presented and the current guidelines for examining
teratology studies, the potential of the test material to
cause developmental toxicity cannot be fully evaluated.
More specifically, certain deficienceis were noted in the
rat study at the time the initial registration standard
was prepared (although occurrances of compound-related terata
were not).   At that time, a request was made for historical
control data with regard to parameters such as delayed
ossification.   Apparently, these data were not submitted.
Since then, new Agency guidelines have stressed that observations be made for all aspects of developmental toxicity,
and not just occurrances of terata.   For these reasons,

complete historical control data for all measured fetal
and maternal parameters (details in MRID 00104532) and
individual litter data for all measured fetal parameters
are now required. The issue of potential developmental
toxicity in the rat needs to be addressed before the status
of the study (now supplemental) can be upgraded.

A teratology study in New Zealand White rabbits was
submitted in which ethoprop technical was administered by
gavage to groups of animals at doses of 0, 0.125, 0.500,
and 2.00 mg/kg. Since the data submitted in the study
were not sufficient to fully evaluate such things as
whether test material adminstration resulted in maternal
toxicity or resulted in an increase in skeletal variations,
a NOEL and LEL for maternal and developmental toxicity
could not be determined. Additional data, including historical control data for fetal and maternal parameters, are
needed for the resolution of these issues (MRID 00161619).

Reproduction

A three-generation study in Fisher 344 rats in which
ethoprop technical was administered at dose levels of 0,
60.5, 131, and 262 ppm was not sufficient to satisfy the data
requirements for reproductive toxicity (MRID 00162164). The
data presented was considered insufficient to determine a NOEL
and LEL for maternal and developmenal toxicity, because questions
were raised with regard to such things as culling procedures and
the appropriateness of other parts of the protocol, animal illness, lack of food consumption and diet analysis data, inadeqate
data presentation, and descrepancies in data reporting.

Mutagenicity: An acceptable battery of mutagenic tests
(gene mutation, negative; chromosomal abberation, positive;
and DNA damage, inactive) were evaluated. An acceptable
bone marrow cytogenetic analysis in rats is needed to provide
in vitro confirmation of in vitro findings in the chromosomal
abberation studies submitted.

Metabolism: Available studies do not fulfil the requirement for metabolism data. However in one study the
metabolites O-ethyl-S-propylphosphorothioic acid and O-
ethyl-phosphoric acid, as well as ethoprop, were detected
in the urine. An acceptable study is required.

Special Studies: The following special studies are
required to address questions generated from the evaluation of the submitted data:

A special study is required in rats for the purpose
of determining a definitive NOEL for plasma, erythrocyte,
and brain cholinesterase inhibition. A protocol must be
submitted to the Agency for approval prior to commencement
of this study.

A subchronic feeding study in dogs is required to
address issues of decrease in animal body weight gain and
the lack of a NOEL for cholinesterase inhibition in the
present 1-year study. This study is required for purposes
of determining a most sensitive species for cholinesterase
inhibition. A protocol must be submitted to the Agency
prior to initiation of this study.

An acceptable rat bone marrow cytogenetic analysis
study is required for in vivo confirmation of in vitro
cytogenetic findings observed in the chromosomal aberration
studies.

Two 60-day mouse studies are required in order to
resolve the question of whether eye lesions observed in
the 78-week mouse oncogeniciy study were systemic effects
of the test material administration. One test must be by
oral gavage in the B6C3F1 mouse, and the other one must
be a dietary study in another mouse strain.

Environmental Characteristics

Data on the metabolism of ethoprop in plants are not
considered adequate. These data identified several metabolites of ethoprop, but did not quantitate them. Given the
nature of this chemical, unusual or exceptionally toxic
metabolites are not expected. Quantitation of known and/or
supposed metabolites will allow a more complete toxicological
evaluation. Additional data form ^{14}C-radiolableled experiments on corn, potatoes, and cabbage are needed. Although
not required previously, metabolism studies on ruminants and
poultry are now needed to elucidate the pathway for metabolism
of ethoprop in animals. Tolerances for ethoprop are currently
expressed in terms of parent compound per se, and will be
reassessed when the additional required studies are submitted
and reviewed.

The Agency is unable to provide a quantitative estimation
of the ethoprop potential for groundwater contamination.
Ethoprop was found to be very mobile in columns of loamy
sand and loam soil types. Additional studies are required to
assess the potential for ground water contamination including
hydrolysis, photolysis in water and soils, leaching of soil
degradates, and field dissipation.

Ecological Characteristics

Birds

Ethoprop technical is highly toxic to bird species on
acute oral, dietary, and dermal bases. Acceptable acute and
subchronic dietary toxicity studies are available. Field
dissipation studies which will more accurately define acute
hazards to bird species inhabiting treated areas are required.
Avian reproduction studies may be needed if it is found that
ethoprop residues remain at significant levels in the field
for an extended period of time. Acute and simulated field
studies show sufficient hazard to wildlife to require the
following field studies: 1) One study with the emulsifiable
concentrate on pineapples. 2) One study with the granular
product (G) on corn or potatoes.

-Ring-neck pheasants; 95% a.i.; oral LD50 = 118 (103-
134 mg/kg).
-Bobwhite quail; 95% a.i.; oral LD50 = 33
(27-40 mg/kg).
-Mallards; LC50 = 287 (215-382 mg/kg).

Aquatic Studies

Technical ethoprop is very highly toxic to aquatic
invertebrates. It is moderately to highly toxic to rainbow
trout and highly toxic to bluegills, crustaceans, and marine
fish species. Ethoprop is slightly toxic to embryo larvae of
oyster species. Further assessment to the potential hazards
to aquatic organisms cannot be made until certain environmental fate data are submitted and reviewed. At that time an
estimated environmental concentration (EEC) will be developed.
Further aquatic data requirements will be reserved until the
EEC and environmental fate data are available.

- Rainbow trout; 95% a.i.; LC50 = 1.02 (0.56-2.10) mg/l.
- Bluegills; 95% a.i.; LC50 = 0.30 (0.23-0.40 mg/l).
- Mysidopsis bahai (shrimp); LC50 = 23 ppb.
- Callinectes sapidus (blue crabs); 100% mortality at
24 hours when exposed to 1 ppm.
- Cyprinidon varieatus (sheepshead minnows); LC50 (static
tank; 96 hours testing) = 748.3 ppb.
- Leiostomus xanthurus (spot); LC50 (static tank; 96 hours
testing) = 32 ppb.
- Cyprinidon varieatus (sheepshead minnows); flow thru

tank; 96 hours testing; LC50 = 232.67 ppb.
- Lagodon rhomboides (pinfish); flow thru tank; 96 hours
  testing; LC50 = 7.2 ppb.
- Oyster species (embryo larvae); 95% a.i.; EC50 = 11.0
  5.6-32 ppm).

Tolerance Assessment

Tolerances for residues of ethoprop in or on food commodities are publisned in 40 CFR §180.262 and §180.31. The tolerances are set at 0.02 ppm for all the listed commodities.
A conclusive tolerance reassessment was not made at this time
due to lack of data for the following: a) The metabolism of
ethoprop in plants and animals. b) Storage stability. c) Residue and toxicity studies. Additional data are required.
A final reevaluation of the tolerances and ADI will be made
as soon as the requested data concerning storage stability,
metabolism, residue, and toxicitiy are reevaluated. No
new tolerances for ethoprop will be granted in the interim.

Data on the metabolism of ethoprop in plants are not
considered adequate. These data indentified several plant
metabolites of ethoprop, but did not quantitate them. Given
the nature of this chemical, unusual or exceptionally toxic
metabolites are not expected. Quantitation of known and/
or supposed metabolites will allow a more complete toxicological evaluation. Additional data from ^{14}C-radiolabled
experiments on corn, potatoes, and cabbage are needed.
Although not required previously, metabolsim studies on
ruminants and poultry are now needed to eludidate the pathway for metabolism of ethoprop in animals. Tolerances for
ethoprop are currently expressed in terms of parent compound
per se, and will be reassessed when the additional required
studies are submitted and reviewed.

Adequate analytical methodologies are available for en
forcement of the present tolerances in terms of ethoprop per
se. Ethoprop is completely recovered by the multiresidue
procedures in the Pesticide Analytical Manual (Vol. 1, protocols II and III), and partially recovered by protocol I. No
data are available for protocol IV. These data are required.
If any metabolites of toxicological concern are identified in
the metabolism studies required, additional validated analytical methodologies may be needed.

4.   SUMMARY OF REGULATORY POSITIONS AND RATONALES

- Ethoprop meets the criteria for resticted use classification.
- Ethoprop is not a candiate for Special Review at this
  time.
- The Agency will not grant any new tolerances or new
  uses for ethoprop until data required under this
  Standard have been received and tolerances reassessed.

- The Agency is unable to provide a quantitative estimation of the ethoprop potential for groundwater
  contamination. Additional studies required are hydro-
  lyis, photolysis in water and soil, leaching of soil
  degradates, and field dissipation.
- In order to remain in compliance, updated label precautions are required to address the hazard to fish and
  wildlife.
- The Agency will reevaluate reentry protection for
  ethoprop when the requested data are received and
  evaluated. In the interim, 24 hours is the established reentry interval for all crops.

5. SUMMARY OF REQUIRED LABEL MODIFICATION

An updated Environmental Hazard Statement is required.

6. SUMMARY OF OUTSTANDING DATA REQUIREMENTS

Data Gaps

The Agency has identified missing data required to
fully evaluate the human and environmental risks associated
wih the use of ethoprop.

Toxicology	Time Frame for Data Submission (Months)
°21 day dermal toxicity (rabbit)...................	12
°90 day feeding (rodent; not required.............. if a chronic rat feeding study is performed).	15
°Chronic toxicity (rodent and nonrodent)...........	50
°Oncogenicity (mouse, rat).........................	50
°Teratogenicity (rat, rabbit)......................	15
°Reproduction (2-generation rat)...................	39
°Mutagenicity testing (cytogenetic analysis).......	14
°Metabolism..	24
°Other Special testing	
rat............................	14
subchronic.....................	18
subchronic (2).................	24

Environmental Fate/Exposure

°Hydrolysis..	9
°Leaching and adsorption/desorption................	12
°Soil dissipation..................................	27
°Soil dissipation, long term......................	50
°Spray drift......................................	27

7. CONTACT PERSON AT EPA

William H. Miller
Product Manager (16)
Insecticide-Rodenticide Branch
Registration Division (TS-767C)
Environmental Protection Agency
Washington, D.C. 20460

Telephone No. (703) 557-2600

FENITROTHION

Reason for Issuance: Registration Standard
Date Issued: July 30, 1987
Fact Sheet Number: 142

1. DESCRIPTION OF CHEMICAL

    Generic Name: O,O-dimethyl O-(4-nitro-m-tolyl)
    (Chemical)   phosphorothioate
    Common Name: Fenitrothion
    Other Chemical
    Nomenclature: O,O-dimethyl O-(3-methyl-4-nitrophenyl)
                     phosphorothioate; O,O-dimethyl O-(4-nitro-m-
                     tolyl) phosphorothioate
    Trade Names: Bayer 41831; Bayer S-5660; Bayer S-1102A;
                  AC-47,300; C 47114; Accothion; Cytel; Cyfen;
                  Folithion; Sumithion; Agrothion; Dicofen:
                  Fenstan; Metathion E-50; Verthion; Cekutrothion;
                  Dybar; Fenitox; Novathion; and Nuvanol.
    EPA Shaughnessy Code: 105901
    Chemical Abstracts Service (CAS) Number: 122-14-5
    Year of Initial Registration: 1975
    Pesticide Type: Insecticide/Acaricide
    Chemical Family: Organophosphate
    U.S. and Foreign Producers: Sumitomo Chemical Company (Japan)

2. USE PATTERNS AND FORMULATIONS

    Application Sites: Ornamentals (including outdoor, greenhouse,
                        and nursery); in forests for spruce
                        budworm and southern pine beetle control;
                        and in and around non-food domestic,
                        commercial, institutional and industrial
                        areas for household pest control.
    Formulation Types: 40% wettable powder (for control of
                        adult anopheline mosquitoes in human
                        dwellings), 4 (45.5%) and 8 (76.8%)
                        pound per gallon emulsifiable concentrates
                        (forestry, ornamental and domestic,
                        commercial, institutional and industrial
                        use), and a 93% soluble concentrate/liquid
                        (forestry use).
    Application Methods: Primarily by ground application equipment;
                         aerial equipment is used for spruce budworm
                         control.

3.   SCIENCE FINDINGS

Summary Science Statement
    Fenitrothion is a moderately acutely toxic cholinester-
aseinhibiting pesticide. It is in Toxicity Category II for
the oral and dermal routes of exposure and Toxicity Category
III for the inhalation route of exposure and is mildly
irritating to the eyes and skin (Toxicity Category III).
It has not been shown to be a dermal sensitizer and does
not demonstrate acute delayed neuotoxic effects. Substantial
chronic toxicology and residue chemistry data gaps exist,
including metabolism, oncogenicity, mutagenicity, teratogenicity, and reproductive effects. Human epidemiological
evidence and a dog chronic feeding study have implicated
fenitrothion in causing human eye effects, such as retinal
degeneration and myopia. Laboratory data show that fenitrothion is potentially highly to very highly toxic to birds,
fish, and aquatic invertebrates, including certain endangered
species. Preliminary data indicate that groundwater contamination probably is not a potential threat; however the
Agency is unable to conduct a full assessment due to data
gaps. The Agency is particularly concerned with potential
exposure to applicators using ground application techniques
to control southern pine beetles; reentry workers in greenhouses and nurseries; and non-target organisms following
forestry uses.

Chemical/Physical Characteristics of the Technical Material
Physical State: oily liquid
Color: Yellow-brownish
Molecular weight and formula: 277.2 – $C_9H_{12}NO_5PS$
Boiling Point: 118 °C at 0.01 mm Hg
Melting Point: 0.3 °C
Specific Gravity: 1.32– 1.34
Vapor Pressure: data gap
Solubility in various solvents: data gap
pH: data gap
Stability: data gap

Toxicology Characteristics (Technical Grade)
Acute Oral: Toxicity Category II (800 and 330 mg/kg in male and
                  female rats, respectively)
Acute Dermal: Toxicity Category II (1200 and 890 mg/kg in female
                  and male rats, respectively)
Acute Inhalation: Toxicity Category III (5.0 mg/L in rats)
Primary Dermal Irritation: Toxicity Category III; mild dermal
                  irritation was reported in a
                  rabbit study.

Primary Eye Irritation: Toxicity Category III; mild irritation
                           after a single application of 0.1 mL
                           into unwashed eyes of albino rabbits.
Skin Sensitization: Not a skin sensitizer
Delayed neurotoxicity: Negative in the hen.
Subchronic Oral (rodent) Testing: Data gap for rodent
                                   species (for plasma
                                   cholinesterase effects)
Oncogenicity: Data gap for the mouse
Chronic Feeding: NOEL for brain and red blood cell
                  cholinesterase in rats is 10 ppm;
                  systemic NOEL for plasma inhibition
                  in the dog is 5 ppm.
Metabolism: Data gap
Teratogenicity: Data gap
Reproduction: Data gap
Mutagenicity: Data gap for point mutation assay in mammalian
               cells, structural chromosomal aberration, and
               other genotoxic effects
Major routes of exposure: Inhalation and dermal exposure to
                           occupants of treated dwellings; dermal
                           and respiratory exposure to applicators
                           and reentry workers.

Environmental Characteristics
Data gaps exist for most studies. Preliminary data indicate that
fenitrothion degrades fairly rapidly in soil with a half-life
of less than a week in non-sterile muck and sandy loam soils.
Preliminary data also suggest fenitrothion is intermediately mobile
in a variety of soils ranging in texture from sandy loam to clay.
The potential for groundwater cannot be assessed until acceptable
environmental fate data are received.

Ecological Characteristics (technical grade)
    Avian oral toxicity: highly toxic to upland gamebirds and
                          slightly toxic to waterfowl (acute
                          oral toxicity value to bobwhite quails
                          and mallards was determined to be 23.6
                          mg/kg and 1190, respectively)

    Avian dietary toxicity: highly toxic to upland gamebirds and
        (8 day)   and slightly toxic to waterfowl (sub-
                            acute toxicity value of 157 ppm for
                            bobwhite quail and 2482 ppm for
                            mallards.)

    Freshwater fish acute toxicity: moderately toxic to both
        (96 hr. LC_{50})   warmwater and coldwater fish
                                     (1.7 ppm for brook trout;
                                     3.8 ppm for bluegill)

Freshwater invertebrate toxicity: very highly toxic to
(48 hr. or 96 hr. EC_{50})   aquatic invertebrates
(3 ppb for <u>Gammarus</u>
<u>fasciatus</u>)

Tolerance Reassessment

There are no domestic uses for fenitrothion on food or
feed commodities. There is one established U.S. food additive
tolerance which covers residues of fenitrothion in wheat gluten
imported from Australia arising from the stored wheat grain
treatment registered in that country (2 CFR 193.156[9]). The
nature of fenitrothion residues in plants is adequately
understood. Submitted data indicate that fenitrothion <u>per se</u> (I),
desmethyl fenitrothion (IV), and p-nitrocresol (VII) are the major
components of the residue. Animal metabolism studies are not
available. In the event that future federal registrations for use
of fenitrothion on plant commodities used for animal feeds are
established, or regulations covering importation of animal products
from countries in which fenitrothion is registered for use are
established, additional animal metabolism studies may be required.
Analytical methodology for determining levels of residues
of fenitrothion, fenitrooxon, and p-nitrocresol in plants is
adequate for data collection and tolerance enforcement purposes,
Storage stability and residue data are required to support
the wheat gluten tolerance. A provisional acceptable daily
intake (PADI), based on a one-year dog study with a NOEL of
0.125 mg/kg/day and using a 30-fold safety factor is calculated
to be 0.004 mg/kg/day. A Theoretical Maximum Residue Contribution (TMRC) for the U.S. population is calculated to be 0.000038
mg/kg/day, which utilizes 0.94 percent of the PADI.

4. Summary of Regulatory Positions and Rationales

    ° Fenitrothion is not being placed into Special Review
at this time. Although the Agency is concerned over the potential adverse impact of fenitrothion on birds and aquatic organisms
resulting from the forestry use pattern, comprehensive aquatic
and terrestrial field studies are needed in order to evaluate
the potential risks to birds and aquatic organisms. The Agency
is also requiring submission of special acute and subchronic rat
studies to provide additional information to confirm the potential
for fenitrothion to cause retinal degeneration and changes in
corneal shape and structure in the human eye. Pending receipt and
evaluation of these data, labeling modifications or other regulatory
action may be warranted.

    ° The Agency is classifying the forestry uses of
fenitrothion (spruce budworm and southern pine beetle) for
restricted-use due to avian and aquatic invertebrate hazards
on an interim basis pending receipt and evaluation of the

aquatic and terrestrial field studies.

° Fenitrothion is highly toxic to honeybees, aquatic invertebrates, and avian species. Endangered species label restrictions are required to protect endangered and threatened species in forest areas.

° Special indoor air residue monitoring studies are required to support continued use of the 40% wettable powder formulation in homes to control adult Anopheline mosquitoes.

° No new tolerances or new food uses will be granted until the Agency has received data sufficient to evaluate the dietary exposure of fenitrothion.

° The Agency is imposing an interim 24 hour reentry interval for the greenhouse and nursery ornamental use pending receipt and evaluation of reentry data.

° Protective clothing statements are required for all products containing fenitrothion.

6.   SUMMARY OF OUTSTANDING DATA REQUIREMENTS

Time Frame*

Toxicology

Subchronic oral toxicity--rodent species (for plasma cholinesterase effects)	12 Months
21-day dermal--rabbit	9 "
90-day inhalation--rat	15 "
Oncogenicity--mouse	50 "
Teratogenicity--rat and rabbit	15 "
Reproduction--rat	39 "
Mutagenicity	12 "
Metabolism study	12 "
Special tests--acute and subchronic tests in rats for eye effects	24 " 9 "

Environmental Fate/Exposure

Hydrolysis study	9 Months
Photodegradation in water, soil and air	9 "
Aerobic soil metabolism study	27 "
Anaerobic aquatic metabolism study	27 "
Lab volatility study	12 "
Leaching and adsorption/desorption	12 "
Soil dissipation study	27 "
Forestry dissipation study	27 "
Fish accumulation study	12 "
Applicator exposure studies	9 "

* based upon receipt of the Standard by the registrant.

Indoor air/surface residue exposure study	12	"
Reentry Data	27	"

Fish and Wildlife

Avian reproduction	24	Months
Actual field testing--birds and aquatic organisms	48	"
Acute toxicity to freshwater invertebrates-- typical end-use product	9	"
Fish early life stage and aquatic invertebrate life cycle	15	"

Plant Testing Requirements

Seed germination/seedling emergence	9	Months
Vegetative vigor	9	"
Aquatic plant growth	9	"

Residue Chemistry

Residue analytical methods	18	Months
Storage stability	18	"
Residue data (wheat gluten)	24	"

Product Chemistry   9-15 Months

7.   CONTACT PERSON AT EPA

     William H. Miller
     Product Manager (16)
     Insecticide-Rodenticide Branch
     Registration Division (TS-767C)
     Office of Pesticide Programs
     Environmental Protection Agency
     401 M Street, S. W.
     Washington, D. C.   20460

     Office location and telephone number:
     Room 211, Crystal Mall #2
     1921 Jefferson Davis Highway
     Arlington, VA   22202
     (703) 557-2400

DISCLAIMER: The information presented in this Chemical Information
Fact Sheet is for informational purposes only and may not be used
to fulfill data requirements for pesticide registration and
reregistration.

FENOXAPROP-ETHYL

Reason for Issuance: New Chemical Registration
Date Issued: February 1988
Fact Sheet Number: 157

1. DESCRIPTION OF CHEMICAL

   Generic Name: (+) ethyl 2-[4-[6-chloro-2-
                                    benzoxazolyl)oxylphenoxylpropanoate

   Common Name: Fenoxaprop-ethyl

   Trade Names: Whip, Acclaim

   EPA Shaughnessy Code: 128701

   Chemical Abstracts
   Service (CAS) Number: 66441-23-4

   Year of Initial
   Registration: 1987

   Pesticide Type: Herbicide

   Chemical Family: Structurally related to
                                    diphenyl ethers

   Producer: American Hoechst Corporation

2. USE PATTERNS AND FORMULATIONS

   Application sites: Used for post-emergent control of annual
   and perennial grasses on terrestrial and aquatic food crops
   (rice and soybeans) and on terrestrial nonfood and domestic
   outdoor use sites (turfgrass, including sod farms, rights-
   of-way, and commercial and residential turf).

   Types of formulations: 93% active ingredient technical grade.
   End use products containing 12.50% active ingredient formulated
   as emulsifiable concentrates.

   Usual carrier: Water. Non-phytotoxic crop oil may also be
   used with the agricultural use formulation.

258

Types and methods of application: Fenoxaprop-ethyl is applied
to turfgrass by pressurized hydraulic sprayers (30-60 PSI)
and hand held pump sprayers. It may also be applied as a
spot treatment. Both ground and aerial application are
permitted for the use on soybeans and rice.

Application rates: Application rates for soybeans range from
.8 to 1.2 pints per acre (.10 to .15 pounds active ingredient)
depending on the target weed species. For rice, application
rates vary from 1.2 to 1.6 pints per acre (.15 to .20 pounds
active ingredient) depending on the target weed species and
stage of growth. Application rates for turfgrass range from
15 to 45 fluid ounces per acre and from .15 to 1.02 fluid
ounces per 1,000 square feet depending on the type of
turf and stage of weed growth.

3. SCIENCE FINDINGS

Summary Science Statement: Fenoxyprop-ethyl induces
developmental toxicity (birth defects) in rabbits based
on studies which showed that dietary administration of
the compound caused an increased incidence of rib anomalies and
diaphragmatic hernias at 200 mg/kg, the highest dose tested.
Margins of safety (MOS) calculated based on a no-observed-
effect-level of 50 mg/kg in the rabbit study are 740,000
and 260,000 for one serving daily of rice and soybean oil,
respectively. Even assuming that four servings of each
could be consumed every day, the MOS's would be 180,000
for rice and 66,000 for soybean oil. MOS values for
mixer/loaders are 1,250,000, for inhalation exposure,
and 1000, for dermal exposure. For applicators, the
MOS for inhalation exposure is 250,000; the MOS for
dermal exposure is 2500. In addition, the label requires
that mixer/loaders wear protective clothing, including
impermeable gloves, and long-sleeved shirts and pants.
Applicators must also wear long-sleeved shirts and pants.
The teratogenic risk to consumers of foods treated with
fenoxaprop-ethyl and to users of the herbicide are therefore
quite low.

Fenoxaprop-ethyl did not induce an oncogenic response in long-
term rat and mouse studies. The chemical did not significantly
impair reproductive ability in a two-generation reproductive
effects study in rats. Four mutagenicity studies with
fenoxaprop-ethyl were negative.

In both short and long term animal studies, fenoxaprop-
ethyl induced toxicologically significant increases and
decreases in lipid enzymes (blood cholesterol). The Agency
has calculated MOS values for these effects based on
the highest dose tested, 6 mg/kg, in a mouse oncogenicity
study, the mouse being the most sensitive species. The
dietary MOS values are, for rice, 35,000, and for soybean
oil, 31,000. For applicators, MOS values are 30,000 for
inhalation exposure and 300 for dermal exposure. For
mixers/loaders, the MOS for inhalation exposure is 150,000
and 120 for dermal exposure. The dermal exposure MOS's
were calculated based on exposure estimations which assumed
that 100% of the chemical would be absorbed through the skin.
However, as indicated above, mixer/loaders/applicators are
required to wear protective clothing, and therefore exposure
will be reduced.

Fenoxaprop-ethyl is not acutely toxic to humans or avian
species. The pesticide is toxic to fish and aquatic invertebrates
Environmental fate studies show that fenoxaprop-ethyl does
not persist significantly in the environment, that it is
relatively immobile and therefore should not pose a risk of
leaching to groundwater.

Chemical Characteristics:

Physical state:	brown crystalline solid
Molecular formula:	$C_{18}H_{16}ClNO_5$
Molecular weight:	361.8 g/M
Solubility:	0.9 mg/l (pH 7 at 25°C) in water, low solubility
Melting point:	85-87°C
Vapor pressure:	0.187 x 10^{-7} mbar at 20°C (non volatile)

Toxicological characteristics:

Acute oral toxicity (rat):	2357 mg/kg (relatively nontoxic)
Acute dermal toxicity (rat):	Greater than 2000 mg/kg (moderately toxic)
Acute dermal toxicity (rabbit):	Greater than 2000 mg/kg (moderately toxic)
Dermal sensitization:	Non-sensitizing

Chronic effects:

2-generation Reproduction (rat):	NOEL = 5 ppm (0.25 mg/kg) based on reduced blood lipids in parents and reduced body weight in offspring

Developmental Toxicity:

Rabbit -- NOEL = 12.5 mg/kg for maternal toxicity, based on decreased food consumption and weight gain; NOEL = 50 mg/kg for developmental effects, based on increased incidence of rib anomalies and diaphragmatic hernia; teratogenic

Rat -- NOEL = 32 mg/kg for maternal toxicity based on reduced body weight gain; NOEL = 32 mg/kg for developmental (fetotoxic) effects -- delayed ossification and slightly impaired growth; NOEL = 100 mg/kg for teratogenic effects

Chronic Feeding/
Oncogenicity:

Rat -- not oncogenic at doses up to and including 180 ppom (9 mg/kg); systemic toxicity NOEL = 30 ppm (1.5 mg/kg) based on decreased serum cholesterol

Mouse -- not oncogenic at doses up to and including 40 ppm (6 mg/kg); dosing not adequate to achieve a maximum tolerated dose (MTD) --- Adequate MOS values based on ratio of highest dose tested/ maximum daily dietary intake.

Dog -- NOEL = 15 ppm (0.37 mg/kg) based on reduced body weight

Mutagenicity:

Negative -- chromosomal aberration, Ames test, Unscheduled DNA Synthesis, and mouse micronucleus

Physiological and biochemical behavior characteristics:

Mode of Activity: Fenoxaprop-ethyl is a systemic herbicide which is rapidly absorbed and translocated throughout leaf and stem tissue. Although the precise mode of activity is unknown, fenoxaprop-ethyl is thought to kill weeds by disrupting lipid metabolism. Effects are seen as general yellowing of the weed followed by death in approximately two to three weeks.

Translocation characteristics: Fenoxaprop-ethyl is highly systemic and rapidly metabolized and therefore there is little potential for the presence of residues in the edible parts of crops treated with the herbicide.

Environmental characteristics: Fenoxaprop-ethyl is stable to hydrolysis at 20°C in pH5 and pH7 solutions, but rapidly hydrolyzes in pH9 solutions. Extensive environmental fate studies show that the chemical does not persist significantly in any medium. Residues in irrigated and rotational crops will not be detectable when label restrictions on using water from rice fields to irrigate grops and on planting· rotational crops are strictly followed. Fenoxaprop-ethyl was slightly mobile in two loamy sand soils, two silt loam soils and an aquatic sediment (clay). Therefore, there is little potential that fenoxaprop-ethyl would leach to ground water.

Ecological characteristics:

Avian Reproduction:	Bobwhite -- 30 ppm Mallard duck -- 180 ppm (NOEL's for reproductive effects; does not impair avian reproduction)
Avian Oral toxicity:	Bobwhite quail -- >2510 mg/kg
Avian dietary toxicity:	Mallard duck -- >5620 mg/kg Bobwhite quail -- >5620 mg/kg
Freshwater fish:	Bluegill -- 310 ppb Pumpkinseed sunfish -- 360 ppb Brown trout -- 480 ppb
Aquatic invertebrates:	Daphnia Magna -- 3.18 ppm

These data indicate that fenoxaprop-ethyl is essentially non-toxic to avian species and that it does not impair avian reproduction; and that fenoxaprop-ethyl is acutely toxic to fish and aquatic invertebrates. The label prohibits use in St. Francis and Cross Counties in Arkansas to avoid impact on the endangered fat pocketbook mussel, Potamilus capax. No other endangered species issues have been identified for the rice and soybean uses.

4.   TOLERANCE ASSESSMENT

Tolerances have been established for the combined residues of fenoxaprop-ethyl and its metabolites on the following raw agricultural commodities (40 CFR 180.   ):

Commodities	Tolerance (ppm)
Rice grain	0.05
Soybeans	0.05

There are no international tolerances/residue limits for fenoxaprop-ethyl.

There are sufficient residue chemistry data available to support these tolerances, including plant and animal metabolism, storage stability (for both the parent compound and its metabolites), field residue studies, and analytical methods. Cattle and poultry feeding studies were not submitted. However, under the proposed conditions of use, measurable residues are not expected to be found in the raw agricultural commodities or fractions. These data are therefore not now necessary.

The Acceptable Daily Intake (ADI) and the Maximum Permissible Intake (MPI) are two ways of expressing the amount of a substance that the Agency believes, on the basis of the results of data from animal studies and the application of "safety" or "uncertainty" factors, may safely be ingested by humans without risk of adverse health effects. The ADI is expressed in terms of milligrams (mg) of the substances per kilogram (kg) of body weight per day (mg/kg/day). The MPI, a related figure, is obtained by assuming a human body weight of 60 kg, and is expressed in terms of mg of substance per day (mg/day).

The Agency has calculated an ADI for fenoxaprop-ethyl of 0.0025 mg/kg/day, based on a NOEL of 0.25 mg/kg/day in the 2-generation rat reproduction study and a 100-fold safety factor. The MPI for a 60 kg person is 0.15 mg/day. These tolerances have a theoretical maximum residue contribution (TMRC) of 0.0011 mg/day in a 1.5 kg diet and would utilize .073 percent of the ADI.

5.   CONTACT PERSON AT EPA

Richard F. Mountfort
U. S. Environmental Protction Agency
TS-767C
401 M Street, S. W.
Washington, D. C.   20460

FENTHION

Reason for Issuance: Issuance of Registration Standard
Date Issued: June 30, 1988
Fact Sheet Number: 169

1. DESCRIPTION OF CHEMICAL

    Generic Name: O,O-dimethyl-O-[4-(methylthio)-m-tolyl] phosphorothioate
    (Chemical)

    Other Chemical
      Names: O,O-dimethyl-O-[3-methyl-4-(methylthio)phenyl
                       phosphorothioate
                     O,O-dimethyl-O-4-methylthio-m-tolyl phosphorothioate

    Common Name: Fenthion (FDA), No American National Standards Institute
                    Common Name

    Trade and Other Names: Baytex; Entex; Bayer 29493; Bayer S-1752;
                              Baycid; Lebaycid; Spotton; Tiguvon;
                              Mercaptophos

    EPA Chemical Code: 053301

    Chemical Abstracts Service (CAS) Number: 55-38-9

    Year of Initial Registration: 1965

    Pesticide Type: Insecticide/Acaricide, Avicide, Mosquitocide (adults and
                     larvae)

    Chemical Family: Organophosphate

    U.S. and Foreign Producers: Mobay Chemical Corp. (United States); and
                                   Bayer AG (Federal Republic of Germany)

2.   PHYSICAL AND CHEMICAL CHARACTERISTICS

Chemical Characteristics of the Technical Material[1]

Physical State: Liquid.

Color: Yellow-tan.

Odor: Slight garlic odor.

Molecular Weight and Formula: 278.3 - $C_{10}H_{15}O_3PS_2$.

Melting Point: $<-25^{\circ}C$

Boiling Point: $105^{\circ}C$ at 0.01 mm Hg.

Vapor Pressure: 3×10^{-5} mm Hg at $20^{\circ}C$.

Density: 1.250 at $20^{\circ}C$.

Solubility in various solvents: Practically insoluble in water (55 mg/l), soluble in methanol, ethanol, ether, acetone, and many other organic solvents (especially chlorinated hydrocarbons).

Physiological and Biochemical Characteristics

Mechanism of Pesticidal Action: Cholinesterase inhibition following ingestion of fenthion or dermal absorption from contact with treated surfaces.

Metabolism and Persistence in Plants and Animals: The metabolism of fenthion in plants and animals is not adequately understood. The available plant metabolism data indicate that the following metabolites were found in the leaves of cotton plants treated with fenthion: fenthion oxygen analog, fenthion oxygen analog sulfoxide, fenthion oxygen analog sulfone, dimethyl phosphorothioic acid and dimethyl phosphoric acid. Fenthion sulfoxide, fenthion sulfone, fenthion oxygen analog sulfoxide, and fenthion oxygen analog sulfone were the metabolites detected in fenthion treated bean plants. The available animal metabolism data indicate that the metabolites fenthion

---

[1]The physical/chemical properties of fenthion listed in this section were obtained from Merck Index 10th Edition.

sulfoxide and fenthion oxygen analog
identified as one entity, and fenthion
sulfone, fenthion oxygen analog sulfoxide, and
fenthion oxygen analog sulfone identified as
another single entity were tentatively
identified in steak, liver, milk, urine, and
feces from cattle. Dimethyl phosphoric acid
and dimethyl thiolophosphoric acid were also
tentatively identified in urine from cattle.
The metabolites fenthion sulfoxide, fenthion
sulfone, fenthion oxygen analog, fenthion
oxygen analog sulfoxide and fenthion oxygen
analog sulfone were identified in muscle,
heart, fat, kidney, and liver from swine.
Swine liver was also found to contain the
additional metabolites fenthion sulfoxide
phenol and fenthion sulfone phenol while in
swine kidney the additional metabolites
detected in swine liver plus the metabolite
fenthion phenol. All of the metabolites found
in swine tissues were individually identified.

3.   USE PATTERNS AND FORMULATIONS

Application Sites: Mosquito and insect control on swamps, standing water,
recreation areas, alfalfa, pasture grass, forests,
barns, poultry houses, nonfood/feed areas of restaurants and commercial buildings, and homes; lice control
on cattle (beef and non-lactating dairy) and hogs;
control of ants, mites, leafhoppers, and aphids on
ornamentals and flowers; bird control; rice to control
mosquitoes (in the State of California only).

Types and Methods of Application: Ground application is used for all uses
of fenthion. Both ground and aerial
applications are used to control
mosquitoes with control measures being
directed to both the aquatic (larval/
pupal) stages and the free-flying adult
stage.

Types of Formulations: Dust, emulsifiable concentrates, soluble
concentrate/liquid, granular, and liquid-ready to
use.

4.   SCIENCE FINDINGS

Summary Science Statement

--Fenthion is moderately toxic (Toxicity Category II) by the oral, dermal and inhalation routes of exposure.

--Fenthion is minimally irritating (Toxicity Category IV) to the skin and eyes.

--Fenthion is not teratogenic in rabbits, and does not denonstrate mutagenic effects. Other subchronic and chronic effects, such as oncogenicity and reproductive effects, are not characterized at this time because of lack of data.

--There are insufficient data to fully assess the environmental fate of fenthion.

--Preliminary data indicate that fenthion degrades in soil rapidly, having a half life of less than one day. Other environmental fate characteristics are undefined, including fenthion's ability to contaminate ground water.

--There are insufficient data to characterize the acceptable dietary exposure to fenthion.

--Fenthion is highly toxic to aquatic organisms, birds, and honey bees. However, field kills of fish and birds have not been documented.

--Fenthion is likely to jeopardize endangered species when used as a mosquito larvicide.

Toxicology Characteristics

--Except for mutagenicity studies, acute oral, dermal, and inhalation studies, acute dermal and eye irritation studies, and a rabbit teratology study, the Agency has no acceptable toxicology studies for fenthion.

--The acute oral LD^{50} for rats is ~250 mg/kg (males) and ~295 mg/kg (females).

--The acute dermal LD^{50} for rats is 1680 mg/kg bw (males) and 2830 mg/kg bw (females).

--The acute inhalation LC^{50} for rats is ~1200 mg/m^3 (males) and ~800 mg/m^3 (females).

--Fenthion is moderately toxic (Toxicity Category II) by the oral, dermal and inhalation routes of exposure.

--Fenthion is minimally irritating (Toxicity Category IV) to the skin and eyes.

--Fenthion is not teratogenic in rabbits.

--The fetotoxic NOEL in the rabbit is 1 mg/kg/day.

--The maternal toxicity NOEL in the rabbit is 6 mg/kg/day.

--The teratogenic NOEL in the rabbit is >18 mg/kg/day.

--Fenthion is non-mutagenic in male mice up to 25 mg/kg bw (the highest dose tested).

--Fenthion's systemic NOEL for mutagenicity is 10 mg/kg bw.

Environmental Characteristics

--Available data are insufficient to fully assess the environmental fate and transport of fenthion. (Data gaps exist for nearly all applicable studies.)

--The available data indicate that fenthion degrades fairly rapidly, with a half-life of less than a day reported in nonsterile silt loam soil in the dark at 75% moisture and room temperature.

--Data also suggest that fenthion degrades to 53% under anaerobic conditions after 60 days incubation.

--The major nonvolatile degradates reported were fenthion sulfoxide, 3-methyl-4(methylsulfonyl) phenol, and 3-methyl-4(methylsulfinyl) phenol.

--Data currently available are insufficient to characterize fenthion's leaching potential for contamination of ground water. (Data to characterize the potential to contaminate groundwater are being required.)

Ecological Characteristics

Avian acute toxicity: Acute toxicity values of 5.94 mg/kg in the mallard duck, < 4.0 mg/kg in the bobwhite quail, and 2.50 mg/kg in doves.

Avian dietary toxicity: Subacute dietary toxicity values ranged from 30 ppm in the bobwhite quail to 231 ppm in the mallard duck. The potential for secondary toxicity was demonstrated when kestrels died after being fed house sparrows that had been killed by an oral dose of 10 mg/kg fenthion.

Freshwater fish acute toxicity: 96-hour acute toxicity values ranged from 3.20 ppm for fathead minnows to 1.58 ppm for cutthroat trout.

Marine fish acute toxicity: LC_{50} value of 1.6 ppb for striped mullet.

Freshwater invertebrate toxicity: The acute toxicity values ranged from 0.62 ppb for Simocephalus (Daphnid) to 0.80 ppb for Daphnia pulex.

Marine invertebrate toxicity: EC_{50} value of 340 ppb for mollusks; 96-hour LC_{50} value of 0.11 ppb for pink shrimp.

These data show that technical fenthion is very highly toxic to birds on an acute oral and dietary basis; moderately to highly toxic to both warmwater and coldwater fish species; very highly toxic to aquatic invertebrates; very highly toxic to pink shrimp; moderately toxic to striped mullet; and highly toxic to mollusks.

TOLERANCE REASSESSMENT

Tolerances have been established for residues of fenthion and its cholinesterase inhibiting metabolites in a variety of raw agricultural commodities (40 CFR 180.214).   The Agency has evaluated the residue and toxicology data supporting these tolerances and has determined that it does not have sufficient data to support the currently established tolerances for residues of fenthion.   Because of the extensive residue chemistry and toxicology data gaps, no significant new tolerances or new food uses will be granted until the Agency has received data sufficient to evaluate the dietary exposure to fenthion.

In addition to United States tolerances, there are also Canadian tolerances, Mexican tolerances, and Codex Maximum Residue Limits (MRLs) established for fenthion.   However, some incompatibility exists between some of the permanent Codex MRLs and the U.S. tolerances.   The issue of incompatibility will be addressed when residue data are submitted and evaluated.

The available toxicity data are insufficient for the Agency to calculate an Acceptable Daily Intake (ADI) for fenthion and therefore the Maximum Permissible Intake (MPI) for a 60 kg human has not been determined.

5.   REQUIRED UNIQUE LABELING SUMMARY

Personal protective equipment and work safety statements must appear on the label of all registered end use products containing fenthion.

All end-use products containing fenthion as an active ingredient with directions for use on agricultural crops, ornamental plants and forest trees, uncultivated agricultural and non-agricultural outdoor areas, aquatic sites, livestock and bird roosting areas must bear the following restricted use labeling statements:

RESTRICTED USE PESTICIDE

Due to Very High Acute Toxicity to Birds, Fish and Aquatic Invertebrates

> For retail sale to and use only by certified applicators or persons under their direct supervision and only for those uses covered by the certified applicator's certification. Certified applicators must also ensure that all persons involved in these activities are informed of the precautionary statements."

Environmental hazard statements and a bee precautionary statement must appear on the label of all end-use fenthion products.

The following reentry interval statement and protective clothing for early reentry statement must appear on the labeling of all fenthion products labeled for use on ornamentals:

> "Reentry into treated area is prohibited for 24 hours (1 day) after the end of application, unless the protective clothing specified on this label for early reentry is worn.

> FOR EARLY REENTRY INTO TREATED AREAS BEFORE SPRAYS HAVE DRIED [OR DUST HAS SETTLED, as applicable] wear all protective clothing specified on this label for an applicator.

> FOR EARLY REENTRY INTO TREATED AREAS AFTER SPRAYS HAVE DRIED [OR DUST HAS SETTLED, as applicable] wear protective suit of one or two pieces covering all parts of the body except head, hands, and feet; chemical-resistant gloves; chemical-resistant shoes (or chemical-resistant shoe covers or chemical-resistant boots).

6.   REGULATORY POSITION SUMMARY

The Agency will not grant any tolerances for significant new food uses[2] until sufficient data (residue chemistry and toxicology) are submitted for the Agency to calculate an Acceptable Daily Intake (ADI) for fenthion.

The Agency is classifying all fenthion end-use products with directions for use on agricultural crops, ornamental plants and forest trees, uncultivated agricultural and non-agricultural outdoor areas, aquatic sites, livestock and bird roosting areas as Restricted Use pesticides, based on avian, fish and aquatic invertebrate toxicity.

The Agency is establishing an interim 24-hour reentry interval for the use of fenthion on ornamentals until adequate data have been submitted and evaluated.

---

[2]"New use" is defined in 40 CFR 152.3(p). In the case of a new food or feed use, the Agency will generally consider as significant an increase in the Theoretical Maximum Residue Contribution (TMRC) of greater than 1%.

The Agency is requiring special studies, in rodent and nonrodent species, on the effects of fenthion on the eye.

The U.S. Fish and Wildlife Service (FWS) has determined that certain uses of fenthion may jeopardize the continued existence of endangered species or critical habitat of certain endangered species. EPA is developing a program to reduce or eliminate exposure to these species to a point where use does not result in jeopardy, and will issue notice of any necessary labeling revisions when the program is developed. No additional endangered species labeling is being required at this time.
A feed additive tolerance must be proposed for residues of fenthion and its cholinesterase-inhibiting metabolites in rice hulls.

While data gaps are being filled, currently registered manufacturing use products and end use products containing fenthion may be sold, distributed, formulated, and used, subject to the terms and conditions specified in the Registration Standard for Fenthion. Registrants must provide or agree to develop additional data in order to maintain existing registrations.

7.   Summary of Major Data Gaps

Toxicology

Dermal Sensitization
Acute Delayed Neurotoxicity
Subchronic 90-Day Feeding, two species (rodent and nonrodent)
Chronic Toxicity, two species (rodent and nonrodent)
Oncogenicity, two species
Teratogenicity (rat)
Reproduction

Environmental Fate/Exposure

Hydrolysis
Photodegradation, water
Photodegradation, soil
Photodegradation, air
Anaerobic Aquatic Metabolism
Aerobic Aquatic Metabolism
Leaching and Adsorption/Desorption
Volatility, laboratory
Volatility, field (based on results of the laboratory volatility study)
Terrestrial Field Dissipation
Aquatic Field Dissipation
Forestry Dissipation
Soil Dissipation, long term (based on the results of the Terrestrial,
   Forestry and Aquatic Field Dissipation studies)
Confined Accumulation, rotational crops
Field Accumulation, rotational crops (may be required depending on results
   of acceptable confined rotational crop accumulation data)
Accumulation, irrigated crops

Fish Accumulation, laboratory (registrant should first submit an Octanol/water Partition Coefficient)
Field Accumulation, aquatic nontarget organisms (may be required depending on the results of laboratory accumulation study)
Foliar Dissipation
Droplet Size Spectrum
Drift Field Evaluation

Ecological effects

Avian Acute Oral (for Degradates)
Avian Subacute Dietary (for Degradates) (upland game bird and waterfowl
Wild Mammal Toxicity (for Technical Grade and Degradates)
Avian Reproduction (upland game bird and waterfowl)
Simulated and Actual Field Testing for mammals and birds
Freshwater Fish LC_{50} (for Degradates) (warmwater and coldwater species)
Freshwater Invertebrate LC_{50} (typical end-use product)
Estuarine and Marine Organisms LC_{50} (Technical Grade and typical end-use product)
Aquatic Organism Accumulation
Simulated or Actual Field Testing (Aquatic organisms)
Honeybee - Toxicity of Residues on Foliage
Special Tests -
  -(Residue Monitoring in Water and on Animals)
  -(Bivalve Toxicity)
  -(Reptile and Amphibian Toxicity)

Residue Chemistry

Nature of Residues (Metabolism [Plants, and Livestock])
Residue Analytical Methods
Residue Storage Stability
Magnitude of Residues in Plants

Product Chemistry

All Product Chemistry Studies

8.   CONTACT PERSON AT EPA

George T. LaRocca
Product Manager (15)
Insecticide-Rodenticide Branch
Registration Division (TS-767C)
Office of Pesticide Programs
Environmental Protection Agency
401 M Street, S. W.
Washington, D. C. 20460

Office location and telephone number:

Room 204, Crystal Mall #2
1921 Jefferson Davis Highway
Arlington, VA 22202
(703) 557-2400

DISCLAIMER: The information in this Pesticide Fact Sheet is a summary only and
may not be used to fulfill data requirements for pesticide registration and
reregistration.   The complete Registration Standard for the pesticide is
available from the National Technical Information Service.   Contact the Product
Manager listed above for further information.

FENVALERATE

Reason for Issuance: Registration Update
Date Issued: September 11, 1987
Fact Sheet Number: 145

1. DESCRIPTION OF CHEMICAL

   Generic Name: (S)-cyano (3-phenoxyphenyl) Methyl-(s)-4-chloro-
                   alpha-(l-methylethyl) benzeneacetate

   Common Name: Fenvalerate (BSI, ISO)

   Trade Names: Pydrin; Sumicidin; Belmark

   Other Names: S-5602; Sanmarton; SD 43775; Sumifly; Sumipower

   EPA Shaughnessy Code: 109301

   Chemical Abstracts Service (CAS) Number: 51630-58-1

   Year of Initial Registration: 1978

   Pesticide Type: Pyrethroid-like; Insecticide/miticide

   Chemical Family: Pyrethroid

   Manufacturers: E. I. Dupont de Nemours; Sumitomo Chemical Co, Ltd (Japan)

2. USE PATTERNS AND FORMULATIONS

   Application Sites: Foliar treatments for control of various insect pests
                        on agricultural crops, fruits, vegetables, ornamentals,
                        lawns. Space and contact spray treatments in and around
                        commercial and residential areas such as hospitals,
                        supermarkets, motels, hotels, homes, transportation
                        equipment (buses, boats, ships, trains, airplanes),
                        utilities, food processing plants, restaurants and other
                        food handling establishments, for control of common
                        premise pests such as cockroaches, crickets, and ticks.
                        Direct spray treatments for control of ectoparasites on
                        pets, horses; cattle eartag; and spot treatment of fire
                        ant mounds, and soil treatment for subterranean termite
                        control in and around buildings and structures.

Types of Formulations: Emulsifiable concentrates; liquids (ready-to-use), ready-to-use, and impregnated, and ULV concentrate.

3. Chemical/Physical Characteristics of the Technical Grade

Physical State: Liquid.

Color: Clear viscous yellow

Odor: Mild chemical odor

Molecular weight and formula: 419.9 - $C_{25}H_{22}ClNO_3$.

Melting Point:

Boiling Point:

Density: 1.17 g/ml at 23° C.

Vapor Pressure: 1.1 x 10^8 mmHg at 25°C.

Solubility in various solvents: In H_2O, <1 mg/l at 20° C.
In acetone, chloroform, cyclohexane, ethanol, and xylene, >1 kg/kg at 23° C.
In hexane, 155 g/kg at 23° C.

Stability: Stable to heat and sunlight; stable to moisture; more stable in acid (pH 4) than alkaline solution.

Toxicology Characteristics of the Technical Grade

- Acute Oral: LD_{50} 1-3 gms/kg for rat.

- Acute Dermal LD_{50} for rabbit = 1-3 gms/kg

- Primary Dermal Irritation (rabbit): none observed

- Primary Eye Irritation (rabbit): none observed

- Skin Sensitization (guinea pig): none observed

- Acute Inhalation(rat); LC_{50} > 101 gms/m3/4 hours

- Subchronic oral: rat - NOEL = 125 ppm;

  dog - NOEL = 500 ppm (HDT)

- Chronic Toxicity (rat): NOEL =250 ppm (HDT)

Toxicology Characteristics of the Technical Grade (continued)

- Chronic Toxicity (dog)[*]: NOEL = 200 ppm (HDT)

- Oncogenicity (24 month - mice): systemic NOEL = 10-50ppm, no
  oncogenic effects at 1,250ppm (HDT)

- Teratogenicity: Teratogenic (mice) NOEL = 50 mg/kg/day (HDT);
  Teratogenic (rabbit) NOEL = 50 mg/kg/day (HDT)

- Reproduction (3-Gen. Rat): NOEL = 250 ppm (HDT)

- Mutagenicity (dominant lethal - mice): negative at 100 mg/kg, (HDT)

- Host medicated - (Mice): Negative at 50 mg/kg (HDT)

- Mutagenic - (Ames); Negative

- Mutagenic - (Chinese hamster, bone marrow): Negative at 25 mg/kg

  [*]MO 70616 containing 75% of the active isomer (A-alpha) was tested
  in place of SD 43775 which contains 18% of its A-alpha isomer.

Physiological and Biochemical Characteristics

- The mode of action in biological systems is stomach and contact,
  exhibiting neuropathological characteristics typical of pyrethroid
  insecticides. Slight repellant effect.

- Foliar absorption: N/A

- Translocation: N/A

Environmental Characteristics

Adequate data are sufficient to define the fate of fenvalerate in the
environment. Fenvalerate is stable to hydrolysis at environmental pH
and temperature and to photolysis. In aqueous solutions exposed to

Environmental Characteristics (continued)

natural sunlight, approximately 52% of the parent compound degraded with
a calculated half-life of 41 days. In the laboratory, in soil maintained
under aerobic conditions the half-life ranged from 65 days to 8 months, in
soil maintained under anaerobic conditions the half life was approximately
6 months. In column-leaching studies where fenvalerate was applied to
different soil types and then saturated with water, results indicated that
fenvalerate and aged degradation products were relatively immobile. Field
studies indicate that residues of fenvalerate remain in the 0-4 inch layer
with a half-life of 1-2 months. After 183 days, residues were at negligible
or undetectable levels. Confined and field rotational crop data show that
residues of fenvalerate are likely to occur in root crops at intervals
of less than nine months. Bioaccumulation factors of 400X were found in
edible portions of rainbow trout after exposure for 30 days. Depuration
was relatively slow with about 40-60% of the residual activity remaining
after 33 days and virtually all as intact parent compound. In catfish,
with a bioaccumulation factor of 62X the depuration was also slow with a
half-life of 46 days in whole fish.

Ecological Effects Characteristics

- Avian acute oral LD_{50} (Mallard): LD_{50} = 9,932 mg/kg.

- Avian dietary LC_{50} (Bobwhite quail): LC_{50} > 10,000 ppm.

- Avian dietary LC_{50} (Mallard): LC_{50} = 5,500 ppm.

- Fish acute 96-hour LC_{50} (Bluegill sunfish): LC_{50} = 0.42 ppb.

- Aquatic invertebrate acute 48-hour LC_{50}
      (Pink shrimp)*: (96-hour) EC_{50} = 1.4 ppb.

Tolerance Assessments

The maximum premissible intake (MPI) was calculated using the rat no-observable effect level (NOEL) of 50 ppm, which was determined in a subchronic (13-week) feeding study. The study was conducted with MO 70616 which containing 75% of the active isomer (A-alpha) instead of the SD 43775 which contains 18% of its A-alpha isomer. This NOEL is equivalent to 2.5 mg/kg/day. A safety factor of 100 results in a calculated acceptable daily intake (ADI) of 0.025 mg/kg/day and an MPI of 1.5 mg/kg/day for a 60 kg human.

No additional data are required to support the current crop tolerances listed in 40 CFR 8180.379

Summary Science Statement

- Fenvalerate, a synthetic pyrethroid is toxic to wildlife and extremely toxic to fish. It is higly toxic to bees exposed to direct treatment on blooming crops or weeds. Fenvalerate has low toxicity to mammals. A 24-month feeding/oncogenic (rat) study demonstrated that fenvalerate is not oncogenic. Mutagenicity data indicate that fenvalerate was negative for all mutagenic tests conducted. Fenvalerate is immobilized in soils. The aged degradation products of fenvalerate do not leach significantly in a sandy loam soil column. It degrades in the soil under field conditions. At this time, there are no concerns for ground-water contamination.

4. Summary of Major Data Gaps
    - Freshwater invertebrate life-cycle test
    - Estuarine invertebrate life-cycle test
    - Simulated and/or actual field study

5. Summary of Tolerances Issued for Fenvalerate

Commodity	Parts per million
Almond hulls	15.0
Almonds	0.2
Apples	2.0
Artichokes	0.2
Beans, dried	0.25
Beans, snap	2.0
Broccoli	2.0
Cabbage	10.0
Cantaloupes	1.0
Carrots	0.5
Cattle, fat	1.5
Cattle, mbyp	1.5
Cattle, meat	1.5
Cauliflower	0.5
Collards	10.0
Corn, grain	0.002
Corn, fodder	50.0
Corn, forage	50.0
Corn, sweet, kernels & cobs	0.1
Cottonseed	0.2
Cucumbers	0.5
Eggplant	1.0
English walnuts	0.2
Filberts	0.2
Goats, fat	1.5
Goats, mbyp	1.5
Goats, meat	1.5
Hogs, fat	1.5
Hogs, mbyp	1.5
Hogs, meat	1.5
Honeydew melons	1.0
Horses, fat	1.5
Horses, mbyp	1.5
Horses, meat	1.5
Milk	0.3
Milk, fat	7.0
Muskmelons	1.0
Peanuts	0.002
Peanut hulls	0.10
Pears	2.0
Peas	1.0
Peas, dried	0.25
Pecans	0.2
Peppers	1.0

5. Summary of Tolerances Issued for Fenvalerate (continued)

Commodity	Parts per million
Potatoes	0.02
Pumpkins	1.0
Radish, roots	0.3
Radish, tops	8.0
Sheep, fat	1.5
Sheep, mbyp	1.5
Sheep, meat	1.5
Soybeans	0.05
Stone fruits	10.0
Sugarcane	2.0
Summer squash	0.5
Sunflower seed	1.0
Tomatoes	1.0
Watermelons	1.0
Winter squash	1.0

6. Contact Person at EPA

George T. LaRocca
Product Manager 15
Insecticide-Rodenticide Branch
Registration Division (TS-767C)
Office of Pesticide Programs
Environmental Protection Agency
401 M Street, S.W.
Washington, DC 20460

Office location and Telephone No.
- Room 204, Crystal Mall #2
- 1921 Jefferson Davis Highway
- Arlington, VA 22202
- (703) 557-2400

FLURPRIMIDOL

Reason for Issuance: New Chemical Registration
Date Issued: February 22, 1989
Fact Sheet Number: 202

DESCRIPTION OF CHEMICAL

Generic Name: alpha-(1-methylethyl)-alpha-[4-(trifluoromethyoxy)
                phenyl]-5-pyrimidinemethanol

Common Name: Flurprimidol

Trade Name: Cutless

EPA Shaughnessy Codes: 125701-3

Chemical Abstracts Service (CAS) Number: 56425-91-3

Year of Initial Registration: 1989

Pesticide Type: Plant Growth Regulator

U.S. and Foreign Producers: Elanco Products Company, Division of
                Eli Lilly and Company

USE PATTERNS AND FORMULATIONS

Application Sites: Turfgrasses and ornamental trees

Types and Methods of Application:

   Boom-type sprayer to turfgrasses and specialized
   injection equipment to ornamental trees.

Application Rates:

   50% wettable powder:

   Cool season turfgrasses
        Late Spring - Early Summer 1.5 - 3 lbs./ai/A
        Late Summer - Early Fall 1.5 - 3 lbs./ai/A

   Warm season turfgrasses
        Late Spring - Early Summer 0.75 - 3 lbs./ai/A
        Late Summer - Early Fall 0.75 - 3 lbs./ai/A

281

Poa annua
          Late Spring - Early Summer 1 - 1.5 lbs./ai/A
          Late Summer - Early Fall 1 - 1.5 lbs./ai/A

99% technical powder:

          Ornamental trees - 0.5 - 1.50 grams per inch
          tree diameter

Types of Formulations:

          50% wettable powder (WP) end-use product marketed in
          water soluble packets and 99% technical powder (TP)
          end-use product marketed in one quart bottles.

Major Uses:

          Turfgrasses: End-use formulation is a plant growth
          regulator which reduces internode and leaf elongation
          in cool and warm season turfgrasses.

          Ornamental trees: End-use formulation is a plant
          growth regulator for reduction of growth and pruning
          frequency.

Usual Carrier:

          50% formulation - water
          99% formulation - alcohol

SCIENCE FINDINGS

Summary Science Statement:

          Chronic feeding/oncogenicity studies were conducted in both
the rat and mouse.   Hepatocellular changes in the males including
enzyme induction, fatty change, hepatocellular eosinophilic
change and focal atypia were observed in the rat study.   A core-
supplementary mouse study showed increased absolute and relative
liver weight in females.   Although both the rat and mouse study
are core-supplementary for oncogenicity due to inadequate dose
selection, they both satisfy the requirement for oncogenicity
testing in one species for the requested non-food uses.   No
oncogenic potential was observed at any dose level in either of
these two studies.   New rat and mouse studies (which achieve the
Maximum Tolerated Dose - MTD) will be required for any food use
registrations.

          A 1-year dog study showed adrenal changes including
decreased plasma cortisol response to adrenal cortico-tropine
hormone (ACTH) stimulation (males), decreased relative

and absolute adrenal weight (males) and degenerative changes of the adrenal cortex (males and females). This study satisfies the requirement for a chronic oral study in one species.

Decreased body weight and food consumption were observed in the rabbit and rat teratology studies. In addition the rat teratology study showed increased mortality, stained perigenital area and snout, chromodacryorrhea, decreased muscle tone, hypoactivity and alopecia. Rat and rabbit teratology requirements have been satisfied.

The requirements for a 90-day feeding study have been satisfied. A subchronic oral rat study showed an increased hepatic enzyme induction in males (significant and dose increases in p-nitroanisol o-demethylase activity). The subchronic oral mouse study indicated an increased incidence of hepatocellular hypertrophy in the males.

A 21-day dermal toxicity study (rabbit) noted slight transient dermal irritation. This data requirement has been fulfilled.

The requirements for a 2 generation reproductive study have been satisfied. The Parental Systemic Toxicity in a 2 generation reproduction study showed increased incidence of non-neoplastic hepatocellular alteration including fatty change and vacuolation (males) and increased susceptibility to stress factors. Decreased mating, fertility, fetal survival (stillbirths), neonatal survival and neonatal body weight in both sexes and in both generations were observed at the Reproductive NOEL. Other parental signs of toxicity included increased susceptibility to stress (pregnant females) resulting in death, increased relative liver weight (males and females), depressed body weight, weight gain and food consumption (males and females).

Subchronic, oncogenicity and teratogenicity studies are not usually required for the requested registration use patterns. However, due to the structural similarity of flurprimidol to compounds of toxicological concern (fenarimol, triarimol, and nuarimol), these studies were required for registration. Oncogenicity studies were requested based on flurprimidol's similarity to an oncogenic compound (triarimol) and teratogenicity studies were requested based on it's similarity to compounds (triarimol and fenarimol) associated with developmental concerns.

Mutagenicity tests for gene mutation, chromosomal aberration and direct DNA damage were evaluated. Flurprimidol had no effect on induction of unscheduled DNA synthesis or chromosomal aberration. It was also negative for mutagenic activity.

The hydrolysis, aerobic and anaerobic soil metabolism, leaching and adsorption/desorption, terrestrial field dissipation, and accumulation studies are acceptable and fulfills the data requirements for proposed turf use. Environmental fate data demonstrates that flurprimidol is stable and moderately mobile. Based upon evaluation of the environmental fate data for the currently proposed uses, the Ground Water Team recommends no prospective ground water monitoring study be required. A study may be required if the use of the chemical is expanded to terrestrial food crops.

The aqueous photodegradation study is considered less than adequate by itself to support registration. However, when evaluated with other acceptable photolysis studies on chemicals of the same class, these data demonstrate a degradation pathway consistent with other studies. This study is not required for the proposed tree injection use. As such, this data requirement is satisfied for turf use since the chemical dissipates rapidly on turf. However, in the case of bare loam soil where crops are generally grown, the chemical may not dissipate rapidly and is therefore subject to photodegradation. Accordingly, a repeat photodegradation study is required for future (food) uses of flurprimidol.

Avian acute, avian dietary, freshwater fish, freshwater invertebrate and acute contact toxicity studies have been fulfilled. Studies indicate that flurprimidol is slightly toxic to birds, aquatic invertebrates, and both warmwater and coldwater species.

Chemical Characteristics of the Technical Material

Physical State: crystalline solid

Color: buff to off-white, white to pale yellow

Odor: none - slightly aromatic

Molecular Weight: 312.3

Empirical Formula: $C_{15}H_{15}F_3N_2O_2$

Boiling Point: 264 degrees Celsius

Melting Point: 93 to 95 degrees Celsius

Vapor Pressure: 3.64×10^{-7} mmHg @25 degrees Celsius

Density: 0.83 to 0.88 g/cc

Octanol/Water Partition Coefficient: K_w = 9.33

pH: In D-H_2O 6.5 to 8.9

Solubility in various solvents:
```
    water, pH 4              120 to 140 ppm @25 degrees Celsius
    water, pH 7              120 to 140 ppm @25 degrees Celsius
    water, pH 10             120 to 140 ppm @25 degrees Celsius
    Acetone                 *700 to 800 mg/mL
    Acetonitrile            *200 to 300 mg/mL
    Toluene                 *75 to 100 mg/mL
    Chloroform              *800 to 900 mg/mL
    Dichloromethone         *800 to 900 mg/mL
    Methanol                *700 to 800 mg/mL
    Heavy Aromatic Naphtha  *25 to 35 mg/mL
    Xylene                  *100 to 200 mg/mL
    1-Chlorobutane          *100 to 200 mg/mL
    n-Hexane                *1 to 2 mg/mL
    Methyl Cellosolve       *700 to 800 mg/mL
    Cyclohexane             *2 to 3 mg/mL
    Ethyl Acetate           *500 to 600 mg/mL
    Cyclohexane             *400 to 500 mg/mL
    Isophorone              *400 to 500 mg/mL
    Acetophenone            *400 to 500 mg/mL
    Monochlorotoluene       *400 to 500 mg/mL
      (90% ortho, 10% para)
```
*Solubilities determined at ambient temperature, 20 to 22 degrees Celsius.

Stability: Stable in glass or polyethylene container

Oxidizing or Reducing Action:

   Oxidizing or reducing action reagent:
   Ammonium dihydrogen phosphate - no gas
   evolution, no temperature rise over 24 hours.

   Reagent: Potassium permanganate - no gas
   evolution, no temperature rise over 24 hours.

   Reagent: Zinc dust 100 mesh - no gas
   evolution, no temperature rise over 24 hours.

Corrosion Characteristics:

   Not corrosive to low density polyethylene
   films and high density polyethylene
   containers.

Explodability:

No positive results were obtained in 10
repetitive drops at 20 inches with an eight
pound hammer.

Toxicology Characteristics

Acute Studies

Acute Oral Toxicity: Toxicity Category III
    Rat: LD_{50} 914 mg/kg (male)
           LD_{50} 709 mg/kg (female)

Acute Dermal Toxicity: Toxicity Category III
    Rat: LD_{50} > 500 mg/kg (male and female)

Primary Dermal Irritation: Toxicity Category IV
    Slight dermal irritant

Primary Eye Irritation: Toxicity Category III
    Moderate eye irritant

Dermal Sensitization:
    Not a sensitizer

Acute Inhalation: Toxicity Category IV
    Rat: LD_{50} > 5.231 mg/L (male and female)

The toxicity base for the technical product supports
registration of this compound for the requested uses.

Subchronic Studies

A 90-day feeding study in rats treated to 0, 1.68, 6.04,
20.39, 68.34 mg/kg/day (males) and 0, 1.98, 7.13, 24.37, 78.47
mg/kg/day (females) of flurprimidol. The systemic No-Observable-
Effect-Level (NOEL) was 1.68 mg/kg/day and the Lowest-Effect-
Level (LEL) was 6.04 mg/kg/day based on increased hepatic enzyme
induction in males (significant and dose increases in p-
nitroanisol 0-demthylase activity). At 24.37 mg/kg/day there was
increased relative and absolute ovarian (female) and relative
liver (male) weight. At 68.34 mg/kg/day, there was increased
absolute liver weight (males).

A 90-day feeding study in the mouse, treated with 0, 15,
67.5 and 300 mg/kg/day of technical flurprimidol. The NOEL was
15 mg/kg/day and the LEL was 67.5 mg/kg/day based on increased

incidence of hepatocellular hypertrophy in the males. At 300
mg/kg/day, there was evidence of enzyme induction, increased
liver weight and hepatocellular hypertrophy in females.

A subchronic dermal study (21-day) was conducted in the
rabbit. Groups of 5 rabbits/sex/group were treated by dermal
exposure to 0, 500, or 1000 mg/kg/day of flurprimidol. The
systemic NOEL was greater than or equal to 1000 mg/kg/day and
the LEL was greater than 1000 mg/kg/day. The NOEL for dermal
irritation was less than 500 mg/kg/day and the LEL was less than
or equal to 500 mg/kg/day based on slight transient dermal
irritation.

Chronic Studies

Rodent Feeding Studies

A 2-year study in rats treated with either 0, 1.0, 3.6, 12.1
and 41.2 mg/kg/day of flurprimidol technical for males and 0,
1.2, 4.4, 14.5, and 49.3 mg/kg/day for females the NOEL was 3.6
mg/kg/day and the LEL was 12.1 mg/kg/day based upon
hepatocellular changes in males including enzyme induction, fatty
change, hepatocellular eosinophilic change and focal atypia. At
41.2 mg/kg/day there was also a transient body weight and weight
gain decrease (males), increased cholesterol and triglycerides
(males and females) increased hepatic enzyme induction and liver
weight, fatty change and hepatocellular eosinophilic change
(females). No oncogenic potential was observed at any dose
level.

A 2-year core-supplementary study in mice treated with
either 0, 1.4, 10.5 or 79.9 mg/kg/day of flurprimidol, the
systemic NOEl was 1.4 mg/kg/day. The LEL was 10.5 mg/kg/day
based on increased absolute and relative liver weight in the
males. No oncogenic potential was observed at any dose level.

Although both the rat and the mouse study are supplementary
for oncogenicity due to inadequate dose selection, they both
satisfy the requirement for oncogenicity testing in one species
for the requested non-food use. There was no indication of
oncogenic potential at any dose level in either study.

New rat and mouse studies will be required for any food use
registrations.

Non-rodent Feeding Study

A 1-year dog study treated with flurprimidol at doses of 0,
0.5, 1.5, 7.0 and 30.0 mg/kg/day, the NOEl was 7.0 mg/kg/day and

the LEL was 30.0 mg/kg/day Highest Dose Tested (HDT) based on
adrenal changes including decreased plasma cortisol response to
ACTH stimulation (males and degenerative changes of the adrenal
cortex (males and females).   The histopathology was limited to
the zona fasciculata of the adrenal cortex and was characterized
by eosinophilic degeneration, vacuolation and cortical atrophy.
There was a slight increase in hepatic p-nitroanisole o-
demethylase activity (males).

Teratology Studies

     A rabbit teratology study using doses of 0, 1.7, 9 and 45
mg/kg/day of flurprimidol had a maternal toxicity NOEl of 9
mg/kg/day and the LEL was 45 mg/kg/day based on decreased body
weight and food consumption.

     A rat teratology study using doses of 0, 2.5, 10, 45 or 200
mg/kg/day of flurprimidol had a maternal toxicity NOEL of 10
mg/kg/day and a LEL of 45 mg/kg/day based on decreased body
weight gain and food consumption. The developmental NOEL was 10
mg/kg/day and the LEL was 45 mg/kg/day based on decreased fetal
weight, increased incidence of hydronephrosis, hydroureter and
numerous developmental skeletal anomalies.

Reproduction Study

     A 2 generation reproduction study in the rat treated with
(time weighted average) 0, 1.8, 7.3, and 74 mg/kg/day of
flurprimidol had a Parental Systemic Toxicity NOEL of 1.8
mg/kg/day and a LEL of 7.3 mg/kg/day based on increased
incidence of non-neoplastic hepatocellular alterations including
fatty change and vacuolation (males) and increased susceptibility
to stress factors.   The Reproductive NOEL was 7.3 mg/kg/day and
the LEL was 74 mg/kg/day based on decreased mating, fertility,
fetal survival (stillbirths), neonatal survival and neonatal body
weight in both sexes and in both generations.   There was an
increased incidence of persistent vaginal estrous and no corpora
lutea.   Additional parental signs of toxicity at 74 mg/kg/day
included increased susceptibility to stress (pregnant females)
resulting in death, increased relative liver weight (males and
females), depressed body weight, weight gain and food consumption
(males and females).

Mutagenicity Studies

     No mutagenic activity was observed in mammalian cells.
There was also no activity when tested in S. typhimurium and E.
coli.   Flurprimidol did not induce chromosome aberrations in

vitro in Chinese Hamster ovary cells. There was no effect on the capacity to induce sister chromatid exchanges in bone marrow cells of Chinese Hamsters. It had no effect on induction of unscheduled DNA synthesis in rat hepatocytes.

Mutagenicity tests for gene mutation, chromosomal aberration and DNA repair were negative.

Environmental Characteristics

Hydrolysis
 stable at pHs 5,7, and 9

Aerobic Soil Metabolism
 degrades with a half-life of > 26 weeks in sandy loam, silt loam and clay loam soils.

Mobility
 Mobile in bare loam soil

Dissipation
 persistent in unvegetated loam (estimated half-life of 1.5 years)

 dissipates rapidly on turf (half-life of 5-21 days)

Accumulation
 displays a low bioconcentration factor in fish with rapid depuration.

The Environmental Fate and Ground Water Branch (EFGWB) have concluded that the hydrolysis, aerobic and anaerobic soil metabolism, leaching and adsorption/desorption, terrestrial field dissipation, and accumulation studies are acceptable and fulfills the data requirements for proposed turf use.

Review of the Aqueous Photodegradation study noted several deficiencies. The study is considered less than adequate by itself to support registration. However, when evaluated with other acceptable photolysis studies on chemicals of the same class, these data demonstrate a degradation pathway consistent with the other studies. This study is not required for the proposed tree injection use. As such, this data requirement is satisfied for turf use since the chemical dissipates rapidly on turf. However, in the case of bare loam soil where crops are generally grown, the chemical may not dissipate rapidly and is therefore subject to photodegradation. Accordingly, a repeat photodegradation study is required for future (food) uses of flurprimidol.

A repeat of the Aqueous Photodegradation study is requested, but not as a condition of registration. A new study will remove any doubt concerning the data and the available study submitted by the registrant is a passe' protocol. Without a free standing study, the question of acceptability will continually arise with every new use and generic review of the chemical.

Ecological Characteristics

Avian acute toxicity:
bobwhite quail LC_{50} > 2000 mg/kg

Avian dietary toxicity:
bobwhite quail LC_{50} > 5000 ppm
mallard duck LC_{50} > 5000 ppm

Freshwater fish acute toxicity:
bluegill sunfish LC_{50} 17.2(16.6 and 17.8) ppm
rainbow trout LC_{50} 18.3(17.5 and 19.0) ppm

Freshwater invertebrate toxicity:
Daphnia magna EC_{50} 11.8(10.9 and 12.9) ppm

Acute Toxicity for Green Algae:
Selenastrum capricornutum EC_{50}.84 ppm

Acute Contact Toxicity:
honey bee LD_{50} > 100 ug/bee

All of the above data requirements have been satisfied. Flurprimidol is practically non-toxic to birds on an acute and dietary basis, slightly toxic to aquatic invertebrates, and both warmwater and coldwater species. The hazard to freshwater algae is expected to be minimal. In addition flurprimidol is relatively non-toxic to honey bees.

CONTACT PERSON AT EPA

Robert J. Taylor
Product Manager (25)
Fungicide-Herbicide Branch
Registration Division (TS-767C)
Office of Pesticide Programs
Environmental Protection Agency
401 M Street, S. W.
Washington, D. C.   20460

Office location and telephone number:
Room 243, Crystal Mall #2
1921 Jefferson Davis Highway
Arlington, VA   22202
(703) 557-1800

DISCLAIMER: The information in this Pesticide Fact Sheet is a summary only and is not to be used to satisfy data requirements for pesticide registration and reregistration.   The complete Registration Standard for the pesticide may be obtained from the contact person listed above.

FORMALDEHYDE AND PARAFORMALDEHYDE

Reason for Issuance: Registration Standard
Date Issued: May 17, 1988
Fact Sheet Number: 167

1. Description of Chemical

	FORMALDEHYDE	PARAFORMALDEHYDE
Generic Name		
	Formaldehyde	Paraformaldehyde
Common Name		
	Formic aldehyde, Methanal, Oxomethane, Oxymethylene, Methylene oxide, Formalin	Polyoxymethylene, Mixed polyoxy-methylene glycols
Trade Name		
	Formaldehyde Solution	Paraformaldehyde
EPA/OPP Pesticide Chemical Code		
	043001	043002
Chemical Abstracts Service (CAS) Number		
	50-00-0	30525-89-4
Year of Initial Registration		
	1948	1953
Pesticide Type		
	Disinfectant, Fungicide Microbiocide	Disinfectant, Fungi-cide, Microbiocide
Chemical Family		
	Aldehydes	Aldehydes

Description of Chemical (cont'd)

FORMALDEHYDE	PARAFORMALDEHYDE

U. S. & Foreign Producers

Celanese Chemical Co., Inc.;
Tenneco, Inc. E.I. duPont De
Nemours & Co. Inc.; Georgia
Pacific Corp., Monsanto Corp.
and The Chemical Supply Co.
(Great Britain)

Celanese Chemical Co., Inc.,
The Chemical Supply Company
(Great Britain)

2. Use Patterns and Formulations

Application Sites:

Formaldehyde

Food and non-food crops; products for processing and industrial uses
(e.g., drilling muds, metalworking cutting fluids and packer fluids);
products for use on hard surfaces (e.g., livestock premises, household
premises and contents, hospital critical equipment, transportation
vehicles); fabrics and textiles (e.g., laundry, carpet); products for
control of microbial pests associated with human and animal wastes
(e.g., toilet bowls, urinals, diaper pails); and preservative of herbicidal, algaecidal, bacteriostatic, disinfectant, sanitizer, fungicidal and insecticidal formulations. There are 329 registered products
that contain formaldehyde (46 products in which formaldehyde has always
been considered as an active and 283 products in which formaldehyde
was previously considered as an inert but has been redesignated as an
active ingredient.)

Paraformaldehyde

Sugar maple tree tapholes; products for processing and industrial
uses (e.g., secondary oil recovery systems, metalworking cutting
fluids, oil recovery drilling muds and packer fluids); products for
use on hard surfaces (e.g., kennels and pet animal quarters, livestock
premises and equipment, household premises and contents, barber and
beauty shop equipment and instruments); fabrics and textiles (e.g.,
laundry, mattresses, pillows and draperies); food handling establishment premise treatment; and preservative of bacteriostatic, algaecidal
and fungicidal formulations. There are 119 registered products that
contain paraformaldehyde (44 products in which the paraformaldehyde
has always been considered as active and 75 products in which the
paraformaldehyde was previously considered as an inert but has now
been redesignated as an active ingredient.)

Number of Products

	Active Ingredient	Formerly Designated Inert Ingredient
Formaldehyde	46	283
Paraformaldehyde	44	75

Types of Formulations

Formaldehyde: gaseous, pelleted/tableted, soluble concentrate/liquid;
37%, 44% and 45% formulation intermediate/manufacturing use products.
End use products range from less than 0.1% to 93%.

Paraformaldehyde: crystalline, wettable powder/dust, pelleted/tableted.
There are not any registered manufacturing use products. End use products
range from less than 0.1% to 100%.

Types and Methods of Applications

Formaldehyde

Spray (pump/electrical); dip; mop; brush; swab; sponge; automatic metering;
proportioning pump; automatic pressure vaporizer; fumigation by (a) wet
sheet method (b) addition of product to permanganate of potash in a bucket
raised off of floor, and (c) fog application with electrical sprayer or
mechanical fogging equipment.

Paraformaldehyde

Dip/immerse; proportioning pump; jet mixer; manual insert; fumigation by
(a) hanging product in desired location, (b) generator, (c) application
from electric hotplate, (d) placement of tablets in open dish and (e)
placement of opened bottle in cabinets.

Application Rates

Formaldehyde

Seed treatment - Dosage rates = 1 pt 37%/30-40 gal. water; 37 lb/pt. - 40
gal. water.

Bacteriostatic - Dosage rates = 100 ppm-21,800 ppm.

Mold/mildew control - Dosage rates = 337 ppm-27,400 ppm.

Algae, bacteria and fungi control - Dosage rates = 1 ppm-1580 ppm.

Sanitizer - Dosage rates = .3 ppm - 2750 ppm.

Disinfectant - Dosage rates = 2 ppm - 370,000 ppm.

Paraformaldehyde

Bacteria control - Dosage rates = 2 ppm - 9010 ppm.

Mold/mildew control - Dosage rates = 4.9 ppm - 301 ppm.

Sanitization - Dosage rates = 6.4 ppm.

Disinfectant - Dosage rates = 2.9 ppm - 6.7 ppm.

3. Science Findings

Formaldehyde has been found to be carcinogenic in animal studies and there is limited evidence of carcinogenicity in humans. The Agency classified formaldehyde as a B1 oncogen. It is estimated that the worker risk from use of formaldehyde as an ingredient in an agricultural pesticide formulation is less than 10^{-6}. Areas of uncertainty exist in dietary exposure and exposure from other uses as an active ingredient, many of which are non-agricultural.

Chemical Characteristics:

	Formaldehyde	Paraformaldehyde
Physical State:	Gas	Crystalline, Wettable powder/Dust, Pelleted/Tableted
Odor:	Pungent	Pungent
Boiling Point:	-19.5°C(-3°F)	NA(technical is solid at room temp.)
Melting Point:	NA(gas at room temp.)	64°C
Unusual Handling Characteristics:	Corrosive to metal	

Toxicology Characteristics:

Most of the toxicological information on formaldehyde is from published sources. The toxicity of paraformaldehyde is believed to be identical to that of formaldehyde because it is the solid polymer of formaldehyde. The Agency does not have access to the raw data supporting these studies. Therefore, none of the studies (published and/or unpublished) is adequate for FIFRA regulatory purposes. However, based on these partially satisfactory studies, the following toxicological characteristics of formaldehyde and paraformaldehyde are expected.

Formaldehyde

Acute oral toxicity:
   Toxicity Category 3; LD_{50} = 800mg/kg (rat)

Acute dermal toxicity:
   Toxicity Category 3; LD_{50}>2g/kg

Acute inhalation toxicity:
   Toxicity Category 3;
   Lowest lethal concentration = 250 ppm (4hr. exposure)

Primary eye irritation: Toxicity Category 1

Formaldehyde (continued)

Primary skin irritation: Toxicity Category 2

Major route of exposure: inhalation, dermal

Paraformaldehyde

Acute oral toxicity
    Toxicity Category 3; LD_{50}>1.6 g/kg (rat)

Acute dermal toxicity:
    Toxicity Category 3; LD_{50}>2 g/kg

Acute inhalation toxicity:
    Toxicity Category 1; LC_{50} about 14 ppm

Primary eye irritation: Toxicity Category 1

Primary skin irritation: Toxicity Category 2

Major route of exposure: inhalation, dermal

Chronic feeding and oncogenicity

Formaldehyde has been found to be carcinogenic by inhalation in rats, and
there is evidence suggestive of carcinogenicity in mice. A recent drinking water study with formaldehyde showed evidence of carcinogenicity in the
rat stomach. The Agency is requiring oral oncogenicity testing. Chronic
toxicity testing (rodent and non-rodent) is also being required.

Developmental Toxicity

Inhalation studies in animals and epidemiological studies in workers have
not demonstrated teratogenic effects. Teratogenicity studies in the rat
and rabbit are required.

Reproduction

In one study, prolonged diestrus, but no impairment of reproductive
function was reported. A 2 generation rat study is required.

Mutagenicity

Formaldehyde has caused genetic changes in Drosophila larvae, fungi,
bacteria and mammalian cells. It is believed that formaldehyde is a weak
mutagen and that it operates by some type of genetic interaction. Gene
metabolism, structural chromosomal aberration and other genotoxic effects
data are required.

General Metabolism

Formaldehyde is a normal metabolite in mammalian systems. In dogs, cats,

rabbits, guinea pigs and rats, the half life of formaldehyde is estimated
to be one minute. General metabolism studies are not required.

Risk assessment results

It is estimated that the worker risks from use of formaldehyde as an
ingredient in an agricultural pesticide formulation are $<10^{-6}$. Further
estimates of the dietary risk and risk to workers from agricultural and
non-agricultural uses will be conducted when the requested data are
received.

Physiological and Biochemical Behavioral Characteristics:

Data are not available.

Environmental Characteristics:

There are no data available to assess the environmental fate of formaldehyde or paraformaldehyde. Data are not available to assess the ground
water contamination potential for pesticidal uses of formaldehyde or
paraformaldehyde. Also, formaldehyde was not included in the Agency's
Data-Call-In Notice for potential ground water leachers. Degradation,
metabolism, mobility, dissipation and accumulation studies are required.
Indoor inhalation exposure reentry studies are required.

Ecological Characteristics:

Freshwater species:

LC_{50} = 100 ppm - bluegill sunfish

LC_{50} = 118 ppm - rainbow trout

Freshwater invertebrates

LC_{50} = 14 ppm - Daphnia magna (Water flea)

Marine and estuarine organisms

96 hr. LC_{50} = 69.1 ppm - Trachinotus carolinus (Florida pompano)

96 hr. LC_{50} = 69 ppm - Menidia menidia (Atlantic silversides)

96 hr. LC_{50} = 18 ppm - Roccus saxatilis (Striped bass)

EC_{50} = 1.8 ppm - Crassotrea virginica (Eastern oyster)

LC_{50} = 143 ppm - Panaeus duorarum (Pink shrimp)

Terrestrial species:

Acute oral LD_{50} = 790 mg/kg - mallard ducks

Terrestrial species (continued)

$LC_{50}>5000$ ppm - bobwhite quail and mallard ducks

A potential hazard may exist to marine and estuarine species from the use of formaldehyde in secondary oil recovery systems and other industrial effluents containing formaldehyde and use of paraformaldehyde in oil well drilling muds. Monitoring of formaldehyde levels in waters receiving residues from the use of the pesticide in secondary oil recovery systems and other industrial effluents (formaldehyde) and oil well drilling muds (paraformaldehyde) is required.

Four use patterns may possibly pose a risk to aquatic endangered and nonendangered species - use of formaldehyde on turf and the use of products containing formaldehyde and paraformaldehyde that are discharged as industrial effluents into shallow or enclosed bodies of water, use of formaldehyde in secondary oil recovery systems and use of formaldehyde in oil well drilling muds. However, risk to aquatic species cannot be characterized at this time because the Agency lacks pertinent chemical and exposure data which are being required.

Tolerance Assessments

No tolerances have been established for residues of formaldehyde or paraformaldehyde in or on plant or animal commodities, with the exception of a food additive tolerance of 2 ppm for residues of formaldehyde in maple syrup resulting from use of paraformaldehyde in maple tree tapholes (21 CFR 193.330.)

Formulations containing 1% or less of formaldehyde and 2% or less of paraformaldehyde are exempt from the requirement of tolerances when used as preservatives in pesticidal formulations applied to growing crops [40 CFR 180.1001(d)].

Formaldehyde is exempt from the requirement of a tolerance for residues in or on the grains of barley, corn, oats, sorghum, and wheat and the forages of alfalfa, Bermuda grass, bluegrass, brome grass, clover, cowpea hay, fescue, lespedeza, lupines, orchard grass, peanut hay, peavine hay, rye grass, soybean hay sudan grass, timothy and vetch resulting from postharvest application of formaldehyde or a mixture of methylene bispropionate (MBP) and oxy(bismethylene)bispropionate (OBMP) when used as a fungicide. These raw agricultural commodities are for use only as animal feeds (40 CFR 180.1032).

Paraformaldehyde is exempt from the requirement of a tolerance for residues in or on sugar beets (roots and tops) when applied to the soil not later than planting (40 CFR 180.1024).

The food additive tolerance (21 CFR 193.330) and current exemptions from the requirements of tolerances (40 CFR 180.1032, 180.1024 and 180.1001(d) are not supported due to the inadequacy of available data. Plant metabolism data along with storage stability data are required. If residues of concern are found in plants, then residue data on crops are required. If residues are found in feed commodities, then livestock feeding studies must be submitted.

An acceptable daily intake (ADI) has not been established for formaldehyde. When the requested toxicological data are received, an ADI will be established. Based on requested residue data, a tolerance reassessment will be performed.

4. Summary of Regulatory Position and Rationale

° Formaldehyde and paraformaldehyde are pesticidally active when used in formulations as a preservative. Label ingredient statements must be revised to include the name and percentage of either when used as a preservative.

° Because the formaldehyde and paraformaldehyde exert their pesticidal effect only in the formulation, and do not affect the efficacy of the product itself, the label must include a statement to this effect.

° Formaldehyde and paraformaldehyde are not being placed in Special Review. Risks to workers using formulations containing these chemicals as preservatives are less than 1×10^{-6}, which does not warrant Special Review. The Agency lacks data on other uses of formaldehyde and paraformaldehyde to assess dietary and non-agricultural risks.

° Protective equipment (respirators) and other risk reduction measures are required for fumigation uses.

° Applications for new registration of products containing formaldehyde or paraformaldehyde must include exposure data to enable the Agency to perform a risk assessment.

° Endangered species labeling will not be required at this time. Four use patterns—turf use, products that are discharged into shallow or enclosed bodies of water, secondary oil recovery systems, and oil well drilling muds—may pose a risk, but the Agency lacks information needed to calculate the environmental concentrations and risks.

° While data gaps are being filled, products containing formaldehyde and paraformaldehyde may continue to be sold and distributed, provided that product labeling is revised as specified in the Registration Standard.

5.   Summary of Major Data Gaps

Study	Due Date
Residue Chemistry	
Metabolism	18 months
Residues (Analytical Method)	15 months
Storage Stability	15 months
Environmental Fate	
Hydrolysis	9 months
Photodegradation	9 months
Metabolism	27 months
Leaching & adsorption/desorption	12 months
Volatility	12 months
Dissipation	27 months
Rotational crops (confined)	39 months
Ecological Effects	
Avian Single Dose Oral LD_{50} (Paraformaldehyde only)	9 months
Avian Dietary LC_{50}	9 months
-upland game bird & water fowl (Paraformaldehyde only)	
Aquatic Organism Testing	9-15 months
Special Test	
Residue Monitoring	12 months
Toxicology	
Acute studies	9 months
Mutagenicity tests	9-12 months
Dermal penetration	12 months
90 day feeding (rat & dog)	15-18 months
21 day dermal (rabbit)	15 months
90 day inhalation (rat)	15 months
Chronic toxicity (rat & dog)	50 months
Oral oncogenicity (rat & mouse)	50 months
Teratogenicity (rat & rabbit)	15 months
Reproduction (rat)	39 months
Re-entry	
Inhalation exposure	27 months

6.   Contact Person at EPA

John Lee, Product Manager 31
Registration Division (TS-767C)
Disinfectants Branch
401 M St., S. W.
Washinton, D. C. 20460
Tel: (703) 557-3675

GLYPHOSATE

Reason for Issuance: Registration Standard
Date Issued: June 1986
Fact Sheet Number: 173

1. Description of chemical

   Generic name: N-(phosphonomethyl)glycine
   Common name: Glyphosate
   Trade names: Roundup, Rodeo, Roundup L&G, Polado, Shackle, Shackle C
   EPA Shaughnessy Codes: Isopropylamine (IPA) salt of glyphosate 103601
                           Sodium Salt of glyphosate   103603
   Chemical Abstracts Service (CAS) number: 38641-94-0
   Year of initial registration: 1974
   Pesticide type: Herbicide; Plant Growth Regulator
   U.S. and foreign: Monsanto in the United States and GENP International
                      Corporation in Taiwan.

2. Use patterns and formulations

   Application sites: Terrestrial food (soybeans, cotton, corn, sorghum,
     wheat, vegetables, citrus fruits, pome fruits, pastures and alfalfa)
     nonfood, noncrop, greenhouse nonfood, aquatic food and nonfood,
     domestic outdoor and forestry.
   Types and methods of applications: Postemergence spray to foliage of
     the vegetation controlled before planting, after planting but prior
     to crop emergence or as a directed spray in established crops.
     Applied using ground equipment, hand-held, recirculating and shielded
     sprayers, wiper applications and with aerial equipment.
   Pests controlled: Emerged annual and perennial grasses, broadleaf weeds,
     and woody brush and trees.
   Application rates: 0.3 to 1.0 pound/acre active ingredient (lb/A ai)
     to control annual weeds and grasses; 1.0 to 4.0 lb/ai to control
     perennial weeds and grasses.
   Types of formulations: 1.04 lb active ingredient (ai)/gal emulsifiable
     concentrate (EC); 0.42, 3, and 4 lb ai/gal soluble concentrate/liquid
     (SC/L); 5% and 6.6% ai SC/L; 0.5%, 0.96%, and 1% ai ready-to-use (RTU);
     and 0.75% and 0.96% ai pressurized liquid (Prl).
   Usual carrier: Water

3. Science findings

Summary Science Statement

   Glyphosate has low acute toxicity (Category III) for acute oral,
acute dermal, and primary eye irritation and is in Category IV for
primary skin irritation. It is not teratogenic to rats or rabbits
and is not mutagenic. The oncogenic potential is not fully defined
at this time. Repeat oncogenic studies are required in mice and rats.

   Glyphosate is no more than slightly toxic to birds, aquatic
invertebrates, freshwater fish and marine/estuarine organisms. It
does pose a hazard to some endangered species.

   Glyphosate is stable to hydrolsis and strongly adsorbed to the
soil and had a low potential to contaminate ground water.

Chemical characteristics:

   The technical isopropylamine salt is a white crystalline solid
with a melting point of 200 °C and a bulk density of 1.74. It is
1% soluble in water at 25 °C, insoluble in ethanol, acetone, or
benzene. The technical sodium salt of glyphosate is a white crystalline
solid which decomposes at 140 °C with a bulk density of 30 pounds/cubic
food (lb/ft^3). It is soluble in water and insoluble in organic
solvents.

   Glyphosate is corrosive to iron and galvanized steel. Spray
mixtures should not be held in galvanized or unlined steel tanks
(except stainless) for an extended period. Glyphosate solutions
should be mixed, stored, and applied on in stainless steel, aluminum,
fiberglass, plastic or plastic lined containers.

Toxicology characteristics:

Acute Toxicology

   Acute Oral Toxicity - rat = 4320 milligrams/kilogram (mg/kg) for
                              males and females - Toxicity Category III

   Acute Dermal Toxicity - rabbit - equal to or greater than ($\geq$) 794 mg/kg
                              for females $\geq$ 5010 mg/kg for males -
                              Toxicity Category III

   Primary Dermal Irritation - rabbit - Primary Irritation Score (P.I.S.)
                              = 0.0/8 (24 hours)
                              Toxicity Category IV - Not an Irritant

Primary Eye Irritation - rabbit - P.I.S. = 12.6/110 at 1 hour
Toxicity Category III - Mild Irritant

Acute Inhalation or Dermal Sensitization studies have not been
submitted and are being required.

Major routes of exposure

The primary potential for exposure from SC/L formulation is
during mixing and loading where both dermal and ocular exposure may
occur via splashing. Inhalation and dermal exposure may occur during
application of the liquid ready-to-use (RTU) and pressurized liquid.

Information from the Agriculture Department of Food and Agriculture
reported incidence of worker poisonings and illness during mixing,
loading and application from exposure to glyphosate. Most illnesses
or poisonings were due to skin or eye irritation. The Agency is
requiring "Worker Safety Rules" including protective clothing to
reduce exposure of workers to glyphosate.

Chronic Feeding/Oncogenicity Studies

In the mouse chronic feeding study non-neoplastic changes noted in
male mice included centrilobular hypertrophy and necrosis of hepatocytes,
chronic interstitial nephritis, and proximal tubule epithelial cell
basophilla and hypertrophy in females. The no-observable effect
level (NOEL) for non-neoplastic chronic effects was the mid dose,
5000 parts per million (ppm).

In the rat chronic feeding study no effects of treatment on the
incidence of non-neoplastic lesions were noted. The NOEL for this
study was 31 mg/kg/day.

The 1-year chronic feeding study in dogs indicated an apparent
decrease in the absolute and relative weights of pituitaries from mid-
and high-dose dogs. Additional data have been requested to better
assess this apparent effect. The tentative NOEL is 20 mg/kg/day,
pending submission of requested data.

The oncogenic potential of glyphosate is not fully understood at
this time. A review of the mouse oncogenicity study noted a slight
increase in renal tubular adenomas. Upon review by the Toxicology
Branch Ad Hoc Oncogenicity Committee and the Science Advisory Panel,
the Agency concluded that the mouse study was not adequate to define
the oncogenic potential of glyphosate. A review of the rat oncogenicity
study determined that a maximum tolerated dose had not been reached;
therefore, the rat study was also inadequate to determine oncogenic
potential. The Agency has required that the mouse and rat oncogenicity
studies be repeated.

Subchronic Toxicology Studies

No acceptable subchronic feeding studies are available for technical glyphosate. No additional studies are required because the rat chronic feeding and 1-year dog studies discussed above fulfill these requirements.

An acceptable 21-day dermal study in rabbits is available for glyphosate. The NOEL for this study was 1000 mg/kg/day. The only effect noted was slight edema and erythema of the skin at 5000 mg/kg/day (highest dose tested).

Teratology and Reproduction Studies

A teratology study in rats indicated no evidence of teratology up to 3500 mg/kg/day (highest dose tested). A fetotoxic and maternal toxic NOEL for this study is 1000 mg/kg/day.

A teratology study in rabbits indicated no evidence of teratology up to 350 mg/kg/day (highest dose tested). The maternal toxic NOEL is 175 mg/kg/day and the fetotoxic NOEL is 350 mg/kg/day for this study.

A three-generation rat reproduction study indicated no effects on fertility or reproductive parameters up to 30 mg/kg/day (highest dose tested). The NOEL for systemic effects is 10 mg/kg/day.

No additional teratology or reproduction studies are required for glyphosate.

Mutagenicity Studies

Glyphosate was negative in all mutagenicity studies submitted to satisfy the Agency's requirements for gene mutation, chromosomal aberrations and primary DNA damage. No additional mutagenicity studies are required.

Metabolism

Available rat metabolism studies are not adequate to fulfill Guidelines requirements, therefore repeat studies are required.

N-Nitroso Glyphosate

The Agency has determined that technical glyphosate contains N-nitroso glyphosate (NNG) as a contaminant at levels of 0.1 ppm or less. The Agency has determined that testing of nitroso contaminants will normally be required only in those cases in which the level of nitroso compounds exceeds 1.0 ppm [See "Pesticide Contaminates with N-nitroso Compound; Proposed Policy 45 FR 42854 (June 25, 1980)"]. No data are required on N-nitroso glyphosate because the amount of N-nitroso glyphosate is less than 1.0 ppm.

Plant Metabolite - Aminomethylphosphonic Acid

The Agency has determined that the metabolite aminomethylphosphonic acid (AMPA) is formed on plants in amounts that can range as high as 28 percent of total residue in the plant. Acceptable acute toxicology studies and a 90-day subchronic feeding study are available for this metabolite. The available data do not suggest that this compound poses any hazard distinct from that of the parent compound. The need for additional testing on this compound will be assessed after submission of an acceptable rat metabolism study with glyphosate.

Physiological and Biochemical Behavior Characteristics

    Foliar Absorption: Readily absorbed through foliage and translocated throughout the plant.

    Translocation: Translocated from treated areas to untreated shoots and roots.

    Mechanism of Pesticidal Action: Inhibition of amino acid biosynthesis resulting in a reduction of protein synthesis and inhibition of growth.

    Metabolism and Persistence in Plants: Metabolism occurs via N-methylation and ultimately yields N-methylated glycines and phosphonic acids. Parent compound and its metabolite AMPA are considered to be residues of concern in plants.

    Metabolism and Persistence in Animals: Glyphosate and AMPA have been identified in the tissues, urine, and feces of rats and rabbits; liver of poultry, swine, and cattle; and the kidney of swine and cattle.

Environmental Characteristics:

    Adsorption and leaching in basic soil types: strongly adsorbed to the soil and very little leaching.

    Microbial Breakdown: Microbial breakdown is the major cause of decomposition of glyphosate in soil. Depending on soil and microfloral population types, varying rates of decomposition occur.

    Loss from photodecomposition and/or volatization: Negligible losses via either route.

Bioaccumulation - Low potential to bioaccumulate in edible and
visceral tissue of catfish or whole body tissue of clams.

Potential to Contaminate Ground Water: Low potential to contaminate
ground water because glyphosate is tightly bound to the soil.

Ecological Characteristics:

Avian Acute Oral Toxicity: Bobwhite quail > 2000 mg/kg

Avian Subacute Dietary Toxicity: Mallard duck > 4640 ppm
                                  Bobwhite quail > 4640 ppm

Avian Reproduction Studies: Mallard duck > 1000 ppm
                             Bobwhite quail > 1000 ppm

Acute Toxicity to Freshwater Fish: Bluegill sunfish = 120 mg/L
                                    Rainbow trout = 86 mg/L

Chronic Toxicity to Freshwater Fish: Fathead minnow with a
maximum threshold concentration (MATC) > 25.7 mg/L

Acute Toxicity to Freshwater Invertebrates: Daphnia magna = 780 ppm
                                             Chironomus plumosus = 55 ppm

Chronic Toxicity to Freshwater Invertebrates: Daphnia magna MATC
                                               > 50 < 96 mg/L

Acute Toxicity to Marine/Estuarine Organisms: Grass shrimp = 281 ppm
                                               Fiddler Crab = 934 ppm
                                               Atlantic Oyster > 10 mg/L

These data indicate that glyphosate is no more than slightly
toxic to birds on an acute or dietary basis and will not cause avian
reproduction impairment.

The available data indicate that glyphosate is practically
nontoxic to both coldwater and warmwater fish, no more than slightly
toxic to freshwater invertebrates, grass shrimp, fiddler crab or
Atlantic oyster.

Hazard to Endangered Species

Minimal hazard is expected to aquatic endangered species.
Previous consultations with Office of Endangered Species have
resulted in jeopardy opinions and labeling for crops (alfalfa,
apples, barley, corn, cotton, pears, and wheat), rangeland and
pastureland, silvicultural sites, aquatic sites, and noncropland
use. Labeling is being imposed to reduce the risk to endangered
species.

Tolerance Assessment

40 CFR 180.364 Glyphosate, tolerances for residues.

Tolerances are established for combined residues of
glyphosate(N-(phosphonomethyl)glycine) and its metabolite
aminomethylphosphonic acid resulting from the application of
isopropylamine salt of glyphosate in or on the following raw
agricultural commodities:

Commodity	Parts per million
Alfalfa	200
Alfalfa, fresh and hay	0.2
Almonds, hulls	1
Asparagus	0.2
Avocados	0.2
Bahiagrass	200.0
Bermudagrass	200.0
Bluegrass	200.0
Bromegrass	200.0
Citrus fruits	0.2
Clover	200.0
Coffee beans	1
Cotton, forage	15
Cotton, hay	15
Cottonseed	15
Cranberries	0.2
Fescue	200.0
Forage grasses	0.2
Forage legumes (except soybeans and peanuts)	0.4
Grain crops	0.1 (N)
Grapes	0.2
Grasses, forage	0.2 (N)
Guavas	0.2
Leafy vegetables	0.2 (N)
Mangoes	0.2
Nuts	0.2
Orchardgrass	200.0
Papayas	0.2
Peanuts	0.1
Peanut, forage	0.5
Peanut, hay	0.5
Peanut, hulls	0.5
Pineapple	0.1
Pistachio nuts	0.2

Pome fruits	0.2
Root crop vegetables	0.2 (N)
Ryegrass	200.0
Seed and pod vegetables	0.2 (N)
Seed and pod. vegetables, forage	0.2 (N)
Seed and pod vegetable, hay	0.2 (N)
Soybeans	6
Soybeans, forage	15
Soybeans, hay	15
Stone fruit	0.2
Timothy	200.0
Wheatgrass	200.0

Tolerances are established for combined residues of glyphosate (N-(phosphonomethyl)glycine) and its metabolite aminomethylphosphonic acid resulting from application of glyphosate isopropylamine salt for herbicidal purposes in or the sodium sesqui salt for plant growth regulator purposes in or on the following raw agricultural commodities:

Commodity	Parts per million
Cattle, kidney	0.5
Cattle, liver	0.5
Fish	0.25
Goats, kidney	0.5
Goats, liver	0.5
Hogs, kidney	0.5
Hogs, liver	0.5
Horses, kidney	0.5
Horses, liver	0.5
Poultry, kidney	0.5
Poultry, liver	0.5
Sheep, kidney	0.5
Sheep, liver	0.5
Sugarcane	2.0

Tolerances are established for the combined residues of glyphosate (N-phosphonomethylglycine) and its metabolite aminomethylphosphonic acid, resulting from the use of irrigation water containing residues of 0.5 ppm following applications on or around aquatic sites, at 0.1 ppm on the crop groupings, citrus, cucurbits, forage grasses, forage legumes, fruiting vegetables, grain crops, leafy vegetables, nuts, pome fruits, root crop vegetables, seed and pod vegetables, stone fruit and the individual commodities cottonseed, hops, and avocados. Where no tolerances are established at higher levels from other uses of glyphosate in or on the subject crops, the higher tolerances should also apply to residues from the aquatic uses cited in this paragraph.

21 CFR 193.235 Glyphosate.

Tolerances are established for the combined residues of the
herbicide glyphosate and the metabolites as indicated when present
the therein as a result of the herbicide application to growing crops.

1. Glyphosate·[N-(phosphonomethyl)glycine] and its metabolite
    aminomethylphosphonic acid resulting from the application of the
    isopropylamine salt of glyphosate for herbicidal purposes and/or
    the sodium sesqui salt for plant growth regulator purposes.

Foods	Parts Per Million
Molasses, sugarcane	30.0

2. Glyphosate [N-(phosphonomethyl)glycine] and its metabolite
    aminomethylphosphonic acid resulting from the application of the
    isopropylamine salt of glyphosate for herbicidal purposes.

Foods	Parts Per Million
Oil, palm	0.1
Olives, imported	0.1
Tea, dried	1.0
Tea, instant	4.0

21 CFR 561.253 Glyphosate

A feed additive regulation is established permitting the combined
residues of the herbicide glyphosate (N-(phosphonomethyl)glycine) and
its metabolite aminomethylphosphonic acid in or on the following feed
commodities:

Feeds	Part Per Million
Citrus pulp, dried	0.4
Soybean hulls	20.0

Canadian tolerances have been established for residues of
glyphosate in or on carrots, sugarbeets, lettuce, cabbage, beans,
peas, soybeans, citrus, pome fruits, stone fruits, grapes, cereals,
grasses, and forage legumes) at 0.1 ppm.

The tolerance assessment indicated several changes in tolerance
listing are needed and several crops and processed commodities require
additional residue data. Refer to the Regulatory Position and
Rationale for listings of crops and changes needed.

The acceptable daily intake (ADI) for glyphosate is currently based on the finding of renal tubular dilation in F_{3b} pups in the rat three-generation reproduction study. The NOEL for this effect was 10 mg/mg/day. Using a hundredfold safety factor, the ADI for glyphosate is 0.1 mg/kg/day, which is equivalent to a maximum permissible intake (MPI) of 6.0 mg/day in a 60 kg individual. Existing tolerances produce a theoretical maximum residue contribution (TMRC) of 1.4238 mg/day from a 1.5 kg diet which occupies 23.73 percent of the ADI.

Reported Pesticide Incidents

Most of the pesticide incidents reported involve illnesses cf workers due to skin or eye irritation during mixing, loading, or ground application of glyphosate. Symptoms include dermal irritation, nausea or dizziness.

4.   Summary of Regulatory Position and Rationale

A review of the data available indicate that no risk criteria listed in 40 CFR 154.7 have been exceeded for glyphosate. A repeat of the mouse and rat oncogenicity studies is required to fully detail the oncogenic potential of glyphosate.

The Agency will issue registrations for substantially similar products and will issue significant new uses on a case-by-case basis. Information available is adequate to assess any potential risk from issuance of new uses or new products.

The Agency is not imposing a ground water advisory statement for glyphoste because the potential of glyphosate to contaminate ground water is very low.

The Agency is not requiring a reentry interval or reentry data for glyphosate because glyphosate has a low toxicity and the "Worker Safety Rules" required are believed to reduce any risk to workers from exposure to glyphosate during mixing, loading, or application of glyphosate.

The Agency is requiring that labeling on all end-use products containing glyphosate bear "Worker Safety Rules" including protective clothing (face shield or goggles, chemical resistant gloves, apron, and shoes, shoe coverings or boots) to be worn when mixing, loading, applying the pesticide, or when handling concentrate.

The Agency is imposing endangered species labeling for crops (alfalfa, apples, barley, cotton, pears, and wheat), rangeland and pastureland, silvicultural sites, aquatic sites, and noncropland areas to protect plant and animal species in jeopardy from application of glyphosate. The Agency will notify the registrants when these statements are to appear on the labeling.

The Agency is imposing a label restriction prohibiting the rotation of food or feed crops in glyphosate-treated soils unless glyphosate is registered for use on those crops. This restriction will be in effect until requested crop rotational data are submitted and reviewed.

The Agency is imposing a label restriction prohibiting the use of glyphosate on rice fields in which crayfish and catfish are included in cultural practice to ensure that catfish and crayfish harvested for human food are not exposed to residues of glyphosate.

The Agency is imposing label restrictions prohibiting the use of water containing glyphosate from rice cultivation for irrigation of food or feed crops not appearing on the glyphosate label or use of water from glyphosate-treated ponds for irrigation purposes for 24 hours after treatment. The restrictions are to ensure that food or feed crops not having tolerances are not exposed to residues of glyphosate.

The Agency is requiring that additional residue data and/or information be submitted on the following raw agricultural commodities: parsnips; turnips, turnip greens; onions; cranberries; sorghum grain; asparagus; coffee; mangoes; peanuts, peanut forage, hay and hulls; and sugarcane. Additional residue data and/or information are required on the following processed commodities: potato granules, chips, and dried potatoes; processed commodities of sugarbeets (dehydrated pulp, molasses, and refined sugar); dried citrus pulp; prunes; processed products of grapes; corn oil (crude and refined); corn-milled products; alfalfa seed; processed olives and olives. Additional residue data and/or information are required on the following crop groups: root and tuber vegetables; leaves of root and tuber vegetables; bulb vegetables group; fruiting vegetables (except cucurbits) group; citrus fruit group; small fruits and berries group; and cereal grains group. All other registered food or feed crops are adequately supported by residue data.

The Agency is requiring that a grazing restriction prohibiting the feeding of sugarcane forage or pineapple forage be added to the label. Alternatively, a petition proposing tolerances for sugarcane forage and pineapple forage with supporting data be submitted.

The Agency is requiring the following changes in the tolerance regulations: 1) the established tolerances on crop groups: root crop vegetables, seed and pods vegetables, and leafy vegetables should be deleted and tolerances established for the individual members of the groups, 2) amend the entry "nuts" to read "tree nuts", 3) the existing tolerances for alfalfa, clover, and forage legumes should be deleted and separate tolerances for glyphosate and AMPA of 100 ppm established for residues of forage and hay of nongrass animal feed, 4) the existing tolerances for forage legumes should be deleted and separate tolerances of 0.2 ppm should be established in or on legume vegetables (except soybeans) and the foliage of legume vegetables, except soybean forage and hay, 5) the established tolerances for forage grasses, grasses, forage; bahiagrass, bermudagrass, bluegrass, bromegrass, fescue, orchardgrass, ryegrass, timothy and wheat grass should be deleted and a separate tolerance of 100 ppm for residues in or on grass forage and hay be established. These changes are based on new tolerance expressions (Refer to 40 CFR 180.34).

The Agency will raise the tolerance for instant tea from 4 ppm to 7 ppm because processing studies available support a tolerance of 7 ppm.

The tolerances for cottonseed hay and forage must be cancelled because cottonseed hay is not considered a raw agricultural commodity of cotton and a grazing restriction exists.

Additional data are required on crops treated with irrigation water containing residues of glyphosate. Once these data are submitted, the crop groupings presently listed in Section (c) of 180.364 must be deleted and tolerances based on requested field irrigation tests must be established for glyphosate residues in or on members of the current, appropriate crop groups and all major irrigated crops which are not included in a crop group such as cottonseed, sugarcane, peanuts, etc.

5. Summary of Major Data Gaps

Product Chemistry	6-12 Months
Repeat of Mouse and Rat Oncogenic Studies	50 Months
Environmental Fate Studies	9-50 Months
Residue Chemistry Data	18-24 Months

6. Contact Person at EPA

Robert J. Taylor
Office of Pesticide Programs, EPA
Registration Division (TS-767C)
401 M Street SW.
Washington, DC 20460
Phone: (703) 557-1800

Disclaimer: The information presented in this Pesticide Fact Sheet is for informational purposes only and may not be used to fulfill data requirements for pesticide registration and reregistration.

HARMONY 75 DF

Reason for Issuance: New Chemical Registration
Date Issued: April 25, 1988
Fact Sheet Number: 162

1. Description of Chemical

    Generic Name: Methyl 3-[[[[(4-methoxy-6-methyl-1,3,5-triazin-2-yl)
    amino] carbonyl]amino]sulfonyl]-2-thiophencarboxylate

    Common Name:

    Trade Name: Harmony 75 DF

    EPA Shaughnessy Code: 128845

    Chemical Abstracts Service (CAS) Number: 79277-27-3

    Year of Registration: 1988

    Pesticide Type: Herbicide

    Chemical Family: Sulfonylurea

    U.S. Producer: E.I. DuPont DeNemours & Company

2. Use Pattern and Formulations

    Application Sites: Terrestrial Food Crops

    Major Crops Treated: Small grains (wheat and barley)

    Types and Method of Application: Foliar, applied broadcast by
    ground equipment or broadcast by aircraft. It is applied
    postemergence in relation to the crop. A selective postemergence
    herbicide for control of certain annual and perennial weeds.

    Application Rates: 0.33 - 0.67 ounces active ingredient/A
    (9.4-19g).

Types of Formulation: 75% water dispersible granule

Usual Carrier: Water

3. Science Findings:

  Summary Science Statement: All data are acceptable to the Agency.
  DPX-M 6316 has low acute toxicity (Category III) for acute dermal,
  primary eye irritation, and is less toxic (Category IV)
  for all other forms of acute toxicity. It was not oncogenic
  to rats or mice, not teratogenic to rabbit, and not mutagenic. It
  is practically nontoxic to birds, fish, aquatic invertebates, and
  honeybees. The pesticide and its degradates will leach in soil
  and have the potential to contaminate groundwater at very low levels
  The nature of the residues in plants and animals are adequately
  understood and adequate methodology is available for enforcement
  of tolerances in wheat and barley grain and straw.

Chemical Characteristics:

  Physical State: Crystalline solid

  Color: White

  Odor: None

  Melting Point: 176-178°C

  Density: 1.49g/cc

  Solubility (25°C):

Water (pH 4.0)	24 milligrams 1 liter /l)
(pH 5.0)	260 mg/l
(pH 6.0)	2400 mg/l
Acetone	11.9 mg/l
Acetonitrile	7.3 mg/l
Ethanol	0.9 mg/l
Methanol	2.6 mg/l
Hexane	<0.1 mg/l
Ethyl Acetate	2.6 mg/l
Methylene Chloride	27.5 mg/l
Xylenes	0.2 mg/l

Vapor Pressure: 2.7×10^{-6}mm Hg/25°C

Dissociation Constant: 4.0 (pKa of the acid)

Octanol/Water Partition Coefficient: 0.027

pH: 4.0 (slurry in water)

Stability: Stable to metals and light. Decomposes on melting.
In solution the compound is very stable to methylene
chloride and ethyl acetate, moderately stable
in methanol, and relatively unstable in acetone
and acetonitrile. The photolytic half-life in an
aqueous solution is expected to be 1 to 5 days.

Toxicological Characteristics:

Acute Toxicology (Technical):

Acute Oral Toxicity greater than (>) 5000 milligrams/kilogram
(mg/kg)

(Rat)   Toxicity Category $\overline{IV}$

Acute Inhalation Toxicity > 7.9 mg/l/4 hour

(Rat)   Toxicity Category III

Acute Dermal Toxicity > 2000 mg/kg

(Rabbit)   Toxicity Category III

Acute Toxicology (75% end-use formulation):

Acute Oral Toxicity > 5000 mg/kg

(rat)   Toxicity Category $\overline{IV}$

Acute Dermal Toxicity > 2000 mg/kg

(rabbit)   Toxicity Category III

Primary eye Irritation - Moderate Eye Irritant

(rabbit)   Toxicity Category III

Skin Sensitization - Not a sensitzer

(guinea pig)

Major Routes of Exposure: The major routes of exposure are through
dermal and eye contact.

Chronic Toxicology:

2-Year Feeding/Oncogenicity Study (Rats)
Systemic no-observable effect level (NOEL)= 1.25 mg/kg/day
Systemic lowest effect level (LEL)= 25 mg/kg/day
No-oncogenic effect noted at 125 mg/kg [highest dose tested
(HD>)

18-Month Oncogenicity Study (Mice)

Systemic NOEL= 3.75 mg/kg/day
Systemic LEL= 112.5 mg/kg/day
No oncogenic effects noted at 1125 mg/kg/day (HDT)

1-Year Feeding (Dog)

NOEL = 18.75 mg/kg/day
LEL = 18.75 mg/kg/day

Teratology (Rat)

Maternal NOEL greater than (>) 725 mg/kg/day/(HDT)
Fetotoxic NOEL = 159 mg/kg/day
Fetotoxic LEL = 725 mg/kg/day
Teratogenic NOEL = 159 mg/kg/day

Teratology (Rabbit)

Maternal NOEL = 158 mg/kg/day
Maternal LEL = 511 mg/kg/day
Teratogenic NOEL > 511 mg/kg/day(HDT)

2-Generation Reproduction (Rat)

Systemic and Reproduction NOEL > 125 mg/kg/day(HDT)
No reproductive effects seen at 125 mg/kg/day (HDT)

Mutagenicity - Reverse Mutation Asssay in Salmonella - not mutagenic
with and without S-9.

Mutagenicity - Gene-Mutation-No increase in mutation frequency was
seen at the highest dose tested of 7 nm the limit of solubility.

Mutagenicity - DNA synthesis/rat hepatocytes in nitro-material
did not induce signficant increase in unscheduled DNA synthesis
(UDS) in primary cultures.

Physiological and Biochemical Behavior Characteristics:

Foliar Absorption: Rapid

Translocation: Translocated within the plant

Mechanism of Pesticidal Action: Selectively inhibit acetolactate synthase.

Metabolism and Persistence In Plants: In plants DPX-M6316 degrades into thiophene and triazine containing entities which appear to be stable and behave as sugars.

Bioaccumulation: Not expected to bioaccumulate in fish.

Environmental Characteristics:

Absorption and Leaching: In unaged column leaching studies, DPX-M6316 showed a high propensity to move in all soil types tested (sandy loam, loamy sand, silt and silt loam). In aged column leaching studies, residues were very mobile in silt loam soil with parent compound and DPX-M6316 acid represnting a major fraction of the leachate. The triazine amine degrades and has a high potential to leach. Because of its potential to leach, the triazine amine has a potential to contaminate groundwater, but because of the low application rate rapid degradation and single application/season, the levels would be very low.

Microbial breakdown: Degrades rapidly in the field via microbial degradation. Aerobic soil metabolism study indicates the DPX-M6316 is degraded to CO_2 via several metabolites in 2-6 days. The triazine amine peaked at 15% of the total residues. The anaerobic metabolism study indicated a similar pattern but at a slower rate.

Loss from Photodecomposition and/or volatilization : Does not volatilize. Degrades rapidly in the field via photolysis.

Exposure of Humans and Nontarget Organisms to Chemical or Degradate: Because of the low toxicity from oral, dermal and inhalation (toxicity category IV, III, and III) and since the chemical is not a skin sensitizer, the risk to humans from exposure should be minimal. Nontarget organisms are not likely to be adversely affected by use of DPX-M6316 because of its low toxicity to birds, fish, and invertebrates.

Exposure During Reentry: Because of low acute toxicity and cultural practices for wheat and barley (little or no reason for field workers to enter field after application) reentry data or labeling are not required.

Ecological Characterists:

Avian Oral Acute Toxicity with Mallard Ducks: > 2510 mg/kg

Avian Dietary Toxicity with Bobwhite quail:> 5620 mg/kg

Mallard Ducks: > 5620 mg/kg

Acute Aquatic Toxicity with Rainbow Trout: > 100 mg/kg

Bluegill Sunfish> 100 mg/kg

Acute Toxicity to Invertebrates: > 1000 mg/kg

Acute Toxicity to Honey Bee:> 12.5 mg/kg

Available data indicate that DPX-M6316 is practically non-toxic to birds, fish, aquatic invertebrates, and honey bees.

Endangered Species: The available data indicate that the proposed use of DPX-M6316 is unlikely to pose a hazard to endangered aquatic or avian species. There may be some hazard to endangered plants.

Tolerance reassessment: Tolerances are established for residues of the herbicide DPX-M6316 (methyl 3-[[[[4-methoxy-6-methyl-1,3,5 Triazin-2-yl) amino]carbonyl] amino] sulfonyl]-2-thiophencarboxylate) in or on the following raw agricultural commodities (40 CFR 180.439).

Commodities	Parts Per Million
barley, grain	0.05
barley, straw	0.1
wheat, grain	0.05
wheat, straw	0.1

The acceptable daily intake (ADI) based on the 2-year rat feeding study (NOEL of 1.25 mg/kg/day) and using a safety factor of 100 is calculated to 0.013 mg/kg/day. The theoretical maximum residue contribution (TMRC) from these tolerances is calculated to be 0.000073 mg/kg body weight/day, which occupies approximately 0.6% of the ADI. There are no other published tolerances for this chemical.

Reported Pesticide Incidents: There are no reported pesticide incidents for the chemical.

4. Summary of Regulatory Position and Rationale

The Agency has decided that the data submitted in support of the registration request is acceptable and fulfills the guidelines requirements. Therefore, the Agency has accepted the use of DPX-M6316 for control of weeds in wheat and barley.

The Agency has determined that DPX-M6316 and its degradates have a potential to leach and therefore contaminate groundwater. Therefore the Agency is requiring a small scale groundwater monitoring study, methodology to determine DPX-M6316 and its degradates in groundwater, and an anaerobic soil metabolism study using radiolabeled material with labeling on the triazine moiety.

5. Summary of Data Gaps

Small scale prospective groundwater monitoring study. Methodology to determine DPX-M6316 and its degradates in water. Anaerobic soil metabolism study with radiolabelled triazine moiety.

6. Contact Person at EPA: Robert J. Taylor
                             Office of Pesticide Programs, EPA
                             Registration Division (TS-767C)
                             401 M Street, S.W.
                             Washington, D.C. 20460
                             Phone: (703) 557-1800

Disclaimer: The information in this Pesticide Fact Sheet is a summary only and may not be used to fulfill data requirements for pesticide registration and reregistration.

HEPTACHLOR

Reason for Issuance: Prohibition of Continued Sale or Use of
                      Heptachlor Products for Seed Treatment
Date Issued: April 26, 1989
Fact Sheet Number: 107.2

1. DESCRIPTION OF CHEMICAL

Generic Name: 1,4,5,6,7,8-8-heptachloro-3a,4,7,7a-tetra-
(Chemical)   hydro-4,7-methano-1H indene
Common Name: Heptachlor
Trade and Other Names: 1,4,5,6,7,8,8-heptachlor-3a,4,7,7a-
  tetrahydro-4,7-methanoindene; E-3314; Velsicol 104.
EPA Shaughnessy Code : 044801
Chemical Abstracts Service (CAS) Number: 76-44-8
Year of Initial Registration: 1952
Pesticide Type: Insecticide
Chemical Family: Chlorinated cyclodiene

2. USE PATTERN - SEED TREATMENTS

ACTION: Notice of PROHIBITION OF CONTINUED SALE OR USE OF HEPTACHLOR

PRODUCTS FOR SEED TREATMENT.

The Administrator has signed a Notice of Determination Pursuant to
Section 6(a)(1) of FIFRA which will be published in the Federal
Register. The Notice will prohibit any further sale or use of
heptachlor products for seed treatment purposes. Any sale or use of
heptachlor products for seed treatment will be a violation of Section
12(a)(1)(A) and/or Section 12(a)(2)(K) of the Federal Insecticide,
Fungicide and Rodenticide Act (FIFRA).

320

3. REGULATORY HISTORY

A. NOTICE OF INTENT TO CANCEL

Prior to 1974, heptachlor (along with a related compound, chlordane) was registered for a wide variety of insecticide uses. On November 18, 1974, the Administrator issued a notice of intent to cancel registrations for most uses of heptachlor (and chlordane). The basis for the notice of intent to cancel was evidence that heptachlor and chlordane had demonstrated toxic effects which may have significant adverse effects on human health, and evidence that both chemicals persist in the environment for many years after application, and as such, are subject to considerable movement from the site of actual application. The evidence on toxicity included a finding that heptachlor and its metabolite, heptachlor epoxide induce tumors in mice and that there was evidence of embryotoxicity in mice and rats.

Because of the persistence and wide application of heptachlor and chlordane products, heptachlor epoxide residues were routinely found in water, food sources, and human adult and fetal tissue. The Administrator therefore proposed to cancel all registered uses of chlordane and heptachlor, except those uses for subterranean termiticide control (see note) and dipping of non-food plants.

NOTE: It should be noted that subsequently on October 1, 1987, EPA issued an Order accepting the voluntary cancellation of chlordane and heptachlor termiticide products. A Notice signed on April 5, 1988, in response to a District Court ruling established limits on the sale and use of existing stock of termiticide products after April 15, 1988.

B. THE SUSPENSION OF HEPTACHLOR PRODUCTS

On July 29, 1975, the Administrator issued a notice of intent to suspend (pursuant to FIFRA Section 6(c)) the registrations of heptachlor and chlordane that were subject to the notice of intent to cancel. The grounds for the notice of intent to suspend were "new evidence ... which confirm [ed] and heighten [ed] the human cancer hazard posed by [chlordane and heptachlor]" and the Administrator's determination that the cancellation proceeding resulting from the notice of intent to cancel would not be complete [ed] in time to "avert substantial additions of these persistent and ubiquitous compounds to an already serious human and environmental burden." The notice of intent to suspend applied to all uses covered by the notice of intent to cancel.

An evidentiary hearing on the proposed suspension took place between August and December of 1975. On December 12, 1975, the hearing examiner published a recommended decision dismissing the notice of intent to suspend. The basis for this recommendation was the hearing examiner's unwillingness to find "conclusively" that heptachlor and chlordane were (are) carcinogens in laboratory animals.

Included in the recommended decision was a discussion of the use of heptachlor for seed treatment. The document noted that inadequate alternatives for seed treatment existed at that time. The hearing examiner recommended that heptachlor for seed treatment not be suspended even if the Administrator were to disagree with the examiner on the question of the hazard posed by chlordane and heptachlor.

On December 24, 1975, the Administrator issued his decision on the proposed suspension of chlordane and heptachlor products. The Administrator ordered a suspension of a number of chlordane and heptachlor uses during the pendency of the cancellation hearing.

As to seed treatment, however, the Administrator found that no adequate alternatives to treatment with heptachlor existed at that time, and therefore found that the benefit from heptachlor for seed treatment exceeded the risks of such use during the time necessary to complete the cancellation hearing. Heptachlor for seed treatment was thus not one of the uses suspended by the Administrator.

C.   SETTLEMENT OF THE CANCELLATION PROCEEDING

The cancellation proceeding continued until November of 1977, at which time the parties entered into settlement negotiations. The negotiations resulted in an agreement which was ratified in a Final Order issued by the Administrator on March 6, 1978. The Final Order resulted in the eventual cancellation of all products subject to the original notice of intent to cancel notice. For seed treatment, the effective date of cancellation was September 1, 1982 for barley, oats, wheat, rye and corn, and July 1, 1983 for sorghum. The Order also contained production limitations; production of heptachlor for seed treatment was limited to 175,000 pounds annually from 1978 to 1982, and to 100,000 pounds in 1983. These production limitations were intentionally less than the use of heptachlor for seed treatment purposes in 1976 (which was 200,000 pounds).

The purpose of the phased cancellations was to provide a "transition period" to allow users to make an orderly adjustment to alternative crops or pest control technologies where possible or to promote development of alternative pest control technologies where none then existed.

D.   EXISTING STOCKS DETERMINATION

The sale and use of existing stocks of pesticide products cancelled after a notice of intent to cancel is issued pursuant to Section 6(b) of FIFRA are controlled by Section 6(a)(1) of FIFRA. It provides in part,"... the Administrator may permit the continued sale and use of existing stocks of a pesticide whose registration is canceled under [Section 6(b)] to such extent as he may specify if he determines that such sale or use is not inconsistent with the purposes of [FIFRA] and will not have unreasonable adverse effects on the environment."

At the time the Agency issued the Final Order, it was expected based upon the use practices at that time that sale and use of existing stocks of cancelled products would cease approximately within one year of the effective cancellation date. The existing stocks allowance and phased cancellation was to result in approximately a six year transition period for users of heptachlor treated seeds to adapt alternative management practices after 1978.

The six year transitional period contemplated in 1984 ended over four years ago. The Agency believes that ten years is more than sufficient time for users to find alternatives to heptachlor seed treatment. Moreover, although some heptachlor continues to be used for seed treatment purposes, the transition away from heptachlor seed treatment has largely been completed (the amount of heptachlor used for seed treatment in 1987 was only 1% of the amount used in 1974).

While the benefits associated with heptachlor seed treatment have greatly diminished in the past ten years, the Agency's general concerns with the use of heptachlor have not diminished.

In addition, in late January and early February of 1986, the Food and Drug Administration (FDA) found very high levels of heptachlor and trans-chlordane in finished livestock feeds.

A fermentation/distillation firm purchased and used obsolete pesticide treated seed grain in their fermentation process. The spent distillers mash was, in turn, used in the manufacture of finished animal feeds and fed to dairy cattle. When FDA tested the milk from dairy herds fed the contaminated feed, the levels of heptachlor epoxide (an animal metobolite of heptachlor) found exceeded, by as much as 75 times, the FDA action level of 0.1 ppm for heptachlor epoxide in the milk fat.

As the result of this one incident, taxpayers have already incurred more than ten million dollars in investigative and indemnification costs. Total losses for all affected parties are expected to exceed sixteen million dollars.

FDA and USDA subsequently carried out an extensive investigation to determine how frequently obsolete pesticide treated seeds were being fed illegally to meat and/or milk producing animals or had entered the livestock feed markets. In over 1000 investigations, well over 100 violations were found. Feeding of obsolete heptachlor treated seed was involved in at least two of these additional violations.

EPA subsequently has determined that sizable inventories of cancelled heptachlor seed treatment products remain in the channels of trade. At the present levels of use, these products would be available for use for the next 70 years.

As previously stated, under Section 6 (a)(1), the Agency may permit the continued sale and use of existing stocks of a cancelled pesticide only if the Agency determines that such sale and use is consistent with FIFRA and does not result in unreasonable adverse effects on the environment.

Under the circumstances, the Agency can no longer find that continued sale or use of heptachlor for seed treatment will not have an unreasonable adverse effect on the environment. The Agency therefore no longer believe that such sale or use is consistent with Section 6(a)(1) of FIFRA.

The Agency accordingly served notice in the Federal Register of _____ that sale or use of stocks of heptachlor for seed treatment is no longer permitted, and that any further sale or use shall be a violation of Section 12(a)(1)(A) and/or Section 12(a)(2)(K) of FIFRA.

While any further use of heptachlor for seed treatment is not permitted, existing stocks of seed grain previously treated with heptachlor may be sold and planted in accordance with good agronomic practices.

4.   GUIDANCE ON THE STATUS OF HEPTACHLOR SEED TREATMENT PRODUCTS AS HAZARDOUS WASTES

Unused quantities of cancelled heptachlor seed treatment products can no longer be used as directed on their label. They, therefore, fit the definition of a solid waste as defined in 40 CFR 261.2 and 261.33 when they are discarded or held with the intent to discard.

A hazardous waste is any solid waste which has been listed as a hazardous waste in 40 CFR Part 261 Subpart D or a solid waste which exhibits any of the characteristics of hazardous waste identified in 40 CFR Part 261 Subpart C ignitability, §261.21; corrosivity, §261.22; reactivity, §261.23; and/or E.P. toxicity, §261.24.

Heptachlor is listed as an acutely hazardous waste (P059) in 40 CFR §261.33(e). Any unused heptachlor seed treatment products, rinsate or containers which have not been properly cleaned (triple rinsed as defined §261.7) are therefore acutely hazardous wastes, as defined in 40 CFR §261.33(e) if they are discarded or intended for discard.

Any person by site who holds cancelled heptachlor seed treatment products when they become wastes is a "generator" of hazardous wastes as defined in 40 CFR Part 261. A generator must comply with the requirements of the Resource Conservation and Recovery Act (RCRA) and any other applicable Federal, State, and local laws and regulations.

Those who hold cancelled heptachlor seed treatment products at the time th become wastes are defined as "generators" and they fall into one of three categories of waste generators. They are:

a.   Conditionally Exempt Generator - one who currently holds or generates no more than 1 kilogram (2.2 pounds) of acutely hazardous waste [heptachlor seed treatment products, a listed acutely hazardous waste (P059)] and who generates no more than 100 kilograms (220 pounds) of other hazardous waste in any calendar month.

A conditionally exempt generator is not required to obtain a permit or interim status (40 CFR Part 261.5). He/she, however, is required to:

- Identify all hazardous waste held or generated, §261.5(c).

- Send the hazardous waste to an authorized facility, §261.5(f)(3).

- Never accumulate more than 1000 kilograms (2200 pounds) of hazardous waste and/or more than 1 kilogram (2.2 pounds) of acutely hazardous waste on his/her property, §261.5(f)(2) and (g)(2).

Acutely hazardous waste (P059) may be held [up to 1 kilogram (2.2 pounds)] in containers which are in good condition (do not leak) and are compatible with the waste.

b. Small Quantity Generator - one who holds or generates no more than 1 kilogram (2.2 pounds) of acutely hazardous waste [heptachlor seed treatment products, a listed acutely hazardous waste (P059)] and generates between 100 and 1,000 kilograms (220 to 2,200 pounds) of other hazardous waste in any calendar month.

A small quantity generator must comply with the requirements of 40 CFR Part 262, Standards Applicable to Generators of Hazardous Waste including obtaining an EPA ID number, using the Uniform Hazardous Waste Manifest, accumulating waste in accordance with § 262.34(d) and complying with recordkeeping and reporting requirements of §262.40(a), (c) and (d); § 262.42(b); and §262.43.

Small quantity generators who choose to store or treat beyond the allowances provided in 262.34(d)-(f) or to dispose of hazardous wastes or acutely hazardous wastes at their own facilities are subject to the full regulatory requirements of 40 CFR Parts §264 through §270 which pertain to the operation, maintenance and permitting of treatment, storage and disposal facilities.

Generators must send heptachlor seed treatment products that are not treated or disposed of on site to a hazardous waste facility permitted to accept them.

c. Generator - one who holds or generates more than 1 kilogram (2.2 pounds) of acutely hazardous waste [heptachlor seed treatement products, a listed acutely hazardous waste (P059)] or more than 1,000 kilograms (2,200 pound) of hazardous waste in any calendar month.

A hazardous waste "generator" as defined above must comply with all applicable hazardous waste management requirements set forth in 40 CFR Part 262, Standards Applicable to Generators of Hazardous Waste. Those who choose to transport their own hazardous waste must comply with 40 CFR Part 263, Standards Applicable to Transporters of Hazardous Waste.

If a generator stores his/her waste for longer than 90 days, then he/she must obtain a RCRA hazardous waste storage permit and comply with the requirements of 40 CFR Part 264 and 40 CFR Part 265. An extension of 30 days may be granted by the Regional Administrator under certain emergency situations.

Generators who choose to store or to treat beyond the allowances provided in § 262.34(a) or to dispose of hazardous wastes or acutely hazardous wastes at their own facilities are subject to the full regulatory requirements of 40 CFR Parts 264 through 270 which pertain to the operation, maintenance and permitting of treatment, storage and disposal facilities.

Generators must send heptachlor seed treatment products that are not treated or disposed of on-site to a hazardous waste facility permitted to accept them.

Obsolete seed, which are no longer viable or suitable for planting and which have been treated with heptachlor are not "listed" hazardous wastes in 40 CFR 261 Subpart D. Their status as "characteristic" hazardous wastes under 40 CFR Subpart C 261.20 through 261.24 and 40 CFR Part 261 Appendix I, II and III must be determined by the generator under 40 CFR 261.11.

It should be kept in mind, however, that some serious environmental inpacts have resulted from the inappropriate disposal of obsolete heptachlor-treated seeds. Every effort should be made to plant existing stocks of heptachlor-treated seeds in accordance with good agronomic practices before they become obsolete.

Should the generator find that obsolete heptachlor-treated seed is a "characteristic" hazardous waste under 40 CFR 262.11, then the seed may be stored, treated or disposed of only at a permitted hazardous waste facility. EPA recommends giving serious consideration to incineration.

On the other hand, if after the aforementioned analysis, the obsolete heptachlor-treated seeds are determined to be non-hazardous, the obsolete heptachlor-treated seeds could be landfilled in accordance with the individual state and local requirements for disposal of solid waste. If landfill of the seed is not viable in your area, then consideration must again be given to incineration as the appropriate means of destruction.

5. CONTACT PERSON AT EPA, OFFICE OF PESTICIDE PROGRAMS:

James G. Touhey
Senior Agricultural Advisor (H-7506C)
Field Operations Division
Office of Pesticide Programs
Environmental Protection Agency
401 M Street, SW
Washington, D.C. 20460

Office location and telephone number:

Room 710
Crystal Mall, Building No. 2
1921 Jefferson Davis Highway
Arlington, VA 22202
(703) 557-5664

6. CONTACT FOR ADDITIONAL INFORMATION REGARDING DISPOSAL

For those states which have RCRA authorization, a concerned individual should contact the hazardous waste management agency of that state for additional information concerning the state disposal requirements (see Appendix for list of authorized states and their addresses and phone number).

For non-authorized states the concerned individual should contact the hazardous waste management division of the EPA region in which his/her state falls (see Appendix for list of states by regions).

In addition, concerned parties may call the RCRA/Superfund Hotline toll free (1-800-424-9346) or may call commercially on (1-202-382-3000) for more detailed information concerning RCRA requirements.

APPENDIX

State Programs Branch, OSW (382-2210)

STATES GRANTED FINAL AUTHORIZATION FOR PRE-HSWA PROGRAM
(As of June 1, 1988)

States	Date Authorized	FR Page Number
1. Delaware	22 June 1984	23837 (June 8)
2. Mississippi	27 June 1984	24377 (June 13)
3. Montana	25 July 1984	28245 (July 11)
4. Georgia	21 August 1984	31417 (August 7)
5. North Dakota	19 October 1984	39328 (October 5)
6. Utah	24 October 1984	39683 (October 10)
7. Colorado	2 November 1984	41036 (October 19)
Revisions Approved	7 November 1986	37729 (October 24)
8. South Dakota	2 November 1984	41038 (October 19)
9. Virginia	18 December 1984	47391 (December 4)
10. Texas	26 December 1984	48300 (December 12)
Revisions Approved	4 October 1985	3952 (January 31)
Revisions Approved	17 February 1987	45320 (December 18)
11. North Carolina	31 December 1984	48694 (December 14)
Revisions Approved	8 April 1986	10211 (March 25)
12. New Hampshire	3 January 1985	49092 (December 18)
13. Oklahoma	10 January 1985	50362 (December 27)
14. Vermont	21 January 1985	775 (January 7)
15. Arkansas	25 January 1985	1513 (January 11)
16. New Mexico	25 January 1985	1515 (January 11)
17. Kentucky	31 January 1985	2550 (January 17)
18. Tennessee	5 February 1985	2820 (January 22)
Revisions Approved	11 August 1987	22443 (June 12)
19. Massachusetts	7 February 1985	3344 (January 24)
20. Nebraska	7 February 1985	3345 (January 24)
21. Louisiana	7 February 1985	3348 (January 24)
22. Maryland	11 February 1985	3511 (January 25)
23. Minnesota	11 February 1985	3756 (January 28)
Revisions Approved	18 September 1987	27199 (July 20)
24. Florida	12 February 1985	3908 (January 29)
25. New Jersey	21 February 1985	5260 (February 7)
26. District of Columbia	22 March 1985	9427 (March 8)
27. Kansas	17 October 1985	40377 (October 3)
28. Nevada	1 November 1985	42181 (October 18)
29. South Carolina	22 November 1985	46437 (November 8)
Revisions Approved	13 September 1987	26476 (July 15)
30. Arizona	4 December 1985	47736 (November 20)
31. Missouri	4 December 1985	47740 (November 20)
32. Guam	27 January 1986	1370 (January 13)
33. Pennsylvania	30 January 1986	1791 (January 15)
34. Illinois	31 January 1986	3778 (January 30)
Revisions Approved	5 February 1988	126 (January 5)
35. Oregon	31 January 1986	3779 (January 30)

State Programs Branch, OSW (382-2210)

STATES GRANTED FINAL AUTHORIZATION FOR PRE-HSWA PROGRAM (con't)
(As of June 1, 1988)

States	Date Authorized	FR Page Number
36. Rhode Island	31 January 1986	3780 (January 30)
37. Washington	31 January 1986	3782 (January 30)
Revisions Approved	23 November 1987	35556 (September 22)
38. Wisconsin	31 January 1986	3783 (January 30)
39. Indiana	31 January 1986	3953 (January 31)
Revisions Approved	5 February 1988	128 (January 5)
40. New York	29 May 1986	17737 (May 15)
41. West Virginia	29 May 1986	17739 (May 15)
42. Michigan	30 October 1986	36804 (October 16)
43. Alabama	23 December 1987	46466 (December 8)
44. Maine	14 March 1988	30192 (February 29)

STATE HAZARDOUS WASTE MANAGEMENT AGENCIES

ALABAMA
Alabama Department of
  Environmental Management
Land Division
1751 Federal Drive
Montgomery, Alabama 36130
(205) 271-7730

ALASKA
Department of Environmental
  Conservation
P.O. Box 0
Juneau, Alaska 99811
Program Manager: (907) 465-2666
Northern Regional Office
  (Fairbanks): (907) 452-1714
South-Central Regional Office
  (Anchorage): (907) 274-2533
Southeast Regional Office
  (Juneau): (907) 789-3151

AMERICAN SAMOA
Environmental Quality Commission
Government of American Samoa
Pago Pago, American Samoa 96799
Overseas Operator
(Commercial Call (684) 663-4116)

ARIZONA
Arizona Department of
  Health Services
Office of Waste and Water Quality
2005 North Central Avenue
  Room 304
Phoenix, Arizona 85004
Hazardous Waste Management:
  (602) 255-2211

ARKANSAS
Department of Pollution Control
  and Ecology
Hazardous Waste Division
P.O. Box 9583
8001 National Drive
Little Rock, Arkansas 72219
(501) 562-7444

CALIFORNIA
Department of Health Services
Toxic Substances Control Division
714 P Street, Room 1253
Sacramento, California 95814
(916) 324-1826

State Water Resources Control Board
Division of Water Quality
P.O. Box 100
Sacramento, California 95801
(916) 322-2867

COLORADO
Colorado Department of Health
Waste Management Division
4210 E. 11th Avenue
Denver, Colorado 80220
(303) 320-8333 Ext. 4364

CONNECTICUT
Department of Environmental
  Protection
Hazardous Waste Management
  Section
State Office Building
165 Capitol Avenue
Hartford, Connecticut 06106
(203) 566-8843, 8844

Connecticut Resource Recovery
  Authority
179 Allyn Street, Suite 603
Professional Building
Hartford, Connecticut 06103
(203) 549-6390

DELAWARE
Department of Natural Resources
  and Environmental Control
Waste Management Section
P.O. Box 1401
Dover, Delaware 19903
(302) 736-4781

DISTRICT OF COLUMBIA
Department of Consumer and
  Regulatory Affairs
Pesticides and Hazardous Waste
  Materials Division
Room 114
5010 Overlook Avenue, S.W.
Washington, D.C. 20032
(202) 767-8414

FLORIDA
Department of Environmental
  Regulation
Solid and Hazardous Waste Section
Twin Towers Office Building
2600 Blair Stone Road
Tallahassee, Florida 32301
RE: SQG's
(904) 488-0300

GEORGIA
Georgia Environmental Protection
  Division
Hazardous Waste Management
  Program
Land Protection Branch
Floyd Towers East, Suite 1154
205 Butler Street, S.E.
Atlanta, Georgia 30334
(404) 656-2833
Toll Free: (800) 334-2373

GUAM
Guam Environmental Protection
  Agency
P.O. Box 2999
Agana, Guam 96910
Overseas Operator
(Commercial Call (671) 646-7579)

HAWAII
Department of Health
Environmental Health Division
P.O. Box 3378
Honolulu, Hawaii 96801
(808) 548-4383

IDAHO
Department of Health and Welfare
Bureau of Hazardous Materials
450 West State Street
Boise, Idaho 83720
(208) 334-5879

ILLINOIS
Environmental Protection Agency
Division of Land Pollution Control
2200 Churchill Road, #24
Springfield, Illinois 62706
(217) 782-6761

INDIANA
Department of Environmental
Management
Office of Solid and Hazardous Waste
105 South Meridian
Indianapolis, Indiana 46225
(317) 232-4535

IOWA
U.S. EPA Region VII
Hazardous Materials Branch
726 Minnesota Avenue
Kansas City, Kansas 66101
(913) 236-2888
Iowa RCRA Toll Free:
(800) 223-0425

KANSAS
Department of Health and
Environment
Bureau of Waste Management
Forbes Field, Building 321
Topeka, Kansas 66620
(913) 862-9360 Ext. 292

KENTUCKY
Natural Resources and
Environmental Protection Cabinet
Division of Waste Management
18 Reilly Road
Frankfort, Kentucky 40601
(502) 564-6716

LOUISIANA
Department of Environmental
Quality
Hazardous Waste Division
P.O. Box 44307
Baton Rouge, Louisiana 70804
(504) 342-1227

MAINE
Department of Environmental
Protection
Bureau of Oil and Hazardous
Materials Control
State House Station #17
Augusta, Maine 04333
(207) 289-2651

MARYLAND
Department of Health and Mental
Hygiene
Maryland Waste Management
Administration
Office of Environmental Programs
201 West Preston Street, Room A3
Baltimore, Maryland 21201
(301) 225-5709

MASSACHUSETTS
Department of Environmental
Quality Engineering
Division of Solid and Hazardous
Waste
One Winter Street, 5th Floor
Boston, Massachusetts 02108
(617) 292-5589
(617) 292-5851

MICHIGAN
Michigan Department of Natural
Resources
Hazardous Waste Division
Waste Evaluation Unit
Box 30028
Lansing, Michigan 48909
(517) 373-2730

MINNESOTA
Pollution Control Agency
Solid and Hazardous Waste Division
1935 West County Road, B-2
Roseville, Minnesota 55113
(612) 296-7282

MISSISSIPPI
Department of Natural Resources
Division of Solid and Hazardous
Waste Management
P.O. Box 10385
Jackson, Mississippi 39209
(601) 961-5062

MISSOURI
Department of Natural Resources
Waste Management Program
P.O. Box 176
Jefferson City, Missouri 65102
(314) 751-3176
Missouri Hotline:
(800) 334-6946

MONTANA
Department of Health and
Environmental Sciences
Solid and Hazardous Waste Bureau
Cogswell Building, Room B-201
Helena, Montana 59620
(406) 444-2821

NEBRASKA
Department of Environmental
Control
Hazardous Waste Management
Section
P.O. Box 94877
State House Station
Lincoln, Nebraska 68509
(402) 471-2186

NEVADA
Division of Environmental Protection
Waste Management Program
Capitol Complex
Carson City, Nevada 89710
(702) 885-4670

NEW HAMPSHIRE
Department of Health and Human
Services
Division of Public Health Services
Office of Waste Management
Health and Welfare Building
Hazen Drive
Concord, New Hampshire 03301-6527
(603) 271-4608

NEW JERSEY
Department of Environmental
  Protection
Division of Waste Management
32 East Hanover Street, CN-028
Trenton, New Jersey 08625
Hazardous Waste Advisement
  Program: (609) 292-8341

NEW MEXICO
Environmental Improvement
  Division
Ground Water and Hazardous
  Waste Bureau
Hazardous Waste Section
P.O. Box 968
Santa Fe, New Mexico 87504-0968
(505) 827-2922

NEW YORK
Department of Environmental
  Conservation
Bureau of Hazardous Waste
  Operations
50 Wolf Road, Room 209
Albany, New York 12233
(518) 457-0530
SQG Hotline: (800) 631-0666

NORTH CAROLINA
Department of Human Resources
Solid and Hazardous Waste
  Management Branch
P.O. Box 2091
Raleigh, North Carolina 27602
(919) 733-2178

NORTH DAKOTA
Department of Health
Division of Hazardous Waste
  Management and Special Studies
1200 Missouri Avenue
Bismarck, North Dakota 58502-5520
(701) 224-2366

**NORTHERN MARIANA ISLANDS,
COMMONWEALTH OF**
Department of Environmental and
  Health Services
Division of Environmental Quality
P.O. Box 1304
Saipan, Commonwealth of
  Mariana Islands 96950
Overseas call (670) 234-6984

OHIO
Ohio EPA
Division of Solid and Hazardous
  Waste Management
361 East Broad Street
Columbus, Ohio 43266-0558
(614) 466-7220

OKLAHOMA
Waste Management Service
Oklahoma State Department of
  Health
P.O. Box 53551
Oklahoma City, Oklahoma 73152
(405) 271-5338

OREGON
Hazardous and Solid Waste Division
P.O. Box 1760
Portland, Oregon 97207
(503) 229-6534
Toll Free: (800) 452-4011

PENNSYLVANIA
Bureau of Waste Management
Division of Compliance Monitoring
P.O. Box 2063
Harrisburg, Pennsylvania 17120
(717) 787-6239

PUERTO RICO
Environmental Quality Board
P.O. Box 11488
Santurce, Puerto Rico 00910-1488
(809) 723-8184
 – or –
EPA Region II
Air and Waste Management Division
26 Federal Plaza
New York, New York 10278
(212) 264-5175

RHODE ISLAND
Department of Environmental
  Management
Division of Air and Hazardous
  Materials
Room 204, Cannon Building
75 Davis Street
Providence, Rhode Island 02908
(401) 277-2797

SOUTH CAROLINA
Department of Health and
  Environmental Control
Bureau of Solid and Hazardous
  Waste Management
2600 Bull Street
Columbia, South Carolina 29201
(803) 734-5200

SOUTH DAKOTA
Department of Water and Natural
  Resources
Office of Air Quality and Solid Waste
Foss Building, Room 217
Pierre, South Dakota 57501
(605) 773-3153

TENNESSEE
Division of Solid Waste Management
Tennessee Department of Public
  Health
701 Broadway
Nashville, Tennessee 37219-5403
(615) 741-3424

TEXAS
Texas Water Commission
Hazardous and Solid Waste Division
Attn: Program Support Section
1700 North Congress
Austin, Texas 78711
(512) 463-7761

UTAH
Department of Health
Bureau of Solid and Hazardous
  Waste Management
P.O. Box 16700
Salt Lake City, Utah 84116-0700
(801) 538-6170

VERMONT
Agency of Environmental
   Conservation
103 South Main Street
Waterbury, Vermont 05676
(802) 244-8702

VIRGIN ISLANDS
Department of Conservation and
   Cultural Affairs
P.O. Box 4399
Charlotte Amalie, St. Thomas
Virgin Islands 00801
(809) 774-3320
   – or –
EPA Region II
Air and Waste Management Division
26 Federal Plaza
New York, New York 10278
(212) 264-5175

VIRGINIA
Department of Health
Division of Solid and Hazardous
   Waste Management
Monroe Building, 11th Floor
101 North 14th Street
Richmond, Virginia 23219
(804) 225-2667
Hazardous Waste Hotline:
(800) 552-2075

WASHINGTON
Department of Ecology
Solid and Hazardous Waste Program
Mail Stop PV-11
Olympia, Washington 98504-8711
(206) 459-6322
In-State: 1-800-633-7585

WEST VIRGINIA
Division of Water Resources
Solid and Hazardous Waste/
   Ground Water Branch
1201 Greenbrier Street
Charleston, West Virginia 25311

WISCONSIN
Department of Natural Resources
Bureau of Solid Waste Management
P.O. Box 7921
Madison, Wisconsin 53707
(608) 266-1327

WYOMING
Department of Environmental Quality
Solid Waste Management Program
122 West 25th Street
Cheyenne, Wyoming 82002
(307) 777-7752
   – or –
EPA Region VIII
Waste Management Division
   (8HWM-ON)
One Denver Place
999 18th Street
Suite 1300
Denver, Colorado 80202-2413
(303) 293-1502

EPA HAZARDOUS WASTE CONTACTS

RCRA/Superfund Hotline 1-800-424-9346 (In Washington, D.C.: 382-3000)	EPA Small Business Ombudsman Hotline 1-800-368-5888 (In Washington, D.C.: 557-1938)	National Response Center 1-800-424-8802 (In Washington, D.C.: 426-2675)

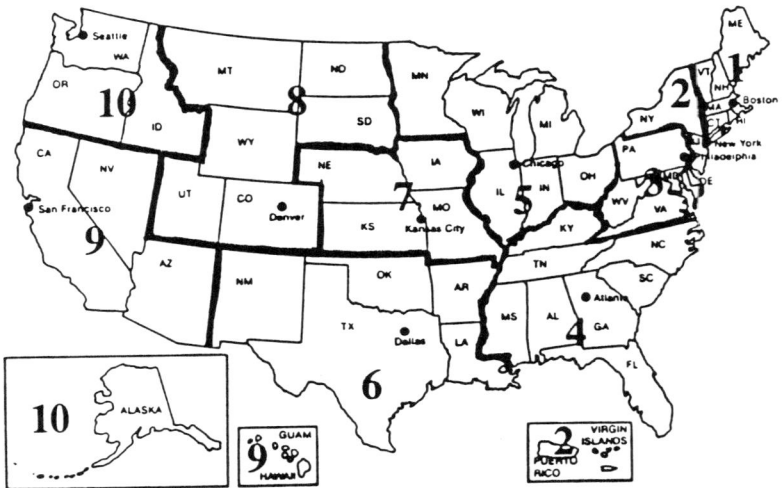

Regions	Regions	Regions	Regions
4 — Alabama	5 — Indiana	9 — Nevada	4 — Tennessee
10 — Alaska	7 — Iowa	1 — New Hampshire	6 — Texas
9 — Arizona	7 — Kansas	2 — New Jersey	8 — Utah
6 — Arkansas	4 — Kentucky	6 — New Mexico	1 — Vermont
9 — California	6 — Louisiana	2 — New York	3 — Virginia
8 — Colorado	1 — Maine	4 — North Carolina	10 — Washington
1 — Connecticut	3 — Maryland	8 — North Dakota	3 — West Virginia
3 — Delaware	1 — Massachusetts	5 — Ohio	5 — Wisconsin
3 — D.C.	5 — Michigan	6 — Oklahoma	8 — Wyoming
4 — Florida	5 — Minnesota	10 — Oregon	9 — American Samoa
4 — Georgia	4 — Mississippi	3 — Pennsylvania	9 — Guam
9 — Hawaii	7 — Missouri	1 — Rhode Island	2 — Puerto Rico
10 — Idaho	8 — Montana	4 — South Carolina	2 — Virgin Islands
5 — Illinois	7 — Nebraska	8 — South Dakota	

U.S. EPA REGIONAL OFFICES

EPA Region I
State Waste Programs Branch
JFK Federal Building
Boston, Massachusetts 02203
(617) 223-3468

Connecticut, Massachusetts, Maine,
New Hampshire, Rhode Island, Vermont

EPA Region II
Air and Waste Management Division
26 Federal Plaza
New York, New York 10278
(212) 264-5175

New Jersey, New York, Puerto Rico,
Virgin Islands

EPA Region III
Waste Management Branch
841 Chestnut Street
Philadelphia, Pennsylvania 19107
(215) 597-9336

Delaware, Maryland, Pennsylvania,
Virginia, West Virginia,
District of Columbia

EPA Region IV
Hazardous Waste Management Division
345 Courtland Street, N.E.
Atlanta, Georgia 30365
(404) 347-3016

Alabama, Florida, Georgia,
Kentucky, Mississippi, North
Carolina, South Carolina, Tennessee

EPA Region V
RCRA Activities
230 South Dearborn Street
Chicago, Illinois 60604
(312) 353-2000

Illinois, Indiana, Michigan,
Minnesota, Ohio, Wisconsin

EPA Region VI
Air and Hazardous Materials Division
1201 Elm Street
Dallas, Texas 75270
(214) 767-2600

Arkansas, Louisiana, New Mexico,
Oklahoma, Texas

EPA Region VII
RCRA Branch
726 Minnesota Avenue
Kansas City, Kansas 66101
(913) 236-2800

Iowa, Kansas, Missouri, Nebraska

EPA Region VIII
Waste Management Division (8HWM-ON)
One Denver Place
999 18th Street, Suite 1300
Denver, Colorado 80202-2413
(303) 293-1502

Colorado, Montana, North Dakota,
South Dakota, Utah, Wyoming

EPA Region IX
Toxics and Waste Management Division
215 Fremont Street
San Francisco, California 94105
(415) 974-7472

Arizona, California, Hawaii,
Nevada, American Samoa, Guam,
Trust Territories of the Pacific

EPA Region X
Waste Management Branch—MS-530
1200 Sixth Avenue
Seattle, Washington 98101
(206) 442-2777

Alaska, Idaho, Oregon, Washington

HEXAZINONE

Reason for Issuance: Registration Standard
Date Issued: September 1988
Fact Sheet Number: 183

DESCRIPTION OF CHEMICAL

Generic Name: 3-cyclohexyl-6-dimethylamino-1-methyl-
1,3,5-triazine-2,4(1H,3H)-dione

Common Name: Hexazinone

Trade Name: Velpar

EPA Shaughnessy Code: 51235-04-02

Chemical Abstracts Service (CAS) Number: 107201

Year of Initial Registration: 1975

Pesticide Type: Herbicide

Chemical Family: Triazine

U.S. Producer: Dupont

USE PATTERNS AND FORMULATIONS

Registered Uses: Terrestrial food crop use on fruit, sugarcane,
alfalfa, pastures/rangeland and fallowland;

Terrestrial nonfood crop use on grasses, rights-of-ways,
and other noncrop areas;

Aquatic nonfood crop uses on drainage ditch banks;

Forestry use on Christmas tree plantation, conifer release,
and conifer forest plantings.

Predominant Uses: A selective herbicide used to control
grasses and broadleaf and woody plants. Approximately
33% is used on alfalfa, 31% in forestry, 29% in
industrial areas, 4% on rangeland and pastures, and
< 2% on sugarcane.

Formulation Types Registered: Technical, formulation
intermediate, granular, pelleted/tableted, dry flowable,
emulsifiable concentrate, soluble concentrate/liquid,
and ready-to-use liquid.

Methods of Application: Postemergence, preemergence, layby,
directed spray, or basal soil treatment using ground
equipment or where appropriate, broadcasted using
aerial equipment.

Rates of Application: lb ai/A = pounds active ingredient per acre
Terrestrial food crop -   0.22 -   6.0 lb ai/A
Terrestrial nonfood crop -   0.67 - 13.5 lb ai/A
Aquatic nonfood crop -   1.0 - 13.5 lb ai/A
Forestry -   0.45 -   6.0 lb ai/A

SCIENCE FINDINGS

Chemical Characteristics:

Physical State: crystalline solid
Color: white
Melting point: 115-117°C
Solubility: soluble in water, chloroform, methanol,
benzene, dimethylformamide, acetone, toluene, and hexane.
Vapor Pressure: 2×10^{-7} mmHg, 25°C.
Stability: Stable in aqueous solutions at pH 5, 7, 9 at
temperatures up to 37°C.

Toxicology Characteristics:

Acute Toxicity:

Oral (rat): 1690 mg/kg (males), toxicity category III.
Testing of females is required.

Dermal (rabbit): > 5278 mg/kg (males), toxicity category IV.
If a gender difference is exhibited in the acute
oral test on females, testing on females may be
required.

Inhalation (rat): 7.48 mg/l (males), toxicity category III.
If a gender difference is exhibited in the acute
oral test on females, testing on females may be
required.

Primary Eye Irritation (rabbit): Corrosive, causing
irreversible eye damage, toxicity category I.

Primary Dermal Irritation: Not an irritant.

Subacute Toxicity:

A 21-day dermal study is required. Two acceptable oral
toxicity studies were reviewed and no additional
data are required. In the 90-day dog study, at the
Lowest Effect Level (LEL) of 125 mg/kg (highest dose tested
(HDT)), there was decreased body weight in both sexes,
increased alkaline phosphatase in both sexes, decreased
albumin /globulin values in both sexes, and increased
absolute and relative liver weight in both sexes.
There were no compound-related histopathological effects.
The No Observed Effect Level (NOEL) was 25 mg/kg.

In the 90-day rat study, at the LEL of 250 mg/kg (HDT),
there was decreased body weight in both sexes. There
were no compound-related effects in mortality, toxic signs,
food consumption, clinical pathology, organ weights,
and histopathology. The NOEL was 50 mg/kg.

Chronic Toxicity:

Chronic Oral Toxicity: One acceptable chronic feeding
study in rats has been submitted. A data gap exists
for a chronic nonrodent (dog) feeding study.

In the rat study, there was increased survival over all
test groups for male rats at 125 mg/kg at 2 years. Survival
at 2 years in female rats was comparable between control
and treated groups. The systemic NOEL was 10 mg/kg.
At the LEL of 50 mg/kg, females had a 5% decreased body
weight and slight decrease in food efficiency. At 125
mg/kg, there were significant toxic effects in both
sexes. Males had a 12% body weight decrease, a 4%
food consumption decrease, increased white blood cells
and eosinophiles, alkaline urine and organ weight changes.
Females had a 19% body weight decrease, slight food
efficiency decrease, alkaline urine and organ weight changes.

Oncogencity: Mouse study - The oncogenic potential is
inconclusive due to a possible ambiguity in the classification of liver neoplasia in both sexes. The systemic
NOEL is 30 mg/kg. The liver slides must be rereviewed
by the Agency.

Rat Study - There were no oncogenic effects up to and
including 125 mg/kg (HDT).

The Agency will reassess hexazinone's oncogenic potential
after rereview of the mouse study.

Teratology: Rat study - At 900 mg/kg/day (HDT), developmental
toxicity was evidenced by decreased fetal body weight, and

partial ossification, kidney anomalies, and misaligned
sternebrae. The LEL was 400 mg/kg and the NOEL was
100 mg/kg/day for developmental. For maternal toxicity,
the NOEL was 100 mg/kg/day.

Rabbit study - NOEL for developmental and maternal
     toxicity was 50 mg/kg/day.

The Agency has determined that hexazinone is not a teratogen.

Reproduction: The Agency is requiring that additional
information be submitted on a rat study that exhibited
a NOEL of 50 mg/kg.

Mutagenicity:

Not a mutagen in a gene mutation, an unscheduled DNA
synthesis or in a chromosomal aberration study in rats.
However, a chromosomal aberration study with the Chinese
hamster showed a positive response. The Agency has
determined that hexazinone is not a mutagenic agent.

Metabolism:

The Agency has reviewed one acceptable rat metabolism
study. C^{14} was excreted as an average of 97% of the
total dosed radioactivity via the urine (ca. 77%) and
feces (ca. 20%) during the collection period. The results
were comparable for each treatment regimen. Very low
levels of radioactivity were detected in the GI tract,
hide, excised organs, muscle, blood, and fat. Hexazinone
was metabolized primarily by hydroxylation and demethyl-
ation resulting in eight major metabolites. No additional
data are required.

Major Routes of Exposure:

The major routes of exposure are dermal and occular
during mixing, loading, and application.

Physiological and Behavioral Characteristics:

Foliar Absorption: Hexazinone is absorbed through the
     roots and/or leaves depending upon the type of
     formulation and method of application.

Translocation: Following root absorption, hexazinone
     translocates upward through the xylem.

Mechanism of Pesticide Action: Hexazinone acts as
     photosynthesis inhibitor.

Environmental Characteristics:

Persistence in Water: Hexazinone is persistent in water at pH 5,7, and 9.

Mobility in soil: Hexazinone is mobile in soil.

Bioaccumulation: Hexazinone does not accumulate in fish.

Environmental Fate and Groundwater:

Contamination Concerns: Hexazinone belongs to the triazine family of pesticides. Some of these pesticides have been found in ground water. Because hexazinone has been identified as being persistent in water and mobile in soils, there is concern for groundwater contamination. Data are required to address this concern.

Nontarget Organisms: The Agency is requiring Droplet Spectrum and Spray Drift Evaluation tests because of the phytotoxicity of hexazinone, its aerial method of application, and the potential exposure of off-site plants to the pesticide.

Exposure of Humans:

Hexazinone has not been reported to be associated with any death or hospitalized cases since 1976. The voluntary accident reporting system reported one accidental ingestion.

Technical grade hexazinone is corrosive to the eye and causes irreversible eye damage. Use of protective goggles, face shield, or safety glasses are required for mixers, loaders, and applicators.

Exposure during Reentry Operations: The Agency has not received adequate toxicological or epidemiological evidence to indicate that residues of hexazinone can cause adverse effects on persons entering treated sites. No reentry data are required and no reentry interval has been imposed.

Ecological Characteristics: LD = lethal dose
                                 LC = lethal concentration
                                 EC = effect concentration

Avian Toxicity: Acceptable data indicate that technical hexazinone is practically nontoxic to birds.

Acute oral (Bobwhite) LD_{50} = 2258 mg/kg
Dietary Toxicity (Mallard) LC_{50} = >10,000 ppm
              (Bobwhite) LC_{50} = >5,000 ppm

Fish Toxicity: Acceptable data indicate that technical
hexazinone is practically nontoxic to fish.

Fish Acute (Rainbow trout) LC_{50} = >320 ppm
(Bluegill sunfish) LC_{50} = >370 ppm
(Fathead minnow) LC_{50} = >274 ppm
(Bluegill sunfish) LC_{50} = >505 ppm

Freshwater Invertebrate Toxicity: Acceptable data indicate
that technical hexazinone is practically nontoxic to
freshwater invertebrates.

Daphnia magna EC_{50} = 145.3 ppm

Estuarine and Marine Organisms Toxicity: Acceptable data
indicate that hexazinone is practically nontoxic to
molluscs and slightly toxic to crustaceans.

Oyster 48-hr EC_{50} = >320 pm
Shrimp 96-hr LC_{50} = 78 ppm
Crab   96-hr LC_{50} = >1000 ppm

Nontarget Insects: There is insufficient information
to determine that use patterns of hexazinone are
nontoxic to honeybees. Therefore a honey bee acute
study is required.

Potential Problems Related to Endangered Species: Because
of the aerial use pattern of hexazinone on forests
and/or rangelands, there is a threat to endangered plant
species around these use sites.

Tolerance Assessment:

Tolerances Established: Tolerance regulations have
been established for residues of hexazinone and
its metabolites in a variety of commodities
(refer to 40 CFR 180.396).

Results of Tolerance Assessment: The nature of the
residue in plants is adequately understood. However,
the metabolism of hexazinone in animals is not.
Studies characterizing the total terminal residue
of hexazinone in ruminants and poultry are required.
Storage stability data and analysis of hexazinone
residues on certain crops must be submitted.
Additional residue analytical data are required.
Processing data are needed for certain commodities.
PHI's must also be established for some commodities.
The adequacy of the established tolerances will be
reassessed when the required data is reviewed.

SUMMARY OF REGULATORY POSITION AND RATIONALE

Hexazinone does not meet any of the criteria specified
in 40 CFR 154.7; therefore a Special Review is not being
initiated at this time.

The Agency will not require restricted use classification
for hexazinone end-use products.

The Agency is not classifying hexazinone as an oncogen
pending rereview of certain data from previously submitted
studies.

Additional data are needed to thoroughly evaluate the
potential of hexazinone to contaminate groundwater.

The Agency is requiring that a tolerance for pasture/
rangeland hay be proposed so that the impractical
restriction against the cutting of hay from these sites
be removed from the label.

The Agency is requiring that the tolerance for alfalfa
hay be revised so that the impractical feeding restriction
of alfalfa hay be removed from the label.

EPA is developing a program to reduce or eliminate
exposure to endangered plant species from the use of
hexazinone to a point where use does not result in
jeopardy and will issue notice of any labeling revisions
when the program is developed. Endangered species
labeling is not required at this time.

While the data gaps are being filled, currently registered
manufacturing-use products and end-use products containing
hexazinone as the sole active ingredient may be sold,
distributed, formulated, and used in the United States,
subject to the terms and conditions specified in this
Standard. Registrants must provide or agree to develop
additional data, required in the Registration Standard.

STATEMENTS REQUIRED ON LABELS:

The following pesticide disposal statement must appear
on hexazinone manfacturing-use products:

"Do not discharge effluent containing this product into
lakes, streams, ponds, estuaries, oceans, or public
waters unless this product is specifically identified
and addressed in an NPDES permit. Do not discharge

effluent containing this product to sewer systems without
previously notifying the sewage treatment plant authority.
For guidance, contact your State Water Board or Regional
Office of the EPA."

The Following Must Appear on End-Use Products:

In the Precautionary Statements:

"Corrosive, causes irreversible eye damage.   Harmful if
swallowed.   Do not get in eyes or clothing.   Mixers,
loaders, and applicators must wear goggles, face shield,
or safety glasses. Wash thoroughly with soap and water
after handling.   Remove contaminated clothing and wash
before reuse."

Environmental Hazard Statement:

"Do not apply directly to water or wetlands (swamps,
bogs, marshes, and potholes).   Do not contaminate water
when disposing of equipment washwaters."

In the Directions for Use Section:

"Do not enter or allow entry into treated areas until
sprays have dried to perform hand tasks.   A person may
enter the areas to perform other tasks only if the
person is wearing the personal protective eye equipment
listed on the label."

As appropriate, the following grazing statements should
appear on the label:

For sugarcane: "Do not feed sugarcane forage to livestock."

For conifer release and forest plantings (reforestation
site preparation):   "Do not graze domestic animals on
treated areas within 30 days after treatment."

SUMMARY OF MAJOR DATA GAPS.   The following data are required for hexazinone.   Specific requirements are detailed in the Data Tables, Appendix I of the Hexazinone Registration Standard, which can be obtained from National Technical Information Service (NTIS) in Springfield, Virginia.

Study	Due Date – From Date of Standard
Product Chemistry	6 – 15 Months
Residue Chemistry	18 – 24 Months
Metabolism Studies	
Analytical methodology	
Magnitude of the Residue	
Storage Stability	
Toxicology	9 – 50 Months
Acute Oral	
Dermal Sensitization	
Subchronic 21 day Dermal	
Chronic Toxicity	
Ecological Effects	9 – 18 Months
Avian Reproduction	
Freshwater Fish LC_{50}	
Fish Early Life Stage	
Freshwater Invertebrate Acute EC_{50}	
Aquatic Invertebrate Life-Cycle	
Honey bee Acute Contact LD_{50}	
Environmental Fate	9 – 50 Months
Photodegradation	
Metabolism Studies	
Mobility Studies	
Soil Dissipation Studies	
Accumulation Studies (Irrigated crop and Rotational Crops)	
Spray Drift	
Groundwater Studies	

CONTACT PERSON AT EPA:

Mr. Richard Mountfort
Product Manager (Team 23)
Fungicide-Herbicide Branch
Registration Division (TS-767C)
Office of Pesticide Programs, EPA
Washington, DC 20460

Telephone: (703) 557-1830

DISCLAIMER: The information presented in this Pesticide Fact Sheet is for informational purposes only and may not be used to fulfill data requirements for pesticide registration and reregistration.

HEXYTHIAZOX

Reason for Issuance: New Chemical—First Food Use
Date Issued: April 13, 1989
Fact Sheet Number: 200

1. Description of Chemical

   Generic Name: Trans-5-(4-chlorophenyl)-4-methyl-2-oxo-3-thiazolidine-
   carboxamide

   Common Name: Hexythiazox

   Trade Name: Savey®

   Other Proposed Names: N/A

   Code Number: DPX-Y5893, NA-73

   EPA Shaughnessy Code: 128849

   Chemical Abstracts Service (CAS) Number: 78587-05-0

   Year of Initial Registration: 1989

   Pesticide Type: Acaricide

   U.S. and Foreign Producers: E.I. du Pont de Nemours & Company, Inc.

2. Use Patterns and Formulations

   Application Sites: Pears (foliar)

   Types and Methods of Application: Ground Application: Use sufficient
   water, 150 to 800 gallons per acre (gal/A) for dilute application,
   50 to 150 gal/A for concentrate applications.

   Application Rates: Apply the lower rates (4.0 oz/A) formulated
   product (form) on low mite egg infestation levels and the high
   rates (6.0 oz form/A) on moderate to high mite egg infestation
   levels or to larger trees. Apply only one application per growing
   season.

Type of Formulation: 50% wettable powder.

Limitations:

   o   Use only in commercial plantings; do not use in home plantings.

   o   Re-entry Statement - Do not treat areas while unprotected humans
       or domestic animals are present in the treatment areas.

       Do not allow re-entry into treated areas without protective
       clothing until sprays have dried.

   o   Do not graze or feed livestock on cover crops growing in
       treated areas.

   o   Do not apply more than a total of 6 oz. form/A per growing season

   o   Do not apply within 28 days of harvest.

   o   Do not apply more than once per growing season.

   o   Do not apply this product through any type of irrigation system.

3.   SCIENCE FINDINGS

Summary Science Statement:

The EPA Peer Review Committee completed its evaluation of hexythiazox
with respect to its oncogenic potential and concluded that the data
available for hexythiazox provide limited evidence of oncogenicity
for the chemical in mice.   According to EPA Guidelines for Carcinogen
Risk Assessment (Federal Register September 24, 1986) the Committee
classified hexythiazox as a Category C oncogen (possible human
carcinogen with limited evidence of carcinogenicity in animals).

The decision supporting a Category C classification (rather than a
Category B classification) was based primarily on the fact that
only one species was affected (mouse), mutagenicity assays did not
support upgrading to a B classification and the structure-activity
relationship of hexythiazox to other compounds supported a C
classification.

In classifying hexythiazox as a Category C oncogen, the Agency
concluded that a quantitative estimation of the oncogenic potential
for humans should be calculated because of the increased incidence
of malignant and combined benign/malignant liver tumors in the
female mouse.   Thus, a Q_1^* of 3.9×10^{-2} (mg/kg/day)$^{-1}$ in human
equivalents has been calculated.   Dietary oncogenic risk to the
general population based on the highly conservative assumption

that all pears are treated with hexythiazox and would bear residues at the proposed tolerance level is estimated to be 10^{-6}. Non-dietary oncogenic risk to the mixer/loader, applicator based on a dermal absorption factor of 2%, use of protective clothing and one application per year is estimated to be 10^{-6}.

Technical hexythiazox exhibits low mammalian acute toxicity. The results of the technical acute toxicity data show a very mild eye irritant. It is not a sensitizer, nor considered to be mutagenic, nor teratogenic. Hexythiazox is readily absorbed by mammals, and the majority of the residue is largely excreted in the feces and urine by 24 hours. The results of the acute toxicity on the end-use formulation (50% WP) indicates that it is of low toxicity (Toxicity Category III and IV).

Sufficient data are available to characterize hexythiazox for pear use from an environmental and ecological effects standpoint. The results of acute testing indicates that hexythiazox is practically non-toxic to birds on both an acute oral and dietary basis. Hexy-thiazox is moderately to highly toxic to aquatic invertebrates. Hexythiazox is highly toxic to fish. Although technical hexythiazox is toxic to aquatic biota, the application rates and physical/chemical properties of the end-use product minimize potential adverse effects for the pear use. Hexythiazox is relatively non-toxic to non-target insects such as honeybees.

No effects to endangered/threatened species are expected, as the trigger for endangered species concern has not been exceeded (the estimated environmental concentration (EEC) is less than 1/20 the LC_{50}).

Adequate data are sufficient to define the fate of hexythiazox in the environment. Hexythiazox is very stable to hydrolysis, with an estimated T 1/2 exceeding 50 days at environmentally expected temperatures and pH values. Hexythiazox undergoes slow photolytic degradation under sunlight, with T 1/2 of 16.6 days in water and 116 days in soil. Hexythiazox degrades in soil under laboratory conditions. The T 1/2 in aerobic soils ranged from 17 to 35 days. Hexythiazox and its soil aged residues do not leach significantly in soil due to its low solubility in water, high soil adsorption characteristics, and slight vertical movement in soils tested. Hexythiazox is not likely to persist in the field, with T 1/2 ranging from 5 to 15 weeks. No crop rotation study is required for this orchard crop use. Both accumulation and depuration of hexythiazox will occur in bluegill sunfish. Bioconcentration factors of approximately 1300X were calculated in the bluegill sunfish flowthrough study. During 28 days of depuration, 97% of the radiolabeled material was eliminated.

Chemical/Physical Characteristics of the Technical Grade Product

Physical State: Crystalline solid
Color: Pale yellow
Odor: Odorless
Melting Point: 105 to 107.5 °C
            108 to 108.5 °C for analytical grade
Vapor Pressure: 2.54×10^{-8} mmHg (20 °C)
Molecular Weight: 352.5
Solubility: Chloroform   137.9 (g/100 mL)
            Acetone   16.0
            n-Hexane   0.39

            Methanol   2.06
            Xylene   36.2
            Acetonitrile   2.86
            Water   0.5 ppm
Specific Gravity: $d_4 20$ 1.289
Bulk Density: 0.50 to 0.70 g/mL
Octanol/Water Partition Coefficient: 340
pH: Stability to Hydrolysis
    (t 1/2 of 0.25 ppm at 22 °C)   pH 5 $> 7 \times 10^4$ (hours)
                                      7   $> 7 \times 10^4$
                                      9   1.21×10^4
Stability to Temperature: Stable after 3 months at 50 °C
Storage Stability: 100% active ingredient stable at room
            temperature and at 50 °C for 180 days

Toxicology Characteristics of the Technical Grade:

o Acute Oral Toxicity – Rat: $LD_{50} > 5000$ mg/kg
      Toxicity Category IV

o Acute Dermal Toxicity – Rat: $LD_{50} > 5000$ mg/kg
      Toxicity Category III

o Acute Inhalation $LC_{50} > 2.0$ mg/L
      Toxicity Category III

o Primary Dermal Irritation – Rabbit: Not a primary skin
      irritant. Toxicity Category IV

o Primary Eye Irritation: Very mild eye irritant.
      Toxicity Category III

o Dermal Sensitization: Non-sensitizer

o 1-Year Feeding - Dog: NOEL = 100 ppm (2.5 mg/kg/day)

o 2-Year Feeding/Oncogenicity - Rat: NOEL (Systemic) = 430
   (21.5 mg/kg/day); oncogenicity-negative at 3000 ppm (HDT)

o 2-Year Feeding/Oncogenicity - Mouse: NOEL (Systemic) = 250
   ppm (37.5 mg/kg/day); Oncogenic in female mouse liver at
   1500 ppm (225 mg/kg/day) HDT

o Reproduction (2 generation) - Rat: Reproductive NOEL > 2400
   ppm (120 mg/kg/day); Maternal NOEL = 400 ppm (20 mg/kg/day)

o Teratology - Rabbit: NOEL > 1080 mg/kg/day for developmental
   toxicity (HDT). Maternal toxicity NOEL > 1080 mg/kg/day
   (HDT)

o Teratology - Rat: Maternal NOEL = 240 mg/kg/day; Fetotoxic
   NOEL = 240 mg/kg/day; Teratogenic NOEL > 2160 mg/kg/day
   (HDT)

o Mutagenicity: Negative in a battery of mutagenicity
        studies.

End-Use Formulation

The stated results for the following acute studies are for the 50
percent wettable powder formulation: Oral (rat), dermal (rat),
inhalation (rat), primary dermal irritation (rabbit), and primary
eye irritation (rabbit) and dermal sensitization (guinea pig).

o Acute Oral - Rat: LD_{50} > 5000 mg/kg (male (M) and female (F)
      Toxicity Category III

o Acute Dermal Toxicity - Rat: LD_{50} > 5000 mg/kg (M&F)
      Toxicity Category III

o Acute Inhalation - Rat: LC_{50} > 2.8 mg/L (M&F)
      Toxicity Category III

o Primary Dermal Irritation - Rabbit: Negative (N)
      Toxicity Category IV

o Primary Eye Irritation - Rabbit: Reddened conjunctivae,
      maximum score of 2, maximum duration 6 days. Chemosis
      ended by day 3 (M); Toxicity Category III

o Dermal Sensitization - Guinea Pig: Not a sensitizer

Physiological and Biochemical Characteristics:

Foliar Absorption: N/A

Translocation: Not translocated.

Mechanism of Pesticide Action: Neurotoxicity Characteristic –
   Controls mites through ovicidal/chemosterilant activity when
   spray mist comes in contact with mite eggs or female mites.

Environmental Characteristics:

The environmental fate data indicate that hexythiazox and its
   soil-aged residues did not have significant vertical mobility
   and thus are not likely to leach and contaminate ground water.
   Field data also indicate that hexythiazox dissipates with half-
   life of 5 to 26 weeks. Hexythiazox underwent slower photolytic
   degradation on soil than in aqueous solution with a half-life
   of 116 days. Hexythiazox did not undergo any noticeable
   hydrolysis under acidic to neutral conditions at 22 °C. At pH
   9 at 22 °C, hexythiazox hydrolyzed very slowly with an estimated
   half-life of 416 days.

^{14}C-DPX-Y5893 (labeled at C-5 of thiazole moiety) was bioaccumulated
in bluegill sunfish under flow-through conditions with a biological
concentration factor (BCF) range of 1000 to 1600 at peak on the
basis of whole fish. The highest accumulation occurred in viscera
with a BCF range of 1.3 to 1.7×10^4. After 14 days of depuration,
about 97 percent of the accumulated radioactivity was removed.
Residue analysis of the 28-day fish samples showed that about 52
to 88 percent of the ^{14}C-residue was present as polar material(s),
5 to 23 percent as parent DPXY5893, 2 to 15 percent as cyclohexane-
hydroxylated metabolites of parent, 4 to 7 percent as conjugated
material, and 2 to 4 percent as tissue bound residues.

Ecological Characteristics:

Technical Formulation

o Avian Oral Toxicity: > 2510 mg/kg (mallard duck LD_{50}).

o Avian Dietary Toxicity (8 days): > 5620 ppm (bobwhite quail
     LC_{50}) and > 5620 ppm (mallard duck LC_{50}).

o Freshwater Fish Acute Toxicity: (96-hr LC_{50}: 0.53 mg/L
     (bluegill) and > 1 mg/L (rainbow trout).

o Freshwater Invertebrate Acute Toxicity (48-hr LC_{50} – Grade:
     1.22 mg/L (Daphnia crinata); (48-hr EC_{50}): 0.74 mg/L.

o   Invertebrate Life Cycle: NOEL 0.5 mg/L (Daphnia magna).

o   Honeybee LC_{50}: >1000 ppm

Tolerance Assessment

A Section 408 tolerance under the Federal Food Drug and Cosmetic Act
has been established for residues of hexythiazox in/on the following
raw agricultural commodity (40 CFR 180.448)

Commodity	Part Per Million
Pears	0.3

The acceptable daily intake (ADI), based on a NOEL of 2.5 mg/kg/day
from a 1-year dog feeding study and a safety factor of 100 is
0.025 mg/kg/body weight/day. The TMRC from the proposed tolerance
is 0.000037 mg/kg body weight/day. This is equivalent to about
7.4 percent of the ADI

The nature of the residue in pears (pome fruit) is adequately defined.
The residue of concern is the parent and its hydroxylated
cyclohexane ring metabolites.

There are no animal feed items with pear orchard use therefore the
nature of the residue in animals is not relevant. Since there
are no feed items involved with pears and the label includes the
restriction "Do not graze or feed livestock or cover crops growing
in treated areas", no secondary residues (meat, milk) are anticipated
from this proposed use.

No processing data have been submitted however none are required
since residue levels in pear juice and nectar will not exceed the
tolerance level on the raw agricultural commodity pears.

There are no Canadian or Mexican tolerances and no Codex Maximum
Residue Limits (MRLS) have been established for hexythiazox and its
metabolites in/on pears. Therefore, no compatibility problem exists.

Reported Pesticide Incidents: None

4. Summary of Regulatory Position and Rationale

A full review of the data indicates that although hexythiazox is an
oncogen in mice the dietary and nondietary risks would be extremely
small from the proposed use on pears. Estimated dietary oncogenic
risk to the general population based on the highly conservative
assumption that all pears are treated with Savey and would bear
residues at the proposed tolerance level is estimated to be 10^{-6}.

The Agency believes that actual exposure and risk would be lower.
The basis for this is that the risk of 10^{-6} reflects a worst-case
dietary exposure because it assumes that 100 percent of the United
States pear crop is treated with Savey and that all quantities of
the food consumed will bear residue levels as high as the proposed
tolerance. In reality, the Agency knows that all pears would not
be treated with this pesticide. Based upon an analysis of the
market penetration of currently registered acaricides, the Agency
expects the percent of crop treated with Savey in a typical year
would be about 30 percent. Likewise, the Agency believes that
residue levels in pear juice and nectar will not exceed the
established tolerance of 0.30 ppm in or on the RAC pears, since
the maximum residue level in pear juice is less than 50 percent
of the residue level in whole fruit. In addition, since there
are no animal feed items involved with pears and the petitioner
has included the label restriction "Do not graze or feed livestock
or cover crops growing in treated areas," no secondary residues
in meat or milk are expected.

Estimated non-dietary oncogenic risk to the mixer/loader applicator
based on a dermal absorption factor of 2%, the use of protective
clothing and 1 application of per year is 10^{-6}. The Agency believes
that this estimate is an overestimation of the lifetime cancer risk
and that actual exposure and risk would be much lower since the
surrogate data base for calculating exposure reflected application
rates of 1 to 7 ai/A whereas the proposed use on pears is for 0.2
lb ai/A. An estimated risk of less than 1×10^{-5} is considered to
be an acceptable risk relative to mixer, loaders and applicators.

Thus, based on the above risk assessment the Agency has characterized
the risk posed to the general public and to pesticide applicators
from the proposed use of Savey as extremely small.

The Agency has determined, based on the available data and use pattern,
that endangered/threatened species would not be adversely affected.

Hexythiazox is not likely to leach and contaminate ground water.

The Agency has reviewed all relevant data and has determined that no
additional data are necessary to make the determination required
by FIFRA sec 3(c)(5). Thus, the Agency is approving this registration under FIFRA sec 3(c)(5).

The Agency has determined that the product will perform its intended
function without unreasonable adverse effects on the environment,
and that when used in accordance with the label directions, the
product will not generally cause unreasonable adverse effects
on the environment.

The Agency has determined that all necessary tolerances have been issued under FIFRA sec. 408.

5. Summary of Data Gaps

    None

6. Contact Person at EPA

    George T. LaRocca
    Product Manager (15)
    Insecticide-Rodenticide Branch
    Registration Division (H7504C)
    Office of Pesticide Programs
    Environmental Protection Agency
    401 M Street S.W.
    Washington, DC   20460

    Office location and telephone number:
    Room 204, Crystal Mall #2
    1921 Jefferson Davis Highway
    Arlington, VA   22202
    Phone: (703) 557-2400

DISCLAIMER: The information presented in this Pesticide Fact Sheet is for informational purposes only and may not be used to fulfill data requirements for pesticide registration and reregistration.

IMAZETHAPYR

Reason for Issuance: New Chemical Registration
Date Issued: March 8, 1989
Fact Sheet Number: 196

1. Description of Chemical

Common Name: Imazethapyr
Chemical Name: (+)-2-[4,5-dihydro-4-methyl-4-(1-methylethyl)-5-oxo-
1H-imidazol-2-yl]-5-ethyl-3-pyridinecarboxylic acid
Trade Name: Pursuit®
OPP (Shaughnessy) No.: 128982
Chemical Abstracts Services (CAS) Number:
Empirical Formula: $C_{15}H_{22}N_4O_3$
Molecular Weight: 306.4
Year of Initial Registration: 1989
Pesticide Type: Herbicide
U.S. Producer: American Cyanamid Company

2. Use Patterns and Formulations

Application Sites: Terrestrial food crops

Major Crops Treated: Soybeans

Types and Methods of Application: Foliar or soil, applied broadcast
by ground equipment for control of broadleaf weeds and grasses.
Applied early preplant, preplant incorporated, preemergence, or
early postemergence.

Application Rate: 0.0625 pounds active ingredient/acre (lb ai/A) or
4 ounces product/A.

Type of Formulations: Emulsifiable concentrate.

Usual Carriers: Water with crop oil or a nonionic surfactant.

3. Science Findings

Summary Science Statement: All data are acceptable. Imazethapyr has
low acute toxicity (Category III) for acute dermal, primary eye
irritation, and acute inhalation and is less toxic (Category IV)

for acute oral toxicity and primary dermal irritation.   Imazethapyr
is not considered a skin sensitizer.   Data does not show imazethapyr
to be oncogenic in rats or mice.   The chemical is not teratogenic
in rats or rabbits, and did not produce any reproductive effects
in rats.   Imazethapyr is not considered mutagenic except at levels
toxic to cells.

Imazethapyr is practically nontoxic to birds, fish, aquatic
vertebrates, and honey bees.

Imazethapyr is persistent regardless of soil type, agricultural
practices, and climatic effects but does not appear to leach under
normal soybean agricultural practices.

The nature of the residue in plants and animals is
adequately understood and adequate methodology is available for
enforcement of the tolerance in soybeans.

Chemical Characteristics:

Technical

Physical State: Solid
Color: Off-white to tan
Odor: Pungent
Boiling Point: N/A
Density: 0.396 g/mL untapped, 0.459 g/mL tapped
Solubility:

Solvent	Solubility at 25 °C (g/100 mL Solvent)
Acetone	4.82
Dimethyl sulfoxide	42.25
Heptane	0.09
Methanol	10.50
Methylene chloride	18.48
2-Propanol	1.73
Toluene	0.50
Water (distilled)	0.14

Vapor Pressure: (pure form) $< 1 \times 10^{-7}$ mm Hg at 60 °C
Dissociation Constant: pka = 3.9
pH: 2.85 at 25 °C

Toxicology Characteristics:

Acute Toxicology - Technical

   o Acute Oral Toxicity - Rat: > 5000 mg/kg/day (male and females)
            Toxicity Category IV

o Acute Dermal Toxicity - Rabbit: > 5000 mg/kg/day (males and
     females) Toxicity Category III

o Primary Dermal Irritation - Rabbit: Imazethapyr is minimally
     irritating. Toxicity Category IV

o Primary Eye Irritation: Imazethapyr is practically nonirritating.
     Toxicity Category III

o Dermal Sensitization - Guinea Pig: Imazethapyr is not a
     dermal sensitizer

o Acute Inhalation: > 3.27 mg/L air (analytical) or 4.21 mg/L
     air (gravimetric). Toxicity Category III

Acute Toxicology - Pursuit Herbicide (22.6% Formulation)

o Acute Oral Toxicity - Rat: > 5000 mg/kg (males and females)
     Toxicity Category IV

o Acute Dermal Toxicity - Rabbit: > 5000 mg/kg (males and
     females) Toxicity Category IV

o Primary Eye Irritation - Rabbit: minimally irritating
     Toxicity Category IV

o Primary Dermal Irritation: Minimally irritating
     Toxicity Category IV

Acute Toxicology - Pursuit Plus (2.39% Formulation)

o Acute Oral Toxicity - Rat: > 5000 mg/kg (males and females)
     Toxicity Category IV

o Acute Dermal Toxicity - Rats: > 2000 mg/kg (males and females)
     Toxicity Category III

o Acute Inhalation Toxicity - LC_{50} (males and females) 1.29 mg/
     imazethapyr and 1.27 mg/pendimethalin. Toxicity Category III

o Primary Eye Irritation - Rabbits: Slight irritation at 24 hours
     Toxicity Category III

o Primary Dermal Irritation - Rabbits: No erythema or edema
     observed. Toxicity Category IV

o Dermal Sensitization - Guinea Pig: not a sensitizer

Subchronic Toxicity

Data are available to satisfy the requirements for sub chronic feeding studies. These data are discussed below.

A 90-day rat subchronic feeding study conducted at dose levels of 0, 1000, 5000, and 10,000 parts per million (ppm) (0, 50, 250, and 500 milligrams/kilogram [mg/kg]/day) resulted in a no-observable effect level (NOEL) of 500 mg/kg/day (highest does tested [HDT]).

A 90-day dog subchronic feeding study conducted at dosages of 0, 1000, 5000, and 10,000 ppm (0, 25, 125, and 250 mg/kg/day) resulted in a NOEL of 250 mg/kg/day (HDT).

A 21-day rabbit dermal study conducted at dosages of 0, 250, 500, and 1000 mg/kg/day resulted in a NOEL of 1000 mg/kg/day (HDT).

Chronic Feeding and Oncogenicity Studies

Data are available to satisfy the requirements for chronic feeding studies and oncogenicity studies in two species. These data are discussed below.

A dog chronic feeding study conducted at dosages of 0, 1000, 5000, and 10,000 ppm (0, 25, 125, and 250 mg/kg/day) resulted in a NOEL of 250 mg/kg/day (HDT) in males and a NOEL for females of 25 mg/kg/day. The lowest-observed-effect level (LOEL) of 125 mg/kg for females was based on decreased packed cell volume hemoglobin and erythrocytes at 250 mg/kg/day.

A 2-year rat chronic feeding/oncogenicity study conducted at dosages of 0, 1000, 5000, and 10,000 ppm (0, 50, 200, and 500 mg/kg) with no oncogenic effects observed up to 500 mg/kg/day (HDT). This study is acceptable as a chronic feeding study with a NOEL of 500 mg/kg/day (HDT) but is classified as Supplemental for an oncogenic study because the maximum tolerated dose (MTD) was not reached (no significant toxic effects at the HDT). A repeat of this study is not required at this time because 1) 500 mg/kg (10,000 ppm) is within 50 percent of the limit dose for an adequately conducted oncogenicity study on a chemical of low toxicity; 2) this pesticide is closely related to two other pesticides ("Scepter" and "Assert") which tested negatively in oncogenicity testing; and 3) there were no positive mutagenicity studies for any of the three pesticides except for "Pursuit," which had a positive result in the in vitro cytogenetics study, but only at levels that were toxic to the cells.

A 78-week mouse oncogenicity feeding study conducted at dosages of 0, 1000, 5000, and 10,000 ppm (0, 50, 750, and 1500 mg/kg/day) produced no oncogenic effects up to 1500 mg/kg/day (HDT). The systemic NOEL was 750 mg/kg/day based on decreased body weight gains in both sexes.

Justification that the highest dose level tested in the rat chronic feeding/oncogenicity study (1500 mg/kg) is sufficient or a new study with an MTD will be required for before additional crop tolerances or registrations can be issued.

Teratogenicity and Reproduction

Data are available to satisfy the requirements for a 2-generation reproduction study and teratology studies in two species. These studies are discussed below.

A rat teratology study conducted at dosages of 0, 125, 375, or 1125 mg/kg/day demonstrated an NOEL for developmental toxicity of 1125 mg/kg/day (HDT). The NOEL for maternal toxicity was 375 mg/kg/day and the LOEl was 1125 mg/kg/day (HDT).

A rabbit teratology study was calculated at dosages of 0, 100, 300, or 1000 mg/kg/day showed no teratogenic effects observed. The NOEL for maternal toxicity was 300 mg/kg/day and the NOEL for developmental (embryo/fetotoxicity was 1000 mg/kg/day (HDT)).

A 2-generation rat reproduction study conducted at dosages of 0, 1000, 5000, and 10,000 ppm (0, 50, 200, and 500 mg/kg). The systemic and reproductive NOEL was 500 mg/kg/day. The study is classified as supplementary because the dose levels were not high enough. A repeat study is not required because it would not likely provide any useful information since the chronic dog study would still be the preferred study.

Mutagenicity

Acceptable data are available for imazethapyr to satisfy the mutagenicity data requirements. These data are discussed below.

Data available to evaluate the potential of imazethapyr to induce gene mutations include reverse mutation assays in S. typhimurium and E. coli, both plate and disc tests. Imazethapyr did not induce gene mutation in either system up to 5000 micrograms (ug)/plate or 1000 ug per disc. Imazethapyr was also tested in a Chinese Hamster Ovary (CHO) cell gene mutation assay (HGPRT locus) up to the limit of solubility with activation (3333 ug/mL) and beyond the limit of solubility without activation (4000 ug/mL). Imazethapyr did not induce gene mutation in this system.

Imazethapyr did not result in a significant increase in chromosomal aberrations in rat bone marrow when tested in an acute in vivo cytogenics assay at dose levels of 0.25, 0.8, and 2.5 grams (g)/kg body weight. When tested in an in vitro cytogenics assay in CHO cells both with and without metabolic activation at dosage levels of 1.14, 1.71, 1.82, 2.05, and 2.25 mg/mL imazethapyr increased chromosomal aberrations without metabolic activation at dosage levels toxic to cells and did not increase chromosomal aberrations with metabolic activation. Imazethapyr was also tested in a dominant lethal rat study at dose levels of 0, 200, 1000, or 2000 mg active ingredient (ai)/kg and did not induce dominant lethal mutations. This study is Supplemental because it was questionable whether or not dose levels were high enough for an adequate negative study. Another study is not required because the other studies satisfy this requirement.

Data available to evaluate the potential of imazethapyr to induce DNA damage includes an in vitro unscheduled DNA synthesis study in primary rat hepatocytes tested at dose levels ranging from 13 to 1333 ug/well. Under the conditions of this test, imazethapyr did not induce unscheduled DNA synthesis in rat hepatocytes.

Metabolism

Two rat metabolism studies were conducted. These studies do not completely characterize the absorption, excretion, retention, and metabolism of imazethapyr. However, a repeat study is not necessary for this use because currently available data give sufficient evidence that most of the chemical will be eliminated via the urine unchanged.

Physiological and Biochemical Characteristics

Translocation and absorption by plants: Absorption through both roots and foliage and translocated rapidly to growing points.

Metabolism/persistence in animals: Eliminated through the urine unchanged.

Environmental Characteristics

Leaching and Adsorption/Desorption: Laboratory studies show imazethapyr to be very mobile in two sand loam and two silt loam soils. The percentage of organic matter and sand do not appear to have an effect on the sorption of imazethapyr.

Field dissipation tests were conducted using preplant incorporation, pre- and postemergence applications. Fields were planted to soybeans. Irrigation was not used because irrigation is not a normal agricultural practice for soybeans. In Kentucky silt loam, imazethapyr was detected at the 0- to 3-inch depth in all methods of application. Half-lives of 31 days for preplant incorporated, 20 days for preemergence, and 17 days for postemergence were calculated. At one sampling for both preemergence and preplant incorporated methods of application, imazethapyr was detected at the 3- to 6-inch depth. Imazethapyr was not detected at other sampling intervals in 3- to 6-inch depths or at 6- to 9- and 9- to 12-inch sampling depths. These plots did not indicate leaching.

In an Illinois silt loam soil, residues of imazethapyr were detected in the 0- to 3-inch depth for all types of application. Concentrations of imazethapyr detected between 30 and 60 days at the 3- to 6-inch depth and persisted until the end of the study when imazethapyr was applied preplant incorporated. The half-life was calculated to be 287 days. When applied pre- and postemergence, imazethapyr was not detected below the 0- to 3-inch depth. The half-life was calculated to be 120 days.

In coarse, sandy soils found in some soils in Georgia and Iowa, movement to 12 inches was noted. Further soil dissipation studies will be needed for future uses with agricultural practices different from soybeans and in soil types similar to soils found in Georgia and Iowa.

Microbial breakdown: Imazethapyr degrades slowly under aerobic (half-life of 33 to 37 months) conditions and very slowly under anaerobic conditions.

Loss from volatilization: None expected because of low vapor pressure.

Degradation from photodecomposition and hydrolysis: Imazethapyr is stable to hydrolysis at pHs 5, 7, 9 in buffered or pond water. Imazethapyr is resistant to degradation in soil when exposed to artificially simulated sunlight. Imazethapyr degrades when exposed to artificially simulated sunlight with a half-life of 46 hours.

Bioaccumulation fish: Does not bioaccumulate in fish.

Potential to contaminate groundwater: Imazethapyr is persistent and mobile as indicated from laboratory studies. In the field, imazethapyr consistently shows persistence, regardless of soil type, agricultural practices, and climatic effects. In the field, imazethapyr does not leach to the same extent in all soil types

and under all agricultural practices. Based on field dissipation
studies performed on soils planted to soybeans under normal
agricultural practices (no irrigation) imazethapyr does not appear
to leach. Therefore, the potential to contaminate groundwater
from use on soybeans is low.

Exposure of humans to pesticides and reentry: Applicator exposure
assessment or reentry exposure are not required because lack of
significant chronic concerns and low acute toxicity (Category III
and IV) result in low exposure to humans from imazethapyr.

Based on available information, only the minimal reentry intervals
required by law should be imposed at this time: Entry into
treated fields shall not be permitted without protective clothing
until sprays have dried and dusts have settled.

Ecological Characteristics

Acceptable data are available to satisfy the requirements for an
avian single dose acute oral toxicity on one species; two subacute
dietary toxicity studies on one species of waterfowl and one
species of upland gamebird; two 96-hour fish acute toxicity
studies on two species of freshwater fish, preferably one coldwater
species and one warmwater species; a 48-hour acute toxicity study
with freshwater invertebrates; and an acute oral toxicity to honey
bees. Studies that satisfy these requirements are listed below.

o Avian Acute Toxicity: Bobwhite Quail LC_{50} > 2150 mg/kg and
      Mallard Duck LC_{50} > 2150 mg/kg

o Avian Dietary Toxicity: Mallard Duck LC_{50} > 5000 mg/kg and
      Bobwhite Quail LC_{50} > 5000 mg/kg

o Freshwater Fish Acute Toxicity: Channel Catfish LC_{50} 240
      mg/L, Bluegill Sunfish LC_{50} 420 mg/L, and Rainbow Trout
      LC_{50} 340 mg/L

o Freshwater Invertebrate Toxicity: Daphnia magna LC_{50} > 1000
      mg/L

o Acute Contact Toxicity: Honey bee LC_{50} > 1000 ug/bee

Based on the above data, imazethapyr is practically nontoxic to birds
on an acute and dietary basis, practically nontoxic to both
warmwater and coldwater fish, practically nontoxic to aquatic
invertebrates, and practically nontoxic to honey bees.

Tolerance Assessment:

The nature of the residue in plants and animals has been adequately defined for the use on soybeans and adequate analytical methods are available for enforcement purposes.

A tolerance is established for residues of the herbicide imazethapyr, ammonium salt, (+)-2-[4,5-dihydro-4-methyl-4-(1-methylethyl)-5-oxo-1H-imidazol-2-yl]-5-ethyl-3-pyridinecarboxylic acid) in or on the raw agricultural commodity soybeans at 0.1 ppm.

The PADI was calculated to be 0.25 mg/kg/day. This value was based on a NOEL of 250 mg/kg/day from the 90-day dog feeding study. An uncertainty factor of 1000 was used because an MTD was not obtained in the rat chronic/ oncogenicity study.

The theoretical maximum residue contribution (TMRC) from this tolerance is calculated to be 0.000034 mg/kg/body weight/day which occupies approximately 0.014 percent of the PADI. There are no other published tolerances for this chemical.

4.   Summary of Regulatory Position and Rationale

The available data submitted to the Agency provide sufficient information to support registration of the use on soybeans. Therefore, the Agency has accepted the use of imazethapyr on soybeans.

5.   Data Gaps

There are no data gaps for the use of imazethapry on soybeans.

6.   Contact Person at EPA

Robert J. Taylor
Product Manager (25)
Fungicide-Herbicide Branch
Registration Division (H7505C)
Office of Pesticide Programs
Environmental Protection Agency
401 M Street SW.
Washington, DC 20460
Office Location and Telephone Number:
Room 243, Crystal Mall #2
1921 Jefferson Davis Highway
Arlington, VA 22202
(703) 557-1800

DISCLAIMER: The information in this Pesticide Fact Sheet is a summary only and is not to be used to satisfy data requirements for pesticide registration and reregistration. The complete Registration Standard for the pesticide may be obtained from the contact person listed above.

KARATE (PP321)

Reason for Issuance: Conditional Registration
Date Issued: May 16, 1988
Fact Sheet Number: 171

1. DESCRIPTION OF CHEMICAL
   *

    Generic Name: PP321; [1 alpha(S), 3 alpha(Z)]-(+)
                   -cyano-(3-phenoxyphenyl) methyl 3(2-chloro-
                   3,3,3-trifluoro-1-propenyl)-2,2-dimethyl-
                   cyclopropanecarboxylate.
    Common Name: PP321
    Trade Name: KARATE
    EPA Shaughnessy Code: 128867(a)
    Chemical Abstracts Service (CAS) Numbers:
    Year of Initial Registration: 1988
    Pesticide Type: Insecticide
    Chemical Family: Synthetic pyrethroid
    U.S. Producers: ICI Americas, Inc.
                    Wilmington, DE 19897

2. USE PATTERNS AND FORMULATIONS

    Application Sites: Agricultural use in/on Cotton

    Method of Application: Foliar ground and aerial
                           application

    Formulation Types: 13% liquid (EC)

    Application Rates: 0.01-0.03 lb.ai/Acre

    Usual Carriers: Organic solvents; Surfactants

    Limitations: RESTRICTED USE pesticide. Use limited to
    certified applicators or persons under their direct supervision.

3. SCIENCE FINDINGS

    Summary Science Statement:

*PP321 comprises one of two diastereomers (enantiomeric pairs)
of Cyhalothrin. Cyhalothrin consists of 4 cis isomers in
the Z configuration (enantiomeric pair A & B) of which PP321
consists of 2 cis isomers (enantiomeric pair B).

Technical PP321 is a synthetic pyrethroid with
moderate acute toxicity. The results of the acute
toxicity on the end-use formulation indicates that
product is of moderate to high acute toxicity. The
end-use product is extremely irritating to the skin
and is a mild sensitizer. Technical PP321 is not
considered to be mutagenic or teratogenic in test
animals. On the basis of structural considerations
and metabolism and subchronic data on both PP321 and
cyhalothrin, the Agency has accepted the long term
data on cyhalothrin in partial fulfillment of the
chronic toxicity requirements for PP321.

Sufficient data are available to characterize PP321
from an environmental fate and ecological effects standpoint.
The results of acute toxicity studies indicate that PP321 is
extremely toxic to fish and other aquatic organisms, but
is practically non-toxic to waterfowl and upland game birds.
However, reproduction data on mallards exposed to cyhalothrin
domonstrated adverse effects on numbers of eggs laid at
doses of 50 ppm, with a no-effect-level (NOEL) of 5 ppm
cyhalothrin. Since technical PP321 is the more biologically
active component of cyhalothrin, it may be more toxic;
therefore, the Agency has determined that the repro
reproduction study be repeated using technical PP321.
An acute contact LD50 study indicated that PP321 is
highly toxic to bees with an LD50 of 0.909 ug/bee.
Formulated PP321 is also highly toxic to honey bees
with a reported LD50 of 0.098 ug/bee and an oral LD50
of 0.483 ug/bee. PP321 is readily degraded by soil
and is virtually insoluble in water. There is little
or no potential for leaching.

The Agency has determined that the registration
of PP321 may effect endangered aquatic and avian species.
Pending a formal consultation with the Fish and Wildlife
Service to determine use limitations with respect to
these species, the product label contsists of language
which will mitigate the risk to endangered species.

A Tolerance Assessment has been conducted by
the Residue Chemistry Branch to provide a dietary
exposure analysis for the use on cotton. The reference dose (Rfd) used to determine this dietary exposure
is calculated to be 0.005 ppm based on a NOEL of 0.5 mg/
kg/day from a 3-generation rat reproduction study and
a safety factor of 100. The Theoretical Maximum
Residue Contribution (TMRC) for the U. S. population is
imately 2.6 % of the Average Daily Intake (ADI).

A. Chemical/Physical Characteristics of the Technical Material

Physical State: liquid
Color: light yellow
Odor: aromatic solvent odor at room temparature
Melting Point: not applicable
Vapor Pressure: 2.1 x 10-8 mbars
Density: 0.830 + .005 g/ml
Solubility: 0.002 mg/ml at 20°C in water
pH: 5.6
Octanol/Water Partition Coefficient: 4.2 x 10^4

B. Toxicological Characteristics:

(Technical PP321)

Acute Oral (Mouse): males: 79 mg/kg   Toxicity Category II
 (LD50)   females: 56 mg/kg

Acute Dermal: males: 632 mg/kg
 (LD50)   females: 696 mg/kg   Toxicity Category II

Primary Dermal Irritation (rabbit): PIS= 0   Toxicity Category IV
(none observed)
Skin Sensitization (guinea pig): not a sensitizer
Acute Inhalation: LD50 (4 hour) males: 0.315 mg/l   Toxicity
(end-use formulation)   females: 0.175 mg/l Category II

Teratology: (rat) Maternal NOEL = 10 mg/kg/day
(Cyhalothrin)   Maternal LEL = 15 mg/kg/day
                    Fetotoxic NOEL = 15 mg/kg/day
                    Teratogenic NOEL = not teratogenic
          (rabbit) Maternal NOEL = 10mg/kg/day
                    Maternal NOEL = 30mg/kg/day
                    Fetotoxic NOEL = 15 mg/kg/day
                    Teratogenci NOEL = not teratogenic

2-Year Feeding/Oncogenicity: (rat) Oncogenic NOEL = 2.5 mg/kg/day
(Cyhalothrin)   Systemic NOEL = 0.5 mg/kg/day
                           (mouse) Oncogenic NOEL = 15 mg/kg/day

Gene Mutation (Ames): negative
Structural Chromosome Aberration: negative
In Vitro Cytogenetics Assay: negative

PP321 formulation- 13.1%

Acute Oral Toxicity in rats:	Toxicity Category II
LD50 = 64 mq/kq	
Acute Dermal Toxicity in rats:	Toxicity Category III
LD50 = >2000 ml/mg	
Acute Inhalation:	Toxicity Category II
LC50 = 0.315 mg/l(M); 0.175 mg/l(F)	
Primary Eye Irritation:	Toxicity Category II
Primary Dermal Irritation:	Toxicity Category I
PIS = 6.7	
Dermal Sensitiztion: mild sensitizer	

C.   Physiological and Biological Characteristics

The mode of action in biological systems is stomach and
contact exhibiting neuropathological characteristics typical
of pyrethroid insecticides.   Slight repellant effect.
Foliar absorption: N/A
Translocation: N/A

D.   Environmental Characteristics

Adequate data are sufficient to define the fate of PP321
in the environment.   PP321 is stable to photolysis
at environmental pH and temperature.   It photodegrades
rapidly and is practically water insoluble.   Under the
conditions of the soil TLC test using various soils,
(aged and unaged) PP321 residues are considered immobile
in soils with a half-life of <14 days in silt and
28-56 days in clay loam, respectively.   Anaerobic conditions
did not alter either degradation rate or products.   It
has a bioaccumulation factor in fish of 858X.   Residues
are depurated radidly in untreated water.   Accumulated
residues are found in non-edible tissue.   PP321 and its
degradates do not leach into the soil.   There are no
concerns at this time in regard to ground water.

E.   Ecological Characteristics

   Avian Acute Oral: Mallard duck − LD50 > 3950 mg/kg
   Avian Subacute Dietary: Mallard duck − LC50 3948 ppm
                    Bobwhite quail − LC50 > 5000 ppm
   Freshwater Fish: Bluegill − LC50 = 0.21 ug/L
                    Rainbow Trout − LC50 = 0.24 ug/L
   Freshwater Invertebrate: Daphnia magna LC50 = 0.36 ug/L
   Marine/Estuarine Invertebrate: Mysid shrimp − LC50 = 4.9 ng/L
                    Eastern Oyster EC50 = 0.59 ng/L
   Marine/Estuarine Fish: Sheepshead minnow − LC50 = 0.807 ng/L

4.   Summary of Regulatory Position and Rationale

   The Agency has determined that it should allow the
conditional registration of PP321 for agricultural use
to control insects in/on cotton. Adequate data are available
to assess the acute and chronic toxicological effects of PP321
to humans. However since certain long-term fish, aquatic
and avian data are missing and required, the registration
is being conditionally approved with an expiration date
of August 30, 1990 which coincides with the dates for
submission of the data required to satisfy the remaining
data gaps listed below.

   In view of the high toxicity of technical PP321 to
aquatic organisms (invertebrates and fish) and the
potential hazard associated with exposure to this
product, the Agency is concerned about exposure
which may result from improper application or use
and so is restricting use of this pesticide.

   The Agency has determined that endangered species
labeling restrictions are necessary to protect endangered
species and is requiring specific limitations on use of
this product to prevent or mitigate exposure.

5.   Summary of Major Data Gaps

   1.   Avian Reproduction − Mallard (71-4)
        A final report is due in May, 1990.

   2.   Fish Life-Cycle (72-5)
        A final report is due in August, 1990.

   3.   Aquatic Invertebrate Lifecycle (72-4)
        A final report is due in August, 1989.

6.   Contact Person at EPA

George T. LaRocca
Product Manager (15)
Insecticide-Rodenticide Branch
Registration Division (TS-767C)
Office of Pesticide Programs
Environmental Protection Agency
401 M Street, S. W.
Washington, D. C.   20460

Office location and telephone number:
Room 211, Crystal Mall #2
1921 Jefferson Davis Highway
Arlington, VA   22202

(703) 557-2400

LACTIC ACID

Reason for Issuance: New Chemical Registration
Date Issued: April 29, 1988
Fact Sheet Number: 163

1. Description of Chemical

   Common Name: Lactic acid

   Chemical Name: 2-Hydroxypropanoic acid

   Trade Names: Propel, SY-83

   OPP (Shaughnessy) Number: 128929

   Chemical Abstracts Service (CAS) Number: 79-33-4

   Empirical Formula: $C_3H_6O_3$

   Molecular Weight: 90.08

   Year of Initial Registration: 1988

   Pesticide Type: Plant growth regulator

   U.S. Producer: Purac Inc.

2. Use Patterns and Formulations

   Types of Effects: Lactic acid hastens ripening, increases
      fruit, seed and pod set and increases shoot and root
      initiation.

   Application Sites: Almonds, apples, beans (green and dry),
      broccoli, cabbage, cauliflower, cherries, citrus, corn
      (sweet and field), cotton, grapes, lettuce, peppers
      (green and chile), pineapples, prunes, strawberries,
      sugarcane, tomatoes and walnuts.

   Types and Methods of Application: Applied by conventional
      aerial or ground equipment to foliage.

   Application Rates: Rates range from 1 to 4 pounds active
      ingredient (ai)/acre. A 1000-2000 parts per million
      (ppm) solution will be used for pineapple and
      sugarcane planting materials.

369

Type of Formulation: 80% ai solution

Usual Carriers: Water

3.   Science Findings

Summary Science Statement: Available acute toxicity data
   indicate Toxicity Category I for lactic acid, based on
   pH.   Lactic acid is a severe dermal and eye irritant.
   However, acute dermal data indicate no systemic
   toxicity.   Protective clothing and eye protection
   statements are required on the label to minimize
   exposure during mixing, loading and application
   operations.   Subchronic/chronic toxicity, mutagenicity
   and environmental fate data requirements are waived.
   The use of lactic acid is not expected to cause adverse
   effects to nontarget organisms including endangered
   species.   The chemical is exempted from the requirement
   of a tolerance when used as a plant growth regulator in
   or on all raw agricultural commodities (RAC).

Chemical Characteristics

Physical State: Nonvolatile liquid

Color: Colorless

Odor: Odorless to weakly acid

Boiling Point: 190 C (760 mm Hg)

Density: 1.195 (at 85.3% purity) grams/ml

Solubility (25 C):   Water - infinite
                     Ether - infinite
                     Ethanol - infinite

Vapor Pressure: 1.3 mm Hg at 90 C

Dissociation Constant: pKa = 3.87

pH: 0.6

Stability: Material is stable at normal temperatures.

Toxicological Characteristics:

Acute toxicity results

Acute oral toxicity (rat):   4,936 mg/kg (male), 3,543
   mg/kg (female), Toxicity Category III

Acute dermal toxicity (rabbit):   greater than 2,000
    mg/kg, Toxicity Category III

Acute inhalation toxicity (rat):   greater than 5.0 mg/l
    (analytical concentration) and 7.94 mg/l (nominal
    concentration) for each sex separately or
    combined.   Toxicity Category III

Primary dermal irritation (rabbit):   Lactic acid is a
    severe dermal irritant.   Toxicity Category I

Primary eye irritation: This data requirement was not
    required since the pH of the chemical is less than
    2.   Toxicity Category I is assigned because of the
    low pH.

Dermal sensitization (guinea pig):   Lactic acid is not
    a dermal sensitizer.

Subchronic, chronic toxicity and mutagenicity data
    requirements are waived.   These waivers are based
    on the following, (1) the natural occurrence of
    lactic acid; (2) residues resulting from the
    registered uses will not be higher than is
    presently allowed in food production and (3) such
    residues will not exceed normal physiological
    lactic acid levels in raw agricultural
    commodities (RACs).

Major Routes of Exposure: Mixers, loaders and
    applicators would receive the most exposure via
    skin/eye contact and inhalation.

Physiological and Biochemical Behavioral Characteristics:

Mechanism of pesticidal action: Preliminary research
    indicates that lactic acid causes temporary
    stimulation of the Krebs cycle after glycolysis.

Metabolism and persistence in plants and animals:
    Lactic acid is a normally occurring compound in
    both plants and animals.   It is formed and
    metabolized from/to pyruvic acid in glycolysis.

Environmental Characteristics:

Lactic acid is naturally occurring and is rapidly
formed and degraded in soil.   The chemical is found in
root exudates of common vegetables.   It is a product
of aerobic and anaerobic soil metabolism of glucose and
plant matter.   Because of the volume of data available
on this compound and its natural occurrence in the
environment, the environmental fate data
requirements are waived.

Ecological Characteristics:

Avian acute oral toxicity (bobwhite quail): greater
    than 2,250 mg/kg

Avian dietary toxicity (bobwhite quail): greater than
    5,620 ppm

Avian dietary toxicity (mallard duck): greater than
    5,620 ppm

Fish acute toxicity (bluegill): 130 ppm

Fish acute toxicity (rainbow trout): 130 ppm

Aquatic invertebrate toxicity (Daphnia magna): 750 ppm

Potential problems related to endangered species:
    Minimal hazard to endangered species is expected
    from the proposed use of lactic acid due to the
    chemical's natural occurrence in the environment
    and living organisms and its low toxicity.

Tolerance Assessment: Lactic acid (2-Hydroxypropanoic acid)
    is exempted from the requirement of a tolerance when
    used as a plant growth regulator in or on all raw
    agricultural commodities (RAC). Lactic acid is
    approved for use as an inert ingredient for
    application to plants (40 CFR 180.1001 (c)). The Food
    and Drug Administration approved the use of lactic
    acid as a generally recognized as safe (GRAS)
    ingredient in human foods. No Mexican or Canadian
    tolerances or Codex Alimentarius maximum residue limits
    have been established for residues of lactic acid in
    any RAC.

4.   Summary of Regulatory Positions

The following Agency positions are summarized from the
Lactic Acid Registration Standard.

None of the risk criteria listed in 40 CFR 154.7 for
initiating a Special Review has been exceeded.

The Agency is requiring protective clothing and eye
protection.

The Agency has waived the subchronic, chronic and
mutagenicity data requirements.

The Agency has waived environmental fate data
requirements for lactic acid.

Use, Formulation or Geographic Restrictions: None

Unique Label Warning Statements: Protective clothing and
     eye protection are required.

5.   Summary of Major Data Gaps

None

6.   Contact Person at EPA

Robert J. Taylor
U.S. Environmental Protection Agency
Registration Division (TS-767C)
401 M Street., S.W.
Washington, D.C.   20460
(703) 557-1800

MALATHION

Reason for Issuance: Registration Standard
Date Issued: January 1, 1988
Fact Sheet Number: 152

1. DESCRIPTION OF CHEMICAL

Generic Name: O,O-dimethyl phosphorodithioate of diethyl
(Chemical)   mercaptosuccinate

Common Name: Malathion

Trade and
Other Names: S-1,2-bis(ethoxycarbonyl)ethyl O,O-dimethyl
                 phosphorodithioate; diethyl(dimethoxy-
                 phosphinothioyl)thiobutanedioate;
                 diethyl mercaptosuccinate S-ester with O,O-
                 dimethyl phosphorodithioate; O,O-dimethyl
                 dithiophosphate of diethyl mercaptosuccinate;
                 [S-(1,2-dicarbethoxyethyl) O,O-dimethyl
                 phosphorodithioate; diethyl mercaptosuccinic
                 acid, S-ester with O,O-dimethyl phosphoro-
                 dithioate; American Cyanamid Co. (USP 2578 652)
                 Code No. EI4049; Calmathion; Celethion; Cython
                 (deodorized grade); Chemathion; Malaspray;
                 Detmol MA 96% (Albert & Co., Germany);
                 Emmatos; Emmatos Extra; For-Mal (Forshaw
                 Chemicals); Fyfanon; Hilthion; Karbofos;
                 Kop-Thion; Kypfos; Malamar; Malaphele;
                 Malathion ULV Concentrate; Malatol; Maltox
                 (All-India Medical); Prentox Malathion 95%
                 Spray; Sumitox; Vegfru Malatox; Zithiol;
                 Malmed.

EPA Pesticide Chemical Code (Shaughnessy Number): 057701

Chemical Abstract Service (CAS) Number: 121-75-5

Year of Initial Registration: 1956

Pesticide Type: Insecticide and Miticide

Chemical Family: Organophosphate

U.S. and Foreign Producers: American Cyanamid Company, A/S
Cheminova, McLaughlin Gormley King Company, Prentiss
Drug and Chemical Corp., Inc., Carmel Chemical Corp.,
Amvac Chemical Corp., Prochimie International Inc.,
Gowan Co., Wesley Industries Inc., Trans Chemic Industries
Inc., Southern Mill Creek Products Co., Inc., Octagon
Process Inc., FMC Corp., and Aceto Chemical Co. Inc.

2.   USE PATTERNS AND FORMULATIONS

Application Sites: <u>Terrestrial food crop</u> use on alfalfa, almond,
anise, apple, apricot, asparagus, avocado, barley, beets,
beets (seed crop), bermudagrass, blackberry, blueberry,
boysenberry, broccoli, brussels sprouts, cabbage, cantaloupe,
carrot, casaba melons, cauliflower, celery, cherry, chestnut,
citrus fruits (nursery stock), clover, collards, corn cotton,
cowpeas (hay), crenshaw melons, cucumber, currant, dandelion,
date, dewberry, eggplant, endive, fig, filbert, flax, garlic,
gooseberry, grapefruit, grapes, grass, grass hay, green beans,
guava, honeydew melons, honey ball melons, horseradish, kale,
kidney beans, kohlrabi, kumquat, leek, lemon, lespedeza,
lettuce, lima beans, lime, loganberry, lupine, macadamia nut,
mango, muskmelons, mustard greens, navy beans, nectarine,
oats, okra, onion, onion (green), onion (seed crop),   papaya,
parsley, parsnip, passion fruit, pasture grasses, peach,
peanuts, pear, peas, pecan, peppermint, peppers, persian
melons, pineapple, rangeland grasses, raspberry, rutabaga,
rye, safflower, salsify, shallot, snap beans, sorgum, soybeans,
spearmint, spinach, squash, strawberry, sugar beets, sweet
potato, swiss chard, tangelo, tangerine, tomato, turnips,
vetch, walnut, watercress, watermelons, wax beans, and wheat;

<u>Terrestrial non-food crop</u> use on tobacco, tobacco (transplant)
beds), ornamental flowering plants, ornamental lawns and turf,
ornamental nursery stock, ornamental woody plants, pine seed
orchards and uncultivated non-agricultural areas;

<u>Greenhouse food crop use</u> on asparagus, beans, beets, celery,
cole crops (including broccoli, cabbage, kale mustard greens,
and turnips), corn cucumber, eggplant, endive, lettuce,
melons, mushrooms, onion, peas, peppers, potato, radish,
spinach, squash, summer squash, tomato, and watercress;

<u>Greenhouse non-food crop use</u> on ornamental plants and Epcot
display crops;

<u>Aquatic food crop uses</u> on cranberry and rice;

Aquatic non-food uses on intermittently fooded areas,
irrigation systems, and sewage systems;

Forestry uses on forest trees (including Douglas fir,
eastern pine, hemlock larch, pines, red pine, spruce, and
true fir);

Indoor uses on stored commodity treatment for almonds,
barley, field corn, field or garden seeds, grapes (raisin),
oats, peanuts, rice rye, sorghum, sunflower, wheat, bagged
citrus pulp, and cattle feed concentrate blocks (non-medicated);
pet and domestic animal uses for beef cattle, cats, chickens,
dairy cattle (lactating and non-lactating), dogs, ducks,
geese, goats, hogs, horses (including ponies), pigeons,
sheep, and turkeys; animal premise uses for dairy and
livestock barns, stables and pens, feed rooms, poultry
houses, manure piles, garbage cans, garbage dumps, kennels,
rabbits on wire, beef cattle feed lots and holding pens,
cat sleeping quarters, dog sleeping quarters, poultry
houses; agricultural premise uses for cull fruit and vegetable
dumps; household uses for indoor domestic dwellings, human
clothing (woolens and other fabrics), mattresses; and
commercial and industrial uses for bagged flour, cereal
processing plants, edible and inedible commercial establishments, dry milk processing plants, edible and inedible eating
establishments, edible and inedible food processing plants,
packaged cereals, pet foods and feed stuff.

Methods of Application: Sprays, aerosols and fogging equipment,
   ground and aerial equipment (including ULV), baits, paints,
   pet collars, dips, soil, bark and foliar application, dormant
   and delayed dormant application, animal dust bags and oilers,
   and cattle feed concentrate blocks.

Formulations: Wettable powders, dusts, granules, emulsifiable
   concentrates, liquids, solids, impregnated materials, and
   pressurized sprays, pellets/tablets, liquids (ready to use).

3. SCIENCE FINDINGS

Summary Science Statement

   Technical malathion is a mildly acutely toxic pesticide,
which is placed in Toxicity Category III based on the oral,
dermal and inhalation routes of exposure. Technical malathion
is non-sensitizing and only mildly irritating to the eyes and
skin (Toxicity Category III and IV, respectively). Additional
data are required to assess the neurotoxic potential of malathion.
Malathion is a cholinesterase inhibitor, reducing plasma and
red blood cell cholinesterase.

Although the Agency possesses a number of studies on the chronic effects of malathion and its principal metabolite malaoxon, several of these studies are deficient scientifically, and must repeated.

Of five studies concerning the oncogenicity of malathion and its metabolite, three are acceptable, and demonstrate that malathion is not carcinogenic in two species of rats, and that its metabolite malaoxon is not carcinogenic in mice. Because of questionable liver findings in the malathion mouse study and the malaoxon rat study, new studies must be conducted in these species.

An acceptable rabbit teratology study demonstrated no teratogenicity at dosages up to 100 mg/kg/day. However, developmental and maternal toxicity were noted at dosages of 50 mg/kg/day. A similar study in rats was unacceptable and must be repeated. A 3-generation reproduction study was also unacceptable.

Laboratory data show that technical malathion is potentially highly toxic to aquatic invertebrates, bees, and aquatic life stages of amphibians; moderately toxic to birds, and slightly toxic to fish. Based on theoretical calculations, both terrestrial and aquatic uses of malathion may pose significant risk to aquatic fauna. Reported fish kills and results of field studies suggest that adverse effects to both aquatic and terrestrial fauna may result from normal use of malathion. However, these studies are not adequately documented to enable EPA to propose restrictions on the use of malathion. EPA will reassess the impacts of malathion use on nontarget organisms after the required environmental fate and ecological effects data have been received and reviewed.

The Agency is unable to assess the potential for malathion to contaminate groundwater because the environmenal fate of malathion is largely uncharacterized. Preliminary data indicate that malathion is very mobile in loamy sand and loam soils. Additional data are needed in order for the Agency to assess its fate in the environment and potential for contaminating groundwater.

A tolerance reassessment of malathion is not possible at this time, since most of the tolerances are not adequately supported, and because there are gaps in the chronic toxicology data base (chronic feeding studies, teratology study, reproduction study, mutagenicity studies, and a metabolism study). The Theoretical Maximal Residue Contribution (TMRC) for the U.S. population average is 0.1014 mg/kg/day and the Provisional Acceptable Daily Intake (PADI) is 0.02 mg/kg/day based on a human study in which plasma and red blood cell cholinesterase were monitored and a 10-fold uncertainty factor was used. The TMRC occupies 507% of the PADI.

Chemical/Physical Characteristics of the Technical Material

Chemical/Physical
Characteristics
(technical grade)

Color: colorless, yellow, amber, or
        brown
Physical state: Liquid
Odor: Mercaptan-like
Specific gravity: 1.2315 at 25°C
Boiling point: 156-157°C at 0.7 mm Hg
Solubility: 145 ppm in water at 25°C;
             completely soluble in most
             alcohols, esters, high
             aromatic solvents, and
             ketones; poor solubility in
             aliphatic hydrocarbons.
Vapor pressure: 0.00004 mm Hg at 30°C
Miscibility: miscible with most organic
              solvents
Stability: may gel in contact with iron,
            terreplate or tinplate

Toxicology Characteristics

Acute Oral: Toxicity Category III (ranges from 1546 to 1945 mg/kg
    in female rats and 1522 to 1650 mg/kg in male rats).

Acute Dermal: Toxicity Category III (>2000 mg/kg in female and
    male rats and rabbits).

Acute Inhalation: Toxicity Category III based on toxicity values
    ranging from 1.7 to >4.0 mg/m^3 in rats.

Primary Dermal Irritation: Toxicity Category IV based on mild
    dermal irritation reported in a rabbit
    study

Primary Eye Irritation: Toxicity Category III based on findings
    of mild conjunctival reactions 72 hours
    post application in rabbits' eyes.

Skin Sensitization: Non-sensitizing

Delayed Neurotoxicity: Data gap.

Subchronic Inhalation: Data gap.

Oncogenicity: Data gaps for mouse (using malathion) and rat
    (using malaoxon).

Chronic Feeding: Data gaps for rodent and nonrodent (using
    malathion) and rodent (using malaoxon).

Metabolism: Data gap.

Teratogenicity: Data gap for rat. Data in rabbit indicated a
NOEL = 25 mg/kg for developmental effects; it
was not teratogenic in any dose group (Highest
Dose Tested was 100 mg/kg).

Reproduction: Data gap.

Mutagenicity: Data gap.

Environmental Characteristics

Data gaps exist for environmental fate. Data reviewed by the
Agency indicate that malathion is very mobile in laomy sand and
loam soils. Adsorption ratios reported (amount adsorbed/initial
concentration) were 0.73 to 0.95. Data are needed before the
Agency can assess the potential for malathion to contaminate
groundwater.

Ecological Characteristics (technical grade)

Avian oral toxicity (8-day LD_{50})	167 ppm for ring-necked pheasant and 1485 ppm for mallard.
Avian dietary toxicity (8-day LC_{50})	Acute toxicity value of 3497 ppm for bobwhite and >5000 ppm for mallard
Freshwater fish acute toxicity (96-hr LC_{50})	200 ppm for rainbow trout and 40 to 103 ppm for bluegill
Freshwater invertebrate toxicity (48-hr EC_{50})	1 ppm for Daphnia magna
Estuarine invertebrate toxicity	>1000 ppm for Eastern oyster

Tolerance Assessment

The available data pertaining to metabolism of malathion
in plants are inadequate. Additional data are required on the
uptake, distribution, and metabolism of malathion in alfalfa,
cotton, soybeans, and either wheat or rice. The data pertaining
to metabolism of malathion in animals are inadequate. Additional
metabolism studies are required that utilize ruminants and
poultry. Metabolism studies using cattle, poultry, and swine
reflecting direct animal treatment are also required.

Analytical methodology for determining the levels of
residues of malathion in plants and animals is adequate.
Malathion is detected by the FDA-USDA multiresidue protocols.

Storage stability data demonstrate that residues of malathion in or on frozen plant commodities are stable up to 185 days after application and in milk stored at -10°C for 98 days after application. No data are currently available for animal tissues and are required. Additional storage stability data are also required in order to evaluate the adequacy of the malathion tolerances.

Insufficient data are available on the magnitude and levels of residues of malathion in or on all commodities listed in 40 CFR 180.111 except flax seed, hops, wild rice, and non-medicated cattle feed concentrate blocks. Processing studies are required.

Tolerances must be proposed and appropriate supporting residue data submitted for the following feed items: beanvines and hay; lentil forage and hay; cowpea seed; soybean straw; barley forage, hay and straw; corn forage and fodder; oat forage, hay and straw; rice straw; rye forage and straw; straw of wild rice; sorghum fodder; lespedeza forage; lupine forage; cotton forage; mint hay; peanut hulls, hay and vines; and pineapple forage.

Feed additive tolerances are required for residues of malathion in or on dried hops and spent hops. A tolerance for residues of malathion in or on anise must be proposed together with supporting residue data. Data are needed to support the use of malathion in food handling establishments. In addition, data reflecting the use of malathion on stored, unfinished tobacco are required.

Based on a study in humans in which red blood cell and plasma cholinesterase activity were inhibited at a dose of 0.34 mg/kg (the lowest effect level or LEL), a NOEL has been extrapolated to 0.2 mg/kg/day. A provisional acceptable daily intake (PADI) of 0.02 mg/kg/day has been calculated using a 10-fold uncertainty factor. The PADI is provisional because the existing data base on malathion is lacking chronic toxicity studies, an acceptable teratology study in rats, an acceptable reproduction study, mutagenicity studies, and a metabolism study.

The Theoretical Maximal Residue Contribution (TMRC) for the U.S. population average is 0.1014 mg/kg/day, occupying 507% of the PADI. For children 1 to 6 years of age, the TMRC occupies 1133% of the PADI. The TMRC is based upon current tolerance levels and an assumption that 100% of the sites are treated. Actual exposure levels are likely to be much lower. When the required data are submitted, the Agency will conduct a full tolerance reassessment.

4. SUMMARY OF REGULATORY POSITIONS AND RATIONALES

    ° No referral to Special Review is being made at this time.

    ° No new tolerances for raw agricultural commodities or significant new uses will be granted until the Agency has received data sufficient to perform a tolerance reassessment. Significant new uses will not be granted until the data gaps have been filled.

    ° The Agency is concerned about the potential hazards to aquatic organisms. However, no regulatory action is being considered at this time for fish and wildlife concerns. EPA will reassess the impacts of malathion use on nontarget organisms after the required environmental fate and ecological effects data have been received and reviewed.

    ° The Office of Endangered Species (OES) in the U.S. Fish and Wildlife Service has determined that certain uses of malathion may jeopardize the continued existence of endangered species or critical habitat of certain endangered species. No additional labeling is required at this time; however, EPA is developing a program to reduce or eliminate exposure to these species, and may require labeling revisions when the program is developed.

    ° In order to meet the statutory standard for continued registration, the Agency has determined that malathion products must bear revised and updated fish and wildlife toxicity warnings.

    ° The Agency is deferring decisions concerning malathion's potential for contamination of groundwater until the environmental fate data have been submitted and reviewed.

    ° The Agency is not restricting the use of malathion products for retail sale only to certified applicators. Malathion does not meet any of the criteria of 40 CFR 162.11 and therefore products containing malathion do not warrant restricted use classification.

    ° The Agency is not establishing a longer reentry interval for agricultural uses of malathion beyond the minimum reentry interval for all agricultural uses of pesticides (sprays have dried, dusts have settled and vapors have dispersed). The Agency will reassess the need for reentry data/reentry intervals upon receipt of the required toxicology data.

5. SUMMARY OF OUTSTANDING DATA REQUIREMENTS

Toxicology	Time Frame
Delayed neurotoxicity	9 mos.
21-day dermal toxicity	9 "
90-day inhalation - rat	15 "
Chronic toxicity (rodent and non-rodent)-- using malathion)	50 "
Chronic toxicity (rodent)--using malaoxon	50 "
Oncogenicity (mouse)--using malathion	50 "
Oncogenicity (rat)--using malaoxon	50 "
Teratogenicity - rat	15 "
Reproductive effects - rat (2-generation)	39 "
Mutagenicity	9-12 mos
Metabolism	24 mos
Domestic animal safety testing	15 "

Environmental Fate/Exposure	
Hydrolysis	9 mos
Aerobic and anaerobic soil metabolism	27 "
Aerobic and anaerobic aquatic metabolism	27 "
Leaching and adsorption/desorption	12 "
Terrestrial field dissipation	27 "
Long-term field dissipation	50 "
Forestry dissipation	27 "
Aquatic (sediment) - field study	27 "
Phhotodegradation in water, soil, air	9 "
Volatility (lab)	12 "
Rotational crops (confined)	39 "
Accumulation in irrigated crops	39 "
Accumulation in fish	12 "
Accumulation in aquatic nontarget organisms	12 "
Spray drift	18 "

Residue Chemistry	
Storage stability data	18 mos
Plant and animal metabolism	18 "
Residue data - raw agricultural commodities	18 "
Processing studies	24 "
Residue data on stored, unfinished tobacco	18 "
Residues in water	15 "
Residue data on food handling establishments	12 "

Product Chemistry	Time Frame
All	9-15 mos

Fish and Wildlife

Acute toxicity to freshwater invertebrates	9 mos
Acute toxicity to estuarine and marine organisms	12 "
Avian reproduction	24 "
Fish early life stage	15 "
Aquatic invertebrate life cycle	15 "
Honeybee - toxicity of residues on foliage	15 "

6. CONTACT PERSON AT EPA

William H. Miller
Product Manager (16)
Insecticide-Rodenticide Branch
Registration Division (TS-767C)
Office of Pesticide Programs
Environmental Protection Agency
401 M Street, SW.
Washington, DC   20460

Office location and telephone number:
Rm. 211, Crystal Mall #2
1921 Jefferson Davis Highway
Arlington, VA

(703) 557-2600

DISCLAIMER: The information presented in this Chemical Information
Fact Sheet is for informational purposes only and may not be
used to fulfill data requirements for pesticide registration
and reregistration.

MALEIC HYDRAZIDE

Reason for Issuance: Registration Standard
Date Issued: June 30, 1988
Fact Sheet Number: 170

1. Description of Chemical

   Generic Name: 1,2-dihydro-3,6-pyridazinedione

   Common Name: Maleic Hydrazide

   Trade Names: Drexel Sucker-Stuff, Super Sucker Stuff, Retard, Burtolin, Decut, Drexel Sprout Stop, Fair 2, Fair Plus, KMH, Maintain 3, Malazide, Mazide, Regulox W, Regulox 50W, Stuntman, Super-De-Sprout, Vendalhyde, Vondrax, Royal MH-30, Royal MH-30 SG, Royal Slo-Gro, Malazide Slo Gro.

   EPA Shaughnessy Codes: Maleic Hydrazide: 051501
   Potassium Salt of Maleic Hydrazide: 051503
   Diethanolamine Salt of Maleic Hydrazide: 051502

   Chemical Abstracts Service (CAS) Number: 123-33-1

   Year of Initial Registration: 1952

   Pesticide Type: Herbicide, Plant Growth Regulator

   U. S. and Foreign Producers: Drexel Chemical
   Trans Chemical Industries
   Uniroyal Chemical, Inc.

2. Use Patterns and Formulations:

   Application Sites: Terrestrial food (potatoes, onions, cranberries), terrestrial nonfood (nonbearing citrus and nonbearing apples, tobacco), terrestrial noncrop.

   Types and Methods of Application: Primarily as a foliar spray with some use by tree injection.

384

Pests Controlled: Sucker control for tobacco, sprout infiltration in
onions and potatoes, growth retardant of quackgrass, wild onions,
and garlic.

Application Rates:
    Terrestrial Food Crops: 0.7 lb acid equivalent (ae)/A to 15 lb ae/A
    Terrestrial Nonfood Crops: 1.5 to 4.5 lb ae/A
    Ornamental Plants and Forest Trees: 0.06 to 6 lb ae/A
    Turf: 0.75 to 6.6 lb ae/A
    Rights-of-Way: 0.66 to 6.6 lb ae/A

Types of Formulation:
    90%, 95%, 97%, and 99% technical grade of the active ingredient
    (TGAI)
    0.66 to 2.25 lb ae/gal emulsifiable concentrate (EC)
    0.6, 1.5, 2.0, 2.25, and 2.5 ae/gal soluble concentrate/liquid
    (SC/L)

Usual Carrier: Water

3. Science Findings

Summary Science Statement: Maleic hydrazide (MH) has low acute toxicity
    (Category III) for primary eye irritation and is in Category IV for
    acute oral toxicity, acute dermal toxicity, and primary dermal irri-
    tation. MH caused no adverse reproductive effects and was not
    oncogenic in mice. The teratology study in rabbits had a teratogenic
    NOEL of 100 mg/kg with malformed scapulae occurring in the mid and
    high dose. Additional information has been requested to clarify
    this effect.

    MH was stable to hydrolysis and photodegradation in soil. It
    photodegraded in buffered aqueous solutions at pH 5, 7, and 9. MH
    was very mobile in five soils and has a low potential to bioaccumu-
    late in fish. Additional persistence and leaching data are needed
    to evaluate MH potential to contaminate ground water.

    MH is considered practically nontoxic to birds, aquatic invertebrates,
    freshwater fish, or honey bees. Endangered animal species are not
    expected to be adversely affected by the use of MH. Since no endan-
    gered plant species are listed for tobacco cropland, citrus, apples,
    or cropland with onions, potatoes, or cranberries, little risk to
    endangered plants is expected from these uses. The hazard evaluation
    for listed plants and the right-of-way uses is being deferred until
    completion of the noncrops cluster.

    Chemical Characteristics:
        Color: White
        Physical state: Crystalline solid

Odor: Faint
Melting point: 292 °C minimum
Bulk density: 30 to 36 lbs cubic feet (cu/ft)
(0.049 grams/milliliter g/ml)

Specific gravity: 1.6 at 20 °C

Solubility:
    90% a.e.- 60 parts per million (ppm) water
        10 ppm isopropyl alcohol
        < 10 ppm xylene
        240 ppm dimethyl formamide

Vapor pressure: < 1 mm Hg at 20 °C

Stability:
    Stable at 45 °C up to 61 days
    Stable at 80 °C up to 30 days

Toxicology Characteristics:
    Existing data are all based on MH (technical) or the potassium salt
    (K salt), further data are requested on the diethanolamine (DEA) salt.

    Acute Toxicology:
        Acute Oral Toxicity (Rat):
            Greater than (>) 5 grams/kilogram (g/kg)
            Toxicity Category IV
        Acute Dermal Toxicity (Rabbit):
            > 20 g/kg
            Toxicity Category IV
        Primary Eye Irritation:
            Primary Irritation Score (PIS) = 0.4
            Toxicity Category III
        Primary Dermal Irritation:
            Slight Irritant
            Toxicity Category IV
        Acute Inhalation and dermal sensitization studies are not
            available and are being required for MH and the DEA salt.

        Subchronic Toxicology Studies: There are no data available for
            subchronic oral, dermal, or inhalation toxicity.
            A 21-day dermal toxicity study is required for both MH
            (technical MH or potassium (K) salt are considered equivalent)
            and the diethanolamine (DEA) salt. Subchronic feeding
            studies for a rodent and nonrodent are not required
            because chronic feeding studies are required.

        Chronic Feeding/Oncogenicity Studies: Available chronic feeding
            studies are inadequate to fulfill guideline requirements
            but are useful to calculate provisional allowable daily intake
            (PADI). The chronic feeding study in rats indicates that
            the no-observable-effect level (NOEL) is less than (<)
            500 milligrams/kilogram (mg/kg). Chronic feeding studies
            on rodents and nonrodents are required for both MH and
            its DEA salt.

Available oncogenic studies in rats are inadequate for maleic hydrazide technical and K salt, no data are available for the DEA salt, therefore data are required for both technical and DEA salt. Two oncogenicity studies are available for mice. Together they indicate that MH is not oncogenic in mice up to 1800 mg/kg. An oncogenic study in mice is required for the DEA salt.

Teratology and Reproduction Studies: A teratogenicity study in rabbits showed that exposure to 300 or 1000 mg/kg resulted in malformed scapulae in offspring, while 100 mg/kg had no effects. Additional information is required on parentage of affected offspring to fully evaluate this effect.

No teratology data are available for rats. Therefore, teratology studies are required for both MH and its DEA salt. A teratology study in rabbits is required for the DEA salt.

A 2-generation rat reproduction study indicated no incidence of adverse reproductive effects up to 2250 mg/kg (highest dose tested [HDT]) with fetal toxic and maternal toxic NOEL of 750 mg/kg based on decreased body weights at the HDT. This study satisfies the requirement for the MH technical. A 2-generation reproduction study in rats is required for the DEA salt.

Mutagenicity Studies: A mutagenicity study of sex-linked recessive lethal gene mutations in Drosophila revealed no sex-linked recessive lethal mutations at cytotoxic doses of 0.4 to 1.0% K salt of MH. All other mutagenicity data are required for MH. A full set of mutagenicity studies for the DEA salt are required.

Metabolism Studies: There are no metabolism studies available for MH technical or the DEA salt; therefore, these studies are required.

Physiological and Behavioral Characteristics:

Foliar Absorption - Absorbed by roots and leaves.

Translocation - Rapidly translocated to leaves and growing shoots.

Mechanism of Pesticidal Action - A uracil antimetabolite which interferes with cell division, plant growth, and maturation.

Metabolism in Plants - Limited available data indicate that the major residues in tobacco are maleic hydrazide and its beta-D-glucoside conjugate. Additional plant metabolism data are required for potatoes, onions and cranberries.

Metabolism in Animals - No data are available for metabolism in animals; therefore, livestock metabolism data is required.

Environmental Characteristics:

Adsorption and Leaching in Basic Soil Types: MH was very mobile
in a silt loam, a sandy clay loam, a sandy loam, and two
sandy soils. Additional leaching and adsorption data are
required.

Microbial Breakdown: Available aerobic/anaerobic soil metabolism
data are insufficient. Therefore, aerobic and anaerobic
soil metabolism studies are required.

Loss from Photodecomposition: Stable to photodegradation in soil.
Photodegraded in buffered aqueous solutions at pH 5, and 7, and
9 with half-lives of 58 days, at ph 5 and 7 and 34 days at pH 9.

Bioaccumulation: Low potential to bioaccumulate in fish.

Potential to Contaminate Ground Water: The available data are
inconclusive for defining potential of MH to leach into ground
water. MH is persistent in water. Once persistence data from
aerobic/anaerobic soil metabolism data and additional leaching
data are submitted, the potential to contaminate ground water
will be evaluated.

Exposure to Humans: Humans may be exposed to maleic Hydrazide by ingestion
of residues on treated crops, and from use of treated tobacco.
The major route of exposure for applicators is expected to be
dermal contact. An exposure assessment was performed for
application on tobacco, potatoes, onions, and rights-of-way.
Exposure ranged from less than 0.5 mg/kg/day to 16 mg/kg/day.
The greatest exposure occurred for the open-pour mixer loader.

Risk to Humans: The major routes of exposures are expected to be
dermal and eye contact. A risk assessment was performed
for dermal exposure to applicators based on a rare
teratogenic effect (malformed scapulae in offspring) seen in
rabbits. The risk assessment indicated that the margin of
safety (MOS) for mixer-loaders was less than 100 and therefore
of concern. Labels for end-use products (EPs) are being amended
to require long sleeve shirts, long pants, chemical resistant
gloves at all times while handling, applying, mixing, or loading
the product.

Reentry: Reentry data are not required at this time because cultural
practices for existing uses indicate little likelihood of exposure or low exposure from these uses. Because MH is placed in
Toxicity Category IV for acute oral and dermal toxicity, minimal
risk to humans is expected.

Ecological Characteristics:

Avian Acute Adult Toxicity:
        Technical: Mallard Duck > 4640 mg/kg
        K salt: Mallard Duck > 2250 mg/kg

Avian Dietary Toxicity:
        Technical:
        Mallard Duck > 10,000 ppm
        Bobwhite Quail > 10,000 ppm
        K salt: Mallard Duck > 5620 ppm

    Acute Toxicity to Freshwater Fish:
        Technical:
        Rainbow Trout = 1435 ppm
        Bluegill = 1608 ppm
        K salt: Rainbow Trout > 1000 ppm

    Acute Toxicity to Freshwater Invertebrates:
        Technical: 107.5 ppm
        K salt: 1000 ppm

        Acute Toxicity to Honey Bee: > 36.26 ug/bee

        These data indicate that MH is considered "very low toxicity"
        to avian species, both warmwater and coldwater fish, freshwater
        invertebrates, and honey bees.

    Hazard to Endangered Species: Endangered animal species are not expected
        to be adversely affected by the use of MH because of its low
        toxicity to mammals, avian species, and aquatic species.

        Since MH is a plant growth retardant endangered plant species
        occurring in areas where MH is used could be at risk. However,
        there are no endangered plant species listed for tobacco crop-
        land, citrus, apples, or cropland planted with potatoes, onions,
        or cranberries, therefore no risk to endangered plant species.
        Evaluation of hazard to indangered plants from use of MH on
        right-of-way will await completion of noncrop cluster.

    Tolerance Assessment: Tolerances are established in 40 CFR 180.175
        for residues of the herbicide and plant growth regulator MH
        (1,2-dihydro-3,6-pyridazinedione) in or on the following raw
        agricultural commodities:

Commodity	Parts per million
Cranberries	15.0
Onions, dry bulb	15.0
Potatoes	50.0

A tolerance is established in 21 CFR 193.270 for residues of the herbicide and plant growth regulator (1,2-dihydro-3,6-pyridazine-dione) on potato chips at 160 ppm as a result of application of the pesticide to the growing potato plant.

A Provisional Acceptable Daily Intake (PADI) for MH is currently based on the finding of renal dysfunction in the rat chronic study. The LOEL for this effect was 500 mg/kg/ day. Using a thousandfold safety factor, the PADI for MH is 0.5 mg/kg/day. Existing tolerances produce a theoretical maximum residue contribution (TMRC) of 0.085 mg/kg/day which occupies 17 percent of the PADI.

The tolerance assessment indicated that additional residue data are needed for onions, potatoes, nonbearing apples, and nonbearing citrus. Additional plant metabolism data are required for MH. No storage stability or animal metabolism data are available; therefore, these data are required.

Available toxicology data that was used to calculate the PADI on technical or K salt include oncogenicity studies in mice, a teratology study in rabbits, and a 2-generation rat reproduction study. Additional data required include chronic feeding studies with a rodent (although the PADI was established based on a supplementary feeding study in rats) and nonrodent oncogenenicity and teratology studies in rats, mutagenicity testing, and rat metabolism data for the technical or K salt. Data required to support tolerances for the DEA salt include acute data, chronic feeding studies with a rodent, non-rodent, oncogenicity studies in rats and mice, teratology studies with rats and rabbits 2-generation rat reproduction study, mutagenicity testing and metabolism data.

Reported Pesticide Incidents

Data from the Pesticide Incident Monitoring System (PIMS) reports and the national study of hospitalized pesticide poisonings (1971-76) show some cases of MH poisonings, mostly occupational. There were an estimated eight persons hospitalized in the United States each year from 1974 to 1976. The circumstances that led to these poisonings are not known.

4. Summary of Regulatory Position and Rationale

A review of the available data indicate that no risk criteria listed in 40 CFR 154.7 have been met or exceeded for MH.

The Agency will not approve any significant new uses of MH until additional residue chemistry data are available to assess existing uses.

The Agency is requiring that labeling on all EPs include a requirement for protective clothing long-sleeve shirt, long pants and chemical resistant gloves at all times when handing, applying, mixing, or loading pesticide.

Products bearing labelling not in compliance with the registration Standard may be released for shipment by the registrant only until July 30, 1989. Such products may be distributed and sold by other persons only until July 30 1990.

The Agency is requiring that additional leaching, aerobic/ anaerobic soil metabolism and dissipation data be submitted to fully define the MH potential to leach and contaminate groundwater.

The Agency is requiring that additional information be submitted on the rabbit teratology study and that a rat teratology be completed within twelve (12) months.

The Agency will continue to require that the acceptable limit of hydrazine occurring in technical products be $\leq$ 15ppm as required by the PD-4

The Agency has determined that all toxicology studies will be necessary for the diethanolamine salt of maleic hydrazide if manufacturers comply with the data requirements which resulted in the suspension of all DEA-MH products in November 1981.

The Agency has determined that all end use product chemistry data for technical MH must be resubmitted and updated.

The Agency has determined that all products containing the DEA salt must be tested for nitrosamines.

The Agency has determined that reentry data or restrictions are not required at this time.

The Agency will not require labeling to protect endangered species at this time for products containing maleic hydrazide.

The Agency has determined that Tier I nontarget area phytotoxicity testing will be required for MH.

The Agency will not require additional residue data on cranberries.

The Agency will require additional residue data on potatoes, onions, tobacco, nonbearing citrus, and nonbearing apples.

The Agency is requiring additional animal and plant metabolism data and storage stability data on all residue data previously submitted and any new residue data requested.

The Agency has determined that certain data essential to the Agency's assessment of this pesticide and its uses and/or that may trigger the need for further data will receive immediate review when submitted. These data include animal and plant metabolism data, data necessary to determine the MH potential to contaminate ground water, and all requested toxicology data.

5. Summary of Major Data Gaps

Requirements	Due Dates
Product Chemistry	6 to 15 months
Residue Data	6 to 24 months
Toxicology Data	9 to 50 months
Environmental Fate	9 to 39 months
Plant Protection	9 months

6.   Contact Person at EPA:

Robert J. Taylor (PM 25)
Fungicide-Herbicide Branch
Registration Division (TS-767C)
Environmental Protection Agency
401 M Street SW.
Washington, D.C.   20460
(703) 557-1800

DISCLAIMER: The information presented in this Pesticide Fact Sheet is for informational purposes only and may not be used to fulfill data requirements for pesticide registration and reregistration.

MANEB

Reason for Issuance: Registration Standard
Date Issued: October 3, 1988
Fact Sheet Number: 182

1. DESCRIPTION OF CHEMICAL

   Chemical Name: Manganese ethylene bisdithiocarbamate

   ANSI Common Name: Maneb

   Principal Trade Names: Dithane M-22", Manzate"

   EPA (Shaughnessy) Code: 014505

   Chemical Abstracts Service (CAS) Number: 12427-38-2

   Year of Initial Registration: Late 1940's

   Pesticide Type: Fungicide

   Chemical Family: Ethylene bisdithiocarbamate (EBDC)

   U.S. and Foreign Producers: Pennwalt

2. USE PATTERNS AND FORMULATIONS

   Registered uses: Terrestrial food crop (fruits, vegetables, seed
   crops, nuts, flax, and grains); terrestrial non-food
   crop (ornamentals, lawns, turf); greenhouse food
   (tomatoes, rhubarb) and non-food crop (ornamentals)

   Predominant uses: Apples, potatoes, tomatoes and sweet corn

   Pests controlled: Foliar fungal diseases of selected fruit, nut,
   vegetable, grain, field and ornamental (including
   turf) crops.

Types of Formulations: Technical, formulation intermediate, dust,
                          granular, wettable powder, wettable powder/
                          dust, flowable concentrate, and ready-to-use.

Types and Method of   Foliar application to vegetable crops and
Application: apples by aerial equipment or ground equipment.
                          Foliar treatment of tobacco or vegetable seed
                          beds, application of sprays or dusts might
                          be by means of hand held compressed air sprayers
                          or dusting equipment. Potato and tomato foliage
                          may be treated by means of solid set, wheel
                          move, or center pivot sprinkler irrigation equipment.

Application Rates: Terrestrial food crop: 0.01 - 8.4 lb. ai/A
                          Terrestrial nonfood crop: 0.8 - 3.2 lb. ai/A
                          Grenhouse food crop: 1.1 - 2.4 lb. ai/100 gal.
                          Greenhouse nonfood crop: 0.8 - 2.4 lb. ai/100 gal.

3. SCIENCE FINDINGS

a. Chemical Characteristics

Physical state: Powder
Color: Yellow
Odor: faint
Molecular Formula: $(C_4H_6MnN_2S_4)x$

Toxicology Characteristics

Acute Oral: LD_{50} = 4,400 mg/kg bw (rat) (Toxicity Category III)

Acute Dermal: LD_{50} > 2 gm/kg bw (rabbit) (Toxicity Category III)

Acute Inhalation: LC_{50} > 2.22 $\pm$ 0.26 mg/l (rat); (Toxicity Category III)

Primary Dermal Irritation: Non-irritating (rabbit)(Toxicity Category IV)

Primary Eye Irritation: Severe eye irritant (rabbit)(Toxicity Category I)

Dermal Sensitization: Sensitizer (guinea pig)

Major Routes of Exposure: Oral, dermal and inhalation

Subchronic Toxicity: No observed effect level (NOEL) = 100 ppm,
                     LEL (increase in thyroid weight in males) = 300 ppm
                     (Monkeys).

Oncogenicity: Studies required

Chronic feeding: Studies required

Metabolism: Studies in rats indicate that maneb is hydrolyzed,
             readily absorbed and excreted in the urine and feces.
             The major metabolite is ETU.

Reproduction: Study required

Teratogenicity & Developmental Toxicity: Studies required

Mutagenicity: Mutagenicity testing showed that maneb was positive
for inducing chromosomal damage in an in vitro SCE
(sister chromatid exchange) assay with metabolic activation.
Evidence showed that maneb is most likely not an
initiating agent and the evidence on promotion
capability was negative. Additional data are required
before the unscheduled DNA synthesis (UDS) assay can
be upgraded to acceptable status. The following
studies were negative: Sister chromatid exchange in
CHO cells in the absence of a metabolic activation and
Host mediated assay in mice.

Physiological and Biochemical Characteristics

Metabolism and Persistence in Plants and Animals:

Metabolism of maneb is not completely understood. Additional
data are being required in plants and livestock. ETU is a
major metabolite of concern.

Environmental Characteristics

Maneb degrades to ETU and other transient degradates in water and
soil. ETU is stable in water at pH 5-9 and under sunlight and the degradation
of ETU on soil is not enhanced by sunlight radiation.

Maneb degrades very rapidly under anaerobic aquatic soil conditions
but ETU is relatively stable under these conditions.

ETU is stable in water at pH 5-9 and under sunlight and the degradation
of ETU on soil is not enhanced by sunlight radiation. ETU is the degradate
of major environmental concern. There are indications that ETU may
leach and enter groundwater. However, additional data are required to
complete the groundwater assessment.

Ecological Characteristics

o Maneb has been found to be practically nontoxic to birds and
mammals.

Avian dietary toxicity: LC_{50} > 9,000 ppm (bobwhite)

o The toxicity of a 80% product to warmwater fish is highly toxic.

4.   TOLERANCE ASSESSMENT

Tolerances, expressed as zinc ethylene bisdithiocarbamate, have been
established for residues of maneb in a variety of raw agricultural commodities
(40 CFR 180.110).

The toxicology data for maneb are insufficient to determine an Acceptable Daily Intake (ADI) or whether the toxicity observed in the studies is due to maneb or ETU. A subchronic study has been used to calculate a Provisional ADI (PADI). Because a subchronic study was used, an uncertainty factor of 1000 was employed. The PADI for maneb is 0.0005 mg/kg/day based on the six month feeding study with a NOEL of 5 mg/kg/day.

The theoretical maximum residue contribution (TMRC), based on the assumption that 100 percent of each crop is treated and contains residues at the tolerance level, is 0.030 or approximately 600 percent of the PADI. Based on a more realistic dietary assessment, using anticipated field residues and estimate of percent crop treated, the estimated average consumption for the U.S. population is 0.0036 mg/kg/day or 70 percent of the PADI.

5. SUMMARY OF REGULATORY POSITIONS

The Agency has initiated a Special Review for maneb along with the other EBDC's in June 1987 because of concern about the oncogenic risk to consumers from dietary exposure to ETU from food treated with these pesticides, and the risks of teratogenicity and adverse thyroid effects to applicators and mixer/loaders from exposure to ETU.

o ETU has been classified as a B_2 oncogen (probable human carcinogen).

o The Agency will not consider establishment of new food use tolerances for maneb because the current residue chemistry and toxicology data are not sufficient to assess existing tolerances and the toxicology data base is insufficient to determine an ADI and also does not allow a decision as to whether observed toxicity is due to maneb or ETU.

° The Agency will consider the need for establishment of tolerances for ETU and any intermediate metabolites when data are sufficient to permit such decisions.

° The Agency will not establish any food/feed additive regulations pursuant to Section 409 of the Federal Food, Drug and Cosmetic Act (FFDCA) and is deferring action on previously established food/feed additive regulations.

° Protective clothing labeling for maneb products, as specified as a result of the 1982 Decision Document should be updated.

° The Agency is requiring reentry data for maneb. In order to remain in compliance with FIFRA, an interim 24-hour reentry interval requirement must be placed on the label of all maneb end-use products registered for agricultural uses, until the required data are submitted and evaluated and any change in this reentry interval is announced.

° The Agency has screened and reviewed the environmental fate data to determine if maneb/ETU and/or its degradate(s) have the potential to leach in to ground water. The Agency has decided that a small-scale retrospective groundwater monitoring study is required to further define the extent of the ground water problem.

o While the data gaps are being filled, currently registered manufacturing-use
   products (MP's) and end-use products (EP's) containing maneb as the
   sole active ingredient may be sold, distributed, formulated and used,
   subject to the terms and conditions specified in this Standard.
   However, new uses will not be registered. Registrants must provide
   or agree to develop additional data, as specified in the Data Appendices
   of the Registration Standard, in order to maintain existing registrations.

6. LABELING REQUIREMENTS

   All maneb products must bear appropriate labeling as specified in
40 CFR 156.10. Appendix II of the Registration Standard contains information
on labeling.

The following are the major labeling specification:

   ° Environmental hazard statement
   ° Protective clothing requirements
   ° Preharvest interval
   ° Worker safety rules
   ° Grazing restrictions for almonds, apples, beans, corn, peanuts, potato,
     sugar beets, ornamental grasses, ornamental turf.

7. SUMMARY OF DATA GAPS

Product Chemistry All

Toxicology

Subchronic dermal (21-Day)
Subchronic inhalation (90-Day)
Chronic toxicity (rodent and nonrodent)
Oncogenicity (rat and mouse)
Teratology (rabbit and rat)
Reproduction (rat)
Mutagenicity (gene mutation) (other genotoxic effects)
Dermal absorption

Residue Chemistry

Nature of the Residue in Plants and Livestock
Analytical Methods
Magnitude of Residue for Variety of Commodities

Environmental Fate

Hydrolysis
Photodegradation studies in water and soil
Aerobic soil studies
Aerobic aquatic
Leaching and absorption/desorption
Aquatic (sediment)

Dissipation Soil Studies
Small-scale retrospective monitoring study
Fish accumulation

Reentry Protection

Reentry Studies on Foliar and Soil Dissipation

Wildlife and Aquatic Organisms

Avian oral toxicity
Avian dietary toxicity
Avian reproduction
Freshwater fish toxicity
Acute freshwater invertebrates
Estuarine and marine organism toxicity
Fish early life stage and invertebrate life-cycle
Aquatic organism accumulation

ETU Data Requirements

Toxicology

Chronic (rodent and non-rodent)
Reproduction

Environmental Fate

Aerobic and anaerobic soil metabolism
Aerobic aquatic
Lab volatility
Degradation (soil)
Aquatic (sediment)
Degradation (soil long-term)
Fish accumulation

8. CONTACT PERSON AT EPA

Lois A. Rossi
Product Manager (21)
Fungicide-Herbicide Branch
Registration Division (TS-767C)
Office of Pesticide Programs
Environmental Protection Agency
401 M St., SW
Washington, D.C. 20460

Office location and phone number:
Room 227, Crystal Mall #2
1921 Jefferson Davis Highway
Arlington, VA
(703) 557-1900

9. DISCLAIMER: The information in this Pesticide Fact Sheet is a summary only and may not be used to satisfy data requirements for pesticide registration and reregistration. The complete Registration Standard for the pesticide may be obtained from the National Technical Information Service. Contact the Product Manager listed above for further information.

MCPA

Reason for Issuance: Registration Standard (SRR)
Date Issued: September 22, 1989
Fact Sheet Number: 208

1. Description of Chemical

   Generic name: 2-methyl-4-chlorophenoxyacetic acid, and its sodium salt, esters and organic amines

   Common name: MCPA

   Trade names: ACME MCPA AMINE 4, AGRITOX, AGRO ONE, BORDERMASTER, BH MCPA, CHIPTOX, DED-WEED, EMPAL, KILSEM, MEPHANAL, METHOXONE, PHOMENE, RHONOX, and WEEDAR

   EPA Pesticide Chemical (Shaughnessy) Number:

   MCPA acid - 030501
   Sodium salt - 030502
   Diethanolamine salt - 030511
   Dimethylamine salt - 030516
   Butoxyethyl Ester - 030553
   Isobutyl Ester - 030562
   Isooctyl Ester - 030563
   Isopropyl Ester - 030566

   Chemical Abstracts Service (CAS) Number:

   MCPA acid - 94-74-6

   Year of Initial Registration: 1973

   Pesticide Type: Herbicide

   Chemical Family: Phenoxy herbicides

   U.S. Registrant: Dow Chemical

2. Use Patterns and Formulations

Application sites: Terrestrial food crop use on small grains
(wheat, oats, barley, rye), peas, beans, pasture grasses,
grain sorghum, alfalfa and clovers.

Terrestrial nonfood crop use on homelawns, ornamental turf,
flax, grass seed crops, noncrop areas, pasture grasses,
rangeland grasses, and forestry.

Aquatic food use on rice.

Aquatic nonfood use on aquatic weeds.

Formulations: Granular (MCPA acid and isooctyl ester);
soluble concentrate/liquid (sodium salt, diethanolamine
salt, dimethylamine salt); Technical (MCPA acid, Butoxyethyl
ester, isooctyl ester); Formulation Intermediate (Butyl
ester, isobutyl ester, isopropyl ester, dimethylamine salt,
MCPA acid); Ready-To-Use (Dimethylamine salt); and
Emulsifiable Concentrate (Butoxyethyl ester, isooctyl
ester). There are 117 federally registered products
containing MCPA as an active ingredient either by itself or
itself or in combination with other pesticides.

Methods of application: Aerial and ground equipment, knapsack
sprayers, pressure and hose-end applicators, and lawn
spreaders.

Application rates: 0.2 - 3.0 lb active ingredient per acre.

Mode of activity: MCPA is absorbed through both leaves and
roots and is readily translocated throughout the plant.
MCPA stimulates nucleic acid and protein synthesis affecting
the activity of enzymes, respiration and cell division.
Broadleaf plants exhibit malformed leaves, stems and roots.

3. Science Findings

This review of MCPA acid, its salts, and esters, is the
second intensive evaluation of the compound. In its original
Registration Standard, issued in 1982, the Agency summarized the
available data supporting the registration of MCPA and concluded
that additional data were needed to fully evaluate the pesticide.
The Agency has since received and reviewed the data on these
compounds.

The Agency has registered 9 salts, esters and amines of MCPA
in addition to the acid. From a toxicological standpoint the
acid and the sodium salt are essentially identical. The

registrants, however, must show that the other derivatives would
be equivalent to MCPA acid under testing conditions.   Otherwise,
a complete set of data must be submitted on each derivative to
support their continued registration.   The following scientific
assessment discusses the Agency's knowledge on MCPA acid.   If the
data refers to a derivative, the derivative is identified.

Chemical/Physical Characteristics of the Technical Material

MCPA is white to light brown and can be a solid,
flakes, crystal powder or liquid.   It has no odor or can be
slightly phenolic smelling.   The melting point is 114 to
119°C and is soluble to varying degrees in various solvents.

Toxicology Characteristics

Acute Oral.   Toxicity Category III (LD_{50} 1.38 g/kg in male
   rats, 0.76 g/kg in female rats).

Acute Dermal.   Toxicity Category III (LD_{50} > 4000 mg/kg in
   the rat).

Acute Inhalation.   Toxicity Category III (LC_{50} > 6.36 mg/L
   in the rat).

Primary Eye Irritation.   Toxicity Category I (corneal
   opacity with irritation of conjunctive observed 21 days
   post-instillation with rabbits).

Skin Sensitization.   Not a skin sensitizer in guinea pigs.
   Testing on other species required.

Acute Delayed Neurotoxicity.   MCPA is not an
   organophosphate, therefore a study is not required.

Subchronic Oral.   Sufficient data are available to satisfy
   the requirements of a subchronic oral study in rodents
   and nonrodents.   In beagles there was evidence of dose-
   related liver and kidney toxicity.   A NOEL for systemic
   effects was set at 1 mg/kg/day.   In rats, the mid- and
   high-dose males exhibited increased kidney weights and
   both sexes had indications of kidney disfunction.
   Hepatotoxicity, based on prolongation of clotting times
   and decreased cholesterol concentrations occurred in high
   dose males.   The NOEL for systemic effects is set at 2.5
   mg/kg/day.

Subchronic Dermal.   Data gap.   A 21-day dermal toxicity
   study is required.

Chronic Toxicity. Sufficient data are available to satisfy
the requirements for the chronic feeding studies in two
species for technical MCPA acid. In beagles, after one
year, kidney and liver toxicity was observed at the mid-
and high dose levels. The systemic NOEL was set at 0.15
mg/kg/day, LDT.

In a 2 year rat study, hepatotoxicity was observed with
elevated triglycerides, decreased cholesterol, and kidney
nephropathy. The NOEL for systemic toxicity was set 1.0
mg/kg/day.

Oncogenicity. Rat and mouse studies were reviewed and
found acceptable. MCPA is considered to be non-
oncogenic. No additional studies are required.

Teratogenic. The studies available to the Agency are
unacceptable under current guideline requirements. While
these studies showed no developmental alterations, the
Agency is requiring teratology studies in two species.

Reproduction. In a two-generation reproduction study with
rats, there were indications of a potential postnatal
growth effect. The NOEL is set at 7.5 mg/kg/day. No
additional reproduction studies are required.

Mutagenicity. The Agency has data to satisfy the
Structural Chromosomal Aberration study and the DNA
Damage and Repair study. MCPA acid was found to be non-
mutagenic in the first study and weakly mutagenic in the
second. The Agency is requiring a gene mutation study.

Metabolism. In a rat study, MCPA did not appear to be
significantly metabolized in vivo (82 and 88% were
recovered in male and female rat urine, respectively).
Repeat dosing did not generally indicate bioaccumulation
in any site except the kidney, primarily in male rats.
Fat appeared to be the site for some MCPA sequestration.
No additional metabolism studies are required.

Special Neurotoxicity Testing. MCPA is structurally
related to 2,4-D which is suspected of causing neuropathy
in humans. Special neurotoxicity testing of 2,4-D is
currently required. Because of this structural
similarity to 2,4-D, the Agency is also requiring special
neurotoxicity testing of MCPA.

Toxicology Profile of Other MCPA Formulations. As stated
previously from a toxicological standpoint one may expect
the acid and sodium salt to be identical. The organic

amines and esters may be significantly different and
lacking data to show otherwise, may have different
toxicological properties. The major exposure to these
compounds is during application therefore acute
toxicology data are required on the various derivatives.
The Agency has some acute testing on these derivatives
and they fall in toxicity Category III. Additional acute
testing is required.

Human Exposure.   The greatest potential for direct human
  exposure comes during mixing and loading operations.
  Exposure would be minimized by wearing of protective
  equipment. The Agency is requiring protective eyewear
  for mixers, loaders, and home users of MCPA acid
  products. No specialized protective language other than
  those required in 40 CFR 156 is required for other MCPA
  derivative homeowner use products. There have been no
  poisoning incidences with MCPA reported in California
  since 1974. Reentry data are not required since MCPA is
  generally in Toxicity Category III for acute studies.

Environmental Characteristics. Preliminary data have shown
  that MCPA degrades under aerobic laboratory conditions
  with a half-life of less than a week to 50 days. Under
  aerobic conditions 89% of parent MCPA remained undegraded
  for 374 days. MCPA is stable to hydrolysis and to
  photolysis in soils. MCPA salts are highly stable in
  water. Available data are insufficient to fully assess
  the environmental fate of MCPA and its various
  derivatives.

Groundwater Concerns. MCPA appears to be mobile from
  preliminary leaching studies. MCPA was found in well
  water in Missouri. Groundwater contamination appears to
  be associated with point sources. A special groundwater
  precautionary statement on labels is being imposed.

Ecological Effects

     Based on available data, MCPA acid has been determined
to be moderately toxic to avian species, slightly toxic to
freshwater fish, practically nontoxic to freshwater
invertebrates and estuarine and marine organisms. Additional
ecological effects data are required on MCPA and its
derivatives. The following data comes from acceptable
studies:

     o   Acute LD_{50} 377 mg/kg for Bobwhite quail

o   Dietary LC_{50} > 2000 ppm for mallard, bobwhite, and ring-necked pheasant

o LC_{50} rainbow trout = 89 ppm

o LC_{50} bluegill = 97 ppm

o LC_{50} Daphnids > 180 ppm

o LC_{50} Atlantic silverside = 179 ppm

o LC_{50} Oyster Larvae = 155 ppm

Effects on Plants.   The Agency has no data for toxicity to nontarget plants. A complete battery of tests are required using MCPA and all its derivatives.

Potential Problems Related to Endangered Species. Because of MCPA's demonstrated toxicity to nontarget species and its intended use pattern, several endangered species could be put at risk from the application of MCPA acid and its derivatives. The Agency has proposed a comprehensive Endangered Species Protection program (Federal Register 54(126) July 3, 1989).

Nontarget Insects. MCPA acid and its sodium salt derivative are relatively nontoxic to honeybees. Data are required on the dimethylamine salt and isooctyl ester as their use patterns allow for significant potential exposure to bees.

Product Chemistry. The Agency is requiring that all product chemistry data be resubmitted. Further, MCPA may be contaminated with dioxins or dibenzofurans, and the amine salts of MCPA may be contaminated with n-nitrosamines. Therefore, analytical data are required for certain products.

Tolerance Assessment

Tolerances for residues of MCPA per se in or on food and feed commodities are published in 40 CFR 180.339(a). Tolerances for residues of MCPA and its metabolite 2-methyl-chlorophenoxyacetic acid in or on animal commodities are published in 40 CFR 180.339(b).

The residue data reviewed in support of these tolerances showed the following:

- Data on metabolism of MCPA in plants available for

review in the 1982 document indicated that MCPA is readily taken up and translocated by plants. Additional data are required on the identity and quantities of residue in or on plants. The nature of the residue is not adequately understood.

- Data pertaining to the residues of MCPA in animals were reviewed for the 1982 document. That document did not require additional data on animal metabolism. Current Guidelines specify that terminal residues in animals be identified and quantified using radioactive material. The metabolism of MCPA in animals is not adequately understood and additional data are required.

- The current residue analytical methods in PAM I are adequate for enforcement of tolerances for residues in plants and animals.

- Data depicting the stability of MCPA residues in storage were not required in the 1982 document. Current Guidelines specify that storage stability data must be submitted in support of established tolerances.

- There are available data to support the established tolerances for MCPA in or on canarygrass seed and straw.

- Additional residue data are required on dried beans, peas (succulent and dry), pea vines and hay, rice grain, sorghum grains, wheat grains, rice straw, sorghum forage and fodder, wheat straw, annual canarygrass, pasture and rangeland grasses, grass hay, alfalfa and alfalfa hay, flaxseed, and flax straw.

- The data requested on wheat grain, forage, and straw may, by translation, support the established tolerances for residues of MCPA in or on the grain, forage, and straw of barley, oats, and rye. The data requested on alfalfa and alfalfa hay may, by translation, support the established tolerances for residues of MCPA in or on clover and clover hay, lespedeza and lespedeza hay, trefoil and trefoil hay, and vetch and vetch hay.

- Processing data are needed on rice grain, sorghum grain, wheat grain, and flaxseed. The requirements for processing data on barley, oat, and rye grain may be satisfied by the data requested on wheat.

- Tolerances need to be proposed for residues of MCPA in or on bean vines and hay, barley hay, oat hay, rye hay, wheat forage and hay, and canarygrass forage.

- Upon receipt of the data requested on animal metabolism and livestock feed items, the established tolerances for the combined residues of MCPA and 2-methyl-4-chlorophenol in the meat, fat, and meat by-products of cattle, goats, hogs, horses, and sheep and in milk will be assessed and the need for tolerances for residues in poultry tissues and eggs will be determined.

- A provisional acceptable daily intake (PADI, RfD) of 0.0015 mg/kg/day for MCPA has been established based on a 1-year feeding study (dog, NOEL 0.15 mg/kg). The value given is a PADI because of the teratology data gaps. However, when the teratology studies are submitted and found acceptable, they are not expected to greatly alter the RfD calculations. A safety factor of 100 was utilized. The dietary exposure was calculated using the published tolerances in 40 CFR 180.339. A dietary exposure for the U.S. population is calculated to be 0.001547 mg/kg/day, corresponding to 103 percent of the RfD. The population subgroups with the highest calculated exposure were nonnursing infants (0.007405 mg/kg/day, 493% of the RfD) and children 1 to 6 years of age (0.004069 mg/kg/day, 271% of the RfD). A dietary exposure was also conducted using the published tolerances factored by the percent of crop treated with MCPA. The dietary exposure for the U.S. population is then 10% of the RfD, for nonnursing infants, 51% of the RfD, and for children 1 to 6 years of age, 27% of the RfD.

4.   Summary of Regulatory Positions and Rationales

   - MCPA is not a candidate for special review.

   - MCPA does not meet the criteria for restricted use classification.

   - Precautionary labeling is required to minimize any hazard to nontarget organisms.

   - The Agency is requiring data on MCPA acid as well as its derivative amines and esters.

   - A special groundwater warning statement is required on the label because of MCPA's potential to contaminate these waters.

   - MCPA does not meet the criteria to require a reentry interval.

   - The Agency will consider establishment of significant new food use tolerances for MCPA.

5. Required Unique Labeling

    A. Groundwater Advisory Statements
    B. Environmental Hazards Statement
    C. Feeding and/or Grazing Restrictions
    D. Protective Clothing Statement
    E. Nontarget Species Precautionary Statements

6. Summary of Outstanding Data Requirements

                                  Timeframe Ranges

    Toxicology   1 - 4 years
    Environmental Fate/Exposure   1 - 4 years
    Ecological Effects   2 years
    Residue Chemistry   2 - 4 years
    Product Chemistry   1 - 2 years

7. Contact Persons at EPA

    Product Specific Inquiries:
    Joanna Miller
    Acting Product Manager (23)
    Fungicide Herbicide Branch
    Registration Division (H-7505C)
    Office of Pesticide Programs
    Environmental Protection Agency
    401 M Street, S.W.
    Washington, D.C. 20460

    Office location and telephone number:
    Room 245, Crystal Mall #2
    1921 Jefferson Davis Highway
    Arlington, VA 2202
    (703) 557-1800

    Reregistration Standard Document Inquiries:
    Philip T. Hundemann
    Review Manager
    Reregistration Branch
    Special Review and Reregistration Division (H-7508C)
    Environmental Protection Agency
    401 M Street, S.W.
    Washington, D.C. 20460

Office location and telephone number:
Room 1124, Crystal Mall #2
1921 Jefferson Davis Highway
Arlington, VA   22202
(703) 557-0933

DISCLAIMER: The information in this Pesticide Fact Sheet is a
summary only and is not be used to satisfy data requirements for
pesticide registration and reregistration.   The complete
Reregistration Document for the pesticide may be obtained from
the National Technical Information Service.   Contact the Review
Manager listed above for further information.

MECOPROP

Reason for Issuance: Registration Standard
Date Issued: December 1988
Fact Sheet Number: 192

DESCRIPTION OF CHEMICAL

Chemical Names:
2-(2-Methyl-4-chlorophenoxy)propionic
acid;
potassium 2-(2-methyl-4-chlorophenoxy)-
propionate;
diethanolamine 2-(2-methyl-4-
chlorophenoxy)-propionate;
dimethylamine 2-(2-methyl-4-
chlorophenoxy)-propionate;
isooctyl 2-(2-methyl-4-chlorophenoxy)-
propionate

ANSI Common Names:
MCPP (includes MCPP acid as well as the
MCPP potassium salt; MCPP diethanolamine
salt; MCPP dimethylamine salt; and
MCPP isooctyl ester)

Other Common Names: **Mecoprop**

Principal Trade Names: MCPP and Mecoprop

EPA (Shaughnessy) Code: 031501 (acid); 031503 (potassium salt);
                           031516 (diethanolamine salt); 031519
                           (dimethylamine salt), 031563 (isooctyl
                           ester).

Chemical Abstracts
Service (CAS) Number: 7085-19-0 (acid)
                           1929-86-8 (potassium salt)
                           1432-14-0 (diethanolamine salt)
                           32351-70-5 (dimethylamine salt)
                           28473-03-2 (isooctyl ester)

Year of Initial
Registration: 1964 (potassium salt)

Pesticide Type(s):   Herbicide; postemergence

Chemical Family: Chlorinated phenoxy

U.S. and Foreign
Producers: Aceto Chemical Co., Inc.;
                           Akzo Zout Chemie Nederland;
                           Dow Chemical USA; Gilmore, Inc.;
                           A.H. Marks Co., Ltd.; Rhone-Poulenc
                           Ag Co.

USE PATTERNS AND FORMULATIONS

Registered Uses: Terrestrial nonfood crop - Golf courses,
                           lawns, ornamental and sports turf, wide
                           area general indoor/outdoor treatments.
                           Aquatic nonfood crop - Drainage ditch
                           banks.
                           Forest trees - site preparation.

Principle Uses: Ninety-six percent of the production
                           is used on turf, including lawns, sport
                           turf and commercial sod production. A
                           small percentage (1-4%) is used in
                           noncrop areas such as rights-of-way,
                           drainage ditch banks and forest site
                           preparation.

Formulations: Granular, wettable powder, emulsifiable
                           concentrate, soluble concentrate/liquid,
                           pressurized liquid.

Methods of Application: Foliar applications by ground equipment
                           including spreader, sprayer, pressure on
                           hose-end sprayer and pressurized
                           containers.
                           Application by air on specific
                           noncropland sites.

SCIENCE FINDINGS

Summary Science Statement: Virtually all data
reviewed were on MCPP acid. Data are needed for the
ester and each salt, with the exception of the
potassium salt, for which the data on the acid will
suffice, before the Agency can completely assess the
effects the forms of MCPP pesticides to organisms
and the environment.

The Agency's Office of Pesticide Programs (OPP) has
classified MCPP as a teratogen, based upon results of
a study in rats. Margins of safety may not be
adequate for some workers; additional data are
required to refine estimates of worker exposure. A
decision on the need for further regulatory action
will be made when these data have been submitted and
evaluated.

MCPP is stable to hydrolysis, had slow
photodegradation in water, and was very mobile in
sand, sandy loam, silt loam, and silty clay loam
soils. The submission of data are required for
terrestrial nonfood and aquatic nonfood uses for all
type formulations.

Freshwater fish toxicity data on MCPP acid is
incomplete, with a rainbow trout study that revealed
very low acute toxicity of MCPP acid of 124 ppm.
There were no acceptable acute freshwater aquatic
invertebrate data submitted.

There were no acceptable data for toxicity of any
MCPP formulations to nontarget plants, endangered
species, or nontarget insects.

It is possible that chlorinated dibenzo-p-dioxins or
dibenzofurans may be formed during the manufacture of
MCPP. It is also possible that nitrosamines may be
formed during the manufacture or storage of the MCPP
amine salts. Thus, manufacturing process information
is required to assess the possibility of the
formation of these contaminants.

Chemical Characteristics

Physical state: Crystalline solid
Color: White to light brown
Odor: Odorless
Molecular Weight: 214.7
Empirical Formula: $C_{10}H_{11}ClO_3$
Boiling Point: Decomposes before reaching boiling point
Vapor Pressure: 1.4×10^{-5} mm Hg at 20°C; 4.7×10^{-5} mm
                   Hg at 30°C
Melting Point: 94°-95°C
Density, bulk: 0.6g/ml, dry uncompacted
Solubility: 0.06% in water at 20°C; readily soluble
                   in alcohol, benzene, acetone,
                   chlorinated hydrocarbons

Toxicology Characteristics

Acute Toxicity		Toxicity Category*
Acute Oral:	558 mg/kg	III
Acute Dermal:	Not lethal to rabbits at 2 g/kg	III
Acute Inhalation:	Data gap	
Primary Dermal Irritation:	No dermal irritation in rabbits exposed for 24 hours to abraded and unabraded skin	IV
Primary Eye Irritation:	Produces persistent corneal opacities, iritis, and conjunctival irritation in both washed and unwashed eyes.	I
Dermal Sensitization:	Data gap.	

*For a description of Toxicity categories, see 40 CFR 156.10

Delayed Neurotoxicity: Data are not available. Testing is not required because MCPP and its derivatives and metabolites are not organophosphates or cholinesterase inhibitors.

Subchronic Toxicity: Oral exposure of rats to technical MCPP caused increased creatinine levels in females and decreased glucose levels in males at 27 mg/kg; increased relative kidney weights in both sexes at 9 mg/kg and 27 mg/kg, respectively. The NOEL was 3 mg/kg.

Chronic Studies

Chronic toxicity: No data are available and none are required because this pesticide does not have food uses.

Oncogenicity: There were no data available nor are any required.

Teratogenicity & Developmental Toxicity: In a rat study, oral exposure to 125 mg/kg of MCPP on days 6-15 of gestation caused increased intra-uterine deaths, decreased crown/rump lengths, and an increased incidence of delayed or absent ossification of the sternebrae. The NOEL was 50 mg/kg/day. MCPP was not teratogenic in rabbits. A reproductive toxicity study is not required.

Mutagenicity

Gene Mutations. In a spot test, no increases in reverse mutations in 4 strains of Salmonella were seen with or without metabolic activation and no forward mutations to streptomycin resistance were seen in S.Coelicolor A3(2) in spot or standard plates, with or without activation.

Other Tests. MCPP caused an increase in sister chromatid exchange after single oral doses of 470 and 3800 mg/kg in Chinese Hamsters. Additional data are required.

Major Route of Exposure: Dermal.

Physiological and Biochemical Characteristics

MCPP is absorbed through leaves and roots and is translocated. It stimulates nucleic acid and protein synthesis and affects the efficacy of enzymes, either stimulating or inhibiting cell division.

Environmental Characteristics

Absorption and Leaching: Unaged MCPP is very mobile in sand, sandy loam, silt loam, and silty clay loam soils. Absorption has been positively correlated with the organic matter of the soils.

Hydrolysis: Technical MCPP at 51-52 ppm did not hydrolyze during 31 days of incubation in sterile buffered solutions (pH 5, 7, and 9) at 25°C ± 2°C.

Bioaccumulation: Available data indicate a low potential of MCPP to bioaccumulate in fish.

Ground Water: Available data indicate that MCPP is mobile in many soils and may potentially leach to ground water.

Ecological Characteristics

Avian acute toxicity: MCPP acid has been shown to have an acute oral LD_{50} of 700 mg/kg to the bobwhite quail and is considered to be of low toxicity.

Avian dietary toxicity: Studies on MCPP acid indicate LC_{50} values of >5620 ppm and 5000 ppm for the mallard and the bobwhite quail, respectively. These values are considered practically nontoxic to avian species on subacute basis.

Freshwater fish acute toxicity: MCPP acid has very low toxicity to the rainbow trout (LC_{50} = 124 ppm).

Freshwater invertebrates toxicity: There were no acceptable freshwater invertebrate studies submitted.

Fish early lifestage and aquatic invertebrate life cycle toxicity: There were no fish early lifestage nor aquatic invertebrate life cycle studies submitted.

Nontarget plant toxicity: There were no acceptable nontarget plant studies submitted.

Nontarget insects: There were no data on effects of MCPP to honeybees and nontarget insects submitted.

Pesticide Incidents

There were no reports of MCPP pesticide incidents.

SUMMARY OF REGULATORY POSITION

1.   Special review: The Agency is deferring a decision on initiating a Special Review of MCPP based on teratogenicity until data needed to refine worker exposure estimates have been submitted and evaluated.

2.   Restricted Use: The Agency will not restrict the use of MCPP products at this time.

3.   Special Data Requirements: The Agency is requiring a submission of special studies to refine exposure estimates to workers. These data will be used to assess the teratogenic risk posed by MCPP to workers or other persons exposed to MCPP.

4.   Basic Data Requirements: The Agency is requiring data specific to the amines and the ester formulations as well as the

acid.   Acceptable data received on the MCPP acid will be
generally acceptable for the potassium salt.

5.   Groundwater concerns: MCPP has the potential to contaminate
groundwater because it is mobile in many mineral soils.   Studies
to evaluate the potential of MCPP and its metabolites to
contaminate groundwater are required.   A groundwater statement
is required (see labeling section below).

6.   Reentry requirement: Reentry data are required because of
the potential teratogenic risk and because some use patterns
involve post-treatment tasks that could result in dermal
exposures of workers to MCPP residues.   Workers are not permitted
to reenter treated areas until sprays have dried and dusts have
settled (see labeling section below).

7.   Protective clothing requirement: The Agency is requiring
mixers/loaders and applicators to use protective clothing and
equipment.   See the protective clothing statement in the labeling
section below.

8.   Endangered species regulatory position: There has been no
determination that MCPP acid or any MCPP formulation would
affect endangered animal or plant species.

9.   Dibenzo-p-dioxins/Dibenzofurans: The Agency is requiring
submission of manufacturing process data on MCPP to assess the
possibility of contamination by dibenzo-p-dioxins or
dibenzofurans.

10.   Nitrosamines: The Agency is requiring chemical analysis of
products containing MCPP amine salts for possible nitrosamine
contamination.

11. Data for Immediate Review: The Agency has identified data
which will be reviewed as soon as they are received: product
chemistry; environmental fate; dermal absorption, applicator
exposure and reentry; toxicity on the amine salts and ester; and
Tier II phytoxicity.

LABELING-REQUIREMENTS

   Statements for Manufacturing-Use Products

   "Do not discharge effluent containing this product into
   lakes, streams, ponds, estuaries, oceans or public waters
   unless this product is specifically identified and
   addressed in an NPDES permit.   Do not discharge effluent
   containing this product to sewer systems without previously
   notifying the sewage treatment plant authority.   For
   guidance, contact your State Water Board or Regional Office
   of the EPA."

Statements for End-Use Products

Ground Water Contamination: "Although highly mobile, in general phenoxy herbicides such as MCPP are not sufficiently persistent to reach ground water from use as specified on product labels. Most cases of ground water contamination involving phenoxy herbicides have been associated with mixing/loading and disposal sites. Caution should be exercised when handling MCPP at such sites to prevent contamination of water supplies."

Protective Clothing: "Mixer/loader and applicators must wear long pants, long sleeve shirts, and gloves. Persons applying liquid MCPP products such as sprays are required to use goggles to cover their eyes."

Worker Reentry:
"Do not reenter or permit workers to reenter treated areas until sprays have dried or dusts have settled."

Nontarget Plant Species:

Aquatic uses:
"Drift or run-off may adversely affect nontarget plants. Do not apply directly to water except as specified on this label. Do not contaminate water when disposing of equipment washwaters."

Nonaquatic uses:
"Drift or run-off may adversely affect nontarget plants. Do not apply directly to water or wetlands (swamps, bogs, marshes and potholes). Do not contaminate water when disposing of equipment washwaters."

SUMMARY OF DATA GAPS

The following data are required as specified. Uses of MCPP potassium salt may be supported with data on either the salt or MCPP acid. Codes for the different formulations are MCPP Acid (1), Potassium Salt (2), Diethanolamine (3), Dimethylamine (4), and Isooctyl Ester (5).

Study	MCPP Formulation	Due Date - From Date of Standard
Product Chemistry:	All	9-12 months
Toxicology:		
Acute Oral, Dermal, Inhalation, Eye	3, 4, 5	9 months

```
Irritation, Dermal
Irritation and
Sensitization

Subchronic
  21-Day Dermal           All                  12 months

Chronic
  Teratogenicity          3, 4, 5              15 months

  Mutagenicity            3, 4, 5               9 months
  Chromosomal Aberration  All                  12 months

  Other (sister
    chromatid exchange)   3, 4, 5              12 months

Special Testing
  Dermal Absorption       All                   9 months
  Applicator Exposure     All                  (protocols due
                                                 in 6 months)
```

Reentry Protection:

```
  Foliar Dissipation      All                  27 months
  Soil Dissipation        All                  27 months
  Dermal Exposure         All                  27 months
```

Environmental Fate:

```
Hydrolysis                All                   9 months
Photodegradation
  Water and Soil          All                   9 months
Metabolism
  Aerobic Soil            All                  27 months
  Anaerobic Aquatic       2, 4, 5              27 months
  Aerobic Aquatic         2, 4, 5              27 months
Mobility
  Leaching and Ad-
    sorption/Desorption   All                  12 months
Dissipation
  Soil                    All                  27 months
  Aquatic (sediment)      2, 4, 5              27 months
  Forestry                4                    27 months
Accumulation
  Irrigated Crops         2, 4, 5              39 months
  Fish                    All                  12 months
  Aquatic Nontarget
    Organisms             2, 4, 5              12 months
```

Wildlife & Aquatic Organisms:

Avian & Mammalian:

Acute Avian Oral	3, 4, 5	9 months
Avian Subacute		
Upland Game Bird	3, 4, 5	9 months
Waterfowl	3, 5	9 months

Aquatic Organism:

Freshwater Fish			
Coldwater Fish	TGAI	3, 4, 5	9 months
Coldwater Fish	TEP	1, 3, 4, 5	9 months
Warmwater Fish	TGAI	All	9 months
Warmwater Fish	TGAI	2, 4, 5	9 months
Freshwater			
Invertebrates	TGAI	All	9 months
Invertebrates	TEP	2, 4, 5	9 months

Plant Protection:

Nontarget Phytotoxicity - Tier II:

Seed Germina-			
tion/Emergence	TGAI	All	9 months
Vegetative Vigor	TGAI	All	9 months
Aquatic Plant			
Growth	TGAI	All	9 months

Nontarget Insect:

Honey Bee Contact	TGAI	2, 3, 4, 5	9 months

CONTACT PERSON AT EPA: Mr. Richard Mountfort
                         Product manager (Team 23)
                         Fungicide-Herbicide Branch
                         Registration Division (TS-767C)
                         Office of Pesticide Programs
                         Environmental Protection Agency
                         401 M St., SW
                         Washington, D.C. 20460

Physical Location: Crystal Mall #2
                    1921 Jefferson Davis Highway
                    Arlington, VA
Telephone number: (703) 557-1830

DISCLAIMER: The information in this Pesticide Fact Sheet is a
summary only and may not be used to satisfy data requirements
for pesticide registration and reregistration. The complete
Registration Standard for the pesticide may be obtained from the
National Technical Information Service. Contact the Product
Manager listed above for further information.

METALAXYL

Reason for Issuance: Registration Standard
Date Issued: September 1988
Fact Sheet Number: 155.1

DESCRIPTICN OF CHEMICAL

Generic Name: N-(2,6-Dimethylphenyl)-N-(Methoxyacetyl) Alanine
Methyl ester.
Common Name: Metalaxyl
Trade Names: Ridomil, Subdue, Apron, Proturf.
EPA Shaughnessy Code: 113501
Chemical Abstracts Service (CAS) Number: 57837-19-1
Year of Initial Registration: 1979
Date of Initial Registration Standard: December 1981
Pesticide Type: Systemic Fungicide
U.S. and Foreign Producers: Ciba-Geigy, O.M. Scott and Son, Co.
Wilbur-Ellis, and Gustafson.

USE PATTERNS AND FORMULATIONS

Application Sites: Metalaxyl is registered for use on over 100
agricultural crops (including more than 30
seed treatment uses). Metalaxyl is also
registered for ornamental and turf uses.

Major Uses: More than 90% of the total poundage of metalaxyl
used domestically is used in the following ten
crops/sites: tobacco, turf, potatoes, ornamentals,
soybean (seed treatment), onions, citrus, cucurbits,
tomatoes, and cotton.

Application Rates: Metalaxyl is applied to soil or foliage.
Application rates range from 0.135 to 8.0 lb
ai/acre for agricultural crops, from 0.25 to
1.12 oz ai/100 lb seed for agricultural seed
treated, from 0.33 to 1.35 lb/ai/acre for
ornamental turf, and from 0.90 to 7.20 lb/ai
acre for ornamental trees and plants. Multiple applications (varying with use) are
approved.

Method of Application: Foliar application; soil application by
incorporation, surface spraying (broadcast
or band), drenching, sprinkler or drip
irrigation; soil mixing; trunk spraying.
For agricultural seed treatment metalaxyl
is applied with conventional slurry or
mist seed treating equipment.

Types Registered: Single active ingredient products containing
metalaxyl are formulated as a granular (G),
pelleted/tableted (P/T) (in fertilizer spikes),
wettable powder (WP), emulsifiable concentrate
(EC), and flowable liquid concentrate (FlC),
as well as a 90% technical product. The
granular, wettable powder and emulsifiable
concentrate formulations are also formulated
as multiple active ingredient products. In
addition, metalaxyl is sold in a combination
with mancozeb, chlorothalonil, pentachloronitrobenzene, captan, and triadimefon.

SCIENCE FINDINGS

Summary of Science Statements

Studies indicate that metalaxyl is not oncogenic or teratogenic. Studies also indicate that metalaxyl does not cause increased incidence of tumors or cause embryotoxic, fetotoxic or
teratogenic effects. Metalaxyl also does not cause reproductive
effects nor did it induce gene mutations in bacteria, yeast, and
mouse lymphoma cells and does not cause chromosomal aberrations
in tests with yeast, hamsters, and mice.

Metalaxyl was found to be practically nontoxic acutely and
subacutely to avian species and to present no adverse effects to
avian and mammalian populations. Metalaxyl poses no hazard to
endangered terrestrial or aquatic animal species or to plant
species.

Physical/Chemical Characteristics

Technical

        Physical State: Crystalline
        Color: White to beige
        Odor: Odorless
        Solubility: Water   - 00.7%
                           Benzene   - 55.0%
                           Hexane   - 00.9%
                           Methanol - 65.0%
                           Isopropanol - 27.0%
                           Methylene Chloride - 75.0%
        Stability: Stable up to 300°C; slight exothermic reaction
                    up to 450°C
        Melting Point: 71 - 72°C
        Vapor Pressure: 2.2×10^{-6} Torr at 20°C
        Density: 1.21 g/cm^3 at 20°C

Toxicology Profile

Acute Toxicity:

o Acute Oral Toxicity   Toxicity Categoty III
                                                     (moderate acute oral)
              (Rat):   669 mg/kg
             (Mice):   788 mg/kg
          (Hamster):   7120 mg/kg

o Acute Dermal Toxicity   Toxicity Category III
                                                     (moderate acute dermal)
           (Rabbit): >6000 mg/kg
             (Rat):   >3170 mg/kg

o Primary Eye Irritation   Toxicity Category II
                                                     (moderate eye irritant)
           (Rabbit): No effect

o Primary Skin Sensitization   Toxicity Category IV
                                                     (slight skin irritant)
           (Rabbit): slight effect

o Primary Dermal Sensitization   Toxicity Category IV
                                                     (not a sensitizer)
        (Guinea Pig): No effect

Subchronic Feeding Studies

Rodent Feeding Studies

    In 90-day feeding studies in rats and mice the liver was
    the target organ for metalaxyl toxicity.

    In female rats a NOEL of 250 ppm (12.5 mg/kg/day) and a
    lowest-observed effect level (LOEL) of 1250 ppm (6.25 mg/
    kg/day) were observed.

    In male mice, a NOEL of <1250 ppm (187.5 mg/kg/day), and
    a LOEL of 1250 ppm (6.25 mg/kg/day) were observed.

Non-Rodent Feeding Study

    A 90-day feeding study in dogs showed no toxicity up to
    1250 ppm in their diet (31.25 mg/kg/day; highest dose tested).

Subchronic Dermal Toxicity

    Metalaxyl had no effect on rabbits when applied to intact
    or abraded skin for 21 days at doses up to 1000 mg/kg/day.

Subchronic Inhalation Toxicity

No effects were observed in rats exposed to smoke from ciga-
rettes containing metalaxyl. The NOEL in this study is
greater than 13,000 ppm (highest dose tested).

Chronic Toxicity

Metalaxyl had minimal effects in chronic feeding studies
with rats and dogs. The NOEL established in rats was 250 ppm
(12.5 mg/kg/day), and the LOEL was 1250 ppm (62.5 mg/kg/day).
In dogs the LOEL was established at 1000 pppm (250 mg/kg/day),
and the NOEL was 250 ppm (62.5 mg/kg/day).

Oncogenicity

The long-term feeding studies in rats and mice showed no increase in the incidence of tumors as a result of metalaxyl.

Teratology

In pregnant rats, NOEL's for maternal and developmental toxicity
were established at 50 mg/kg/day, and LOEL's for both types of
toxicity were established at 250 mg/kg/day.

Metalaxyl caused no embryotoxic, fetotoxic, or teratogenic
effects in treated rabbits. The NOEL for maternal toxicity
was 300 to 500 mg/kg/day, and the developmental toxicity was
greater than 300 mg/kg/day (highest dose tested in the main
study).

Reproduction

In a multi-generation reproduction study with rats, no dose-
related effects were observed throughout the three generations
with respect to toxicity or reproductive parameters. The NOEL
for reproductive and developmental toxicity is greater than
1250 ppm (12.5 mg/kg/day).

Mutagenicity

Metalaxyl did not induce gene mutations in bacteria, yeast and
mouse lymphoma cells in vitro with or without metabolic activation. The fungicide also caused no structural or numerical
chromosomal aberrations as indicated by yeast, hamsters or mice.
No DNA damage was observed in bacteria and no unscheduled DNA
synthesis was noted in rat primary hepatocytes or human fibro-
blasts in vitro as the result of exposure to metalaxyl. These
results suggest that metalaxyl is not genotoxic.

Metabolism

Metabolism studies in rats showed that single oral doses of
metalaxyl are readily absorbed. Approximately 62 to 65% of

the administered radioactivity is recovered in the urine and feces within 24 hours after dosing, and 96% is recovered during the 48 hours after dosing. The major route of excretion in males was the feces after oral and dermal doses, while that for females was the urine for both routes of administration. The distribution of radiolabel observed six days after dosing did not indicate that metalaxyl residues were stored in tissues following a single dose.

Dermal Absorption

The absorption $T_{1/2}$ values for metalaxyl in tetrahydrofuran (THF) ranged from 12 hours (in male rats receiving a 1 mg/kg dermal dose) to 20 hours (male rats given a dermal dose of 10 mg/kg). The value for females was 13 hours for both doses.

ECOLOGICAL CHARACTERISTICS

Metalaxyl has been found to be practically nontoxic acutely and subacutely to avian species and to present no adverse effects to avian and mammalian populations. There is no indication of detrimental effects on aquatic plant species and the technical pesticide is practically nontoxic to freshwater aquatic animal species. The most sensitive organism appears to be Daphnia magna, having LC_{50}s of 28 and 12.5 ppm with technical and formulated pesticides, respectively.

Chronic toxicity assays on aquatic species and fish accumulation testing do not suggest that metalaxyl presents a long-term risk in the aquatic environment. It appears very unlikely that metalaxyl could accumulate in water or sediments to concentrations that would pose a risk to aquatic populations.

Although use of metalaxyl presents little risk to freshwater populations it cannot be assumed that the same holds true for marine/estuarine species which may be exposed in connection with several of the registered uses. The Agency is requesting data on marine/estuarine species such as oysters and shrimp.

Endangered Species

The registered uses of technical metalaxyl and a widely used formulation, Ridomil® 27.9% ai EC (which appears to be more toxic to aquatic species than the technical), do not present a hazard to endangered terrestrial or aquatic animal species or plant species.

Environmental Fate

Metalaxyl was found to be moderately stable under normal environmental conditions. Fish accumulation was found not to exceed 7X when fish were exposed to metalaxyl at 1 ppm in water, and the residues were found to accumulate in the nonedible portions over the edible portions. Residues declined rapidly during depuration.

The rotational crop data demonstrated the need for a 12-month rotational crop restriction because some crops will take up metalaxyl residues of concern when planted 12 months or more after treatment of a prior crop. Confined studies are needed to identify all residues of concern plus field tests are necessary to determine the need for additional tolerances.

In addition, ground water monitoring studies were required early in the registration process for metalaxyl. While subsequent submissions were judged to be sufficient at the time, these studies are no longer adequate and further data are required.

TOLERANCE REASSESSMENT

Tolerances have been established for residues of metalaxyl in numerous varieties of raw agricultural commodities (40 CFR 180.408) and also in food and feed commodities (21 CFR 193.277 and 21 CFR 561.273, respectively). In addition, tolerances have also been established for indirect or inadvertant residues of metalaxyl.

The acceptable daily intake (ADI) is based on the six month feeding study in dogs (NOEL of 6.3 mg/kg/of body weight/day) and a 100-fold safety factor. Therefore, the ADI is calculated to be 0.063 mg/kg/day.

REGULATORY POSITION

This review of metalaxyl is the second intensive evaluation of the compound. A Registration Standard was developed in 1981 in conjunction with its initial registration. At that time metalaxyl was registered for non-food uses on tobacco, conifers, ornamentals, and turf and was not registered for any food or feed uses. The only additional data needed to support the registered non-food uses in 1981 were groundwater monitoring, subchronic inhalation toxicity, phytotoxicity and storage stability. Since the issuance of the 1981 Registration Standard, registrations have been approved for use on over 100 agricultural crops. These registrations were granted based on adequate supporting data (including residue, acute and chronic data) at the time of application for registration.

In 1984, the Agency promulgated general rules at 40 CFR Part 158, which set forth the range of data which must be submitted to EPA to support the registration or reregistration of each pesticide under FIFRA. Based on these revised and expanded data requirements, the toxicity data base for metalaxyl is still virtually complete and in most cases is adequate to support continued registration of existing uses. However, some data determined to support registration in the past only partially fullfill current data requirements. As a result, several studies primarily in the disciplines of residue chemistry and environmental fate, must be conducted and submitted to the Agency. In addition, several new data requirements are being imposed to characterize potential adverse effects to marine/estuarine species.

The following Agency positions are based on the substantially complete data base currently available for metalaxyl:

o   Metalaxyl is not being placed in Special Review at this time because none of the risk criteria listed in 40 CFR 154.7 prescribing a Special Review have been met.

o   The Agency is requesting rotational crop studies and, in order to meet the statutory standard for continued registration, product labeling must bear a 12-month rotational crop restriction as an interim measure.

o   Additional ground water monitoring and laboratory leaching studies are being required. The Agency has determined that data submitted on ground water monitoring are inadequate.

o   Ground water monitoring data will be reviewed when submitted in order for the Agency to determine whether further regulatory action is warranted based on this concern.

o   The Agency is requiring further data on potential adverse effects to marine/estuarine species to determine if currently registered uses will result in exposure levels of concern to these populations.

o   While data gaps are being filled, currently registered manufacturing-use products (MP's) and end use products (EP's) containing metalaxyl as the sole active ingredient may be sold, distributed, formulated, and used, subject to the terms and conditions specified in this Standard. However, registrants must provide or agree to develop additional data, as specified in the Data Appendices, in order to maintain existing registrations.

LABELING REQUIREMENTS

All metalaxyl products must bear appropriate labeling as specified in 40 CFR 156.10. Appendix II of the Standard contains information on label requirements.

In order to remain in compliance with FIFRA, no pesticide product containing metalaxyl may be released for shipment by the registrant after October 30, 1989, unless the product bears amended labeling which complies with the specifications in the Standard.

In order to remain in compliance with FIFRA, no pesticide product containing metalaxyl may be distributed, sold, offered for sale, held for sale, shipped, delivered for shipment, or received and (having been so received) delivered or offered to be delivered by any person after October 30, 1990, unless the product bears amended labeling which complies with the specifications of the standard.

In addition to the above, in order to remain in compliance with FIFRA, the Agency is requiring:

   o Revised environmental hazard labeling

   o 12-month rotational crop statement

SUMMARY OF MAJOR DATA GAPS

40CFR§158.120 - Product Chemistry
   o Description of Beginning Materials and
      Manufacturing Process
   o Discussion of the Formation of Impurities
   o Preliminary Analysis
   o Certification of Ingredient Limits
   o Analytical Methods to Verify Certified Limits

40CFR§158.125 - Residue Chemistry
   o Nature of the Residue in Livestock
   o Residue Analytical Method
   o Storage Stability Data
   o Magnitude of the Residue (potatoes; sugar beet roots;
      soy beans; cereal grains; forage, fodder and straw of
      cereal grains; cottonseed; hops; peanuts; pineapples;
      sunflower seed)

40CFR§158.130 - Environmental Fate
   o Photodegradation Studies in Water
   o Terrestrial Field Dissipation Studies
   o Confined Accumulation Studies on Rotational Crops
   o Ground Water Monitoring and Laboratory Leaching Studies

40CFR§158.135 - Toxicology
   o Acute Inhalation Toxicity
   o Metabolism Studies

40CFR§158.145 - Ecological Effects
   o Acute LC_{50} Estuarine/Marine Organisms (shrimp and oyster)

CONTACT PERSON AT EPA: Lois A. Rossi
                          Product Manager (21)
                          Fungicide-Herbicide Branch
                          Registration Division (TS-767C)
                          Office of Pesticide Programs, EPA
                          Washington, D.C.   20460

                          Telephone: (703) 557-1900

DISCLAIMER: The information in this Pesticide Fact Sheet is a summary only and is not to be used to satisfy data requirements for pesticide registration and reregistration. The complete Registration Standard for the pesticide may be obtained from the contact person listed above.

METALDEHYDE

Reason for Issuance: Registration Standard
Date Issued: December 23, 1988
Fact Sheet Number: 191

1. DESCRIPTION OF CHEMICAL

   Common Name : Metaldehyde
   Generic Name: 2,4,6,8-tetramethyl-1,3,5,7-
       tetraoxycyclo-octane
   Trade Name : Meta, Metason, Halizan, Antimilace,
       Namekil, Cekumeta
   OPP Chemical Code: 053001
   Chemical Abstracts Service (CAS) Number: 9002-91-9
       (homopolymer); 108-62-3 (tetramer)
   Year of Initial Registration: 1962
   Pesticide Type: Molluscicide
   Chemical Family: Hydrocarbon
   U.S. Producer: Lonza, Incorporated

2. USE PATTERNS AND FORMULATIONS

   Application Sites: Terrestrial (food and nonfood crop);
       greenhouse (food and nonfood crop); domestic and
       nondomestic outdoor; indoor.

   Predominate Uses: Home orchards; gardens and orna-
       mental; and commercial vegetables (artichokes,
       broccoli, cauliflower, lettuce, cabbage, and
       tomatoes) and fruit (citrus, stone, pome and
       strawberries).

   Methods of Application: Hand, ground or aerial
       equipment.

   Types of Formulations: Wettable powders, dusts,
       emulsifiable concentrates, flowable concentrates,
       granules, ready-to-use products (liquids) and
       pelleted/tableted.

3. SCIENCE FINDINGS

   Summary Science Statement: Metaldehyde is of moderate
       to low acute toxicity and is not a dermal irritant.

430

Subchronic and chronic toxicity data are not available to assess this chemical. Poisoning incident reports show that the use of metaldehyde has been associated with dog poisonings, resulting in death.

Data are not available to assess the ecological effects or environmental fate of metaldehyde.

Chemical Characteristics:

Physical State: Crystalline powder
Color: White
Odor: Aldehyde odor
Melting Point: 246°C
Bulk Density: 65g/100cc
Solubility: at 20°C, 0.2 g/L in water and very
    slightly soluble in alcohol
Vapor Pressure: 1mm Hg at 20°C
Flammability: 36-40°C

Toxicology Characteristics: The only acceptable toxicological study is a partially acceptable mutagenicity study. Acute studies are available, however, the purity of the test material must be established before they can fulfill guideline requirements.

Acute Toxicology:

Oral--Rat: 383 mg/kg (females), Toxicity
    Category II[1]; 734 mg/kg (males), Toxicity
    Category III

Dermal--Rat: >5 g/kg, Toxicity Category III

Inhalation--Rat: >13.5 mg/L/4 hr, Toxicity
    Category III

Primary Dermal Irritation--Rabbit: Not a
    dermal irritant

Mutagenicity: In the acceptable portion of a
    paritally acceptable study, metaldehyde was
    not found to be mutagenic.

Environmental Characteristics: Data are not available to assess the environmental characteristics of metaldehyde.

---

[1]Refer to 40 CFR 156.10 for a discussion of toxicity categories.

Ecological Characteristics: None of the available data
meet Agency requirements but they do suggest that
metaldehyde is of low toxicity to birds and is
practically nontoxic to aquatic organisms.

Tolerance Assessment: Although tolerances had not been
generally required for food/feed uses of baits,
there is a food additive regulation of zero parts
per million (ppm) for strawberries (40 CFR
185.4025). This is not supported by adequate data.
The toxicological data are inadequate to allow an
assessment of any dietary exposure to metaldehyde.

4.   SUMMARY OF REGULATORY POSITION AND RATIONALE. The
Agency, historically, did not consider the use of
baits in-field to protect fruits and vegetables to
be a food use and tolerances were not required.
However, updated guidelines, issued in October
1982, state that if residues could occur in foods
or feed, the use is considered to be a food use and
a petition for tolerance or exemption from
tolerance is required. The Agency has no basis for
concluding that residues will not occur in the food
and feed crops treated with metaldehyde.
Therefore, the Agency will require data on
metaldehyde to support food uses.

The Agency will also continue to require
precautionary labeling statements to protect dogs
and other pets.

5.   SUMMARY OF MAJOR DATA GAPS. Essentially all data,
including toxicological data required to support
food uses, are required for metaldehyde.

6.   EPA CONTACT

Dennis Edwards, Product Manager
Office of Pesticide Programs
Registration Division (TS-767C)
401 M Street SW.
Washington, DC 20460

Telephone: (703) 557-2386

DISCLAIMER: The information presented in this Pesticide Fact
Sheet is for informational purposes only and may not be used
to fulfill data requirements for pesticide registration and
reregistration.

METHOMYL

Reason for Issuance: Issuance of Registration Standard
Date Issued: April 1989
Fact Sheet Number: 201

1. Description of Chemical

-Generic Name: S-methyl N-[(methylcarbamoyl)oxy]-
                  thioacetimidate

-Common Name: Methomyl

-Trade Name and
  other names: methyl N-[[(methylamino)carbonyl]oxy]-
                  ethanimidothioate, methyl N-[(methylcar-
                  bamoyl)oxy]-thioacetimidate, Lannate,
                  Lanox, and Nudrin

-EPA Shaughnessy Number: 090301

-Chemical Abstracts Service (CAS) Number: 16752-77-5

-Year of Initial Registration : 1968

-Pesticide Type: Insecticide

-Chemical Family: Carbamate

-U.S. Producer: E.I. duPont de Nemours and Company

2. Use Patterns and Formulations

-Application sites: Terrestrial food and non-food crops,
                     Greenhouse food and non-food crops,
                     Aquatic food crops, Forestry (ground
                     only), and Indoor- human and animal
                     premise

-Types and method of applications: Foliar and broadcast
 soil application by both ground and aircraft equipment

-Application rates: 0.1 to 1.5 lb a.i. per acre

-Types of formulations: Wettable powders, emulsifiable
                          and soluble concentrates, gran-
                          ulars, baits, and dusts

-Usual carriers: Petroleum and clay carriers

3. Science Findings

Physicochemical Characteristics:

Technical methomyl is a white crystalline solid with a melting
point of 78-79 C. Methomyl is soluble in water and most
organic solvents. Methomyl's empirical formula is
$C5H10N2O2S$ and its molecular weight is 162.2.

Toxicological Characteristics:

-Acute Oral: 17 to 24 mg/kg (rat)   Toxicity Category I
-Acute Dermal: >5000 mg/kg (rabbit)   Toxicity Category III
-Primary Eye Irritation: Data gap
-Acute Inhalation: 0.30 mg/liter/4 hours Toxicity Category III
-Primary Skin Irritation: Data gap
-Dermal Sensitization: Data gap
-Acute Delayed Neurotoxicity: Data shows no potential for
 this effect.
-Subchronic dermal (21 day):   Data gap
-Oncogenicity: A rat chronic/oncogenicity study showed no
 oncogenic effects at the highest dose tested (HDT-400 ppm).
 A mouse oncogenicity study showed no oncogenic effects at
 HDT 200 ppm.
-Metabolism: Data Gap- Preliminary data suggests that
 methomyl may be metabolized to the possible human oncogen
 acetamide. Additional metabolism data in the rat and mon-
 key are required to detect the possible presence of acetamide
 in the tissues.
-Teratology: Not teratogenic or embryotoxic at HDT 400 ppm
 in rats. Not teratogenic or embryotoxic at HDT in rabbits
 (Maternal No Observed Effect Level [NOEL] of 2 mg/kg/day).
-Reproduction: No observed effects with a NOEL of 100 ppm.
-Mutagenicity: Not a mutagen in all the required tests.

Physiological and Biochemical Characteristics:

-Mechanism of Pesticide Action: Methomyl kills by poisoning
 the insects' nervous system.
-Metabolism and persistence in plants and animals:
 Plants: Available data demonstrate that methomyl is rela-

tively persistent on leaf surfaces and fruit.   There has
been some concern in the recent past that acetamide, a
suspected human carcinogen, may occur in plants following
treatment with methomyl.   However, recently available data
on thiodicarb, a related insecticide that breaks down initially
to methomyl in plants, reveal that acetamide will not occur
in plants treated with methomyl.   Furthermore, any acetamide
formed from acetronitrile, a known metabolite of methomyl, will
be hydrolyzed to form acetic acid and ammonium ion.

Animals: The available data are not adequate to assess the
nature of methomyl in animals.   Additional metabolism data
(ruminants and poultry) are required to detect the possible
presence of acetonitrile and acetamide.   These data were
requested in a FIFRA 3(c)(2)(B) letter dated March 23, 1987.

Environmental Characteristics:

-Available data are insufficient to fully assess the environmental fate of methomyl.   The Registration Standard issued
in 1981 did not address the aquatic uses of methomyl.   Data
must now be submitted for the following: aquatic aerobic,
aquatic anaerobic, aquatic field dissipation and irrigated
crops.   Vapor pressure data indicate the need for volatility
data.   Monitoring data are needed to assess the potential of
this pesticide to contaminate groundwater.

Ecological Characteristics:

-Acute avian oral toxicity: LD50 24.2 mg/kg for bobwhite
quail (highly toxic).
-Avian dietary toxicity: LC50 of 1100, 1975, and 2883 ppm
respectively for bobwhite quail, ring-necked pheasant, and
mallard duck (slightly toxic).
-Freshwater fish acute toxicity: LC50 = 1.6 ppm for Rainbow
trout.   LC50 = 0.5 ppm for channel catfish (moderate to
highly toxic for fish).
Freshwater aquatic invertebrate toxicity: LC50 values of
0.0698 to 0.343 ppm suggest that is very highly toxic to
freshwater invertebrates.

Major Routes of Exposure:

-Dermal followed by inhalation. Human exposure occurs from mixing,
loading and application.   Exposure can be reduced by the use of
goggles or face shield and protective clothing.

Tolerance Assessment:

-Tolerances have been established for residues of methomyl in
a variety of raw agricultural commodities (Refer to 40
CFR 180.253 for listing of tolerances).   Methomyl's tolerances

have been reassessed using the Tolerance Assessment System
(TAS). The Acceptable Daily Intake (ADI) for this chemical
is 0.025 mg/kg/day. The Theoretical Maximum Residue Contribution (TMRC) for the U.S. population is 0.016188 mg/kg/day,
which occupies 65% of the ADI.

4.   Summary of Regulatory Position

   This review of methomyl is the second intensive evaluation
of the compound. In its original Standard, issued in 1981,
the Agency summarized the available data supporting the registration of methomyl and concluded that additional data
were needed data to fully evaluate the pesticide.

   The Agency has since received and reviewed the data
and has revised its scientific and regulatory conclusions in
relative to these data. Additionally, other information cn the
chemical (for example the acetimide issue) and the expanded
data requirements promulgated in 1984 at 40 CFR Part 158,
have added new data requirements. This Standard, which
supercedes the 1981 document, is the Agency's updated assessment of the pesticide and the data needed to support its
continued registration.

A.   Methomyl is not being placed in Special Review at this
time because none of the criteria for initiation of Special
Review listed in 40 CFR 154.7 have been met. The Agency
believes that the water soluble bag use restriction and the
increased reentry intervals provide mixer/loader and field-
worker protection. The Agency intends, however, to monitor
State pesticide incidents monitoring systems to determine
the effectiveness of labeling changes identified in the
Standard. The Agency may impose further regulatory actions
if these incidents reports indicate that these labeling changes
are inadequate.

B.   The Agency is requiring the submission of acute aquatic
toxicity data and aquatic and non-aquatic field monitoring
data on the end-use formulations and aquatic life stage data
and avian reproduction data on the technical formulation in
order to complete the wildlife risk assessment.

C.   Based on methomyl's use pattern and toxicity data, the
Agency has determined that methomyl may trigger the endangered
species criteria for fish, aquatic organisms and insects.
No endangered species labeling is required at this time. A
program is being developed by the Agency to reduce or eliminate exposure of this chemical to these species. After this
program is developed, the Agency will notify registrants of
any additional labeling that may be required to remain in
compliance with FIFRA. The labeling requirements affecting

methomyl, e.g. those listed in PR Notices 87-4 and 87-5, have been withdrawn pending reissuance.

D. Various methomyl formulations were classified as restrict-ed use products by regulation in 1978 (see 40 CFR 162.13). The Agency has now determined that the 90% water soluble bag formulation should also be classified as a restricted use pesticide. Labeling language for each of the restricted use products must specify that the restriction is based on high acute toxicity to humans.

E. The following reentry intervals are being imposed at this time based on the submitted reentry data: one day for beans, cabbages, roses grown outdoors and carnations, whether grown outdoors or in a greenhouse; three days for cotton, nectarines, and oranges/citrus; four days for peaches; and seven days for grapes. Because of the similarity in crops and in the work tasks performed in those crops, a three day reentry interval is being established for apples, and a one day reentry interval for alfalfa, asparagus, broccoli, brussel sprouts, carrots, cauliflower, celery, collards, cucumbers, lettuce, melons, onions, peanuts, peas, peppers, potatoes, sorghum, soybeans, summer squash, spinach, sugar beets, tobacco, and tomatoes. Additional data are being requested to set reentry intervals for mint, corn, roses grown in greenhouses, and chrysanthemums grown in greenhouses or outdoors. Until these data are received and evaluated, an interim seven day reentry interval is being established for corn, and a one day reentry interval is being established for these crops and all other crops and sites not specifically listed above.

F. The following labeling is required for all manufacturing use products:

This pesticide is toxic to fish. Do not discharge effluent containing this product into lakes, streams, ponds, estuaries, oceans, or public waters unless this product is specifically indentified and addressed in the NPDES permit. Do not discharge effluent containing this product to sewer systems without previously notifying the sewage treatment plant authority. For guidance, contact your State Water Board or Regional Office of the EPA.

G. The following labeling is required for all end use pro-ducts:

PERSONAL PROTECTIVE EQUIPMENT:

USE ONLY WHEN WEARING THE FOLLOWING PERSONAL PROTECTIVE EQUIPMENT DURING MIXING/LOADING, APPLICATION, REPAIRING AND CLEANING OF MIXING, LOADING, AND APPLICATION EQUIPMENT, AND

DISPOSAL OF THE PESTICIDE: longsleeve shirt; long-legged pants; shoes and socks, chemical resistant gloves; face shield or goggles; NIOSH or MSHA approved respirator. During equipment repair and cleaning, the respirator need not be worn.

IF APPLICATION IS PERFORMED USING AN ENCLOSED CAB OR COCKPIT, THE FOLLOWING PROTECTIVE CLOTHING AND EQUIPMENT MAY BE WORN AS AN ALTERNATIVE: long-sleeve shirt and long-legged pants; shoes and socks. All other protective clothing and equipment required for use during application must be available in the cab and must be worn when exiting the cab into treated area. When used for this purpose, contaminated clothing may not be brought back into the cab unless in an enclosure such as a plastic bag.

IMPORTANT! If pesticide comes in contact with skin, wash off with soap and water and contact a physician immediately. ALWAYS WASH HANDS, FACE, AND ARMS WITH SOAP AND WATER BEFORE USING TOBACCO PRODUCTS, EATING, DRINKING, OR TOILETING.

AFTER WORK: Before removing gloves, wash them with soap and water.   Take off all work clothes and shoes. Shower using soap and water. Wear only clean clothes when leaving job--do not wear contaminated clothing. Personal clothing worn during work must be stored and laundered separately from protective clothing and household articles. Store protective clothing separately from personal clothing. Clean or launder protective clothing after each use. Respirators must be cleaned and filters replaced according to instructions included with the respirators. Protective clothing and protective equipment heavily contaminated or drenched with methomyl must be destroyed according to state and local regulations. HEAVILY CONTAMINATED OR DRENCHED CLOTHING CANNOT BE ADEQUATELY DECONTAMINATED.

DURING AERIAL APPLICATION, HUMAN FLAGGERS ARE PROHIBITED.

PROTECTIVE CLOTHING LABEL STATEMENTS FOR 1% BAITS

Use only when wearing the following personal protective equipment during loading, application, repairing and cleaning of mixing, loading, and application equipment, and disposal of the pesticide: longsleeve shirt; long-legged pants; shoes and socks; gloves.

IMPORTANT! If pesticide comes in contact with skin, wash off with soap and water.

ALWAYS WASH HANDS, FACE, AND ARMS WITH SOAP AND WATER BEFORE USING TOBACCO PRODUCTS, EATING, DRINKING, OR TOILETING.

OTHER LABEL STATEMENTS FOR BAITS

Do not contaminate feed and foodstuffs. Do not apply
where poultry or other animals, especially dogs and young
calves, can pick it up or lick it.

Do not use in edible product areas of food processing
plants, restaurants, or other areas where food is commercially
prepared or processed. Do not use in serving areas while
food is exposed.

REENTRY INTERVALS:

The following reentry intervals are required in the
directions for use section of all labels with terrestrial and
greenhouse food and non-food uses: three days for cotton,
nectarines, citrus and apples; four days for peaches; seven
days for grapes and corn; all other crops, one day.

The following fish and wildlife statements are required
to appear under the "Environmental Hazards" heading:

1. Granulars (including baits): This pesticide is
toxic to birds. Collect, cover or incorporate granules spilled
on the soil surface. Do not apply directly to water or wetlands
(swamps, bogs, marshes, and potholes). Do not contaminate
water when disposing of equipment washwaters.

2. Non-Granular:

a. Aquatic (Watercress): This pesticide is toxic to
fish. Drift and runoff from treated areas may be hazardous
to aquatic organisms in neighboring areas. Do not contaminate
water when disposing of equipment washwaters.

b. Terrestrial: This pesticide is toxic to fish and
wildlife. Do not apply directly to water or wetlands (swamps,
bogs, marshes, and potholes). Drift and runoff may be hazardous
to aquatic organisms in neighboring areas. Do not contaminate
water when disposing of equipment washwaters.

H. End use products (except granulars and baits) with outdoor
crop uses must have the following bee caution: This product
is highly toxic to bees exposed to direct treatment on blooming
crops or weeds. Do not apply this product or allow it to
drift to blooming crops or weeds while bees are actively
visiting the treatment area.

Products with lentil use must include a pregrazing interval
of three days and a preharvest interval of seven days.

I. Pursuant to the data requirements in 40 CFR Part 158,

the Agency has determined that the following data are essential
to the Agency's assessment and should receive a priority
review when they are received by the Agency:

40 CFR 158.240 Residue Chemistry

171-4 Nature of the Residue (Metabolism-Livestock),
      Meat/Milk/Poultry and Eggs

40 CFR 158.390 Reentry Protection

   132-1 Foliar Dissipation (Reentry)
   201-1 Droplet Size Spectrum
   201-1 Drift Field Evaluation

40 CFR 158.340 Toxicology

   85-1 General Metabolism (Rat and Monkey)
   82-2 Subchronic Dermal (21-Day)

40 CFR 158.290 Environmental Fate

   162-3 Anaerobic Aquatic Metabolism
   162-4 Aerobic Aquatic Metabolism
   162-2 Laboratory Volatility
   164-2 Aquatic (Sediment)
   165-5 Accumulation in Non-Target Organisms
   - Groundwater Monitoring

40 CFR 158.490 Ecological Effects

   71-4 Avian Reproduction
   71-5 Simulated and Actual Field Testing- Birds
   72-1 Freshwater Fish Acute Toxicity
   72-2 Acute Toxicity- Freshwater Invertebrate
   72-3 Acute Toxicity- Aquatic Estuarine and Marine Organism
   72-5 Fish Life-Cycle
   72-7 Field Testing for Aquatic Organism

5.   Summary of Major Data Gaps

-The following studies are required to assess the toxicological
 characteristics of technical methomyl: Eye irritation, dermal
 sensitization, 21-day dermal toxicity, and general metabolism
 testing (rat and monkey).
-The following data are required to fully characterize
 methomyl's environmental fate: Reentry, volatility (lab),
 aquatic sediment dissipation, accumulation studies in irrigated crops and in fish, and groundwater monitoring.
-Additional residue and processing studies in certain commodities, are required to support existing tolerances.
-The following data are required to complete a wildlife risk

assessment: Avian subacute dietary toxicity, avian reproduction, freshwater fish toxicity, acute toxicity to freshwater invertebrates, acute toxicity to estuarine and marine organisms, fish early life stage and aquatic invertebrate life cycle, and simulated or actual field testing for aquatic organisms and mammals and birds.
-Product chemistry and acute toxicity data are required.

Contact person at EPA:

Dennis Edwards
Product Manager (12)
Insecticide-Rodenticide Branch
Registration Division (H7505C)
Office of Pesticide Programs
Environmental Protection Agency
401 M. Street, S.W.
Washington,D.C. 20460

Office location and telephone number:
Room 202, Crystal Mall Building #2
1921 Jefferson Davis Highway
Arlington,Va. 22202
703-557-2386

Disclaimer: The information in this Fact Sheet is a summary only and is not to be used to satisfy data requirements for pesticide registration and reregistration. The complete Registration Standard for methomyl may be obtained from the Information Services Section, Program Management and Support Division (TS-757C), EPA, 401 M St., SW, Washington, D.C. 20460.

METHOXYCHLOR

Reason for Issuance: Registration Standard
Date Issued: December 1988
Fact Sheet Number: 187

1. DESCRIPTION OF CHEMICAL

    Generic name: 1,1,1-trichloro-2,2-bis(4-
                   methoxyphenyl) ethane

    Common name: Methoxychlor

    Trade Names: Marlate, Prentox, and Methoxcide

    Other Chemical Nomenclature: 1,1,1-trichloro-2,2-di(4-
                                    methoxyphenyl)ethane; 1,1-
                                    (2,2,2-tri chloroethylidene)-
                                    bis[4-methoxybenzene]; 1,1,1-
                                    trichloro-2,2-bis(p-
                                    methoxyphenyl)ethane; 2,2-
                                    bis)p-methoxyphenyl)-1,1,1-
                                    trichloroethane

    CAS Registry No.:   72-43-5

    EPA Pesticide Chemical Code (Shaughnessy Number): 034001

    Empirical Formula: $C_{16}H_{15}Cl_3O_2$

    Molecular Weight: 345.7

    Year of initial registration: 1948

    Pesticide type: Insecticide/Acaricide

    Chemical family: Chlorinated Hydrocarbon

    U.S. Registrants: Chemical Formulators; Prentiss Drug &
                       Chemical Co., J.R. Simplot Co.;
                       Dynachem Industries; Clover Chemical
                       Co.; Drexel Chemical Co.; Kincaid
                       Enterprises; and Wesley Industries

Enterprises; and Wesley Industries

2. USE PATTERNS AND FORMULATIONS

Registered
Uses:

TERRESTRIAL FOOD CROP use: (1) **seed treatment only use**
on grains and various vegetables; (2) **foliar application
(including seed treatment) use** on vegetables and fruits;
and (3) **foliar application only use** on vegetables and
fruits

TERRESTRIAL NON-FOOD CROP use on grasses, ornamentals and
trees

GREENHOUSE FOOD CROP use on mushrooms

DOMESTIC AND NON-DOMESTIC OUTDOOR use around dwellings
and for garbage and sewer areas, general urban outdoor
use

AQUATIC FOOD use on cranberry

AQUATIC NON-FOOD use for mosquito larvae control in
aquatic sites, such as beaches, lakes, marshes and
rivers

FORESTRY use on forest trees

INDOOR use on: (1) **postharvest stored grain commodity
and premise treatment**; (2) **direct animal treatment** for
dogs, cats, and farm animals; (3) **agricultural premise**
use; (4) **kennels, dog sleeping quarters and cat** sleeping
quarters; (5) **indoor domestic dwellings** for use on
household contents such as human clothing (including
woolens); (6) **direct application to humans**; (7)
commercial and industrial use in food processing, storage
transportation areas and equipment

Pests Controlled: various nuisance species (some of public
health significance) including cockroaches,
mosquitoes, flies and chiggers; various
arthropods attacking field crops,
vegetables, fruits, ornamentals, stored
grain, livestock and domestic pets

Methods of Application: sprays, fogs, paints,
ground and aerial equipment,
animal dust-bags, dips, sprays
and back-rubbers

Formulations: Wettable powders, dusts, emulsifiable

concentrates, flowable concentrates,
liquid soluble concentrates, granules,
ready-to-use products (liquids) and
pressurized liquids

3. SCIENCE FINDINGS

Summary Science Statement

With the exception of one mutagenicity study, there are no
acceptable acute, subchronic, or long-term
toxicology/oncogenicity studies available to support technical
methoxychlor. In the acceptable mutagenicity study, an
unscheduled DNA synthesis assay in mammalian cells in culture, no
abnormal DNA synthesis was noted at any of the dose levels
tested.

Based on acceptable laboratory data, technical methoxychlor
is characterized as very highly toxic to fish and aquatic
invertebrates, and practically nontoxic to birds and bees. Based
on theoretical calculations, both terrestrial and aquatic uses of
methoxychlor may pose a hazard to aquatic organisms, although
there is no field evidence to support this. The impacts of
methoxychlor use to nontarget organisms will be assessed upon
receipt of ecological effects and environmental fate data.

The environmental fate of methoxychlor cannot be
characterized because acceptable data are lacking. Preliminary
data suggest that methoxychlor is unlikely to contaminate
groundwater because of its low solubility and high rate of
adsorption to soil particles.

The nature of the residues of methoxychlor in plants and
animals is not adequately understood. None of the tolerances
for methoxychlor is adequately supported. Plant and animal
metabolism studies, residue studies, analytical methodology,
processing studies, and storage stability data are needed before
the Agency can determine the adequacy of current tolerance
levels.

Chemical/Physical Characteristics of the Technical Material

Chemical/Physical
Characteristics:
          Color: Data Gap
          Physical State: Crystalline solid (Farm
          Chemicals, 1987)
          Odor: Data Gap
          Melting Point: 89 $^{\circ}$C (Farm Chemicals,
          1987)
          Specific Gravity: Data Gap
          Solubility: Very soluble in aromatic
                     chlorinated, or ketonic solvents,

somewhat soluble in paraffinic
types; essentially insoluble
in water (Farm Chemicals, 1987)

Vapor Pressure: Data Gap
Flammability: Data Gap
pH: Data Gap

Toxicology Characteristics

With the exception of one mutagenicity study, there are no
acceptable acute, subchronic or long-term toxicology/oncogenicity
studies available to support technical methoxychlor. In the
mutagenicity study, a mammalian cell in culture unscheduled DNA
synthesis assay (UDS assay), no increase in abnormal DNA
synthesis was noted.

Environmental Characteristics

The Agency is unable to assess the environmental fate of
methoxychlor because acceptable data are lacking. Preliminary
data indicate that methoxychlor is stable to hydrolysis (half-
life > 200 days); photodegradation in water (half-life of 4.5
months); and aerobic soil metabolism (half-life > 3 months in
sandy loam soil). The half-life for anaerobic soil metabolism is
reported at less than 1 month in sandy loam soil. Preliminary
data also indicate that methoxychlor has a high adsorption rate
to soil sediment (K_d value is 620).

Ecological Characteristics

Based on acceptable laboratory data, technical methoxychlor
is characterized as practically nontoxic to birds on both an
acute oral and subacute dietary basis and very highly toxic to
fish and aquatic invertebrates on an acute basis.   There is
sufficient information to characterize methoxychlor as
relatively nontoxic to honey bees. The acute toxicity value = 24
ug/bee.

- Acute LD50 (bobwhite):
    >2510 mg/kg
- Dietary LC50
    >5620 ppm (upland gamebird)
- Freshwater invertebrates toxicity (96-hr LC50) for
    daphnid .78 ppb
- Fish acute toxicity (96-hr LC50) for rainbow trout:
    1.31 ppm
- Fish acute toxicity (96-hr LC50) for brook trout:
    0.009 ppm

Tolerance Assessment

Tolerances have been established for residues of

methoxychlor in a variety of raw agricultural commodities, in
meat, fat and meat byproducts (40 CFR 180.120). Tolerances are
expressed in terms of methoxychlor per se.

The nature of the residues of methoxychlor in plants and
animals is not adequately understood. None of the tolerances
for methoxychlor is adequately supported. Plant and animal
metabolism studies, residue studies, analytical methodology,
processing studies, and storage stability data are needed before
the Agency can determine the adequacy of current tolerance
levels.

The Preliminary Limiting Dose (PLD) of methoxychlor is .005
mg/kg/day. This is based on a rabbit teratology study with a No
Observed Effect Level (NOEL) of 5 mg/kg/day for increased loss
of litters and an uncertainty factor of 1000 to account for
inter- and intraspecies differences, poor quality of the study
used and total incompleteness of the subchronic and chronic
toxicity data base.   The study is not considered to be adequate
to define a NOEL for purposes of setting an Acceptable Daily
Intake, since the experimental design was considered to be
inadequate.   It is being used on an interim basis for calculation
of the PLD.   The Agency is unable to complete a tolerance
assessment of methoxychlor because of the incompleteness of the
toxicology and residue chemistry data bases.

4. SUMMARY OF REGULATORY POSITIONS AND RATIONALES

-    Methoxychlor is not being placed into Special Review at
this time. Since there are so few acceptable studies available
to support registration of products containing methoxychlor, the
Agency is not yet able to make a determination as to whether any
of the criteria of 40 CFR 154.7 have been met or exceeded.

-    The Agency will not approve any new food uses, including
minor uses for this chemical since none of the tolerances are
adequately supported.

- The Agency is unable to assess methoxychlor's potential for
contaminating groundwater. When data required in the Standard
have been received and evaluated, the Agency will assess the
potential for methoxychlor to contaminate groundwater.

- Updated worker safety rules are required for end-use product
labels.

-The Agency is not establishing a longer reentry interval for
agricultural uses of methoxychlor beyond the minimum reentry
interval (sprays have dried, dusts have settled, and vapors have
dispersed).

- Revised and updated fish and wildlife statements are

required for end-use product labels. Since methoxychlor is
practically nontoxic to bees, the bee statement imposed under
under PR Notice 68-19 is no longer appropriate. Registrants must
remove the bee statement from the labeling.

- The Agency is not classifying methoxychlor as a restricted
use pesticide at this time, since it is unable to determine if
this pesticide meets any of the risk criteria of 40 CFR 152.170.
Upon receipt of data required under this Standard, the Agency
will apply the criteria of 40 CFR 152.170 to determine if any
uses of methoxychlor warrant restricted use classification.

- Since methoxychlor is an analogue of DDT, the Agency is
requiring specific analysis of methoxychlor for the potential
impurities 1,1,1-trichloro-2,2-bis(p-chlorophenyl)ethane (DDT)
and other structurally similar compounds.

SUMMARY OF OUTSTANDING DATA REQUIREMENTS

Toxicology	Time Frame
Acute Oral Toxicity	9 Months
Acute Dermal Toxicity	9 Months
Acute Inhalation Toxicity	9 "
Eye Irritation	9 "
Dermal Irritation	9 "
Dermal Sensitization	9 "
21-Day Dermal Toxicity	9 "
Chronic Testing (rodent)	50 "
Chronic Testing (non-rodent)	50 "
Oncogenicity (rat)	50 "
Oncogenicity (mouse)	50 "
Teratogenicity (rat)	15 "
Teratogenicity (rabbit)	15 "
Reproduction	39 "
Gene Mutation	9 "
Other Mechanisms of Mutagenicity	12 "
Metabolism	24 "

Environmental Fate/Exposure	
Hydrolysis	9 Months
Photodegradation in Water	9 "
Photodegradation on Soil	9 "
Aerobic Soil Metabolism	27 "
Anaerobic Soil Metabolism	27 "
Anaerobic Aquatic Metabolism	27 "
Aerobic Aquatic Metabolism	27 "
Leaching and Adsorption/ Desorption	12 "

Aquatic Dissipation   27 "
Forestry   27 "
Soil, Long-term   39 "
Confined Rotational Crop   39 "
Accumulation in Irrigated Crops   39 "
Accumulation in Fish   12 "
Accumulation in Aquatic
Nontarget Organisms   12 "

Fish and Wildlife

Avian Reproduction   24 Months
Freshwater Fish LC_{50} Studies (TEP)   9 "
Freshwater Invertebrate LC_{50} Studies (TEP) 9 "
Estuarine and Marine
Organisms LC_{50} Studies (TEP)   12 "
Fish Early Life Stage
and Invertebrate Life Cycle   15 "
Simulated or Actual Field
Testing-Aquatic Organisms   24 "
Seed Germination/Seedling
Emergence   9 "
Aquatic Plant Growth   9 "

Residue Chemistry

Residue data - Raw Agricultural Commodities 18 Months
Processing Studies   24 "
Plant and Animal Metabolism   18 "
Storage Stability   15 "
Residue Analytical Methods   15 "

Product Chemistry

All Data   9 -15 Months

6.   Contact Person at EPA

     Dennis H. Edwards Jr.
     Product Manager (12)
     Insecticide-Rodenticide Branch
     Registration Division (TS-767)
     Environmental Protection Agency
     Washington, DC 20460

     Tel. No. (703) 557-2386

DISCLAIMER: The information presented in this Chemical
Information Fact Sheet is a summary only and may not be used to
fulfill data requirements for pesticide registration and
reregistration.

METIRAM

Reason for Issuance: Registration Standard
Date Issued: October 3, 1988
Fact Sheet Number: 181

1. DESCRIPTION OF CHEMICAL

   Chemical Name: Mixture of 5.2 Parts by Weight (83.9%) of
   Ammoniates of [ethylenebis (dithiocarbamate)]
   zinc with 1 part by weight (16.1%) ethylenebis
   [dithiocarbamic acid], bimolecular and
   trimolecular cyclic anhydrosulfides and
   disulfides

   Common Name: Metiram

   Principal Trade Names: Polyram™, Polyram-Combi™

   EPA (Shaughnessy) Code: 014601

   Chemical Abstracts Service (CAS) Number: 9006-42-2

   Year of Initial Registration: late 1940's

   Pesticide Type: Fungicide

   Chemical Family: Ethylene bisdithiocarbamate (EBDC)

   U.S. and Foreign Producers: FMC and BASF

2. USE PATTERNS AND FORMULATIONS

   Registered uses: Terrestrial food crop uses on apples, asparagus,
   celery, corn (sweet), cotton, cucumber, peanuts,
   pecans, potatoes (including seed pieces), sugar beets,
   and tomatoes; Terrestrial nonfood crop uses on tobacco
   (field and transplants) and roses

   Predominant uses: Apples and potatoes

   Pests controlled: Foliar fungal diseases of selected fruit, nut,
   vegetable, field and ornamental crops.

Types of Formulations: Formulation intermediate, dust, and wettable powder.

Types and Method of Application: Foliar application to fruits, vegetables and nuts by aerial equipment, as well as ground equipment. For ground equipment metiram suspensions typically made from a wettable powder, would be applied by means of air blast sprayers or in the case of some row crops by means of tractor mounted boom sprayers.

Application Rates: Terrestrial food crop: 0.3 - 6.4 lb ai/A
Terrestrial nonfood crop: 1.2 - 2.4 lb ai/A

3. SCIENCE FINDINGS

a. Chemical Characteristics

Physical state: solid
Color: Light yellow
Odor: odorless
Vapor Pressure: $<1 \times 10^{-7}$ mbar at 20 C
Molecular Formula: $(C_{16}H_{33}N_{11}S_{16}Zn_3)$

Toxicology Characteristics

Acute Toxicity: All studies required

Major Routes of Exposure: Dermal, Inhalation and oral by ingestion of food residues

Subchronic Toxicity: Inhalation study is adequate, other studies required

Oncogenicity: Studies required

Chronic feeding: Studies required

Metabolism: Studies in rats indicate that the polymer is hydrolyzed and readily absorbed and eliminated in the urine and feces. ETU was one of the metabolites in the urine and bile of rats.

Reproduction: Study required

Teratogenicity & Developmental Toxicity: Studies required

Mutagenicity: Considering only the acceptable studies, the majority of mutagenicity studies on metiram were negative. However, the in vitro sister chromatid exchange assay in Chinese hamster ovary cells was positive and is considered a sensitive test for chromosomal effects. According to the present data, metiram is considered positive for chromosomal damage. A gene mutation assay is required.

Physiological and Biochemical Characteristics

Metabolism and Persistence in Plants and Animals:

Metabolism of metiram is not completely understood. Additional data are being required in plants and livestock. ETU is a major metabolite of concern.

Environmental Characteristics

Presently only the hydrolysis and photodegradation in soil and in water data requirements on both metiram and ETU were fully satisfied. Metiram has a very limited solubility in water. Metiram in water solution degrades primarily to ETU and other transient degradates. ETU is also a soil degradate of metiram and its formation on soil is enhanced by sunlight. ETU is stable in water at pH 5-9 and under sunlight and the degradation of ETU on soil is not enhanced by sunlight radiation. ETU is the degradate of major environmental concern. There are indications that ETU may leach and enter groundwater. However, additional data are required to complete the groundwater assessment.

Ecological Characteristics

° Metiram has been found to be slightly toxic to birds. Formulated metiram showed that LC50 values for mallard duck and bobwhite quail are both greater than 3712 ppm.

° Based on an acute contact honeybee toxicity study, there is sufficient information to characterize metiram as practically nontoxic to honeybees.

4. TOLERANCE ASSESSMENT

Tolerances, expressed as zinc ethylene bisdithiocarbamate, have been established for residues of metiram in a variety of raw agricultural commodities (40 CFR 180.217 and 180.319).

The toxicology data for metiram are insufficient to determine an Acceptable Daily Intake (ADI) or whether the toxicity observed in the studies is due to metiram or ETU. A three generation rat reproduction study has been used to calculate a Provisional ADI (PADI). Because a NOEL was not reached in the three generation study, an uncertainty factor of 1000 was employed. The PADI for metiram is 0.0003 mg/kg/day.

The theoretical maximum residue contribution (TMRC), based on the assumption that 100 percent of each crop is treated and contains residues at the tolerance level, is 0.009 or approximately 3000 percent of the PADI. Based on a more realistic dietary assessment, using anticipated field residues and estimate of percent crop treated, the estimated average consumption for the U.S. population is 0.00038 mg/kg/day or 122 percent of the PADI.

5. SUMMARY OF REGULATORY POSITIONS

The Agency initiated a Special Review for metiram along with the
other EBDC's in June 1987 because of concern about the oncogenic risk to
consumers from dietary exposure to ETU from food treated with these pesticides,
and the risks of teratogenicity and adverse thyroid effects to applicators
and mixer/loaders from exposure to ETU.

o ETU has been classified as a B$_2$ oncogen (probable human carcinogen).

o The Agency will not consider establishment of new food use tolerances
   for metiram because the current residue chemistry and toxicology data
   are not sufficient to assess existing tolerances and the toxicology
   data base is insufficient to determine an ADI and does not allow a
   decision as to whether observed toxicity is due to metiram or ETU.

° The Agency will consider the need for establishment of tolerances for
   ETU and any intermediate metabolites when data are sufficient to permit
   such decisions.

° The Agency will not establish any food/feed additive regulations pursuant
   to Section 409 of the Federal Food, Drug and Cosmetic Act (FFDCA) and is
   deferring action on previously established food/feed additive regulations.

° Protective clothing labeling for metiram products, as required as a result
   of the 1982 Decision Document, must be updated.

° The Agency is requiring reentry data for metiram. In order to remain in
   compliance with FIFRA, an interim 24-hour reentry interval requirement
   must be placed on the label of all metiram end-use products registered
   for agricultural uses, until the required data are submitted and evaluated
   and any change in this reentry interval is announced.

° The Agency has screened and reviewed the environmental fate data to
   determine if metiram/ETU and/or its degradate(s) have the potential to
   leach into ground water. The Agency has decided that in addition to
   environmental fate data requirements, a small-scale retrospective
   ground water monitoring study is also required to define the extent of
   the ground water problem.

o While the data gaps are being filled, currently registered manufacturing-use
   products (MP's) and end-use products (EP's) containing metiram as the
   sole active ingredient may be sold, distributed, formulated and used,
   subject to the terms and conditions specified in this Standard.
   However, new uses will not be registered. Registrants must provide
   or agree to develop additional data, as specified in the Data Appendices
   of the Registration Standard, in order to maintain existing registrations.

6. LABELING REQUIREMENTS

All metiram products must bear appropriate labeling as specified in
40 CFR 156.10. Appendix II of the Registration Standard contains information
on labeling requirements.

In addition to the above, in order to remain in compliance with FIFRA, the Agency is requiring:

° Protective clothing requirements
° Environmental hazard precautions
° Worker safety rules
° Reentry interval
° Grazing restrictions for apples, pecans, corn (sweet), cotton, peanuts, sugar beets and potato (seed pieces).

7. SUMMARY OF DATA GAPS

Product Chemistry All - Due within 6 months

Technical Grade:
Preliminary analysis of product samples

MUP:
Analysis & certification of product ingredient
Oxidizing or reducing action
Flammability
Explodability
Storage stability
Corrosion characteristics

Toxicology - The last studies are due 12/90

Acute testing
Dermal sensitization
90-Day feeding (rodent and nonrodent)
21-Day subchronic dermal
Chronic toxicity (rodent and nonrodent)
Oncogenicity (rat and mouse)
Teratology (rabbit and rat)
Reproduction (rat)
Mutagenicity (point gene mutation)

Residue Chemistry - Data due 10/88 and 4/89

Nature of the Residue in Plants and Livestock
Analytical Methods
Magnitude of Residue for Variety of Commodities

Environmental Fate - Last studies are due 7/90

Leaching and adsorption/desorption
Field dissipation
Degradation soil
Degradation (soil long-term)
Small-scale retrospective ground water monitoring
Confined rotational crops
Fish accumulation

Reentry Protection   - Data due 7/89

Reentry Studies on Foliar and Soil Dissipation

Wildlife and Aquatic Organisms - Last data are due in 12 months

Avian oral toxicity
Freshwater fish toxicity
Acute freshwater invertebrates
Estuarine and marine organism toxicity

ETU Data Requirements

Toxicology

Chronic (rodent and non-rodent) Data due 5/90
Reproduction   Data due 12/90

Environmental Fate - Last studies due 7/90

Aerobic and anaerobic soil metabolism
Aerobic aquatic
Lab volatility
Degradation (soil)
Aquatic (sediment)
Degradation (soil long-term)
Small-scale retrospective ground water monitoring study
Fish accumulation

8. CONTACT PERSON AT EPA

Lois A. Rossi
Product Manager (21)
Fungicide-Herbicide Branch
Registration Division (TS-767C)
Office of Pesticide Programs
Environmental Protection Agency
401 M St., SW
Washington, D.C.   20460

Office location and phone number:
Room 227, Crystal Mall #2
1921 Jefferson Davis Highway
Arlington, VA
(703) 557-1900

9. DISCLAIMER: The information in this Pesticide Fact Sheet is a summary only and may not be used to satisfy data requirements for pesticide registration and reregistration. The complete Registration Standard for the pesticide may be obtained from the National Technical Information Service. Contact the Product Manager listed above for further information.

MEVINPHOS

Reason for Issuance: Issuance of Registration Standard
Date Issued: March 31, 1988
Fact Sheet Number: 156

1. DESCRIPTION OF CHEMICAL

   Generic Name: 3-[(Dimethoxyphosphinyl)oxy]-2-butenoic acid methyl ester; 3-hydroxycrotonic acid methyl ester dimethyl phosphate.

   Common Name: Mevinphos

   Trade and Other Names: Apavinphos, Duraphos, Menite, Mevinox, OS-2046, Phosdrin, and Phosfene

   EPA Shaughnessy Code: 015801

   Chemical Abstracts Service (CAS) Number: 7786-34-7

   ENT Registry Number: 22374

   Year of Initial Registration: early 1950s

   Pesticide Type: Insecticide

   Chemical Family: Organophosphate

   U.S. and Foreign Producers: E. I. DuPont (U.S.); Amvac Chemical Corp. (U.S.); APA Spa (Italy); Comlets Chemical Industrial Co., Ltd. (Taiwan); GEMP International Corp. (Taiwan); KenoGard VT AB (Sweden); and Shell International Chemical Co., Ltd. (England)

2. PHYSICAL AND CHEMICAL CHARACTERISTICS

Chemical Characteristics of the Technical Material

Physical State: Liquid.

Color: Light yellow to orange.

Odor: Mild to none.

Molecular Weight and Formula: 224.1 - $C_7H_{13}O_6P$.

Boiling Point: 99 - 103°C at 0.03 mm Hg.

Vapor Pressure: 0.003 mm Hg at 21°C.

Density: 1.24 at 16°C.

Solubility in various solvents: Miscible with water, acetone, benzene, carbon tetrachloride, chloroform, ethanol, isopropanol, methanol, toluene, and xylene; slightly soluble in carbon disulfide and kerosene; insoluble in hexane.

Physiological and Biochemical Characteristics

Mechanism of Pesticidal Action: Cholinesterase inhibition following contact with, or ingestion of, mevinphos treated surfaces.

Metabolism and Persistence in Plants and Animals: The metabolism of mevinphos in plants and animals is not adequately understood. The available plant metabolism data indicate that mevinphos is absorbed by plant roots, translocated readily to leaves and growing shoots, and is degraded rapidly. The major residues identified in plants are the alpha and beta isomers of mevinphos and dimethyl phosphate. Mevinphos acid is a minor metabolite in plants and it has been demonstrated that mevinphos acid is converted to desmethyl mevinphos acid. It has been suggested that the end products of mevinphos metabolism are methanol, acetone, and carbon dioxide. The available animal metabolism data indicate that twelve hours after dosing 57 to 65% of the 32p-residues was excreted in the urine and feces (45 to 50%) in urine; 12 to 15% in feces). Only mevinphos hydrolysis products were present.

3. USE PATTERNS AND FORMULATIONS

Application Sites: Alfalfa, anise (fennel), apples, artichoke, beans, beets (including tops), Bermudagrass (seed crop), birdsfoot trefoil, broccoli, broccoli raab, Brussels sprouts, cabbage, carrots, cauliflower, celery, cherries (sour), Chinese broccoli (gai lon), Chinese cabbage (including napa or nappa cabbage, bok choi, pak choi, gai choy, and mizuna), clover, collards, corn, cucumbers, eggplant, grapes, grapefruit, kale, lemons, lettuce, melons (including cantaloupes, honeydew melons, muskmelons, and watermelons), mint, mustard greens, okra, onions (including green onions), oranges, peaches, pears, peas, peppers, plums, potatoes, raspberries, red chicory [also known as radicchio] (tops), sesame (seed crop), sorghum, spinach, squash (summer), strawberries, tomatoes, turnips (including tops), walnuts, ornamental flowering plants, greenhouse agricultural crops (lettuce), watercress, and sewage disposal plants.

Types and Methods of Application: Foliar application using conventional ground or aerial equipment. Water treatments are permitted in sewage disposal plants

Application Rates: Recommended application rates range from 0.125 to 3.0 pounds of active ingredient per acre.

Types of Formulations: Dust, emulsifiable concentrates, soluble concentrate/liquid, and ready to use.

4. SCIENCE FINDINGS

Summary Science Statement

Mevinphos is an organophosphorus compound and is a potent cholinesterase inhibitor. The Agency has virtually no acceptable toxicity data for mevinphos. However, a high incidence of physician-treated poisonings and hospitalizations due to mevinphos have been noted. Poisoning Reports from California indicate that mevinphos was among the top five pesticides causing occupational poisoning in that state during 1981-1985. The major routes of applicator exposure are through dermal exposure, with some inhalation exposure. Results of a preliminary review of recently received studies are consistent with a Toxicity Category I classification for acute oral and acute dermal toxicity for mevinphos. Mevinphos was not teratogenic in rats at a maternally toxic dose. Data gaps exist for acute toxicity, subchronic effects, chronic feeding, oncogenicity, reproduction, mutagenicity, and metabolic effects. A partial data gap exists for teratogenicity. Mevinphos is very mobile in sandy loam, silt loam, loam, and clay loam soils, however, insufficient data are available for the Agency to fully assess the environmental fate of the compound. Tolerances have been established for a number of raw agricultural commodities, however additional data are required to support many of them. The Agency has insufficient data to establish an Acceptable Daily Intake (ADI) for mevinphos at this time. Available ecological effects data show that mevinphos is very highly toxic to aquatic organisms and birds. However, no documented field kills of fish or birds have been noted. Because of its demonstrated toxicity to nontarget species and its intended use pattern, mevinphos has been identified by the Office of Endangered Species (OES), U.S. Fish and Wildlife Service (FWS), as being likely to jeopardize endangered species when used on corn and sorghum. Based on this determination, OES specified reasonable and prudent alternatives to avoid jeopardizing the continued existence of the identified species by these uses. EPA is developing a program to reduce or eliminate exposure to these species to a point where use does not result in jeopardy, and will issue notice of any necessary labeling revisions when the program is developed.

Toxicology Characteristics

Except for a teratology study conducted in the rat, the Agency has no acceptable toxicology data for mevinphos. In the rat teratology study 24 mated female Sprague-Dawley rats were dosed by gastric intubation with 0, 0.20, 0.75, 1.00, and 1.25 mg/kg/day of technical mevinphos in water on gestation days 6 through 15. Based on data from this study, the Agency determined that the fetotoxic, embryotoxic, and teratogenic NOELs were >1.00 mg/kg/day and concluded that mevinphos does not induce developmental effects in rats.

Environmental Characteristics

Available data are insufficient to fully assess the environmental fate of mevinphos and transport of mevinphos. Data gaps exist for nearly all applicable studies. The available data indicate that mevinphos residues are very mobile in sandy loam, silt loam, loam, and clay loam soils. Freundlich K_{ads} values ranged from 0.392 to 1.92 and Freundlich K_{des} values ranged from 1.16 to 3.53. Data currently available are insufficient to characterize mevinphos's leaching potential for contamination of ground water. Data to characterize the potential to contaminate groundwater are being required. Treated areas should not be re-entered for at least 96-hours (4-days) for citrus, grapes, nectarine and peach orchards; and 48-hours (2-days) for all other mevinphos treated crops, unless protective clothing is worn.

Ecological Characteristics

Avian acute toxicity: Acute toxicity values of 1.34 mg/kg in the sharp-
                         tailed grouse and 4.63 mg/kg in the mallard duck.

Avian dietary toxicity: Subacute dietary toxicity values ranged from 236
                         ppm in the Japanese quail to 1991 ppm in the
                         mallard duck.

Freshwater fish acute toxicity? 96-hour acute toxicity values ranged from
                         11.9 ppb for rainbow trout to 22.5 ppb for
                         bluegill sunfish.

Marine fish acute toxicity: LC_{50} value of 640 ppb for sheepshead minnow.

Freshwater invertebrate toxicity: The acute toxicity values ranged from 0.18 ppb for Daphnia pulex to 5.00 ppb for Pteronarcys californica.

Marine invertebrate toxicity: 96-hour shell deposition EC_{50} value of greater than 1000 ppb for the Eastern oyster; EC_{50} value of 150 ppb for brown shrimp.

These data show that technical mevinphos is very highly toxic to birds on an acute oral basis; highly toxic to upland game birds and slightly toxic to waterfowl on a dietary basis; very highly toxic to both warmwater and coldwater fish species; very highly toxic to aquatic invertebrates; and highly toxic to the brown shrimp and sheepshead minnow and moderately toxic to the Eastern oyster.

TOLERANCE REASSESSMENT

Tolerances have been established for residues of mevinphos in a variety of raw agricultural commodities (40 CFR 180.157). The Agency has evaluated the residue and toxicology data supporting these tolerances and has determined that a full tolerance reassessment for mevinphos cannot be made at this time because of extensive residue chemistry and toxicology data gaps. Because of these extensive data gaps, no significant new uses, including group tolerances, will be granted until the Agency has received data sufficient to thoroughly evaluate the dietary exposure to mevinphos.

In addition to United States tolerances, there are also Canadian tolerances, Mexican tolerances, and Codex Maximum Residue Limits (MRLs) established for mevinphos.   However, some incompatibility exists between some of the permanent Codex MRLs and the U.S. tolerances.   The issue of incompatibility will be addressed when residue data are submitted and evaluated.

The Agency does not have sufficient data to support the established tolerances for residues of mevinphos alpha and beta isomers in or on all raw agricultural commodities (RACs).

Processing studies are required for apples, citrus fruits, grapes, plums, potatoes, tomatoes, corn, and sorghum.

The initial Acceptable Daily Intake (ADI) for mevinphos was based on a 2-year dog feeding study.   On the basis of plasma and erythrocyte cholinesterase inhibition, the No Observed Effect Level (NOEL) in this study was defined as 0.025 mg/kg/day. A safety factor of 10 was used to calculate the ADI of 0.0025 mg/kg/day.   The cholinesterase inhibition Lowest Effect Level (LEL) in this study was defined as 0.075 mg/kg/day.   When the Agency rereviewed this study in the reregistration review for mevinphos, serious report deficiencies were noted that made the study unacceptable and the Agency found that it was not possible to define the NOEL and LEL doses. The Agency reviewed seven chronic toxicity studies during the mevinphos reregistration review process and all but one of them (a rat teratology study) were found to be unacceptable. Therefore, the Agency concludes that there is insufficient data to establish an ADI or PADI for mevinphos at this time.

5. REQUIRED UNIQUE LABELING SUMMARY

The following revised environmental hazard statement must appear on the label of all manufacturing-use products containing mevinphos: "This pesticide is toxic to fish and wildlife. Do not discharge effluent containing this product into lakes, streams, ponds, estuaries, oceans, or public waters unless this product is specifically identified and addressed in an NPDES permit. Do not discharge effluent containing this product to sewer systems without previously notifying the sewage treatment plant authority. For guidance, contact your State Water Board or Regional Office of the Environmental Protection Agency."

Personal protective equipment and work safety statements for mixer/loaders and applicators are required to be included on the label of all end-use mevinphos products.

All end-use products containing mevinphos as an active ingredient, must bear the following restricted use labeling statements:

"RESTRICTED USE PESTICIDE

Due to Very High Acute Toxicity to Humans
and
Residue Effects on Avian, Mammalian and
Aquatic Species

For retail sale to and use only by certified applicators or persons under their direct supervision and only for those uses covered by the certified applicator's certification. Direct supervision for this product is defined as the certified applicator being physically present during application, mixing, loading, repair and cleaning of application equipment. Certified applicators must also ensure that all persons involved in these activities are informed of the precautionary statements."

Environmental hazard statements and a bee precautionary statement must appear on the label of all end-use mevinphos products.

The following reentry interval statement and protective clothing for early reentry statement must appear on the labeling of all products except aquatic non-food use products:

"Reentry into treated citrus groves, grape vineyards, nectarine and peach orchards is prohibited for 96 hours (4 days) after the end of application, unless the protective clothing specified on this label for early reentry is worn. Reentry into all other treated areas is prohibited for 48 hours (2 days) after the end of application, unless the protective clothing specified on this label for early reentry is worn.

FOR EARLY REENTRY INTO TREATED AREAS AFTER SPRAYS HAVE DRIED: Use protective suit of one or two pieces covering all parts of the body except head, hands, and feet; chemical-resistant gloves; chemical-resistant shoes (or chemical-resistant shoe covers or chemical-resistant boots).

FOR EARLY REENTRY INTO TREATED AREAS BEFORE SPRAYS HAVE DRIED, wear all protective clothing specified on this label for an applicator.

Written or oral warnings must be given to workers who are expected to be in treated areas or in an area about to be treated with this product. (Indicate specific oral warnings which inform workers of areas or fields that may not be entered without specific protective clothing, period of time field must be vacated and appropriate actions to take in case of accidental exposure.) When oral warnings are given, warnings shall be given in a language customarily understood by workers. Oral warnings must be given if there is reason to believe that written warnings cannot be understood by workers. Written warnings must include the following information: **DANGER:** Area treated with MEVINPHOS on (Date) _____ . **Do not enter without appropriate protective clothing until** (insert date/time reflecting end of reentry interval set forth on this label). In case of accidental exposure see STATEMENTS OF PRACTICAL TREATMENT found on the MEVINPHOS product label."

The following statement must appear on the labeling of all products, except aquatic non-food use products, which permit aerial application: "HUMAN FLAGGERS ARE PROHIBITED during aerial application of this product unless in totally enclosed vehicles."

The telephone number of the National Pesticide Telecommunications Network must appear on the label of all end-use mevinphos product.

6. REGULATORY POSITION SUMMARY

The Agency will not grant any tolerances or any new uses for mevinphos until sufficient data are submitted for the Agency to calculate the Theoretical Maximum Residue Contribution (TMRC) for mevinphos and then will not grant any tolerances or significant new food uses until all of the required residue chemistry and toxicology data have been submitted and reviewed.

The Agency is continuing the restricted use classification of all mevinphos products.

The Agency is establishing the following reentry intervals for end-use mevinphos products: four days (96-hours) for citrus, groves, grape vineyards, nectarine and peach orchards; and two days (48-hours) for all other crops treated with mevinphos.

The Agency is prohibiting the use of human flaggers during aerial application of mevinphos unless the flaggers are in enclosed vehicles.

The agency is requiring the addition of the telephone number of the National Pesticide Telecommunications Network to all end-use mevinphos product labels.

The Office of Endangered Species (OES) in the U.S. Fish and Wildlife Service (FWS) has determined that certain uses of mevinphos may jeopardize the continued existence of endangered species or critical habitat of certain endangered species. EPA is developing a program to reduce or eliminate exposure to these species to a point where use does not result in jeopardy, and will issue notice of any necessary labeling revisions when the program is developed. No additional endangered species labeling is being required at this time.

While data gaps are being filled, currently registered manufacturing use products and end use products containing mevinphos may be sold, distributed, formulated, and used, subject to the terms and conditions specified in the Registration Standard for Mevinphos. Registrants must provide or agree to develop additional data in order to maintain existing registrations.

7. Summary of Major Data Gaps

Toxicology

Acute Oral Toxicity
Acute Dermal Toxicity
Acute Inhalation Toxicity
Primary Eye Irritation
Primary Dermal Irritation
Dermal Sensitization
Acute Delayed Neurotoxicity
Acute Dermal (to define lethality, toxicity, and ChE NOELs)
Subchronic 90-Day Feeding, two species (rodent and nonrodent)[1]
Subchronic 90-Day Inhalation (to define toxicity from greenhouse
    exposure)
Subchronic 21 Day Dermal
Subchronic Neurotoxicity (conditionally in hen and/or mammal)
Chronic Toxicity, two species (rodent and nonrodent)
Oncogenicity, two species
Teratogenicity (rabbit)
Reproduction
Mutagenicity (full battery - Gene Mutation, Chromosomal Aberration and Other
    Mechanism of Mutagenicity Studies)
Metabolism

---

[1]This requirement is waived since chronic studies are required.

Environmental Fate/Exposure

Hydrolysis
Photodegradation, water
Photodegradation, soil
Photodegradation, air
Aerobic Soil Metabolism
Anaerobic Soil Metabolism
Anaerobic Aquatic Metabolism
Volatility, laboratory
Volatility, field (pending results of the laboratory volatility study)
Terrestrial Field Dissipation
Aquatic Field Dissipation
Soil Dissipation, long term (reserved pending the results of the laboratory
   fish accumulation study)
Confined Accumulation, rotational crops
Field Accumulation, rotational crops (deferred pending receipt of
   acceptable confined rotational crop accumulation data)
Accumulation, irrigated crops
Fish Accumulation, laboratory
Field Accumulation, aquatic nontarget organisms (deferred pending the
   receipt of acceptable accumulation in laboratory fish data)
Leaching and Adsorption/Desorption
Droplet Size Spectrum
Drift Field Evaluation

Ecological effects

Wild Mammal Toxicity
Avian Reproduction (upland game bird and waterfowl)
Freshwater Fish LC_{50} (warmwater and coldwater species) (test material:
   typical end-use product)
Freshwater Invertebrate LC_{50} (test material: typical end-use product)
Estuarine and Marine Organisms LC_{50} (fish and shrimp)
Fish Early Life Stage and Invertebrate Life Cycle
Aquatic Organism Accumulation
Honeybee - Toxicity of Residues on Foliage
Special Test - (Terrestrial Residue Monitoring of
   (Avian and Mammalian Food Items)

Residue Chemistry

Nature of Residues (Plants, Livestock)
Storage Stability
Magnitude of Residues

Product Chemistry

All product chemistry studies

8.   CONTACT PERSON AT EPA

William H. Miller
Product Manager (16)
Insecticide-Rodenticide Branch
Registration Division (TS-767C)
Office of Pesticide Programs
Environmental Protection Agency
401 M Street, S. W.
Washington, D. C.   20460

Office location and telephone number:

Room 211, Crystal Mall #2
1921 Jefferson Davis Highway
Arlington, VA   22202
(703) 557-2600

DISCLAIMER: The information presented in this Chemical Information Fact Sheet
is for informational purposes only and may not be used to fulfill
data requirements for pesticide registration and reregistration.

MONOCARBAMIDE DIHYDROGEN SULFATE

Reason for Issuance: New Chemical Registration
Date Issued: December 30, 1987
Fact Sheet Number: 151

1. Description of Chemical:

   Chemical Name: Monocarbamide dihydrogen sulfate

   Other Name: Monourea sulfuric acid adduct

   Trade Name: ENQUIK

   OPP Internal Control Number: 078001

   Year of Initial Registration: 1987

   Pesticide Type: Herbicide/Desiccant

   U.S. and Foreign Producers: Union Chemicals Division, Unocal Corp.

2. Use patterns and formulations:

   Application sites: Herbicide uses include; Onions (green & dry bulb),
      leeks, shallots, spring onions, garlic, peanuts, peas, lentils,
      dry beans, garbanzos and grass seed. Desiccant uses include; dry
      beans, peas, lentils, peppermint and potato vines.

   Types of formulations: liquid, 81.6% active ingredient

   Types and methods of application: Applied postemergence to crops and weeds
      as a contact pesticide. Diluted with water and applied as a spray for
      broadcast, directed or shield spray treatments, using ground equipment.
      A wetting agent or non-ionic surfactant may be added.

   Application rates: 20 to 309 lbs. a.i./A, depending upon application site

   Usual carriers: water

3. Science findings:

Summary science statement:

Acute toxicology data indicate that the chemical is extremely
caustic and corrosive. The appropriate toxicity category is I
(DANGER), based on primary eye and skin irritation studies.
Chronic toxicology and environmental fate data requirements are
waived. Avian dietary studies indicate that the formulated product
is practically nontoxic to mallard ducks and bobwhite quail.
Diluted monocarbamide dihydrogen sulfate is less hazardous to wildlife
than the undiluted chemical. Application is restricted to use of
the diluted chemical only. Potential hazard to fish and wildlife
including endangered species is minimized due to label restrictions
against aerial application and limiting use to the following states;
Alabama, Georgia, Idaho, Michigan, Oregon and Washington.

Chemical characteristics:

Physical state: Viscous liquid

Color: Colorless

Odor: Odorless

Density: 13.14 pounds per gallon

Miscibility: Limited miscibility in polar organic solvents.
                   Not miscible with hydrocarbons and lipids.

Vapor pressure: Negligible up to decomposition temperature of 80°C.

Crystallization point: 2°C

pH: 1 (aqueous solution)

Unusual handling characteristics: Corrosive to nylon, cotton, leather,
                                  aluminum and copper alloy such as
                                  brass. Corrosive to skin and eyes.
                                  Explosively decomposes above 160°F.
                                  Do not mix with nitrogen fertilizers
                                  as explosive mixtures may result.

Toxicology characteristics:

Acute toxicology results:

Acute oral toxicity (rat): 1,200 mg/kg (male)
                             350 mg/kg (female)
                           Toxicity category II

Acute dermal toxicity (rabbit): greater than 2 g/kg. Study was terminated at 48 hours because of extreme caustic action and dermal necrosis. Toxicity category was not determined.

Primary skin irritation (rabbit): Caustic on intact and abraded skin at 24 hours. Study was terminated at 24 hours after similar results were obtained with 1:4 v/v dilution with water. Toxicity category I

Primary eye irritation (rabbit): Severe corneal involvement, grade 4 chemosis of conjunctivae at 24 hours. Because of severity of reaction, study was terminated at 24 hours. Toxicity category I

Acute inhalation toxicity (rat): greater than 10.8 mg/l
Toxicity category III

Additional data not required on acute dermal and primary eye irritation studies, in accordance with EPA-540/9-82-025, October 1982, Pesticide Assessment Guidelines Subdivision (§81-2(a)(1)) and Subdivision F (§81-4(d)(2)), respectively.

Chronic toxicology results: Additional toxicology studies (including a three-generation reproduction study) were waived in accordance with provisions of 40 CFR 162.45 (c). Toxicological concerns over heavy metal impurities in the sulfuric acid, have been adequately resolved. Recycled (spent) sulfuric acid will not be used to manufacture the active ingredient.

Major routes of exposure: Mixers, loaders and applicators would be expected to receive the most exposure via skin contact and inhalation.

Physiological and biochemical behavioral characteristics:

Mechanism of pesticidal action: Monocarbamide dihydrogen sulfate is a contact herbicide and desiccant. The chemical disrupts cell membrane structures in a catalytic non-acid consumptive reaction. The rate of activity is dependent upon the amount of waxy cuticle present on the surface of treated plants. Plants with a significant wax coating are less susceptible to damage than others.

Environmental characteristics: The environmental fate of urea and sulfuric acid are well known from the literature. Environmental fate data requirements are waived.

Ecological characteristics:

Hazards to fish and wildlife:

Avian dietary toxicity (Mallard): greater than 5620 ppm (49% a.i.)

Avian dietary toxicity (Bobwhite): greater than 5620 ppm (49% a.i.)

Simulated avian field study: Results indicate that eye and foot damage may be pronounced for birds exposed to direct applications of undiluted monocarbamide dihydrogen sulfate. Application of a diluted solution caused some but much less damage.

Avian re-entry field study: Results indicate that birds that can escape during ground application are unlikely to be harmed if they enter treated fields after application.

pH study in ephemeral ponds and irrigation water: The study indicates that under most conditions the acidifying effect of ENQUIK will be temporary. Only shallow ponds receiving an over-spraying of ENQUIK would be affected by a lowering of pH.

Potential problems related to endangered species: Contamination of habitat areas is not likely to occur as a result of runoff, since the chemical, once it reaches the soil, is expected to be neutralized. Potential hazard may exist for those endangered species receiving a direct spraying of ENQUIK. However, geographical restrictions and a prohibition against aerial application should minimize any potential hazard to endangered species.

Tolerance assessments: Monocarbamide dihydrogen sulfate is exempted from the requirement of a tolerance when used as a herbicide or desiccant in or on all raw agricultural commodities (40 CFR 180.1084).

4. Summary of regulatory position and rationale:

Use, formulation, geographical restrictions: Pesticidal use is limited to the following states; Alabama, Georgia, Idaho, Michigan, Oregon and Washington. Aerial application is prohibited.

Unique label warning statements:

End-Use Product:

Physical or Chemical Hazards

"ENQUIK is corrosive to nylon, aluminum, and any copper alloy such as brass. Do not use pumps or fittings containing nylon, mild steel, aluminum, brass, leather, natural rubber or buna N. Non-nylon plastic and 316-L

stainless steel are recommended for application equipment.
Diluted ENQUIK is more corrosive to steel than the
concentrate. Do not allow ENQUIK to be heated above
160°F as explosive decomposition may occur. Do not weld
equipment containing ENQUIK.

Do not mix with nitrogen fertilizers as explosive mixtures
may result. Do not mix with other materials without
specific authorization as hazardous combinations may result.

Clothing: ENQUIK can attack cotton, nylon, and leather
clothing. If ENQUIK contacts clothing of this type,
flush with plenty of water to minimize damage. Wear non-
nylon plastic protective clothing."

Human Hazards

"Corrosive. Causes irreversible eye damage and burns to skin.
Face shields or goggles must be worn. Wear suitable protective
equipment to protect skin, such as synthetic rubber or non-nylon
plastic apron, gloves, pants and boots. Wash after contact with
skin. Shower at the end of the working day. Do not wear contaminated
clothing. Avoid spray. Avoid breathing spray mist."

Environmental Hazards

"This product may be harmful to wildlife directly sprayed. Keep out
of lakes, ponds and streams. Do not apply directly to water or
wetlands. Do not contaminate water by cleaning of equipment or
disposal of wastes. Do not apply near waters already damaged by
acid pollution or in areas with soils of poor buffering capacity
if important aquatic resources are adjacent."

5. Summary of major data gaps: None

6. Contact person at EPA:

Robert J. Taylor
Product Manager (25), (TS-767C)
Environmental Protection Agency
401 M Street, S.W.
Washington, D.C. 20460
(703) 557-1800

DISCLAIMER: The information presented in this Pesticide Fact Sheet
is for informational purposes only and may not be used to fulfill data
requirements for pesticide registration and reregistration.

OXYDEMETON-METHYL

Reason for Issuance: Registration Standard/Initiation of Special
Review
Date Issued: October 1987
Fact Sheet Number: 144

1. Description of Chemical

    Chemical name: S-[2-(Ethylsulfinyl)ethyl]
                    O,O-dimethyl phosphorothioate

    Common name: Oxydemeton-methyl

    Other Chemical Names: Demeton-o-methyl; s-2-ethylsulphinylethyl-
                           o,o-dimethyl phosphorothioate; metilmerkapto-
                           fosoksiol; o,o-dimethyl-s-[2-(ethylsulfinyl)
                           ethyl]phosphorothioate; demeston-s-methyl-
                           sulfoxid;

    Trade Names: R2170; Metasystemox; Metasystox-R; and Bay 21097

    OPP Shaughnessy No.: 058702
    Chemical Abstracts Service (CAS) No.: 301-12-2
    Year of Initial Registration: 1961
    Pesticide Type: Insecticide, Acaricide.
    Chemical Family: Organophosphate
    U.S. and Foreign Producers: Mobay Chemical Corp., U.S.A.; Bayer AG,
                                West Germany.

2. Use Patterns and Formulations

    Application sites: terrestrial food crops (vegetable, field tree fruit,
                        and nut crops), terrestrial nonfood crops (ornamentals
                        and forest trees), greenhouse nonfood crops (research
                        crops not for consumption), domestic indoor and
                        outdoor, and forestry (trunk injection only).

    Types of Formulation: Oxydemeton-methyl is formulated in combination
                           with trichlorfon, methoxychlor, carbaryl, dinocap,
                           dicofol, or tolpet. Single ai formulations
                           consist of 0.48 to 4.0 lb/gal EC, 2%, 2.45%,
                           and 13.5% EC and 50% P/T.

Types/Methods of Application: applied by broadcast spray using ground
                                 equipment or aircraft, soil injection,
                                 tree trunk injection.

3. Science Findings

Summary Science Statement:

Oxydemeton-methyl is an organophosphate compound. It is classified as
Toxicity Category I due to its dermal toxicity, Toxicity Category II due
to its inhalation and oral toxicity. Toxicity Category IV based on
primary dermal irritation potential, and Toxicity Category III because
of its ability to cause primary eye irritation. A 1-year dog feeding
study conducted with the 50% concentrate of the technical material,
instead of the technical grade, at dietary concentrations of 0,
0.0125, 0.125 and 1.25 mg/kg/day, showed inhibition of RBC and brain
cholinesterase at the mid-dose. A new study with the technical
grade is required. Submitted oncogenicity data are inadequate to
allow an assessment of the oncogenic potential of oxydemeton-methyl.
Based on available data, the Agency has determined that oxydemeton-methyl
has the potential to induce reproductive effects in humans, and that
that effect is of concern because of the potential exposure to applicators,
mixers, and loaders who use products containing oxydemeton-methyl,
and to field workers who may enter treated fields. Additional data
are required to fully assess this concern. Additional toxicology
and residue chemistry data are required for a reassessment of the
tolerances of oxydemeton-methyl and its cholinesterase-inhibiting
metabolites. Avian acute oral studies indicate toxicity ranges from
moderately toxic (mallards) to highly toxic (rock dove). Eight-day
dietary studies also show a wide range of toxicity from practically
non-toxic (mallards) to highly toxic (bobwhite quail). Available
data indicate oxydemeton-methyl is very highly toxic to daphnids,
moderately toxic to isopods, highly toxic to amphipods, moderately
to highly toxic to coldwater fish, slightly to moderately toxic to
warmwater fish, and moderately toxic to estuarine crustaceans.
Additional data are required for the Agency to complete its assessment
of potential risks to avian and aquatic species.

Chemical Characteristics:

Physical/chemical properties of pure oxydemeton-methyl, and of the
53.1% FI[a] and 90% unregistered T:

  Color: Colorless (P), yellow-brown (T), Yellow to amber (FI)
  Physical state: liquid (P, FI)
  Odor: Odorless (P), Typical sulfur-containing (T)
  Melting Point: <10° C (P), <10° C (T)
  Boiling Point: 106° C at 0.013 mbar (P)
  Specific Gravity: 1.289 at 20° C (P), 1.03 at 20° C (FI)
  Solubility: 10-100 g/100 ml of xylene, cyclohexanone, 2-propanol,
               and methylene chloride at 20° C (P). Miscible in water
               at 20° C (P).

---

a FI = 53.1% Formulation Intermediate; T = unregistered 90% Technical;
  P = Pure Active Ingredient;

Vapor Pressure: 3.8 x 10^{-5} mbar at 20° C (P)
Dissociation Constant: N/A
Octanol/Water Partition Coefficient: 0.18 (P)
pH: N/A

Toxicology Characteristics

Acute oral: Toxicity Category II
- Supplementary data indicates 50-80 mg/kg in the rat

Acute dermal: Toxicity Category I
- 112 and 152 mg/kg in female and male rats, respectively

Acute inhalation: Toxicity Category II
- 0.51 mg/L and 1.5 mg/L in mice and rats,
respectively

Primary dermal irritation: Toxicity Category IV
- No irritation observed in rabbit

Primary eye irritation: Toxicity Category III
Reversible mild irritation of conjunctivae
in the rabbit

Acute delayed neurotoxicity: Data gap

Subchronic oral (rodent): Data insufficient to assess subchronic
toxicity
Chronic feeding: Data gap

Oncogenicity: Data gap, rat and mouse

Metabolism: Data gap

Teratogenicity: Rat – Maternal NOEL = 1.5 mg/kg/day
Maternal LEL = 4.5 mg/kg/day
Developmental Toxicity NOEL = 1.5 mg/kg/day
Maternal ChE NOEL = 0.5 mg/kg/day (lowest
dose tested)

Rabbit – Maternal NOEL = 0.2 mg/kg/day
Maternal ChE LEL = 0.8 mg/kg/day
Developmental Toxicity LEL = 0.05 mg/kg/day
(lowest dose tested)
Maternal ChE NOEL = 0.2 mg/kg/day

Reproductive Effects: Rat – Reproductive NOEL = 1 ppm
Reproductive LEL = 10 ppm (decreased viability
index and pup body weight, histopathological
changes in epididymis)
Systemic NOEL = 1 ppm (decreased testes weight)

Mutagenicity: Data insufficient to assess mutagenic potential

Physiological and Biochemical Characteristics:

Metabolism and persistence in plants and animals

The metabolism of oxydemeton-methyl in plants is not adequately understood. Because of the structural similarity between disulfoton, demeton, and oxydemeton-methyl it is possible that certain metabolites of oxydemeton-methyl will be common to those of disulfoton and/or demeton. However, in the absence of adequate data delineating the nature of residues of oxydemeton-methyl in plants, no comparisons with demeton or disulfoton plant metabolism can be made at this time. Presently, oxydemeton-methyl and its cholinesterase-inhibiting metabolites are the residues of concern in plants. The tolerance definition may be revised to specifically include the metabolites which constitute residues of concern upon receipt of the requested data. The metabolism of oxydemeton-methyl in ruminants and poultry is not adequately understood. The only available in vivo ruminant metabolism study did not indicate what percentages of the total radioactivity in tissues and milk were present in the extracts of these samples. Additional metabolism studies in plants and animals are required.

Environmental Characteristics:

The environmental fate of oxydemeton-methyl cannot be fully assessed. With the exception of a hydrolysis study, available data reviewed by the Agency are not sufficient to fulfill the environmental fate data requirements. Available data are not adequate to fully assess leaching potential (e.g., field dissipation). Oxydemeton-methyl and its degradation products are soluble in water. Under aerobic conditions, as indicated by preliminary studies, oxydemeton-methyl breaks down in soils of different textures to dimethyl phosphate, 2-sulfonic diethyl sulfoxide, 2 sulfonic diethyl sulfone, and oxydemeton-methyl sulfone. The Agency cannot conclusively determine the potential for ground water contamination at this time.

Ecological Characteristics:

Avian Species

The avian acute oral studies with 50% ai oxydemeton-methyl indicate that the acute toxicity ranges from moderately toxic to highly toxic, depending on the species of bird tested. The acute toxicity values range from 14 mg/kg (rock dove) to 47.6 mg/kg (California quail) for the highly toxic range and from 53.9 mg/kg (mallard) to 120 mg/kg (Chuka) for the moderately toxic range. Likewise, 8-day dietary studies demonstrate that there is a wide range of toxicity from practically nontoxic (>5000 ppm for mallards) to highly toxic (434 ppm for bobwhite quail). Additional data are not required.

No acceptable avian reproduction studies are available. Data are
required to support uses of oxydemeton-methyl on crops with multiple
applications such as cotton, alfalfa, grapes and sugar beets.

Avian terrestrial field studies are not required. A comparison of
the acute toxicity values with estimated pesticide residues on food
items indicate that birds will not be exposed to lethal concentrations
of oxydemeton-methyl in the field at application rates up to 0.75 lb
ai/A (the highest rate currently registered).

Aquatic Organisms

The acute toxicity tests with freshwater invertebrates indicate that
oxydemeton-methyl is very highly toxic to daphnids (3.3 ppb), moderately
toxic to isopods (1400 ppb), and moderately to highly toxic to amphipods
(190 to 1000 ppb). However, none of the available studies satisfies
the requirement for an acute toxicity study for freshwater invertebrates
since older specimens were used in the tests. A new study is required.

The acute fish studies indicate that oxydemeton-methyl is moderately
to highly toxic to coldwater fish and slightly to moderately toxic
to warmwater fish. The 96-hour acute toxicity values ranged from
0.73 ppm to 6.4 ppm for rainbow trout and from 1.22 ppm to 31.5 ppm
for several species of warmwater fish. Additional data are not
required.

Evaluation of tests on a 25% ai oxydemeton-methyl concentrate
product, indicates this formulation is moderately toxic to estuarine
crustaceans. The 96-hour acute toxicity value for pink shrimp (Penaeus
duorarum) was 1.2 (0.9-1.8) ppm, and the 96-hour acute toxicity
value for fiddler crab was 8.6 (6.6-11) ppm. Based on currently
registered uses, oxydemeton-methyl is not expected to enter estuarine
environments in significant concentrations. Additional acute
toxicity testing on estuarine organisms is not required.

Nontarget Insects

Studies indicate that oxydemeton-methyl is moderately to highly
toxic to honey bees exposed to direct application. However, toxicity
of foliar residues appears to be short lived. Toxicity to honey
bee larvae is highly variable depending upon larval development at
the time of exposure. It is unlikely that these nontarget insects
would be affected by application of oxydemeton-methyl made in
accordance with label directions.

Tolerance Reassessment

Tolerances have been established for residues of oxydemeton-methyl
expressed in terms of combined residues of oxydemeton-methyl and its
cholinesterase-inhibiting metabolites on various raw agricultural
commodities (40 CFR 180.330).

A feed additive tolerance of 2.0 ppm is established for the combined
residues of oxydemeton-methyl and its cholinesterase inhibiting

metabolites in the milled fractions of sorghum (except flour) for animal feed when present therein as a result of application to growing sorghum (21 CFR 561.234).

Because of the extensive residue chemistry and toxicology data gaps, the Agency cannot complete all tolerance reassessments.

4. Required Labeling and Regulatory Position Summary

Oxydemeton-methyl is being placed into Special Review: The Agency is initiating a Special Review on all uses of pesticide products containing oxydemeton-methyl, based on concerns regarding reproductive effects. The Agency has determined that oxydemeton-methyl has the potential to induce reproductive effects in humans, and that that effect is of concern because of the potential exposures of applicators, mixers and loaders who use products containing oxydemeton-methyl, and to field workers entering treated fields.

The following precautionary measures are being required for all products containing oxydemeton-methyl either as a single active ingredient or in combination with other active ingredients, pending receipt and evaluation of additional data which will allow the Agency to fully assess the potential of oxydemeton-methyl to induce adverse effects.

Use Classification: All products containing oxydemeton-methyl are classified for Restricted Use under this Standard, on the basis of the potential for reproductive effects.

Protective Clothing/Equipment and Closed Systems: The use of maximum full body protective clothing/equipment by mixer/loaders/applicators, and by field workers entering treated fields prior to the end of the 48 hour reentry interval is required.

Reentry Interval: The 48-hour reentry interval established under 40 CFR 170 for all agricultural uses of oxydemeton-methyl which may involve hand labor tasks will continue to be required. In addition, the use of maximum protective clothing and equipment is required if treated fields are entered prior to the end of the reentry interval. Reentry data are being required to reassess the adequacy of the 48 hour reentry interval.

Flaggers: The use of human flaggers during aerial application is prohibited, unless they are in totally enclosed vehicles.

Tolerances: No new tolerances and food uses will be granted until the Agency has received data sufficient to evaluate the dietary exposure to oxydemeton-methyl.

5. Summary of Major Data Gaps

    Toxicology: Date Due *

        Acute Oral   9 Months
        Dermal Sensitization (Guinea pig)   9 Months
        Acute Delayed Neurotoxicity (Hens)   12 Months
        Subchronic Oral - Rat   15 Months
                       - Non-rodent   18 Months
          (Not required if chronic data are acceptable)
        Subchronic Dermal (21-Day; Rabbit)   12 Months
        Subchronic Neurotoxicity   15 Months
          (Required if acute delayed neurotoxicity
          is positive)
        Chronic Toxicity (Rat, dog)   50 Months
        Oncogenicity (Rat, Mouse)   50 Months
        Mutagenicity Tests (Chromosomal and DNA)   12 Months
        Metabolism (General)   24 Months
        Male Short Term Exposure Study   10/31/88
        Male Reproductive System Toxicity Study   12/31/87**

    Environmental Fate/Exposure

        Photodegradation
          In water   9 Months
          In soil   4 Months
          In air (condiditonal)
        Metabolism - Lab
          Aerobic Soil   27 Months
          Anaerobic Soil   7/14/88
        Mobility
          Leaching and Adsorption/Desorption   12 Months
          Laboratory Volatility   12 Months
          Field Volatility (conditional)
        Terrestrial Field Dissipation
          Field Dissipation - Soil   27 Months
          Long-term Field Dissipation (conditional)
          Confined Rotational Crop   07/14/89
          Field Rotational Crop (conditional)
        Spray Drift Studies
          Droplet Size Spectrum   27 Months
          Drift Field Evaluation   27 Months
        Reentry
          Foliar Dissipation   27 Months
          Soil Dissipation   27 Months
          Dermal Exposure (conditional)
          Inhalation Exposure (conditional)

---

*   Due date is based on date of receipt of Standard by the registrant, unless
    otherwise indicated.
** Study required by the June, 1987 Data Call-in and is to be submitted by
    either December 31, 1987 (if an ongoing study being conducted by Mobay for
    the California Dept. of Food and Agric. is used) or October 31, 1988 (if a
    new study is undertaken).

Ecological Effects

Avian Reproduction
Aquatic Invertebrate Acute Toxicity
Fish Early Life Stage and
   Aquatic Invertebrate Life Cycle
Aquatic Organism Accumulation

Date Due*

04/14/88
9 Months

15 Months
27 Months

Residue Chemistry/Product Chemistry

Product Chemistry
Plant and Animal Metabolism
Storage Stability
Magnitude of Residue Studies

6 - 12 Months
10/14/87
24 Months
4/14/88

6. Contact Person at EPA

William H. Miller, (PM-16)
Insecticide-Rodenticide Branch (TS-767)
401 M Street SW.
Washington, DC   20460.
Tel. No. (703) 557-2600

DISCLAIMER: The information presented in this chemical Information Fact
Sheet is for informational purposes only and may not be used to fulfill
data requirements for pesticide registration and reregistration.

OXYTETRACYCLINE

Reason for Issuance: Registration Standard
Date Issued: December 1988
Fact Sheet Number: 188

1. DESCRIPTION OF CHEMICAL

    Generic Name: Oxytetracycline
    Oxytetracycline calcium complex
    Oxytetracycline hydrochloride

    Chemical Name: 4-(dimethylamino)-1,4,4a,5,5a,6,11,12a-octahydro-
    3,5,6,10,12,12a-hexahydro-6-methyl-1,11-dioxo-2-
    naphthacenecarboxamide and its calcium complex and
    hydrochloride salts.

    Trade Names: Glomycin, Terrafungine, Riomitsin, Hydroxy-
    tetracycline, Berkmycin, Biostat, Impercin, Oxacycline, Oxyatets,
    Mycoshield, Agricultural Terramycin, Terramycin Hydrochloride,
    Terramycin.

    Chemical Class: Antibiotic (produced by the actinomycete
    Streptomyces rimosus)

    Pesticide Type: Plant Fungicide/Bactericide and Algicide

    CAS Registry Number: 79-57-2 (oxytetracycline)
                         7179-50-2   (calcium complex)
                         2058-46-0 (hydrochloride)

    EPA Shaughnessy Codes: 006304 (oxytetracycline)
                           006321 (calcium complex)
                           006308 (hydrochloride)

    Empirical Formulae: $C_{22}H_{24}N_2O_9$ (oxytetracycline) Mol. Wt. 460.44
    $C_{22}H_{22}N_2O_9$ Ca (calcium complex) M.W. 498.52
    $C_{22}H_{24}N_2O_9$ HCl (hydrochloride) M.W. 496.9

    Year of Initial Registration: August 1974

    U.S. and Foreign Producers: Pfizer, Inc.

478

2.   USE PATTERNS AND FORMULATIONS

Type of Pesticide: Plant Fungicide/Bactericide and Algicide.

Pests Controlled: Bacterial and fungal diseases and slime-forming microorganisms.

Registered Uses:

1. Calcium oxytetracycline [17% WP]:
Nectarines, peaches, pears, and creeping Bentgrasses.

2. Oxytetracycline hydrochloride [21.6% Soluble Concentrate.
-(Tree Trunk Injection) pears, peaches, and ornamental palms)
- Marine antifoulant paint additive.

3. Oxytetracycline hydrochloride [21.6% Soluble Concentrate.
Formulation Intermediate for Marine antifoulant additive.

Predominant Uses: Pears and peaches (98%).

Minor Uses: Ornamental palms and Bentgrasses.

Annual Usage: 21,350 pounds/ai

Method of Application: Foliar, tree injection, and brush on (marine use).

3.   SCIENCE FINDINGS

Summary Science Statement

1. Oxytetracycline hydrochloride oncogenicity data indicates equivocal evidence of oncogenicity in male and female rats. The Agency concludes that, although the findings were termed "equivocal" by the National Toxicology Program, they do not represent positive evidence of carcinogenicity in the rat. A similar study in mice indicated no evidence of oncogenicity.

2. Tolerances for oxytetracycline are limited to peaches (which includes nectarines) and pears, 0.1 ppm and 0.35 ppm respectively. A Reference Dose (RfD) of 1.0 mg/kg/day has been established based on several chronic studies. The Theoretical Maximum Residue Contribution (TMRC) for the U.S. population is 0.000065 mg/kg/day, corresponding to 0.006% of the RfD. A proposed increase in the peach tolerance from 0.1 to 0.35 ppm would result in a TMRC of 0.000118

mg/kg/day. The largest subgroups, nursing and non-nursing infants, represent 0.061% and 0.076% respectively of the current RfD.

3. The potential for development of oxytetracycline resistance due to increased background levels from pesticidal uses to applicators and field workers appears minimal.

4. The Agency is unable to assess the potential for oxytetracycline to contaminate ground water because the environmental fate of oxytetracycline is uncharacterized.

5. The Agency is unable to assess the ecological effects of oxytetracycline on terrestrial or aquatic wildlife, because no data are available.

Toxicology Characteristics

Oxytetracycline has been used extensively for over 37 years in animals and in man. It is one of a group of broad spectrum antibiotics known as tetracyclines which were developed for control of bacterial diseases in man and animals. As a result of its human drug use, there is an extensive body of toxicological data available on oxytetracycline. Thus, all toxicological data requirements have been waived.

Chronic toxicity

Due to various deficiencies, the available studies do not fulfill current requirements. Additional data are not required, based upon availability of both animal and human data from oxytetracycline's drug uses.

Two 2-year chronic toxicity studies in rats are available. In one study, Osborne-Mendel rats were fed diets containing 0, 100, 1000 and 3000 ppm, oxytetracycline hydrochloride in the diet for 24 months. The NOEL was determined to be 3000 ppm, approximately 150 mg/kg/body weight/day, highest dose tested.

In a second study, Sprague Dawley rats were fed diets containing 0,100, and 1000 ppm oxytetracycline hydrochloride in the diet for 24 months. The NOEL for oxytetracycline hydrochloride was 1000 ppm, 50 mg/kg/day, highest dose tested.

Two chronic toxicity studies in dogs are available. In the first study, dogs were fed diets containing 0, 100, 3000, and 10000 ppm of oxytetracycline hydrochloride in the diet for 24 months. A yellow discoloration of the long bones and brownish discoloration of the thyroid was observed in all dosed animals at necropsy. The NOEL was determined to be 10000 ppm, approximately 250 mg/kg/day, highest dose tested.

In a second study, mongrel dogs were fed diets containing 0, 5000, and 10000 of oxytetracycline hydrochloride in the diet for 12 months. The NOEL was determined to be 10000 ppm, approximately 250 mg/kg/day, highest dose tested.

Oncogenicity

NCI/NTP Oxytetracycline Oncogenicity Study in the F344N/Rat

In this study, oxytetracycline hydrochloride (purity 98.8%) was administered to groups of F344/N rats fed 0, 25000, and 50000 ppm in the diet for 103 weeks. Fatty metamorphosis of the liver was increased in rats in the low dose group. The National Toxicology Program concluded that ". . . . there was equivocal evidence[1] of carcinogenicity for male F344/N rats as indicated by increased incidences of pheochromocytomas of the adrenal gland.   There was equivocal evidence of carcinogenicity for female F344/N rats as indicated by increased incidences of adenomas of the pituitary gland in the high dose group."

NCI/NTP Oxytetracycline Oncogenicity Study in the B6C3F1 Mouse

In this study, oxytetracycline hydrochloride (purity 98.8%) was administered to groups of B6C3F1 mice fed 0, 6300, and 12500 ppm in the diet for 103 weeks. The National Toxicology Program concluded that "... there was no evidence of carcinogenicity for male or female B6C3F1 mice fed diets containing 6300 or 12500 ppm of oxytetracycline hydrochloride for 2 years."

Teratogenicity

Female Charles River CD rats were dosed during days 6 through 15 of gestation with 1200, 1350, or 1500 mg/kg of oxytetracycline hydrochloride. There were dose-related decreases in maternal survival and body weight gain, and increases in the incidence of respiratory difficulties and rough coat.   In addition, there were significant dose-related decreases in the percent of treated dams found pregnant. There was also a dose-related decrease in fetal body weight. The high incidence of maternal deaths and the fetotoxicity noted in all dose levels tested did not allow for an establishment of a NOEL.   The LEL was 1200 mg/kg/day (lowest dose tested).

The significant findings discussed in this study can be attributed to the excessive dose levels used, and the overly stressing of the treated dams.

Female CD-1 mice were dosed during day 6 through 15 of gestation with 0, 1325, 1670, and 2100 mg/kg oxytetracycline hydrochloride.   No adverse effects were demonstrated, due probably to the low dose levels used.   The NOEL for maternal and developmental toxicity in this study was 2100 mg/kg (highest dose tested).

---

[1] The NCI/NTP uses five levels of interpretative evaluations in animal carcinogenesis studies; in decreasing order of strength (not potency or mechanism) of the experimental evidence, these are: (i) clear evidence ofcarcinogenicity (ii) some evidence of carcinogenicity, (iii) equivocal evidence of carcinogenicity, (iv) no evidence of carcinogenicity, and (v) inadequate study of carcinogenicity.

Antibiotic Microbial Resistance

Mature beagles were fed a diet containing 0, 2, or 10 ppm, approximately of oxytetracycline for 44 days. The 10 ppm (0.25 mg/kg/day) diet resulted in a shift from a predominantly drug-susceptible population of enteric lactose-fermenting organisms to a multiple antibiotic-resistant population. A shift to drug-resistance did not occur in the group fed 2 ppm approximately 0.05 mg/kg/day. The NOEL was 0.05 mg/kg/day.

4. TOLERANCE ASSESSMENT

Tolerances have been established for residues of oxytetracycline in two raw agricultural commodities (40 CFR 180.337). Use of oxytetracycline as a drug in food animals is regulated by the FDA according to 21 CFR 520, 522, 524, and 558. The FDA has established tolerances for oxytetracycline in or on meat, fat, meat byproducts, and in uncooked edible tissues of salmonoid fish and catfish (21 CFR 556.500).

No data are available to evaluate the nature of the residue of oxytetracycline in plants. The Agency has assessed the need for data reflecting the metabolism of oxytetracycline in plants and has concluded that these data are not required because of the drug uses of oxytetracycline.

No data are available to evaluate the nature of the residue of oxytetracycline in animals. However, data on the metabolism of oxytetracycline in food animals are not required: residues of oxytetracycline in meat and milk are unlikely since there are no registered uses of animal feed items at the present time.

The available microbiological assay method for the determination of oxytetracycline residues in or on peaches, nectarines and pears is adequate for data collection and for tolerance enforcement. The Agency will not require any additional analytical methods at this time. The method is similar to Final Action Microbiological Methods I and II in the AOAC Official Methods of Analysis (1984;42.293-42.298).

5. Summary of Regulatory Positions

Oxytetracycline is not a candidate for Special Review at this time.

Oxytetracycline does not meet the criteria for restricted use classification.

The Agency will continue to grant new uses for oxytetracycline.

The Agency will propose that the tolerance level for peaches be increased from 0.1 ppm to 0.35 ppm.

The Agency will not propose the establishment of crop group tolerances for pome fruits or stone fruits.

Current tolerances are sufficient to cover the actual residues resulting from tree injections (pears only) and foliar applications.

The Agency is deferring its decision concerning the potential of oxytetracycline to contaminate groundwater until information on its environmental fate has been submitted and evaluated.

The Agency believes that the potential for development of resistant microorganisms in applicators and/or field workers as a result of exposure are negligible.

Potential for development of oxytetracycline resistant microorganisms as a result of dietary exposure is minimal.

6. Summary of Major Data Gaps

	Timeframe for Submission	
Environmental fate/Exposure:[2]		
Hydrolysis	9	Months
Photodegradation in water and in soil	9	Months
Metabolism Studies (lab)		
-Aerobic Soil	27	Months
-Anaerobic Soil	27	Months
-Anaerobic Aquatic	27	Months
-Aerobic Aquatic	27	Months
Leaching and Adsorption/Desorption	12	Months
Dissipation Studies (field)		
-Soil	27	Months
-Aquatic (Sediment)	27	Months
Accumulation in Fish	12	Months
Fish & Wildlife:		
Avian Acute Oral LD50	9	Months
Avian Dietary LC50	9	Months
Freshwater Fish LC50 (TGAI)[3]	9	Months
Freshwater Invertebrate (TGAI)	9	Months
Product Chemistry		
All product chemistry studies	9	Months

---

[2]   Environmental Fate data requirements only for calcium oxytetracycline.

[3] TGAI: Technical grade of the active ingredient

7. **CONTACT PERSON AT EPA**

    Larry Schnaubelt
    Product Manager (21)
    Herbicide/Fungicide Branch
    Registration Division (TS-767C)
    Office of Pesticide Programs
    Environmental Protection Agency
    401 M Street, S. W.
    Washington, D. C. 20460

    Office location and telephone number:

    Room 227, Crystal Mall #2
    1921 Jefferson Davis Highway
    Arlington, VA 22202
    (703) 557-1900

DISCLAIMER: The information in this Pesticide Fact Sheet is a
summary only and is not to be used to satisfy data requirements for
pesticide registration and reregistration. The complete Registration
Standard for the pesticide may be obtained from the National Technical
Information Service. Contact the Product Manager listed above for
further information.

PHENMEDIPHAM

Reason for Issuance: Registration Standard
Date Issued: March 30, 1987
Fact Sheet Number: 172

1. Description of Chemical

The following chemical is covered by this Registration Standard:

Common name: Phenmedipham
Chemical name: 3-methoxycarbonylaminophenyl -3-methylcarbanilate
CAS Number: 13684-63-4
OPP (Shaughnessy) Number: 098701
Empirical Formula: $C_{16}H_{16}N_2O_4$
Trade Names: Betanal®, GBP-1-127050®, Kemipham®, Morton EP-452®, Phenmediphame®, Schering 4072®, Schering 4075®, Schering 38584®, SN 38584®, Spinaid®.
Chemical Family: Carbanilate
Pesticide Type: Herbicide
Year of Initial Registration: 1970
Registrants of Technical Products: Schering AG, Nor-AM Chemical Co.

2. Use Patterns And Formulations

Application Sites: Terrestrial food uses (sugar beets, table beets, spinach & spinach seed).
Methods of Application: Applied postemergence to foliage of weeds and crops. Primarily applied as broadcast or banded spray by ground or aerial equipment.
Application Rates: Table beets, spinach: 0.5 to 1 lb, sugar beets: 0.17 to 0.61 lb (lb ai/Acre).
Types of Formulations: Emulsifiable concentrate [EC] (8.0% and 15.9% a.i.)
Usual Carrier: Water

3. Science Findings

Phenmedipham did not induce neoplastic effects in chronic toxicity studies of rats. Except for weight changes in a 3 generation rat study, this chemical does not induce reproductive
effects in rats. This pesticide has shown low acute oral and
dermal toxicity in test animals. It is low in avian toxicity
and moderately toxic to fish and aquatic invertebrates.

It does not cause the destruction of habitat of animals or nontarget organisms.

Although the present data base for phenmedipham does not indicate
major toxicological concerns, there are still toxicology data
gaps: acute inhalation toxicity, primary eye irritation, primary
dermal irritation, dermal sensitization, subchronic dermal toxicity
(21-day), oncogenicity in two species, teratology in two species,
mutagenicity and general metabolism have not been characterized
for phenmedipham.

In addition, the environmental fate, metabolism in food crops,
metabolism in ruminants and poultry for this compound have not
been characterized.

Chemical Characteristics:

Phenmedipham is an odorless white to slightly colored powder at
room temperature. Its melting point is 140-144° C, and its
molecular weight is 300.3. Its water solubility is 3.1 mg/liter
in pH 4 water at 25°C.

Toxicological Characteristics:

   Acute Oral: Toxicity Category IV - > 8,000 mg/kg (male &
                 female rats)
   Acute Dermal: Toxicity Category III - > 4,000 mg/kg (male &
                 female rats)

   Chronic Toxicity:

   A rat chronic feeding study with phenmedipham showed a No-
   Observable-Effect-Level (NOEL) for oncogenicity greater than
   500 ppm (25 mg/kg), highest dose tested (HDT). Although an
   increase in neoplastic findings was not observed in this study
   up to and including a dosage level of 500 ppm (HDT), the
   oncogenic potential of phenmedipham cannot be ascertained in
   the absence of a maximum tolerated dose. Therefore, this study
   does not fulfill the data requirement for oncogenicity (rat).
   Oncogenicity studies in two species are required. The body
   weight gain changes noted in the high dose females during the
   second year are minimal evidence of systemic toxicity (less
   than 10% decrease). Therefore, the NOEL for systemic effects
   is 100 ppm with a lowest observed effect level (LOEL) for
   systemic effects at 500 ppm (HDT).

In a dog chronic feeding study, phenmedipham showed a NOEL of 1000 ppm (25 mg/kg, HDT). Neoplastic findings were not observed in this study up to and including a dosage level of 1,000 ppm (HDT).

Major routes of exposure: Applicators (mixer/loader) handling this pesticide.

Physiological & Biochemical Chracteristics:

Translocation: Phenmedipham is absorbed by the plant leaves and translocated to other portions of the plant.

Mechanism of pesticidal action: It is a strong inhibitor of the Hill reaction in photosynthesis.

Metabolism & Persistence in Plants & Animals: The available data are inadequate to evaluate the persistence of phenmedipham in plants and animals.

Environmental Chracteristics:

Available data are insufficient to fully assess the environmental fate and potential exposure of humans and non-target organisms to phenmedipham. Additional data are required to characterize the potential for phenmedipham to reach ground water supplies.

Ecological Characteristics:

There is sufficient information to characterize phenmedipham as having very low toxicity on an avian dietary basis for Bobwhite quail and Mallard duck. However, an avian single-dose oral toxicity study is required. There is sufficient acute toxicity information to characterize technical phenmedipham as moderately toxic to freshwater invertebrates and to both warm and cold water fish. In an acute contact study, phenmedipham was shown to be very low in toxicity to honey bees.

	(Low toxicity)
Avian dietary toxicity:	Bobwhite quail - > 10,000 ppm
(8 day diet)	Mallard duck - > 10,000 ppm
Freshwater fish toxicity:	(Moderate toxicity)
(96-hour exposure in water)	Bluegill - 3.98 ppm
	Rainbow trout - 1.41 ppm
Aquatic invertebrate toxicity:	(Moderate toxicity)
(48-hour exposure in water)	Daphnia magna - 3.2 ppm.

Endangered species:

No endangered species labeling is required because of the low to moderate toxicity of phenmedipham products to avian and aquatic species, and the fact that this pesticide will not readily drift from the application sites to endangered plants. In addition, treated crops are not planted in the habitat of these plants.

Tolerance Assessment

Sufficient data are available to ascertain the adequacy of the established tolerances for residues of phenmedipham in or on table beet roots and tops, sugar beet roots and tops, and spinach (40 CFR 180.278).

Tolerances for Phenmedipham have been approved for the raw agricultural commodities (RAC's) listed below.

Crop	Tolerance(ppm)	Food Factor	mg/day(1.5kg)
Beets	0.20	0.17	0.00052
Sugar, beet roots	0.10	3.64	0.00546
Sugar, beet tops	0.10	3.64	0.00546
Spinach	0.50	0.05	0.00038

The provisional acceptable daily intake (PADI) for phenmedipham is based on a 104 week rat feeding study. The systemic NOEL for for this study is 100 ppm (5 mg/kg). The systemic lowest observed effect level (LOEL) is 500 ppm (25 mg/kg/day, HDT). There is also a 104 week dog feeding study with a NOEL of greater than 1000 ppm (25 mg/kg, HDT). Utilizing a safety factor of 100, the PADI was set at 0.050 mg/kg/day. This is equivilant to a maximum permitted intake (MPI) of 3.00 mg/day for a 60 kg individual. The theoretical maximum residue contribution (TMRC) for phenmedipham in the daily diet based on the total tolerances above and a daily food intake of 1.5 kg is 0.0064 mg/day. Under these conditions, 0.21% of the PADI has been utilized.

However, data gaps exist for plant and animal metabolism, analytical methods, and storage stability. Processing studies are required for sugar beets. Since the data required for individual commodities are dependent on metabolism data, the Agency recommends that metabolism data be obtained and submitted prior to any required residue data.

There are Canadian tolerances of 0.10 ppm for phenmedipham residues in or on sugar beets and tops but no tolerances on spinach and table beets. There are no Mexican tolerances, Codex Maximum Residue Levels, or exemptions for phenmedipham on sugar beets, spinach and table beets.

Reported Pesticide Incidences

The Pesticide Incident Monitoring System (PIMS) does not have
any incident involving phenmedipham at this time.

4. Summary of Regulatory Positions & Rationale:

-- The Agency has determined that phenmedipham does not exceed any
of the risk criteria for adverse effects in 40 CFR, Section 154.7
at the present time. Available data indicate that phenmedipham
does not promote neoplastic findings in chronic toxicity studies
of rats. Except for weight changes in a 3-generation rat study,
this compound does not induce reproductive effects in rats. It
does not pose a risk of serious acute injury to humans, domestic
animals or non-target organisms and, it does not cause the
destruction of habitat of non-target organisms.

-- The Agency has determined that certain toxicological studies
are required to support the reregistration of phenmedipham
products: acute inhalation, primary eye irritation, primary
dermal irritation, dermal sensitization, subchronic dermal toxicity
(21-day), oncogenicity in two species, teratology in two species,
mutagenicity and metabolism studies.

-- The Agency has determined that present precautionary statements
for persons handling or applying phenmedipham products are sufficient for the labels of manufacturing-use and end-use products.
Available data indicate that phenmedipham causes low oral (Category
IV) and dermal (Category III) toxicities in test animals. Therefore, the labeling of these products contain statements that
caution persons applying or handling this compound, give first
aid instructions, and require the use of precautionary measures
to ensure safe handling of the pesticide products.

-- The Agency has determined that reentry intervals for workers
are not required for phenmedipham products. The low acute toxicity of this chemical does not warrant significant concern about
exposure of workers reentering treated areas, according to the
criteria in 40 CFR Part 158.140.

-- At this time, the Agency will defer action on ground water
issues until receipt and evaluation of environmental fate data.

-- The Agency is requiring labeling which warns of potential
hazards to aquatic organisms. Available acute toxicity data
indicate that phenmedipham is moderately toxic to fish and
aquatic invertebrates.

--The available data indicate that this pesticide, when applied
at recommended rates, does not present unreasonable hazards to
birds. Existing studies on this compound indicate that its dietary toxicity to birds is very low.

-- The Agency has determined that specific labeling to mitigate
potential hazards to endangered species are not required.
Available data indicate that phenmedipham exhibits low toxicity
to avian species, and moderate toxicity to fish and aquatic
invertebrates. Crops treated with this pesticide are not planted
in the habitat of endangered plants. Also, this chemical will
not readily drift from its application sites to the endangered
plants.

-- The Agency will not require phytotoxicity data.
Phenmedipham has low volatility and will not readily drift from
its application sites to non-target plants and it has limited
terrestrial use on minor crops.

5. Precautionary Statements

    a. Manufacturing-Use Product Statements

        All phenmedipham products intended for formulation into
        end-use products must bear the following statements:

        "This pesticide is toxic to fish and aquatic organisms.
        Do not discharge effluent containing this product into
        lakes, streams, ponds, estuaries, oceans, or public
        water unless this product is specifically identified and
        addressed in an NPDES permit. Do not discharge effluent
        containing this product into sewer systems without
        previously notifying the sewage treatment plant authority.
        For guidance contact your State Water Board or Regional
        Office of the EPA."

    b. End-Use Product Statements

        The following precautionary statements must appear on all
        EP labels:

        "This pesticide is toxic to fish and aquatic organisms.
        Do not apply directly to water or wetlands (swamps, bogs,
        marshes, and potholes). Drift and runoff from treated areas
        may be hazardous to fish and aquatic organisms in adjacent
        aquatic sites. Do not contaminate water by cleaning
        of equipment or disposal of wastes."

6. Summary of Major Data Gaps

    158.120 Product Chemistry data are required during 1987.

    158.125 Residue Chemistry:

171-4	Nature of Residue (Plant & Animal Metabolism)	October 30, 1988
171-4	Residue Analytical Methods	July 30, 1988
171-4	Storage Stability	July 30, 1988
171-4	Residue Studies on Crops, Processed Food/Feed Commodities	April 30, 1989

    158.135 Toxicology:

81-3	Acute Inhalation Toxicity (Rat)	January 30, 1988
81-4	Primary Eye Irritation Toxicity (Rabbit)	January 30, 1988
81-5	Primary Dermal Irritation (Rabbit)	January 30, 1988
81-6	Dermal Sensitization (Guinea pig)	January 30, 1988
82-2	Subchronic Dermal (21-day)	April 30, 1988
83-2	Oncogencity (Two species)	
	(Mouse)	May 30, 1988
	(Rat)	June 30, 1991
83-3	Teratogenicity (Two species)	July 30, 1988
84-2	Mutagenicity	April 30, 1988
85-1	Metabolism	April 30, 1989

    158.130 Environmental Fate

161-1	Hydrolysis	January 30, 1988
161-2	Photodegradation in Water	January 30, 1988
162-2	Anaerobic Soil Metabolism	July 30, 1989
163-1	Mobility Studies (Leaching & Adsorption/Desorption)	April 30, 1988
164-1	Soil Dissipation	July 30, 1989
165-1	Rotational Crops (Confined)	July 30, 1990
165-4	Accumulation in Fish	April 30, 1988

    158.145 Wildlife and Aquatic Organisms

71-1	Avian Oral Toxicity	January 30, 1988

7. Contact Person at EPA

Robert J. Taylor, PM-25
Office of Pesticide Programs, EPA
Registration Division (TS-767C)
401 M Street, S.W.
Washington, DC 20460
Phone (703) 557-1800

DISCLAIMER: The information presented in this Pesticide Fact Sheet is for informational purposes only, and may not be used to fulfill data requirements for pesticide registration or reregistration.

PHORATE

Reason for Issuance: Registration Standard
Date Issued: December 1988
Fact Sheet Number: 34.2

1. DESCRIPTION OF CHEMICAL

    Generic name: O,O-diethyl S-[(ethylthio)methyl]
                      phosphorodithioate

    Common name: Phorate

    Trade Names: Thimet, AAstar and Rampart

    Other Chemical
    Nomenclature: O,O-diethyl-S-(ethylthio)
                 methyl phosphorodithioate
                 and O,O-diethyl S-
                 [(ethylthio)methyl] ester

    EPA Pesticide Chemical (Shaughnessy) Number: 057201

    Chemical Abstracts Service (CAS) number: 298-02-2

    Year of initial registration: 1959

    Pesticide type: Insecticide

    Chemical family: Systemic Organophosphate

    U.S. Registrants: American Cyanamid Company, Uniroyal
                      Chemical Company Inc., Aceto Chemical
                      Company Inc., Wilbur Ellis Co.,
                      Riverside Terra Corp., Farm Bureau
                      Cooperative, and Platte Chemical Co.

2. USE PATTERNS AND FORMULATIONS

    Application sites: Terrestrial food crop use on beans,
                      corn (field and sweet), cotton,
                      hops, peanuts, potatoes, sorghum,
                      soybeans, sugar beets, sugarcane,
                      barley and wheat.

Terrestrial non-food crop use on
lilies (bulb production).

Formulations: Granular

Pests Controlled: Various leaf-feeding insects, mites,
and soil insects

Methods of application: Soil and foliar applications
(band, broadcast, in-furrow and
drilling) using conventional
ground and aerial equipment

3. SCIENCE FINDINGS

Summary Science Statement

Technical phorate is in Toxicity Category I by the oral,
dermal and inhalation routes. The acute oral administration of
phorate to hens did not cause a delayed neurotoxic effect.
Based on results of acceptable subchronic and chronic feeding
studies with rats and dogs, cholinesterase (plasma, blood or
brain) is the primary target for phorate. Phorate does not
produce oncogenic effects, based on results of acceptable
chronic studies in rats or dogs. Phorate does not induce
teratogenic effects, based on results of acceptable teratology
and reproduction studies. Phorate did not cause a mutagenic
response in several in vitro (microbial and mammalian cells)
studies or in an in vivo dominant lethal study. Results of an
acceptable metabolism study using male rats indicated that a
large proportion of phorate labeled metabolites were excreted in
urine and feces within 24 hours of dosing. The total
radioactivity levels in tissues was low. The oxidative,
phosphorylated products (metabolites of phorate which may be more
potent anticholinesterase compounds through oxidative
desulfuration and/or sulfide oxidation) represented a minor
proportion of the phorate metabolites measured.

Based on acceptable laboratory data, technical phorate is
characterized as very highly toxic to birds on an acute oral
basis; highly toxic to birds on a dietary basis; very highly
toxic to mammals on a dietary basis and very highly toxic to
freshwater fish and aquatic invertebrates and estuarine and
marine organisms on an acute toxicity basis. Results of the
terrestrial field studies (Level 1) showed mortalities to avian
and mammalian species. An aquatic field study (required in the
1984 Standard) is still in progress and is due in 1991.

Many of the tolerances for phorate are still not adequately
supported. Additional data (residue studies, residue analytical
methods, processing and cooking studies, poultry metabolism study
and storage stability data) are needed before the Agency can

determine the adequacy of current tolerance levels and perform a tolerance reassessment.   Based on the NOEL for brain cholinesterase in a one year dog study (0.05 mg/kg) and applying an uncertainty factor of 100, the Agency has calculated the Anticipated Residue Contribution (ARC) for the U.S. population average to be 0.000491 mg/kg/day, corresponding to 98% of the RfD.   The ARC assumes residues are present at tolerance levels, but takes into account percent of crop treated, where possible. For children 1 to 6 years of age, the ARC occupies 235% of the RfD, and for non-nursing infants, the ARC occupies 331% of the RfD.   The Agency is requiring processing and cooking studies to assess anticipated residue levels in meat and milk.   The Agency expects that cooking and processing of meat and milk will reduce residues of phorate to levels which will be of little or no concern.

Chemical/Physical Characteristics of the Technical Material

Chemical/Physical
Characteristics: Color: pale straw to light brown
(TGAI, 2749-106); color-
less to very light yellow
(TGAI, 241-212 and 241-213)

Physical state: liquid

Odor: characteristic of mercaptan
containing compounds

Boiling Point: 118-120 ^{0}C, 0.8 mm Hg

Specific Gravity: 1.15 at 20 ^{0}C (TGAI,
2749-106); 1.17 at
25 ^{0}C (TGAI for 241-
212 and 241-213)

Solubility: 50 mg/l in water;
miscible with carbon
tetrachloride, vege-
table oils (unspecified),
xylene, and unspecified
alcohols, ethers and
esters

pH: 5-7 (TGAI, 2749-106);
3.56-3.81 (TGAI, 241-212 and
241-213)

Viscosity: 80 cps at 21 ^{0}C

Corrosion: non-corrosive to steel,
aluminum, porcelain,

fiberglass, and phenolic
resins

Toxicology Characteristics

Acute Oral: Toxicity Category I (LD_50 of 3.7 and 1.4 mg/kg
in male and female rats, respectively)

Acute dermal: Toxicity Category I ($LD50$ of 9.3 and 3.9 mg/kg
in male and female rats, respectively)

Acute inhalation: Toxicity Category I ($LD50$ of 60 and 11
mg/m3 for male and female
rats, respectively)

Primary dermal irritation: None Available. Not required
since the toxicity of phorate
prohibits the administration of
appropriate dosage levels.

Primary eye irritation: None Available. Not required since
the toxicity of phorate prohibits
the administration of appropriate
dosage levels.

Skin sensitization: None available. Not required due to the
high acute toxicity of the chemical.

Delayed Neurotoxicity: Did not induce delayed neurotoxicity in
an acceptable study in hens.

Subchronic non-rodent study: None available. Not required
since acceptable chronic
data for the non-rodent are
available.

Subchronic rodent study: A rat study is available. The LEL
in this study was 2.0 ppm (0.1
mg/kg/day); the NOEL was 0.66 ppm
(0.033 mg/kg/day).

Chronic toxicity: Dog study is available (NOEL and LEL for
systemic toxicity were 50 and 250
ug/kg/day, respectively). Mouse study is
available (NOEL and LEL were .45 and .9
ug/kg/day, respectively). Rat study is
available (LEL was 0.05 mg/kg/day, NOEL
was not determined)

Oncogenicity: The rat combined chronic toxicity and oncogenicity
study did not reveal any evidence that phorate

was oncogenic under the condition of that study. Based on a reevaluation of the mouse study, the evidence does not show that an MTD was attained. Confirmatory data are required.

Mutagenicity: Phorate was negative in all areas of mutagenicity testing (gene mutation, structural chromosome aberration and tests for other genotoxic effects)

Teratogenicity: A rat study is available (LEL for developmental toxicity, based on embryotoxicity, and maternal toxicity was 0.50 mg/kg and the corresponding NOEL for each was 0.25 mg/kg). A rabbit study is available (LOEL and NOEL for maternal toxicity was 0.5 and 0.15 mg/kg, respectively. The NOEL for developmental toxicity was 1.2 mg/kg, the highest dosage administered).

Reproduction: A mouse study is available (LEL was .45 mg/kg/day and the NOEL was .23 mg/kg/day).

Metabolism: A study in male rats is available. Results indicated that a large proportion of the administered 14C was recovered in urine and feces. Oxidative, phosphorylated products only represented a minor proportion of the metabolites measured.

Environmental Characteristics

Based on the results of an acceptable leaching study, ^{14}C phorate was reported to be very mobile to mobile in loamy sand, sandy loam, silt loam, and loam soils. The 1984 Registration Standard indicated that phorate has some potential to leach through the soil and contaminate groundwater. Based on recently submitted data, phorate does not appear to be a potential leacher. However, its sulfone and sulfoxide degradates show greater persistence and mobility in soil, and therefore may have a greater leaching potential. Since data are still outstanding, the Agency cannot fully assess phorate's potential for contaminating groundwater.

Ecological Characteristics

Based on acceptable acute data, technical phorate is characterized as very highly toxic to birds on an acute oral basis, highly toxic to birds on a dietary basis, very highly toxic to mammals on a dietary basis, and very highly toxic to freshwater fish and aquatic invertebrates and estuarine and marine organisms on an acute toxicity basis.

- Acute LD50 (mallard):
  0.62 mg/kg
- Acute LD50 (chukar):
  12.8 mg/kg
- Dietary LC50
  248 ppm (waterfowl)
  441 ppm (upland gamebids)
  28 ppm (small mammals)
- Freshwater invertebrates toxicity (96-hr LC50) for
  amphipods: 0.68 ppb to 9 ppb
- Fish acute toxicity (96-hr LC50) for rainbow trout: 6
  to 13 ppb
- Fish acute toxicity (96-hr LC50) for bluegill sunfish:
  2 ppb; 5 ppb for bass
- Estuarine fish and invertebrates (LC_{50})
  0.11 to 1.9 ppb for shrimp and 1.3 to 5.0 ppb for
  spot and sheepshead minnow; for mollusks (900 ppb)

Tolerance Assessment

    Tolerances for residues of phorate in or on food and feed
commodities are published in 40 CFR 180.206. A tolerance for
residues of phorate on the processed feed commodity, dried
sugarbeet pulp, is published in 21 CFR 180.590. Tolerances are
expressed in terms of phorate and its cholinesterase-inhibiting
metabolites.

    Based on data submitted in response to the 1984 Standard,
the nature of the residue in plants is adequately understood.
The nature of the residue in animals not adequately understood.
A poultry metabolism study is required. The available data
support the established tolerances for the combined residues of
phorate and its cholinesterase-inhibiting metabolites in or on
potatoes, sugar beets, sugar beet tops, sugar beet pulp, and
soybeans. Additional data (residue studies, processing and
cooking studies, residue analytical methods, poultry metabolism
study and storage stability data) are needed before the Agency
can determine the adequacy of current tolerance levels and
perform a tolerance reassessment.

    The Agency has performed a preliminary dietary exposure
analysis using tolerance level residues and percent of crop
treated where possible. The ARC for phorate for the U.S.
population average is 0.000491 mg/kg/day. For the U.S.
population average, the ARC occupies 98% of the ADI. For
children 1 to 6 years of age, the ARC occupies 235% of the ADI,
and for non-nursing infants, it occupies 331% of the ADI. The
ARC is based on current tolerance levels, and, where possible, on
percent of crop treated. Due to the significant contribution
made by milk to the diet of children and non-nursing infants,

data regarding the reduction of residues through cooking and processing are required.

4. SUMMARY OF REGULATORY POSITIONS AND RATIONALES

- Based on a high acute toxicity of phorate to avian species and the current registered uses of phorate, there exists a high potential for adverse effects to avian species from exposure to phorate granules at or near the soil surface. This potential for exposure to phorate is demonstrated from results of Level I studies and is confirmed by bird kill incidents. The Agency is currently evaluating these data in the context of a comparative risk assessment of granular pesticides which may pose a risk to birds. Based on this assessment, regulatory action may be taken.

- The Agency is not placing phorate into Special Review at this time for hazards to aquatic organisms. Available field reports and laboratory data indicate that the concentrations of phorate in the aquatic environment resulting from the registered uses of phorate might expose aquatic species to residue levels exceeding risk criteria for Special Review. Upon receipt and evaluation of the aquatic field study (due in 1991), a determination will be made regarding further regulatory action.

- Unique warning statements required include revised and updated fish and wildlife toxicity statements, reentry statements, and protective clothing statements.

- The Agency is requiring special acute and subchronic eye studies to evaluate phorate's effect on the eye.

- The Agency will not approve significant new uses for this chemical since many of the tolerances are still not adequately supported.

- The Agency will continue to restrict the use of products containing phorate. Phorate meets the risk criteria of 40 CFR 152.170 due to acute oral and dermal toxicity and bird toxicity.

- The Agency is still unable to fully assess phorate's potential for contaminating groundwater. Upon receipt of a terrestrial field dissipation study, and other requested environmental fate data, the need for groundwater monitoring will be determined.

SUMMARY OF OUTSTANDING DATA REQUIREMENTS

Toxicology	Time Frame
Special Testing for Eye Effects	9-18 Months
Metabolism	24 "
Mouse Short-term Study	12 "

Environmental Fate/Exposure

Hydrolysis	9 Months
Photodegradation in Water	9 Months
Photodegradation on Soil	9 "
Aerobic Soil Metabolism	27 "
Anaerobic Soil Metabolism	27 "
Lab Volatility	12 "
Soil Dissipation	27 "
Confined Rotational Crop	39 "
Accumulation in Fish	12 "
Foliar Dissipation	18 "
Soil Dissipation	18 "

Fish and Wildlife

Avian Reproduction	24 Months
Estuarine/Marine Organism Testing (TEP)	12 "
Freshwater and Estuarine Fish Early Life Stage Life-Cycle Study	15 "
Freshwater and Estuarine Invertebrate Life-Cycle Study	15 "
Aquatic Organism Field Testing	January, 1991

Residue Chemistry

Residue data - Raw Agricultural Commodities	18 Months
Processing Studies	24 "
Poultry Metabolism	18 "
Cooking Studies	24 "
Storage Stability	15 "
Residue Analytical Methods	15 "

Product Chemistry

Majority of Data	9 -15 Months

6. Contact Person at EPA

     William H. Miller
     Product Manager (16)
     Insecticide-Rodenticide Branch
     Registration Division (TS-767)
     Environmental Protection Agency
     Washington, DC 20460

     Tel. No. (703) 557-2600

DISCLAIMER: The information presented in this Chemical
Information Fact Sheet is for informational purposes only and
may not be used to fulfill data requirements for pesticide
registration and reregistration.

PHOSALONE

Reason for Issuance: Registration Standard
Date Issued: November 30, 1987
Fact Sheet Number: 148

1. DESCRIPTION OF CHEMICAL

Generic Name: S-6-chloro-2,3-dihydro-2-oxobenzoxazol-3-yl
(Chemical)   methyl 0,0-diethyl phosphorodithioate

Common Name: Phosalone (ANSI)

Other Chemical Nomenclature:
  °S-[(6-chloro-2-oxo-3(2H)benzoxazolyl)methyl]
    0,0-diethylphosphorodithioate;
  °0,0-diethyl phosphorodithioate S-ester with 6-chloro-
    3-(mercaptomethyl)2-benzoxazolinone;
  °0,0-diethyl-S-[(6-chloro-2-oxobenzoxazolin-3-yl)
    methyl]phosphorodithioate;
  °S-[6-chloro-3-(mercaptomethyl)-2-benzoxazolinone]
    0,0-diethylphosphorodithioate;
  °0,0-diethyl-S-[6-chloro-3-(mercaptomethyl)-2-benzoxazolinone]
    phosphorodithioate.

Trade Names: Azonfene, Benzofos, RP11974, Rubitox, and Zolone.

EPA Shaughnessy Code: 097701

Chemical Abstracts Service (CAS) Number: 2310-17-0

Year of Initial Registration: 1974

Pesticide Type: Insecticide/Acaricide

Chemical Family: Organophosphate

U.S. and Foreign Producers: Rhone-Poulenc, Inc.

2. USE PATTERNS AND FORMULATIONS

Application Sites: Nut crops, citrus crops, pome fruits,
                  stone fruits, grapes, potatoes, artichokes,
                  roses, and arborvitae.
Formulation Types: Technical, Emulsifiable Concentrate and
                  Wettable Powder.
Application Methods: Foliar: ground and aerial application.

3.   SCIENCE FINDINGS

Summary Science Statement

Phosalone may have an adverse impact on birds and aquatic organisms resulting from all use patterns, excluding ornamentals. Aquatic and terrestrial field studies are required to determine the potential risks to these organisms. Laboratory data show that technical phosalone is highly toxic to fish and aquatic invertebrates.

The toxicological profile of the end-use products places them in Toxicity Category II for primary eye irritation, acute oral and dermal exposures. The end-use products were only mildly irritating to the skin when tested for dermal irritation. The products were characterized as weak dermal sensitizers. Phosalone can cause adverse effects to persons entering treated fields and to persons involved in the preparation and application of this pesticide. Preliminary data show that groundwater contamination is unlikely, but the Agency is unable to conduct a full assessment due to data gaps.

Chemical/Physical Characteristics of the Technical Material

Physical State: Crystalline solid

Color: White

Molecular weight and formula: 367.8 gms-$C_{12}H_{15}ClNO_4PS_2$

Melting Point: 45-47°C

Density: 1.391 g/ml at 20°C

Vapor Pressure: $\leq$ 0.5 x 10^{-6} mm Hg at 24°C and 16.4 x 10^{-6} at 60°C

Solubility: At 20°C: 1.7ppm in water; 20 g/100 ml in methanol
and ethanol; and 10g/100ml in acetone, benzene,
cyclohexanone, acetonitrile, xylene, toluene,
dioxane, chloroform, and methyl chloride.

Stability: Very stable under normal laboratory conditions for
a period of 2 years.

Toxicology Characteristics (Technical Material)

Acute Oral: Toxicity Category II (90 and 125 mg/kg in female and
              male rats respectively).

Acute Dermal: Toxicity Category II (LD50 > 350 mg/kg for both
                males and females).

Acute Inhalation: Waived for technical, particles analyzed were
                    determined not to be within the respirable
                    range (≤ 15 microns).

Primary Dermal Irritation: Toxicity Category IV (PDIS-0.77, mildly
                             irritating to intact and abraded skin).

Primary Eye Irritation: Data gap.

Skin Sensitization: Data gap.

Delayed neurotoxicity: Negative in hens.

Subchronic Oral (non-rodent): Data gap.

Oncogenicity: Data gap.

Chronic Feeding: NOEL for RBC inhibition in the rat is 25 ppm.

Metabolism: Data gap

Teratogenicity: Data gap

Reproduction: Data gap

Mutagenicity: Data gap for point mutation assay in mammalian
                cells, structural chromosomal aberration, and
                other genotoxic effects.

Major routes of exposure: Dermal and respiratory exposure to
                            mixers, loaders, applicators, and
                            fieldworkers.

Environmental Characteristics

     Phosalone is stable at pH 5 and 7, but is hydrolyzed at a
pH of 9 with a half-life of 9 days. Artificial light accelerated degradation in buffered solution at a pH of 5 and in
soil. Aerobic soil metabolism studies demonstrate half-life
values of 1-7 days. Field dissipation studies showed half-life
values of 1-9 weeks. Phosalone was essentially immobile in
a soil column test. Based upon this preliminary data phosalone
appears unlikely to contaminate ground water. It exhibited
moderate accumulation in the bluegill sunfish, with rapid
dissipation in untreated waters.

Ecological Characteristics (Technical grade)

Avian Oral Toxicity: Slightly toxic to waterfowl (acute oral toxicity value: mallards > 2150 mg/kg).

Avian Dietary Toxicity: Slightly toxic to waterfowl and upland-
(8-Days)   game birds (subacute toxicity values: mallards 1659 ppm and bobwhite quail 2033 ppm).

Freshwater Fish Acute: Very highly toxic to warmwater fish and
Toxicity (96-Hours)   and highly toxic to coldwater fish (acute LC values: 0.05 ppm bluegill and 0.63 ppm rainbow trout).

Freshwater Invertebrate: Very highly toxic to aquatic invertebrate
Toxicity (48-Hours)   (acute EC value: 0.0012 ppm <u>Daphnia</u> <u>magna</u>)

4. Tolerance Reassessment

Tolerances have been established for phosalone on a variety of raw agricultural commodities, in meat, fat and meat byproducts (40 CFR 180.263) and in processed food (21 CFR 193.340) and feed (21 CFR 561.300).

Tolerances for the following commodities are adequately supported: potatoes, citrus, apples, pears, apricots, cherries, nectarines, peaches, plums (fresh prunes), dried prunes, tree nuts, almond hulls, grapes, raisins, artichokes, and the fat, meat, and meat byproducts of cattle, goats, hogs, horses and sheep.

Additional data are required to assess the need for food/feed additive tolerances for the following products processed from raw agricultural commodities bearing measurable, weathered residues: potato granules or flakes, chips, and wet and dry peels, raisin waste and grape juice.

The nature of the residues in both plants and animals are not adequately understood. If, on receipt of the required metabolism data, the Agency determines that residues in addition to the parent require regulation, additional methods for data collection and enforcement may be required.

Crop group tolerances may be proposed for the Pome Fruit Group at 10 ppm and for the Stone Fruit Group at 15 ppm.

The Agency has established a Provisional Acceptable Daily Intake (PADI) at 0.0025 mg/kg/day based on a 6-month dog feeding study in which plasma cholinesterase activity was depressed. Because a NOEL was not established in this study a 100 fold uncertainty factor has been used rather than the 10 fold factor normally applied for cholinesterase inhibition.

5. Summary of Regulatory Positions and Rationales

°Phosalone is not being placed into Special Review at this time. Although the Agency is concerned about the potential adverse impact of phosalone on birds and aquatic organisms resulting from the agricultural use patterns, aquatic and terrestial field studies are needed in order to evaluate the potential risks to these species.

°The Agency has sufficient data (analysis of pesticides with similar uses were found to be in jeopardy) to indicate that the current use patterns of phosalone may affect endangered species. Endangered species labeling is reserved pending concurrence from the Fish and Wildlife Service.

°The Agency is classifying all Phosalone end use products as Restricted Use Pesticides; except for products packaged and labeled solely for use around the home. Products containing phosalone for use on cherries and citrus are restricted due to avian hazards. All use patterns are restricted due to aquatic toxicity. The estimated environmental concentrations exceed the LC50 value for fish and the EC50 value for aquatic invertebrates, and the NOEL values for avian species.

°The Agency is imposing a 6-month rotational crop restriction for small grains and a 12-month rotational crop restriction for leafy vegetables and root crops.

°No significant new tolerances or new food uses will be granted until the Agency has received sufficient data to evaluate the dietary exposure of phosalone.

°The Agency will retain the 24-hour reentry interval imposed in the 1981 Registration Standard. This reentry interval will be retained until the required reeentry data are received and evaluated. Data will be reviewed on a priority basis because of reported poisoning incidents in California.

°The Agency will require that end use products bear label statements to protect mixers, loaders, applicators, flaggers, and fieldworkers.

°Preliminary data indicate that groundwater contamination is unlikely. The Agency is requiring environmental fate studies to fully characterize phosalone's fate in the environment.

6. SUMMARY OF OUTSTANDING DATA REQUIREMENTS

Toxicology

	Time Frame
Primary Eye Irritation	9 Months
Dermal Sensitization	9 "
Acute Inhalation	9 "
Subchronic oral toxicity--Dog (for cholinesterase effects)	12 "
Oncogenicity--rat	50 "
Teratogenicity--(rat and rabbit)	15 "
Reproduction--(rat)	39 "
Mutagenicity	12 "
Metabolism study	12 "
Dermal Absorption (rat)	9 "

Environmental Fate/Exposure

Aged leaching study	12 "
Rotational Crop (Confined)	12 "
Foliar dissipation study	15 "
Spray Drift	12 "
Soil Dissipation	27 "

Fish and Wildlife

Avian reproduction	24 Months
Actual field testing- birds and mammals (citrus)	30 "
Simulated or Actual field testing-aquatic organisms	48 "
Acute toxicity to Estuarine and Marine organisms	12 "
Fish early life stage and aquatic invertebrate life cycle	15 "
Acute toxicity to Freshwater Invertebrate	9 "

Residue Chemistry

Animal and Plant metabolism	18 Months
Storage stability	15 "
Processing studies for potatoes and grapes	24 "
Residue data (tea)	18 "

Product Chemistry   6-15 Months

7. ## CONTACT PERSON AT EPA

Dennis H. Edwards, Jr.
Product Manager (12)
Insecticide-Rodenticide Branch
Registration Division (TS-767C)
Office of Pesticide Programs
Environmental Protection Agency
401 M Street, S. W.
Washington, D. C.   20460

Office location and telephone number:
Room 211, Crystal Mall #2
1921 Jefferson Davis Highway
Arlington, VA   22202
(703) 557-2386

DISCLAIMER: The information presented in this Chemical Information Fact Sheet is for informational purposes only and may not be used to fulfill data requirements for pesticide registration and reregistration.

PHOSPHAMIDON

Reason for Issuance: Registration Standard
Date Issued: January 1, 1988
Fact Sheet Number: 154

1. Description of Chemical

   Common Name: Phosphamidon

   Chemical Name: 2-chloro-3-(diethylamino)-1-methyl-3-
                   oxo-1-propenyl dimethyl phosphate

   Other Chemical Nomenclature: dimethyl phosphate ester
                                 2-chloro-N,N-diethyl-3-
                                 hydroxycrotonamide;

                                 2-chloro-2-diethylcarbamoyl-
                                 1-methylvinyl dimethyl
                                 phosphate;

                                 O,O-dimethyl-O-(2-chloro-
                                 2-diethylcarbamoyl-1-
                                 methylvinyl) phosphate

   Trade Names: Apamidon; C570; Ciba 570; Dimecron; Dixon;
                 Dimenox

   Chemical Class: Organophosphate
   Empirical Formula: $C_{10}H_{19}ClNO_5P$
   Chemical Abstracts Service (CAS) Nos.:
                   13171-21-6 [(E)-+(Z)-isomers]
                   23783-98-4 [(Z)-isomer]
                   297-99-4 [(E)-isomer]
   OPP Shaughnessy No.: 018201
   Pesticide Type: Systemic insecticide/acaricide
   Year of Initial Registration: 1963
   U.S. and Foreign Producers: Ciba-Geigy Corp., USA

2. Use Patterns and Formulations

   Application sites: terrestrial food crops -- apple; apple
                       (nonbearing); broccoli; cantaloupe;
                       cauliflower; cotton; cucumber; grapefruit;
                       lemon; orange; peppers; potato; sugarcane;
                       tangerine; tomato; walnut; watermelon.

507

Types of Formulation: Phosphamidon is marketed as an 8
                        lb ai/gal SC/L (soluble concentrate/
                        liquid), and is formulated from a
                        single 89.5% technical product.

Types/Methods of Application: Phosphamidon is applied
                                by ground and aerial
                                application.

3. Science Findings

    Summary Science Statements:

    Phosphamidon is an organophosphate compound. The Agency
    has no valid acute toxicity studies for phosphamidon,
    although the Agency believes that phosphamidon is highly
    toxic. No valid subchronic studies are available for
    phosphamidon. A 2-year rat feeding study showed toxic
    signs of cholinesterase inhibition activity in serum and
    brain, decreased body weights, erythrocyte counts, hemoglobin
    levels and, necrotic changes in the stomach and other
    organs. Studies on the toxic effects of phosphamidon in
    nonrodent species are not available. The submitted
    oncogenicity data are considered inadequate to assess the
    oncogenic potential of phosphamidon. Sufficient developmental
    toxicity data are available to satisfy the regulatory
    requirements in two species. Based on these data the
    Agency concluded that phosphamidon did not demonstrate any
    significant developmental toxic effects. The available
    2-generation rat reproduction study satisfies this requirement.
    No data are available to evaluate the mutagenic potential
    of phosphamidon. No data on the metabolic pathway of
    phosphamidon are available. Phosphamidon is acutely and
    subacutely very highly toxic to a variety of avian species
    and can be lethal to birds through dermal exposure. In
    addition, available information indicate that delayed
    mortality of birds occurs after applications of phosphamidon
    (in some cases up to several weeks). Since available
    information suggests that phosphamidon may have a relatively
    short half-life, it is possible that some degradate is
    toxic to birds. Based upon the fish and wildlife data
    available, it is observed that technical phosphamidon is
    very highly toxic to both coldwater and warmwater fish
    species, aquatic invertebrates, and mammals. Available
    data indicate that phosphamidon is highly toxic to honey
    bees, predaceous mites, parasitic wasps, and predaceous
    beetles.

Chemical Characteristics:

Physical/chemical properties of pure phosphamidon and of
the technical phosphamidon.

   Color: Colorless (PAI*), slightly amber oil (T*)
   Physical State: Liquid (PAI)
   Odor: Odorless, faint, mild (PAI)
   Melting Point: Liquid at room temperature
   Boiling Point: 160 °C, 1.5 mm (PAI)
   Density: 1.2 at 20 °C (PAI)
   Solubility: Miscible with water, alcohol, and ketones.
               Highly soluble in aromatic and chlorinated
               hydrocarbons, esters and ethers. Solubility
               in hexane is 3.23 g/100 g 25°C (PAI)
   Vapor Pressure: 2.5×10^{-5} mm Hg at 20 °C (PAI)
   Octanol/Water Partition Coefficient: Log P = 0.8 (PAI)
   Stability in water: Half-life at 2 ppm and 38 °C is
                        70 hours at pH 9.1 and > 300
                        hours at pH 1.1 (PAI). Half-life
                        (in days) of phosphamidon in
                        buffered media (T):

	Temperature	
pH	23 °C	45 °C
4	74	6.6
7	13.8	2.1
10	2.2	0.14

Toxicology Characteristics:

Acute oral: Data gap
Acute dermal: Data gap
Acute inhalation: Data gap
Primary dermal irritation: Data gap
Primary eye irritation: Data gap
Acute delayed neurotoxicity: Data gap
Subchronic oral (nonrodent): Data gap **
Chronic feeding (nonrodent): Data gap
Oncogenicity : Data inadequate to assess oncogenic potential,
               Study required in the mouse.
Metabolism: Data gap

Developmental Toxicity:
               Rabbit - Maternal NOEL = 3 mg/kg
                        Maternal LEL = 10 mg/kg
                        Developmental NOEL = > 10 mg/kg

               Rat - Maternal NOEL = 0.5 and 1 mg/kg
                     Developmental NOEL = 2 mg/kg

*PAI = Pure Active Ingredient, T = Technical
** Not required if an acceptable chronic study is submitted.

Reproductive effects:
               Rat - Parental NOEL = 30 ppm (1.5 mg/kg/day)
                     Parental LEL = 50 ppm (2.5 mg/kg/day)
                     Reproductive/Developmental
                         NOEL = 5 ppm (0.25 mg/kg/day)
                     Reproductive/Developmental
                         LEL = 30 ppm (1.5 mg/kg/day)
Mutagenicity: Data gap

Physiological and Biochemical Characteristics:

Metabolism and persistence in plants and animals

The metabolism of phosphamidon in plants and animals is
not adequately understood. Residues of [^{14}C] phosphamidon
were identified and quantitatively determined only in
immature bean plants. No edible livestock tissues (other
than milk) were analyzed for residues, and residues were
not characterized sufficiently in livestock. No poultry
data were submitted. The available plant metabolism data
indicate that phosphamidon degrades rapidly when applied
directly to the leaves of very young "two-leaf stage"
bean plants. The limited available ruminant metabolism
data indicate that phosphamidon is degraded rapidly in
animals, with most of the metabolic compounds being
excreted in the urine. Available data support the
established tolerances for residues of phosphamidon
including all of its related cholinesterase-inhibiting
compounds (as currently known) in or on the raw agricultural
commodities (RACs) potatoes, tomatoes, cucumbers, cottonseed,
and sugarcane. Note, however, that these tolerances,
including the residue definition, will be reassessed upon
receipt of the requested plant metabolism studies.
Ultimately, the tolerance definition will be changed to
list specific metabolites.

Environmental Characteristics:

Available data are not sufficient to allow the Agency to
fully assess the environmental fate of phosphamidon. The
available data suggest that phosphamidon is readily susceptable
to hydrolysis. Phosphamidon appears to be relatively
short lived in aerobic soil. N,N-diethyl-2-chloroacetoacetamide
and N-ethyl-2-chloroacetamide were identified as the two
major nonvolatile degradates, but the characterization of
degradates has not been completed. Phosphamidon residues
are considered to be highly mobile in soil. However, the
relative mobilities of the parent compound and its degradates
have not been adequately defined. Available data are not
adequate to fully assess the potential of phosphamidon to
contaminate ground water.

Ecological Characteristics:

Avian Species

Technical phosphamidon is very highly acutely toxic to birds as demonstrated by both acute and dietary studies. Subacute dietary studies demonstrate that there is a range of toxicity from 24 ppm (bobwhite quail) to 712 ppm (mallard duck). Available data indicate that phosphamidon can be toxic to birds through contact with head, feet or through contact with sprayed foliage. Data indicate that small doses picked up from perches or applied to the feet of birds can be lethal. Special tests are being required to determine the dermal toxicity of phosphamidon to birds. Avian reproduction studies for phosphamidon are not available. There are no data available on the toxicity of degradates of phosphamidon to birds. Because delayed mortality is an indicated adverse effect of phosphamidon on birds, and because substantial reduction of populations of songbirds have occurred several weeks after phosphamidon applications to forests, it is possible that degradation products of phosphamidon can result in such effects.

Aquatic Organisms

Acceptable acute toxicity tests with technical phosphamidon indicate that phosphamidon is highly toxic to fish. The acute toxicity tests with freshwater invertebrates indicate that phosphamidon is very highly acutely toxic to aquatic invertebrates. There are no acceptable data evaluating the toxicity of technical phosphamidon to estuarine and marine organisms. Phosphamidon may reach estuarine environments from its use on citrus orchards. Acute studies on the toxicity of phosphamidon to estuarine and marine invertebrates are required. Aquatic invertebrate life cycle studies are required to support the agricultural use applications.

Wild Mammal Toxicity

There are no adequate data with which to assess the toxicity of phosphamidon to mammals. The only study available, which suggests a lethal dose of 18 mg/kg in the deer mouse, indicates that phosphamidon may be highly toxic to wild mammals. Although the Agency can draw no conclusions regarding the potential toxicity of phosphamidon to mammals, a wild mammal toxicity study is not being required at this time. However, if the acute mammalian studies required in the Toxicology Section indicate a rat acute toxicity $\leq$ 5 mg/kg then a wild mammal toxicity study will be required.

Non-target Insects

Available data indicate that phosphamidon is highly toxic to honey bees, predaceous mites, parasitic wasps, and predaceous beetles.

Tolerance Reassessment

Tolerances (expressed as phosphamidon) for residues of the insecticide phosphamidon (2-chloro-2-diethylcarbamoyl-1-methylvinyl dimethyl phosphate) including all of its related cholinesterase-inhibiting compounds have been established in or on various raw agricultural commodities (40 CFR 180.239).

Because of the extensive residue chemistry and toxicology data gaps, the Agency cannot complete a tolerance reassessment.

4. ## Required Labeling and Regulatory Position Summary

Restricted Use Classification

All currently registered products containing phosphamidon have been classified for Restricted Use (40 CFR 162.31) due to the acute dermal toxicity to humans and residue effects on avian and mammalian species.

Endangered Species Labeling

Refer to PR Notice 87-5, issued May 1, 1987, for endangered species labeling statements for all end-use products containing phosphamidon for use on cotton.

It is the Agency's position that in order to remain in compliance with FIFRA, all products containing phosphamidon must include labeling which requires compliance with the following precautionary measures, pending receipt and evaluation of additional data which will allow the Agency to fully assess the potential to induce adverse effects.

Protective Clothing, Equipment and Work Safety Statements

The use of maximum full body protective clothing/equipment by mixers/loaders/applicators, and by field workers entering treated fields prior to the end of the 48 hour reentry interval is required.

Reentry Interval

A 43-hour reentry interval is required for all agricultural
uses of phosphamidon. In addition, the use of protective
clothing is required for early reentry into treated
areas.

Tolerances

No tolerances or significant new food uses will be granted
until the Agency has received sufficient data to evaluate
the dietary exposure to phosphamidon.

5.   Summary of Major Data Gaps

Toxicology   Date Due *

    Acute Oral Toxicity   9 Months
    Acute Dermal Toxicity   9 Months
    Acute Inhalation Toxicity   9 Months
    Primary Eye Irritation   9 Months
    Primary Dermal Irritation   9 Months
    Dermal Sensitization   9 Months
    Acute Delayed Neurotoxicity (hen)   12 Months
    Subchronic 90-day feeding (nonrodent)**   18 Months
    Subchronic 21-Day Dermal   12 Months
    Chronic Toxicity (nonrodent)   50 Months
    Oncogenicity (mouse)   50 Months
    Mutagenicity (Gene Mutation, Chromosomal
        Aberration and Direct DNA Damage and
        Repair Studies)   9 Months
    Metabolism (rats)   24 Months

Environmental Fate/Exposure

    Hydrolysis   9 Months
    Photodegradation
      In water   9 Months
      In soil   9 Months
      In air   9 Months
    Metabolism
      Aerobic Soil   27 Months
      Anaerobic Soil   27 Months
    Mobility
      Leaching and Adsorption/Desorption   12 Months
      Laboratory Volatility   12 Months
    Dissipation Studies - Field
      Soil Dissipation   27 Months
      Confined Accumulation Study   39 Months
      Fish Accumulation Study   12 Months

----

 * Due date is measured from the date of receipt of Standard by
   the registrant, unless otherwise indicated.
** Not required if an acceptable chronic study is submitted.

Environmental Fate/Exposure (cont'd)	Date Due*
Spray Drift	
Droplet Size Spectrum	24 Months
Drift Field Evaluation	24 Months

Ecological Effects

Acute Avian Oral Toxicity (Degradate)	9 Months
Acute Avian Dietary Toxicity (Degradate)	9 Months
Avian Reproduction	24 Months
Avian Field Testing (Mammals, Birds)	24 Months
Special Avian Testing (Dermal Toxicity)	6 Months – (acceptable protocol)
Acute Toxicity to Estuarine and Marine Organisms	12 Months
Fish Early Life Stage and Aquatic Invertebrate Life Cycle	15 Months
Aquatic Residue Monitoring	6 Months – (acceptable protocol)

Residue Chemistry

Nature of Residues (Plants, Livestock)	18 Months
Residue Analytical Method (Plant, Animal)	15 Months
Storage Stability	18 Months
Magnitude of Residues (Field Crops)	18 Months

6. Contact Person at EPA

William H. Miller, (PM-16)
Insecticide-Rodenticide Branch (TS-767)
401 M Street S.W.
Washington, D.C. 20460

Tel. No. (703-557-2600

DISCLAIMER: The information presented in this chemical information Fact Sheet is for informational purposed only and may not be used to fulfill data requirements for pesticide registration and reregistration.

---

* Due date is measured from the date of receipt of Standard by the registrant, unless otherwise indicated.

PICLORAM

Reason for Issuance: Registration Standard
Date Issued: October 24, 1988
Fact Sheet Number: 48.1

1. Description of Chemical

    Chemical Name: 4-amino-3,5,6-trichloropicolinic acid
    Common Name: Picloram
    OPP Chemical Code: 005101
    Chemical Abstracts Service (CAS) Number: 1918-02-1
    Year of Initial Registration: 1964
    Pesticide Type: Herbicide
    Chemical Family: Picolinic acid
    U.S. Producer: Dow Chemical, U.S.A.

    Chemical Name: Potassium salt of 4-amino-3,5,6-trichloropicolinic
                    acid
    Common Name: Picloram, Potassium (K) salt
    OPP Chemical Code: 005104
    CAS No.: 2545-60-0

    Chemical Name: Isooctyl ester of 4-amino-3,5,6-trichloropicolinic
                    acid
    Common Name: Picloram, Isooctyl ester (IOE) of picloram
    OPP Chemical Code: 005103

    Chemical Name: Triisopropanolamine salt of 4-amino-3,5,6-trichloro-
                    picolinic acid
    Common Name: Picloram, TIPA salt
    OPP Chemical Code: 005102

    Chemical Name: Triethylamine salt of 4-amino-3,5,6-trichloropicolinic
                    acid
    Common Name: Picloram, TEA salt
    OPP Chemical Code: 005105

    Chemical Name: Isopropanolamine salt of 4-amino-3,5,6-trichloropico-
                    linic acid
    Common Name: Picloram, IPA salt

515

2. Use Patterns and Formulations

Application Sites:

Picloram, potassium salt: Terrestrial food crop use on small grains,
flax, pastures and rangeland grasses; Terrestrial noncrop use
on noncrop agricultural areas and rights-of-way; Forestry use cn
forest lands site preparation.

Picloram, isooctyl ester: Terrestrial noncrop use on industrial
sites and rights-of-way; Forestry use on forest trees site
preparation.

Picloram triisopropanolamine salt: Terrestrial food crop use on
small grains and pastures and rangeland; Terrestrial nonfood crop
use on uncultivated agricultural areas, rights-of-way, and industrial
sites; Aquatic noncrop use on drainage ditch banks; Forestry
use on forest trees.

Picloram, triethylamine salt: Terrestrial food crop use on pastures
and rangelands.

Types and Methods of Application: By ground: broadcast or spot
treatment as foliar or soil spray; as a basal spot treatment,
broadcast as pelletized spray; as tree injection, as frill treatment; as a stump treatment, as basal bark treatment, as a wick
application, and as a low-volume dormant stem spray. By air:
broadcast and low-volume dormant spray.

Pests Controlled: Broadleaf weeds and woody plants.

Application Rates: (Section 3 registrations)

Picloram, TIPA salt: 0.27 to 3.00 pounds acid equivalent (lb ae)
per acre (A)
Picloram, IOE: 0.5 to 3.0 lb ae/A(mixtures or MAF)
Picloram, K salt: 1.0 to 8.5 lb ae/A
Picloram, TEA salt: 0.25 to 1.0 lb ae/A(SLNS)

Types of Formulations: [represented Sec 3 registrations of the potassium
salt(K)]: Formulation Intermediate: 30% ae or 34.7% acvtive
ingredient (ai), Pelleted: 2% ae or 2.3 ai, 5 ae or 5.8% ai, 10% ae
or 11.6 ai; Soluble Concentrate, Liquid: 2 lb ae or 24% ai, 24.4%
ai or 29.9% ai

Usual Carrier: Water

3. Science Findings

Summary Science Statement: Technical picloram is in Toxicity
Category I with respect to acute inhalation and Categories III and

IV with respect to other acute toxicities. Picloram has been classified as Group D Oncogen (not classifiable to human carcinogenicity).
Repeat oncogenicity, teratology, and reproduction studies are
being required. Picloram does not appear to be mutagenic based
on available data.

Picloram is stable to hydrolysis, does not photodegrade under light,
and is relatively stable in anaerobic loam soils and under anaerobic
aquatic conditions and does not accumulate in fish. Picloram is
intermediately to very mobile in soils ranging in texture from
clay to loam. Picloram has been identified as a chemical with a
potential to contaminate groundwater.

The Agency is requiring that residue data depicting residues of HCB
in plant and animal commodities be submitted.

Picloram is practically nontoxic to avian species, slightly to
moderately toxic to freshwater fish, and slightly toxic to freshwater
invertebrates.

Chemical Characteristics:

Technical Picloram (Acid):
    Color: White
    Physical State: Powder
    Odor: Chlorine like
    Melting Point: 215 °C (decomposes)
    Bulk Density: 19.7 lb/cu ft
    Solubility at 25 °C:
        0.043 g/100 mL — Water
        0.55 g/100 mL — Isopropanol
        1.05 g/100 mL — Ethanol
        1.98 g/100 mL — Acetone
        0.12 g/100 mL — Diethyl ether
        0.16 g/100 mL — Acetonitrile
        1.85 g/100 mL — Methanol
        0.06 g/100 mL — Methylene dichloride
        0.02 g/100 mL — Benzene
        0.001 g/100 mL — Kerosene

    Vapor Pressure: 6.16×10^{-7} millimeters (mm) Hg at 35 °C
                     1.07×10^{-6} mm Hg at 45 °C

    Storage Stability: Stable under normal conditions.

Picloram, Potassium (K) Salt (34.7% ai)
    Color: Dark brown
    Physical State: Liquid
    Odor: Alcoholic
    Bulk Density: 1.320 at 20 °C

Toxicology Characteristics:

Existing data are all based on picloram (technical) or K salt.
    Further data are requested for the IOE, TIPA salt, TEA salt, and
    IPA salt.

Acute Toxicology – Technical (Acid):

o Acute Oral Toxicity (Rats): Greater than (>) 5000 mg/kg body
       weight for males – Toxicity Category IV; = 4012 mg/kg for
       females – Toxicity Category III

o Acute Dermal Toxicity (Rabbits): > 2000 mg/kg for males and
       females, Toxicity Category III

o Acute Inhalation (Rat): > 0.035 mg/L for males and females,
       Toxicity Category I

o Primary Eye Irritation (Rabbit): Moderate eye irritation,
       Toxicity Category III

o Primary Dermal Irritation (Rabbit): Not an irritant, Toxicity
       Category IV

o Dermal Sensitization (Guinea Pig): Not a skin sensitizer

Acute Toxicology (K Salt):

o Acute Oral Toxicity (Rat): > 5000 mg/kg for males,
       Toxicity Category IV; = 3536 mg/kg for females, Toxicity
       Category II

o Acute Dermal Toxicity (Rabbit): > 2000 mg/kg for males and
       females, Toxicity Category III

o Acute Inhalation Toxicity (Rat): > 1.5 mg/L for males and
       females, Toxicity Category II

o Primary Eye Irritation (Rabbit): Moderate eye irritation,
       Toxicity Category III

o Primary Dermal Irritation (Rabbit): Not a skin irritant,
       Toxicity Category IV

o Dermal Sensitization (Guinea Pig): Skin Sensitizer*

---

*Requires statement for skin sensitization: "May Cause
Allergic Skin Reaction."

Acute Toxicology (IOE):

o   Acute Oral Toxicity (Rat): > 3500 mg/kg for males and females,
       Toxicity Category III

o   Acute Dermal Toxicity (Rabbit): > 2000 mg/kg for males and
       females, Toxicity Category III

o   Acute Inhalation Toxicity (Rats): > 0.35 mg/L for males and
       females, Toxicity Category II

o   Primary Eye Irritation (Rabbits): Moderate eye irritation,
       Toxicity Category III

o   Primary Dermal Irritation (Rabbit): Mild skin irritation,
       Toxicity Category III

Subchronic Toxicology Studies:

An acceptable 13-week subchronic feeding study in rats is available
    for picloram. The no-observed-effect level (NOEL) for this study
    was 50 mg/kg. A dose dependent increase in absolute and relative
    liver weights was seen at 150 mg/kg.

An acceptable 6-month feeding study with dogs is available for
    picloram. The NOEL for this study was 7 mg/kg. A decrease in
    food consumption and increase in liver weights was noted at the
    highest dose.

No subchronic feeding studies are available for the TIPA, TEA, IPA, or
    IOE forms of picloram. Subchronic feeding studies are required in
    a rodent and a nonrodent for each form of picloram.

A 21-day subchronic dermal study is not available for picloram. This
    study is required for all forms of picloram.

Chronic Feeding Studies:

An acceptable 2-year chronic feeding study with rats is available for
    picloram. An increase in size and altered tinctorial properties
    of centrilobular hepatocytes occurred in males and females at the
    high (200 mg/kg/day) and mid (60 mg/kg/day) dose resulting in a
    NOEL of 20 mg/kg/day for this study.

A chronic feeding study in nonrodents is not available for picloram and is required.

Oncogenicity Studies:

The available oncogenic data for picloram include mouse and rat studies performed by the National Toxicology Program (NTP) and a rat study performed by Dow Chemical U.S.A.

An oncogenic effect (neoplastic nodules) was seen in female rats at the highest dose in the NTP study. This study was unacceptable based on experimental design (too short exposure limit, insufficient information to determine if a maximum tolerated dose [MTD] was attained).

No oncogenic effects were noted in either the mouse study done by NTP or the rat study done by Dow. These studies were not acceptable because the available information was insufficient for determining if an MTD had been reached.

The test material in all of these studies contained the contaminant hexachlorobenzene (HCB), which is classified by the Agency as a Group B2 oncogen (probable human carcinogen). Picloram was classified as a Group D oncogen (not classifiable as to human carcinogenicity). The Agency is requiring that both the mouse and rat oncogenicity studies be repeated.

Teratogenicity and Reproduction:

A teratogenicity study in rabbits is available for picloram. A small number of fetuses showed abnormalities such as missing ribs, omphalocele, and hypoplastic tail. Historical control data are required to evaluate the observed abnormalities.

A teratology study in rats is available for picloram. No teratogenic effects were noted. Some fetotoxicity was present at the lowest dose. Because a NOEL cannot be set for the study a repeat teratology study in rats is required.

A multigeneration reproduction study in rat is available for picloram. No reproductive effects were observed however, too few test animals were used and no toxicity was observed at the highest dose. Therefore, a 2-generation reproduction study is required for picloram.

No teratology or reproduction studies are available for the ester and amine forms of picloram. Teratology studies in rats and rabbits are required for all ester and amine forms of picloram. Reproduction studies are not required at this time.

Mutagenicity:

Picloram did not show evidence of chromosomal changes in a cytogenetic bone marrow study exposing rats up to 2000 mg/kg of picloram.

No other acceptable mutagenicity studies are available for picloram, its salt, ester, or amine forms. Additional mutagenicity data are required for picloram, its esters, and its amines.

Metabolism:

Available rat metabolism data are not adequate to fulfill Guideline requirements; therefore, additional studies are required.

Manufacturing Contaminants:

Technical picloram is contaminated with HCB, classified as a probable human carcinogen (Group B2). Dietary and nondietary risk assessments were performed by the Agency. The Agency considered the dietary and nondietary risk from HCB to be acceptable at this time.

Nitrosoamines are a potential contaminant of tertiary amines (TEA) and alkanolamines (TIPA) forms of picloram. Testing is required to show that the level of 1 ppm nitrosoamine contamination is not exceeded.

Physiological and Biochemical Characteristics:

Foliar Absorption and Translocation: Picloram translocates from both the roots and leaves of plants and accumulates in the new growth. Picloram is both foliar-absorbed and root-absorbed.

Mechanism of Pesticidal Action: Alters nucleic acid and protein synthesis.

Metabolism and Persistence in Plants: Available plant metabolism data indicate that picloram degrades to CO_2, oxalic acid, and the metabolites 4-amino-2,3,5-trichloropyridine and 4-amino-3,5-dichloro-6-hydroxypicolinic acid.

Metabolism and Persistence in Animals: Available metabolism data
indicate that animals excrete 82 to 98 percent of the [^{14}C]picloram
used in dosing the animals as picloram.

Environmental Characteristics:

Absorption and Leaching in Basic Soil Types: Available data indicate
that picloram was intermediately mobile to very mobile in soils
ranging in texture from clay to loam. Adsorption of picloram
pH. Addition of inorganic salts to the soil did not affect
adsorption of picloram.

Microbial Breakdown: Picloram degraded with half-lives of 100 to
200 days in loam soil, 200 to 300 days in silt loam soil, and
greater than 300 days in loamy sand, commerce loam, clay, and
sandy loam soils under aerobic conditions. Picloram was relatively stable in anaerobic loam soil under anaerobic aquatic
soil conditions.

Loss from Photodecomposition: Does not degrade.

Bioaccumulation in Fish: Does not accumulate in fish.

Potential to Contaminate Groundwater: Picloram has been previously
identified as a pesticide with a propensity to leach into groundwater. Picloram has been reported as detected in seven States.
Picloram is persistent and mobile and has a high potential to
reach groundwater.

Exposure to Humans: Based on available acute toxicology data the
major routes of exposure appear to be through inhalation and
dermal sensitization. Although technical picloram (free acid) is
in Toxicity Category I based on inhalation, there is little chance
of exposure to mixer/loaders or applicators because there are
currently no products registered containing the free acid form of
picloram.

Risk to Humans: The major risk to humans appears to be from the
contaminant HCB. Both dietary and nondietary risk assessments
were performed. The dietary exposure to HCB occurs from the use
of pesticides containing picloram on small grains and secondary
residues on animal commodities. The oncogenic risk for the U.S.
population based on dietary exposure was calculated to be 6. x
10^{-7}.

Potential nondietary exposure to HCB is to workers, mixer/loaders
and applicators from use of picloram on wheat, forests, rights-
of-way, and pasture/rangeland. The estimated nondietary risk
to mixer/loaders and applicators ranged from 5.0×10^{-5} to
10^{-8}.

Reentry: Reentry intervals are not required because cultural practices
for existing uses indicate little likelihood that field workers
would be exposed to acutely toxic levels of picloram from agricultural
applications.

Ecological Characteristics:

Avian Acute Oral Toxicity (Technical): Bobwhite quail > 2250 mg/kg/day;
   K Salt: Bobwhite quail > 2510 mg/kg/day

Avian Subacute Dietary Toxicity (Technical): Bobwhite quail > 5000 ppm,
   mallard duck > 5000 ppm; K Salt, bobwhite quail > 5620 ppm; IOE,
   bobwhite quail > 5620 ppm

Acute Toxicity to Freshwater Fish (Technical): Rainbow trout = 4.3
   to 19.3 ppm, bluegill sunfish = 14.5 to 23.0 ppm; K Salt, rainbow
   trout = 13 ppm, catfish = 14 ppm, bluegill sunfish = 24 ppm; IOE,
   rainbow trout = 4.0 ppm, catfish = 1.4 ppm, bluegill sunfish =
   6.3 ppm

Fish Embryolarvae Study: Rainbow trout with a maximum acceptable
   threshold concentration (MATC) = 0.55 < MATC < 0.88 mg/L

Acute Toxicity to Freshwater Invertebrates Studies: (Daphnids,
   Gammarus, Pteronarcella, and Pteronarcys) = 10 to 68.3 ppm

Chronic Aquatic Invertebrate Study (Daphnids): 11.8 < MATC
   < 18.1 mg/L

Acute Toxicity to Honey Bees (Technical): = 14.5 micrograms per bee

Technical picloram appears to be moderately to slightly toxic to
   freshwater fish, slightly toxic to aquatic invertebrates, and
   practically nontoxic to birds. Chronic fish testing showed that
   picloram caused a reduction in rainbow trout larval survival at
   2.02 mg/L and a reduction in growth at 0.88 mg/L. Picloram
   affected the growth and survival in cutthroat trout at 0.29 mg/L.

The isooctyl ester form of picloram is moderately toxic to fish, and
   practically nontoxic to birds.

The potassium salt of picloram appears to be slightly toxic to
freshwater fish and practically nontoxic to birds.

Phytotoxicity and Endangered Species

Picloram has been shown to be a highly phyotoxic herbicide.

Because of picloram's demonstrated toxicity to nontarget plant species
and its intended use pattern, picloram has been identified as
being likely to jeopardize endangered plant species when used on
pastures/rangeland and forests. A program is being developed by
the Agency to reduce or eliminate exposure to these species to a
point where use does not jeopardize these species.

Tolerance Assessment:

Tolerances are established under 40 CFR 180.292 for residues of the
picloram. Food and Feed additive tolerance are established
under 40CFR 185.4580 and 40 CFR 186.4580. These replace old
Section 21 CFR 183.350 and 21 CFR 561.305.

The Agency has established a R.F.D or a provisional acceptable daily
intake at 0.07 mg/kg/day based on a 6-month dog feeding study
(NOEL of 7.0 mg/kg/day) using a safety factor of 100. The theoretical maximum residue contribution (TMRC) is calculated to
be 0.001847 mg/kg/day, which utilizes 2.6 percent of the PADI.

The tolerance assessment indicated that additional residue data are
needed for wheat grain, wheat forage, wheat straw, pasture, rangeland grasses, and flax. Data are required depicting residues of
HCB in or on wheat grain, wheat straw, pasture and rangeland
grasses, flax seed, and flax straw. Additional plant and animal
metabolism data are needed.

Reported Pesticide Incidents

Most of the reported pesticide incidents involve crop damage and
damage to other nontarget plants resulting from drift and from
soil contaminated with picloram.

4.   Summary of Regulatory Position and Rationale

A review of available data indicates that none of the risk criteria
listed in 40 CFR 154.7 have been exceeded. Therefore, no referral
to Special Review is being made at this time.

The Agency will continue to require that EPs containing picloram
retain the "Restricted Use" classification and the groundwater
advisory against the use of picloram on well-drained soils.

The Agency is requiring that the rat and mouse oncogenicity studies
be repeated using Osborne-Mendel rats and $B_6C_3F_1$ mice of both

sexes using a commercially available technical grade picloram
uncontaminated with potentially tumorigenic levels of HCB.

The Agency has determined that basic toxicology studies are needed
for the organic esters and amines of picloram in addition to the
complete toxicological testing of the acid and/or K salt.

The Agency will continue to require manufacturers to limit the level
of HCB in the technical to a maximum of 200 ppm.

The Agency is requiring nitrosamine testing for the tertiary amine
and alkanoloamine forms of picloram. The level of nitrosoamines
permitted in these forms is a maximum of 1 ppm.

The Agency is requiring that a prospective groundwater monitoring
study be submitted for picloram.

The Agency is requiring that Tier II phytotoxicity testing be performed
with the technical picloram, its salts, ester, and amine forms.

The Agency is requiring that droplet size spectrum and drift field
evaluation data be submitted for picloram.

The Agency is requiring that additional residue data be submitted for
wheat grain, wheat forage, wheat straw, pasture and rangeland
grasses, and flax. The data must include residues of HCB and the
results of analysis for HCB levels.

The Agency is requiring that additional plant metabolism data be
submitted providing complete identification and quantitation of
all terminal residues.

The U. S. Fish and Wildlife Service has determined that picloram
is likely to jeopardize endangered plant species when used on
rangeland/pastureland and forests. The Agency is developing a
program to reduce or eliminate exposure of this chemical to these
species. After the program is developed, notification of any
additional labeling requirements will be made.

The Agency is requiring that the labels of products containing
picloram determined to be skin sensitizers include the statement "May cause allergic skin reaction after multiple exposure"
on the labels

The Agency will not approve any significant new food uses for picloram
while major data gaps exist. When additional data are evaluated
the Agency will determine whether significant new uses may be
established.

5. Summary of Data Gaps

Requirements	Due Dates
Product Chemistry	6 to 15 months
Residue Data	6 to 24 months
Toxicology Data	9 to 40 months
Environmental Fate	9 to 39 months
Groundwater Monitoring	9 months
Plant Protection	9 months

6. Contact Person at EPA

Robert J. Taylor
Product Manager 25
Fungicide-Herbicide Branch
Registration Division (TS-767C)
Office of Pesticide Programs
Environmental Protection Agency
401 M Street SW.
Washington, DC 20460
Phone: (703) 557-1800

DISCLAIMER: The information presented in this Pesticide Fact Sheet is
for informational purposes only and may not be used to fulfill data
requirements for pesticide registration and reregistration.

PROPANIL

Reason for Issuance: Registration Standard
Date Issued: December 23, 1987
Fact Sheet Number: 149

1. Description of Chemical

   Common Name: Propanil

   Chemical Name: 3',4'-dichloropropionanilide

   Other Names: Surcopur, Supur, Propanex, Propanilo, Supernox, FW-734, Stam, Stampede, DPA, Herbax, Riselect, Erban, Chem Rice, Rogue, S-10165, Strel, Bay 30130

   OPP (Shaughnessy) Number: 028201

   Chemical Abstracts Service (CAS) Number: 709-98-8

   Empirical Formula: $C_9H_9Cl_2NO$

   Molecular Weight: 218.1

   Year of Initial Registration: 1962

   Pesticide Type: Herbicide

   U.S. and Foreign Producers: Rohm & Haas Company, Bayer AG, CIFA Laboratori Chemici, C.I.K. Australia Pty., Ltd., Crystal Chemical Inter-America, Cumberland International Corporation, Sintesul, Tifa, Ltd., Visplant-Chimiren S.r.l.

2. Use Patterns and Formulations

   Application Sites: Rice, spring barley, oats and durum wheat.

   Percent of Crop Treated With the Pesticide: Rice, 70 to 80 percent of total U.S. rice crop.

   Percent of Pesticide Applied to Crop: (1) Rice, approximately 95 percent [9,000,000 to 11,000,000 pounds of active ingredient (ai)] of total domestic usage of propanil, (2) Wheat, barley oats, approximately 2 to 5 percent (less than 1,000,000 pounds ai) of total domestic usage of propanil.

527

Types and Methods of Application: Applied by conventional
aerial or ground equipment as a postemergence application.

Pest Controlled: Grasses, broadleaf and aquatic weeds.

Application Rates: Rates range from 1.13 to 6 lb ai/A.

Types of Formulations:

Manufacturing-Use Products: 85, 90, and 96% ai.

End-Use Products: 33, 33.8, 35, and 35.9% ai (3 lb ai/gal)
emulsifiable concentrate; 43.48, 43.5, 44.5,
45, and 45.4% ai (4 lb ai/gal) emulsifiable
concentrate; 35% ai (3 lb ai/gal) soluble
concentrate/liquid.

Usual Carriers: Water

3. Science Findings

Summary Science Statement: There are no acceptable data to assess
acute toxicity. However, two invalid acute oral studies indicate
Toxicity Category III for propanil. The chronic toxicity data
were developed from testing three technicals of varying purity.
Pending review of the impurity profiles of these technicals, certain
studies may need to be repeated. Maternal and developmental toxicity
no-observable-effect levels (NOEL) were established at 20 mg/kg/day
in separate rat and rabbit teratology studies. In a rat reproduction
study, a reproductive and systemic NOEL was established at 300 parts
per million (ppm). Results of these three studies do not indicate
any toxicological concerns. A supplementary mouse oncogenicity
study conducted with 85.4% and 98% ai propanil was negative for
oncogenic potential at 180 ppm (highest dose tested). Dose-related
histologic findings were observed in the male liver and the NOEL
for this lesion was 30 ppm for the 98% technical and the LEL was
180 ppm for both technicals. Bilateral retinal degeneration in
male and female mice and thyroiditis in female mice were observed
at 180 ppm with the 85.4% ai technical only. NOELs for these effects
were not established and the mouse oncogenicity study did not employ
the maximum tolerated dose. Therefore, this study is classified as
supplementary data and further information, or a new mouse oncogenicity
study is necessary. Propanil is not acutely toxic to birds on a
dietary basis. The chemical is slightly to moderately toxic to
estuarine and marine organisms and freshwater invertebrates.
Propanil is moderately to very highly toxic to freshwater fish.
The chemical may pose a risk to some endangered species. The
environmental fate of propanil is not adequately understood. Available
data indicate that propanil is stable to hydrolysis at pH 7 and 9.

Leaching studies indicate that the chemical is mobile to very mobile on sand loam, silt loam, clay loam, clay and sand soils. A ground water monitoring study may be required pending the results of mobility and soil field dissipation studies.

Chemical Characteristics:

Physical State: Solid

Color: Medium to dark grey

Odor: Mildly acrid

Melting Point: 89-92 °C

Density: 1.25 g/ml at 25 °C

Solubility: 0.002 g/ml in water, readily soluble in ketones, alcohols and chlorinated solvents.

Vapor Pressure: Less than 0.001 mmHg at 50 °C (97% ai), 9.1×10^{-7} torr at 25 °C (85% ai).

Octanol/Water Partition Coefficient: 193 (analytical grade, 99% ai)

Stability: Stable at room temperature. Strong acid or alkali will hydrolyze propanil to 3',4'-dichloroaniline and propionic acid.

Toxicological Characteristics:

The chronic toxicity data base was developed from testing three technicals ranging in purity from 85 to 98% ai. Pending review of the impurity profile of these technicals certain studies may need to be repeated.

°   Teratology - Rat: Maternal and developmental toxicity NOEL is 20 mg/kg/day (85.4% ai)

°   Teratology - Rabbit: Maternal and developmental toxicity NOEL is 20 mg/kg/day (85.4% ai)

°   Reproduction - Rat: Reproductive and systemic NOEL is 300 ppm. There were no compound-related effects on fertility, gestation, pup viability and lactation, or sex ratios for each generation.

°   Mutagenicity: Propanil was not mutagenic in gene mutation and chromosomal aberration assays, and in all but one direct DNA damage assay.

Major Routes of Exposure:

Mixers, loaders, and applicators would receive the most exposure via skin/eye contact and inhalation.

Physiological and Biochemical Behavioral Characteristics:

Foliar Absorption: Propanil is absorbed through the leaves.

Translocation: The chemical is translocated to the growing point and back to other leaves.

Mechanism of Action: Propanil inhibits a number of biochemical reactions, especially photosynthesis.

Environmental Characteristics:

Available data are insufficient to fully assess the environmental characteristics of propanil. Leaching studies indicate that the chemical is mobile to very mobile on sandy loam, silt loam, clay loam, clay and sand soils. Propanil is stable to hydrolysis at pH 7 and 9. Pending the results of anaerobic/ aerobic soil metabolism, aquatic metabolism, and field dissipation studies, a ground water monitoring study may be required.

Ecological Characteristics:

° Avian Dietary Toxicity: 1924 ppm (bobwhite quail), greater than 5000 ppm (mallard duck)

° Freshwater Fish Acute Toxicity: Less than 3.7 to 5.36 ppm (bluegill sunfish), 2.3 ppm (rainbow trout)

° Aquatic Invertebrate Acute Toxicity: 6.7 ppm (Daphnia magna)

Potential problems related to endangered species: The use of propanil on rice may pose a hazard to endangered aquatic species. The Agency is consulting with the U.S. Fish and Wildlife Service to determine whether the fat pocketbook pearly mussel (Potamilus capax) may be at jeopardy from the use of propanil on rice. After U.S. Fish and Wildlife Service review, the Agency may impose labeling requirements to protect endangered species.

Tolerance Assessment:

Tolerances have been established for residues of propanil in or on the
following raw agricultural commodities (RACs):

Commodity	Tolerance (ppm)
Barley, grain	0.2
Barley, straw	0.75
Cattle, fat	0.1 (N)[1]
Cattle, mbyp[2]	0.1 (N)
Cattle, meat	0.1 (N)
Eggs	0.05 (N)
Goats, fat	0.1 (N)
Goats, mbyp	0.1 (N)
Goats, meat	0.1 (N)
Hogs, fat	0.1 (N)
Hogs, mbyp	0.1 (N)
Hogs, meat	0.1 (N)
Horses, fat	0.1 (N)
Horses, mbyp	0.1 (N)
Horses, meat	0.1 (N)
Milk	0.05 (N)
Oats, grain	0.2
Oats, straw	0.75
Poultry, fat	0.1 (N)
Poultry, mbyp	0.1 (N)
Poultry, meat	0.1 (N)
Rice	2
Rice, straw	75 (N)
Sheep, fat	0.1 (N)
Sheep, mbyp	0.1 (N)
Sheep, meat	0.1 (N)
Wheat, grain	0.2
Wheat, straw	0.75

A feed additive tolerance of 10 ppm is established for propanil in
or on rice bran, rice hulls, rice polishings and other milling
fractions resulting from application of the herbicide to the
growing RAC rice.

---

[1] The designation "(N)", standing for negligible residues, will be
deleted from this entry since it is no longer used by the Agency.
[2] Meat byproducts

Dietary Assessment: The provisional acceptable daily intake (PADI) was based on a 2-year rat feeding study. The NOEL for the rat study was 5 mg/kg/day. An uncertainty factor of 1000 was used to account for the inter- and intraspecies differences and the toxicology data gaps. The PADI was obtained by dividing the NOEL by the uncertainty factor of 1000. The resultant PADI was 0.005 mg/kg/day. The theoretical maximum residue contribution (TMRC) of propanil to the daily diet of the U.S. population is 0.0015 mg/kg/day, based on the existing tolerances, with 29% of the PADI being utilized.

Reported Pesticide Incidents: A well documented drift incident occurred in California where propanil was applied to extensive acreage of rice in the Sacramento Valley and resulted in phytotoxicity to plum trees up to 40 miles away. California has disallowed use of propanil in specified geographic areas, requires permits for application in other areas, and has imposed application equipment requirements and meteorological conditions for application. Propanil is considered a "restricted herbicide" in California State Regulations.

4. Summary of Regulatory Positions

The following Agency positions are summarized from the Propanil Registration Standard.

° The Agency is not initiating a Special Review of propanil at this time.
° The Agency will not establish significant new food uses until residue and chronic health effects data are submitted and reviewed.
° The Agency is not imposing reentry requirements at this time.
° Tolerances for residues in catfish and crayfish must be proposed and supporting data submitted. In lieu of proposing tolerances for catfish and crayfish, label restrictions may be implemented.
° The Agency has determined that label restrictions on rice drainage water must be implemented.
° If available rice processing studies are determined to be adequate, then the registrant must propose a method to make consistent, the feed additive tolerance and the expected residues in rice bran.
° The registrant must propose a method to make consistent, the tolerance for residues in or on barley, oats and wheat straw and the expected residues for these commodities if available straw data for these commodities are determined to be adequate.

° The Agency will initiate steps to delete the designation "(N)" from the tolerances listed in 40 CFR 180.274.
° The Agency is requiring product chemistry data on each technical grade propanil product.
° The Agency has determined that product chemistry and metabolism data will be immediately reviewed upon receipt.

Use, Formulation or Geographic Restrictions: No significant new food or feed uses of propanil will be permitted until residue chemistry and chronic toxicology data are available to assess existing uses.

Unique Label Warning Statements: End-use products shall bear the following statements as applicable:

a. End-use products with aquatic uses:

(1) This pesticide is toxic to fish. Drift and runoff from treated areas may be hazardous to aquatic organisms in neighboring areas. Do not apply directly to water except as specified on this label. Do not contaminate water by cleaning of equipment or disposal of wastes.

(2) Water drained from treated rice fields must not be used to irrigate other crops or released within 1/2 mile upstream of a potable water intake in flowing water (i.e., river, stream, etc.) or within 1/2 mile of a potable water intake in a standing body of water such as a lake, pond, or reservoir.

b. The following statements must appear on end-use products labeled for aquatic uses, if the registrant of the product chooses not to conduct tests and propose tolerances for catfish and crayfish:

(1) Do not drain water from treated fields into areas where catfish farming is practiced.

(2) Do not apply to fields where commercial crayfish farming is practiced and do not drain water from treated fields into areas where crayfish farming is practiced.

c. End-use products with nonaquatic uses:

> This pesticide is toxic to fish. Drift and
> runoff from treated areas may be hazardous to
> aquatic organisms in neighboring areas. Do not
> apply directly to water or wetlands (swamps,
> bogs, marshes, and potholes). Do not contaminate
> water by cleaning of equipment or disposal of wastes.

5. Summary of Major Data Gaps

Product Chemistry

All Product Chemistry data.

Residue Chemistry

Metabolism Studies (plant, livestock)
Residue Analytical Method (plant and animal residues)
Storage Stability
Residue Studies

Toxicology

Acute Oral
Acute Dermal
Acute Inhalation
Eye Irritation
Dermal Irritation
Dermal Sensitization
21-Day Dermal
Chronic Toxicity (rodent and nonrodent)
Oncogenicity (rat and mouse)
General Metabolism

Environmental Fate

Hydrolysis
Photodegradation (water and soil)
Aerobic Soil Metabolism
Anaerobic Soil Metabolism
Anaerobic Aquatic Metabolism
Aerobic Aquatic Metabolism
Leaching, Adsorption/Desorption
Soil Dissipation

Aquatic (Sediment) Dissipation
Rotational Crop (Confined) Accumulation
Irrigated Crops Accumulation
Fish Accumulation

Fish and Wildlife

Avian Acute Oral
Avian Subacute Dietary (upland game bird and waterfowl)
Freshwater Fish Toxicity [TEP] (coldwater fish and warmwater fish)
Freshwater Invertebrates Acute Toxicity [TEP]
Estuarine and Marine Organisms Acute Toxicity
Fish Early Life Stage and Aquatic Invertebrate Life Cycle
Fish Life Cycle

Plant Protection

Seed Germination/Seedling Emergence
Vegetative Vigor
Aquatic Plant Growth

6. Contact Person at EPA

Robert J. Taylor
U.S. Environmental Protection Agency
Registration Division (TS-767C)
401 M Street, S.W.
Washington, DC 20460
(703) 557-1800

DISCLAIMER: The information presented in this Pesticide Fact Sheet is
for informational purposes only and may not be used to fulfill data
requirements for pesticide registration and reregistration.

PROPAZINE

Reason for Issuance: Registration Standard
Date Issued: December 20, 1988
Fact Sheet Number: 189

1. ## DESCRIPTION OF CHEMICAL

   Generic Name: 2-chloro-4,6-bis(isopropylamino)-s-triazine
   Common Name : Propazine
   Trade Name : Milogard,® Gesamil,® Milo-Pro, Pramitol, Prozinex
   OPP Chemical Code: 080808
   Chemical Abstracts Service (CAS) Number: 139-40-2
   Year of Initial Registration: 1974
   Pesticide Type: Herbicide
   Chemical Family: S-Triazine
   U.S. and Foreign Producers: Ciba-Geigy, Drexel, Makhteshim-Agan,
      Griffin Corp., I.Pi.Ci.

2. ## USE PATTERNS AND FORMULATIONS

   Application Sites: Propazine is registered for use on the
      terrestrial food crop sorghum and for noncrop areas.

   Percent of Pesticide Applied: 99+% of propazine is used on
      sorghum.

   Types and Methods of Application: Propazine is used as a
      selective preemergent herbicide to control broadleaf and
      grass weeds. Propazine is applied as a spray, at the
      time of planting, prior to planting or immediately follow-
      ing planting by ground or aerial equipment.

   Application Rates: Propazine is applied generally from 1 to
      2 pounds active ingredient per acre; however, as much as
      3.2 pounds active ingredient per acre may be used on
      certain fine textured or highly organic soils for sorghum
      and from 1.6 to 13.3 pounds per acre for non-crop areas.

   Types of Formulations: Wettable powders (90 to 26.67% active
      ingredient); flowable concentrates (44.5 to 18.7% active
      ingredient); soluble concentrates (43% active ingredient)

   Usual Carrier: Water. Agitation in the spray tank is necessary
      to keep the chemical in suspension.

3. SCIENCE FINDINGS

Summary Science Statement: Propazine has low acute oral
   toxicity and is classified in Toxicity Category III*.
   Propazine is not considered to be teratogenic in rats.
   Propazine did not induce tumors in mice but an increased
   incidence of mammary gland tumors was observed in female
   rats. Based on the rat study, the Agency has classified
   propazine at a Group C oncogen (potential human carcinogen)
   but has concluded that quantitative risk assessment is
   not warranted because tumors in the rat study occurred in
   only one sex, were mostly benign and were significantly
   increased only at the highest dose tested.

Propazine can be characterized as slightly toxic to coldwater fish and practically nontoxic to waterfowl. It will
   not pose a hazard to endangered plant or wildlife species.
   Propazine does have the potential to contaminate groundwater.

Chemical Characteristics:

   Physical State: Solid
   Color: Colorless, white
   Odor: Odorless
   Melting Point: 212-214 °C
   Density: 1.16 $\pm$ 0.002 g/cm^3 at 20 °C
   Solubility: 8.6 ppm; water 20-22 °C
   Vapor Pressure: 2.9 x 10^{-8} mmHg at 20 °C
   Stability: Minimum of 3 years at room temperature

Toxicology Characteristics:

   Acute Toxicity:

      Acute Oral--Rat: > 5 g/kg (Toxicity Category IV)

      Acute Dermal--Rabbit: > 2 g/kg (Toxicity Category III)

      Acute Inhalation--Rat: > 2.1 mg/L/4 hr (Toxicity
         Category III)

      Primary Eye Irritation--Rabbit: No corneal opacity at
         24 hours (Toxicity Category III)

      Primary Skin Irritation--Rabbit: Score of 3.9/8.0
         with erythema, eschar, and edema with improvement
         within 72 hours (Toxicity Category III)

   Subchronic Toxicological Results: No acceptable studies
      are available. However, because an acceptable chronic
      rat study is available and a nonrodent chronic study
      is required, subchronic studies are not required.

* Refer to 40 CFR 156.10 for a discussion
   of the toxicity categories.

Chronic Feeding Results: In a chronic feeding study in rats, the NOEL was 100 ppm. A chronic feeding study in nonrodents is required.

Oncogenic Testing Results: Propazine was not oncogenic in a mouse study. In a rat study, however, it did produce an increased incidence of mammory gland tumors in female rats at the highest dosage level tested (1,000 ppm).

Developmental and Reproductive Study Results: Propazine did not induce terata in a rat developmental toxicity study. The reproductive NOEL was 100 ppm. An additional developmental study in a second species is required.

Major Route of Exposure: Dermal (mixers, loaders and applicators)

Physiological Characteristics:

Absorption Characteristics: Propazine is absorbed through plant roots.

Translocation: Propazine is absorbed by plant roots and is translocated upwardly in the plant to the leaves. It accumulates in the growing parts and leaves of plants.

Mechanism of Action: Inhibition of cell division and photo-synthesis

Environmental Characteristics: Propazine is persistent, moderately mobile and stable to hydrolysis, photolysis and microbial degradation, demonstrating a potential to con-taminate groundwater. It has been detected in groundwater samples in 8 states with maximum concentrations of 20 ppb in surface water and 300 ppb in groundwater. Available data are insufficient to fully assess the environmental fate and transport of propazine.

Ecological Characteristics: Propazine is slightly toxic to coldwater fish with a toxicity value (LC_{50}) of 16.5 ppm for rainbow trout. It is practically nontoxic to waterfowl with a toxicity value (LC_{50}) of 32000 ppm for Mallards. Based on use, estimated concentrations and the available toxicity data, there is no threat to endangered wildlife or plant species.

Tolerance Assessment: Tolerances are established for negli-gible residues of propazine in or on sweet sorghum, its grain, fodder and forage at 0.25 ppm (40 CFR 180.243). The provisional acceptable daily intake (PADI) for propazine is 0.02 mg/kg/day, based on a 2-year rat feed-ing study in which the systemic NOEL was set at 100 ppm

(5 mg/kg). The safety factor used was 300 based on an
uncertainty factor of 100 to account for inter- and intraspecies differences with an additional factor of 3 to
account for the incompleteness of the chronic data base.
The theoretical maximum residue contribution (TMRC) for
the U.S. population average is 0.0003 mg/kg/day, equivalent
to 1.7 percent of the PADI.

4. SUMMARY OF REGULATORY POSITION AND RATIONALE. As the result
    of a Data Call-In Notice, issued in April 1988, for a
    groundwater monitoring study, all propazine registrations
    have been either cancelled or suspended. Therefore, the
    Agency has determined, at this time, that it is not
    necessary to formulate specific regulatory positions
    regarding propazine. If a registrant commits to generate
    the required data and complies with the requirements of
    FIFRA, the Agency will then address specific regulatory
    positions for this chemical.

5. SUMMARY OF MAJOR DATA GAPS

        Product Chemistry
        Toxicology
            Dermal Sensitization
            21-Day Dermal
            Chronic Toxicity (Nonrodent)
            Teratogenicity (Rabbit)
            General Metabolism
        Residue Chemistry
        Environmental Fate
        Ecological Effects

6. EPA CONTACT

        Robert J. Taylor, Product Manager (25)
        Office of Pesticide Programs
        Registration Division (TS-767C)
        Environmental Protection Agency
        401 M Street, SW.
        Washington, DC 20460
        Telephone: (703) 557-1800

DISCLAIMER: The information presented in this Pesticide Fact Sheet
is for informational purposes only and may not be used to fulfill
data requirements for pesticide registration and reregistration.

PSEUDOMONAS FLUORESCENS EG-1053

Reason for Issuance: New Chemical
Date Issued: March 3, 1988
Fact Sheet Number: 160

1. Description of Chemical

      Generic Name: Pseudomonas fluorescens EG-1053

      Trade Name: DAGGER™

      EPA Shaughnessy Code: 064188

      Year of Initial Registration   1988

      Pesticide Type: Biofungicide

      U.S. and Foreign
         Producers: Ecogen Inc.
                                       2005 Cabot Boulevard West
                                       Langhorne, PA 19047-1810

2. Use Patterns and Formulations:

      Application Sites: Pseudomonas fluorescens is proposed for
         use on cotton for control of Pythium/Rhizoctonia.

      Types of formulations: 20% Granular End-Use Product

      Types and methods application: End-use product is applied
         in the furrow at planting time with use of a granular
         pesticide applicator.

      Application rates: The proposed application rate is 15 pounds
         per acre.

      Usual carrier: Water

3. Science Findings

    Summary Science Statement:

    The biofungicide was not toxic to test animals when
    administered via the oral, dermal, or pulmonary routes.
    The active ingredient, P. fluorescens was not infective or
    pathogenic for test animals when administered via the oral,
    pulmonary, or intravenous routes. Dagger™ G biofungicide
    did not elicit a delayed type hypersensitivity reaction in
    guinea pigs. The formulated product was mildly irritating
    to the eye. All toxicology studies were considered
    acceptable.

    Avian testing on the active ingredient, P. fluorescens, by
    both the oral and injection route showed no adverse effects
    at levels of 6.9×10^6 and 4.6×10^6 colony forming units
    per bird. The Agency believes that there will not be any
    adverse effects to avian species through the use of Dagger™
    on cotton.

    Likewise, injection studies using mice show no
    pathogenic or toxic signs attributable to P. fluorescens.
    Therefore, the Agency believes that use of this product
    will not cause adverse effects to feral mammals.

    The major concern in aquatic systems is fish. P. fluorescens
    has been associated with disease in fish. The Agency is
    not aware of disease occurrences with Pseudomonas in aquatic
    invertebrates. Because no acceptable fish studies have
    been submitted, the Agency cannot complete an aquatic risk
    assessment at this time.

    Testing on nontarget plants showed that, in most cases, the
    active ingredient in Dagger™ caused an increase in growth
    after germination. The exceptions were cabbage, which
    showed a significant decrease in seedling height, and
    lettuce, which showed a significant decrease in fresh
    weight at the end of four weeks. No phytotoxic effects
    were seen in any of the species tested. The data taken as
    a whole indicate that Dagger™ should not pose a threat to
    most plants. The Agency does not feel that there will be a
    "may effect" situation for endangered terrestrial wildlife,

insects or plants through the use of Dagger™ on cotton.
A final determination of "may affect" can be made for
endangered fish when the Agency receives a pathogenicity
study for fish.

Chemical Characteristics

Color: Light tan (MUP)
           Charcoal grey (EP)
Physical State:
    Aqueous cell suspension (MUP)
    Granular solid (EP)
Odor: Slight musty (MUP)
          Earthy-like (EP)
Density/specific
   gravity: 1.01 at 20 C (MUP)
pH: 6.9-7.1
Stability: Stable for 77 days in distilled water (pH 6.6)
            (MUP)
Storage stability: Loses 60-71% of viability within 1
            month at 40-50 F and 1000-fold loss within 5 months

Toxicology Characteristics

Acute effects

Acute oral toxicity/pathogenicity (rat):
    P. fluorescens was not toxic to, or infective in, or
    pathogenic to rats when administered orally in a
    single dose at 5×10^8 colony forming units (CFU) per animal
    Dagger™ G biofungicide is not toxic to, or infective in,
    or pathogenic for rats when administered at 2×10^7
    CFU/animal in a single oral dose.

Acute intravenous toxicity/pathogenicity (mouse):
    P. fluorescens was not toxic to, or infective in, or
    pathogenic for mice when administered at 5×10^6
    CFU/animal by the intravenous route.

Acute pulmonary toxicity/pathogenicity(rat):
    P. fluorescens was not toxic to, or infective in, or
    pathogenic for rats when administered to rats at
    1.5×10^8 CFU/animal via the pulmonary route.
    Dagger™ G biofungicide was not toxic to, or infective
    in, or pathogenic for rats when administered at
    5×10^7/animal via the pulmonary route.

Acute dermal toxicity (rabbit):
    P. fluorescens was not toxic to rabbits when applied
    to the skin at 2.8×10^8 CFU/animal. The bacterium

also was not a skin irritant.
Dagger" G biofungicide was not toxic to rabbits when
applied at the skin at a dose level of 1 gram/animal.
The test substance also was not a skin irritant.

Primary eye irritation (rabbit):
Dagger" G biofungicide is a mild eye irritant when
applied to the eye of rabbits at 0.1 g/eye. Irritation
signs observed were slight to moderate redness, and
slight chemosis and discharge associated with the
conjunctivae.

Hypersensitivity study (guinea pig):
Dagger" G biofungicide is not a skin sensitizer in
guinea pigs.

Ecological Characteristics

Avian Injection Pathogenicity test (Mallard Duck):
The administration of TGAI Pseudomonas strain 1053
at a dose of 10,000 mg/Kg by intraperitoneal
injection showed no apparent pathogenicity or
effect on the survival of young mallards.

Avian Acute Oral LD_{50} Pathogenicity test (Mallard Duck):
The administration of TGAI-Pseudomonas Strain 1053
(Washed) at a dose of 15,000 mg/Kg showed no
pathogenicity or effect on survival of young mallards.

4. Benefits

Dagger" G, a dry granular formulation is applied at planting time
for controlling cotton seedling disease.

5. Tolerance Assessment

Pseudomonas fluorescens is exempt from the requirement of a
tolerance in or on the raw agricultural commodities cottonseed
and cotton forage.

6. Summary of Major Data Gaps

Data required for conditional registration:

Guideline Number	Study	Time Generally Allowed for Response to Data Request
Ecological Effects		
154-20	Toxicity/Pathogenicity to Freshwater Aquatic Invertebrates	9 months
154-19	Freshwater fish testing	9 months
121-1	Target Area Phytotoxicity Tests (2) [The previously submitted tests may be upgraded to core if the percent active ingredient is reported and the test microorganism is confirmed as the same one registered.]	9 months
122-1	Seedling Germination/Vegetative Vigor - Tier I. [The previously submitted test may be upgraded to core if the microorganism under test is confirmed as the same as the one registered].	9 months
123-1	Seed Germination/Seedling Emergence	9 months

Contact Person at EPA

Lois A. Rossi
Product Manager (21)
Registration Division (TS-767C)
Environmental Protection Agency
401 M Street SW.
Washington, D.C.   20460

DISCLAIMER: The information presented in this Pesticide Fact Sheet
is for informational purposes only and may not be used to
fulfill data requirements for pesticide registration and reregistration.

QUIZALOFOP ETHYL

Reason for Issuance: Registration of New Chemical
Date Issued: June 10, 1988
Fact Sheet Number: 168

1. Description of Chemical

   Generic Name: Ethyl 2-[4-(6-chloroquinoxalin-2-yl oxy)phenoxy]propanoate
   Common Name: Quizalofop Ethyl, DPX-Y6202
   Trade Name: DuPont Assure Herbicide
   EPA Shaughnessy Code: 128201
   Chemical Abstracts Service (CAS) Number: 76578-14-8
   Year of Initial Registration: 1988
   Pesticide Type: Herbicide
   Chemical Family: Phenoxy Propionic Ester
   U.S. and Foreign Producers: E.I. du Pont de Nemours & Company, Inc.
                               Nissan Chemical Industries, Ltd.

2. Use Patterns and Formulations

   Application Sites: Terrestrial food crops

   Major Crops Treated: Soybeans

   Types and Methods of Application: Foliar, applied broadcast by air
      or ground equipment for control of annual and perennial grasses.
      It is applied postemergence to both crops and grasses.

   Application Rates: Up to 4 ounces active ingredient per acre.

   Types of Formulations: 9.5% EC

   Usual Carrier: Water

3. Science Findings

   Summary Science Statement: The submitted data are acceptable to the
      Agency. Quizalofop ethyl has low acute toxicity, Category III for
      acute oral and acute inhalation (technical) and primary eye
      irritation (formulation), and Category IV for acute oral, acute

545

dermal, and primary dermal (formulation), acute dermal, primary dermal, delayed hypersensitivity, and primary eye irritation (technical). It was not oncogenic to rats or mice [increases in liver tumors occurred at highest dose which exceeded the maximum tolerated dose (MTD)] and not teratogenic to rats or rabbits and not mutagenic. Quizalofop ethyl is practically nontoxic to birds or honey bees; highly toxic to freshwater fish, very highly toxic to invertebrates, moderately toxic to marine fish and very toxic to marine invertebrates. It has a low potential to leach and contaminate ground water and does not accumulate in fish. The nature of the residues in plants and animals is adequately understood and adequate methodology is available for enforcement of tolerances in soybeans, processed soybean food/feed items and meat, milk, poultry, and eggs.

Chemical Characteristics:

Physical State: Technical, solid; 9.5% EC, liquid
Color: Technical, white; 9.5% EC, amber
Odor: Technical, None; 9.5% EC, aromatic/petroleum
Melting Point: 91 °C
Specific Gravity: Technical, 1.35 grams per cubic centimeter
(g/cc^3); 9.5% EC, 1.01 g/cc^3
Molecular Weight: 372.81
Solubility: In distilled water at 25 °C, .3 milligrams/liter (mg/L); in 0.05 N sodium phosphate buffer at 25 °C as a function of pH:

```
pH  5.4    0.34 mg/L
pH  7.1    0.31 mg/L
pH  9.0    0.29 mg/L
pH 10.4    0.31 mg/L
```

In organic solvents at 25 °C:

```
Acetone        110 g/L
Acetonitrile    86 g/L
Ethanol          9 g/L
Benzene        290 g/L
Xylene         120 g/L
N-hexane         2.6 g/L
Dioxane        350 g/L
```

Unusual Handling Characteristics: No special handling needed.

Toxicology Characteristics:

Acute Studies (9.5% EC)

Acute Oral Toxicity - Rat: 6600 mg/kg (males), 5700 mg/kg (females); Toxicity Category IV

Acute Dermal Toxicity - Rabbit: > 5000 mg/kg; Toxicity Category IV

Primary Dermal Irritation - Rabbit: Not a primary skin irritant;
   Toxicity Category IV

Primary Eye Irritation - Rabbit: Mild irritant; Toxicity
   Category III

Acute Studies (Technical)

Acute Oral Toxicity - Rat: 1670 mg/kg (males), 1480 mg/kg
   (females); Toxicity Category III

Acute Dermal Toxicity - Rabbit: > 5000 mg/kg; Toxicity
   Category IV

Acute Inhalation - Rat: 4.8 to 5.8 mg/L; Toxicity Category III

Delayed Hypersensitivity - Guinea Pig: Not a sensitizer.

Primary Eye Irritation - Rabbit: Not an eye irritant; Toxicity
   Category IV

Primary Dermal Irritation - Not a skin irritant; Toxicity
   Category IV

Chronic Toxicology:

90-Day Feeding Study - Rat:
   No-observed-effect level (NOEL) = 2 milligram/kilogram/day
   (mg/kg/day) lowest dose tested (LDT)

90-Day Feeding Study - Mouse:
   NOEL less than (<) 15 mg/kg/day (LDT)

6-Month Feeding Study - Dog:
  NOEL = 2.5 mg/kg/day

2-Year Chronic Feeding/Oncogenicity Study - Rat:
   Systemic NOEL = 0.9 mg/kg/day
   Systemic lowest effect level (LEL) = 3.7 mg/kg/day
   No oncogenic effects up to and including 15.5 mg/kg/day [highest
   dose tested (HDT)]

18-Month Chronic Feeding/Oncogenicity Study - Mouse:
   Systemic NOEL = 12 mg/kg/day
   Systemic LEL = 48 mg/kg/day
   No oncogenic effects up to and including 12 mg/kg/day; an
   effect (increase in combined benign and malignant liver tumors)
   occurred at the HDT which exceeded the MTD.

1-Year Feeding Study - Dog:
   NOEL = 10 mg/kg/day (HDT)

Teratology Study - Rat:
   NOEL > 300 mg/kg/day (HDT)
   No teratogenic effects at 300 mg/kg/day (HDT)

Teratology Study - Rabbit:
   Maternal toxic NOEL = 20 mg/kg/day
   Maternal toxic LEL = 60 mg/kg/day (HDT)
   Developmental NOEL = 60 mg/kg/day (HDT)

2-Generation Reproduction - Rat:
   Fetotoxic NOEL = 1.25 mg/kg/day
   Maternal NOEL = 5 mg/kg/day
   Developmental NOEL = 1.25 mg/kg/day
   No reproductive effects up to 20 mg/kg/day (HDT)

Mutagenicity (Salmonella typhimurium): Negative

Mutagenicity, Chromosomal Aberrations (CHO) In Vitro: Negative

Mutagenicity, Unscheduled DNA Synthesis (Rat): Negative, not
   mutagenic

Metabolism - Rat: Readily absorbed from gastrointestinal tract
   and excreted rapidly.

Major Routes of Exposure:

The major route of exposure is via eye contact (formulation) and acute
oral and acute inhalation (technical).

Physiological and Biochemical Characteristics:

Foliar Absorption: Rapid.

Translocation: Systemic after absorption by foliage.

Mechanism of Pesticidal Action: Inhibition of fatty acid biosynthesis
   in susceptible plants. Potential to Contaminate Ground Water:
   Based on the low potential for leaching, quizalofop ethyl has a
   low potential to contaminate ground water.

Metabolism in Plants and Animals: Metabolized to several
   nonherbicidal compounds.

Persistence in Animals and Plants: Does not persist in either
   animals or plants.

Environmental Characteristics:

Absorption and Leaching in Basic Soil Types: DPX-Y6202 was poorly
   absorbed in two sandy loam and two silt loam soils and characterized
   as immobile. DPX-Y6202 and its acid metabolite are immobile to
   moderately mobile and have a low potential to leach.

Microbial Breakdown: Rapidly metabolized to DPX-acid which further
   degraded to phenols and CO_2.

Loss from Decomposition and Volatilization: None expected.

Bioaccumulation in Fish: Does not accumulate in fish.

Resultant Average Persistence: Half-life of 139 and 145 days in
   silty clay loam and silt loam, respectively.

Exposure of Humans and Nontarget Organisms to Pesticide or
   Degradates: Human risk from exposure is minimal because of low
   acute toxicity (Categories III and IV). Nontarget organism risk
   is minimal because maximum expected residues on soil and water do
   not approach the toxicity values for organisms tested.

Ecological Characteristics:

Avian Acute Oral Toxicity - Mallard Duck: > 2000 mg/kg.

Avian 8-Day Dietary Toxicity - Mallard Duck: > 5000 ppm.

Avian 8-Day Dietary Toxicity - Bobwhite Quail: 5620 ppm.

Fish Acute 96-Hour Toxicity - Rainbow Trout: 870 ppb.

Fish Acute 96-Hour Toxicity - Bluegill Sunfish: 460 ppb.

Fish Acute 96-Hour Toxicity - Mysid Shrimp: 0.15 ppm.

Fish Acute 96-Hour Toxicity - Mysid Shrimp: 0.15 mg/L.

Acute Toxicity (96-Hour) Test for Shrimp, Static - Mysid Shrimp:
   0.25 mg/L.

Acute Toxicity Test Mollusc, 96-Hour Flowthrough - Eastern Oyster:
   187 mg/L.

Acute Toxicity Test for Mollusc, 48-Hour Embryo and Larvae - Eastern
   Oyster: 0.079 ppm.

Acute Toxicity Test for Estuarine Fish, Static - Sheepshead Minnow:
   1.4 mg/L.

Acute Toxicity Test for Estuarine Fish, Static - Sheepshead Minnow:
   1.76

Freshwater Fish Early Life Study - Fathead Minnow: Maximum allowable
   toxic concentration (MATC):   11 to 30 ppb.

Acute Toxicity to Honey Bee: > 50 ug/bee.

Quizalofop ethyl is practically nontoxic to birds, highly toxic to
   freshwater fish, very highly toxic to invertebrates, moderately
   toxic to marine fish, very toxic to marine invertebrates, and
   relatively nontoxic to honey bees.

Tolerance Assessment:

The nature of the residues in plants and animals is adequately
   understood and adequate analytical methods are available for
   enforcement purposes.

Tolerances are established for the combined residues of quizalofop
   (2-[4-(6-chloroquinoxalin-2-yl oxy)phenoxy]propanoic acid and
   quizalofop ethyl (ethyl-2-[4-6-chloroquinoxalin-2-yl oxy)phenoxy]
   propanoate) all expressed as quizalofop ethyl in or on the raw
   agricultural commodity soybeans at 0.05 part per million (ppm).

Tolerances are established for the combined residues of quizalofop,
   quizalofop ethyl, and quizalofop methyl (methyl 2-[4-(6-chloro-
   quinoxalin-2-yl oxy)phenoxy]propanoate), all expressed as quizalofop
   ethyl in or on the following raw agricultural commodities.

Commodity	ppm
Cattle, fat	0.05
Cattle, meat	0.02
Cattle, mbyp	0.05
Eggs	0.02
Goats, fat	0.05
Goats, meat	0.02
Goats, mbyp	0.05
Hogs, fat	0.05
Hogs, meat	0.02
Hogs, mbyp	0.05
Horses, fat	0.05
Horses, meat	0.02
Horses, mbyp	0.05
Milk	0.01
Milk, fat	0.05
Poultry, fat	0.05
Poultry, meat	0.02
Poultry, mbyp	0.05
Sheep, fat	0.05
Sheep, meat	0.02
Sheep, mbyp	0.05

Tolerances are established for the combined residues of quizalofop and quizalofop ethyl, all expressed as quizalofop ethyl in or on the following processed food/feed commodities.

Commodity	ppm
Soybean soapstock	1.00
Soybean hulls	0.02
Soybean meal	0.50
Soybean flour	0.50

The acceptable daily intake (ADI) based on the 2-year rat feeding/ oncogenicity study (NOEL of 0.9 mg/kg/day) and using a hundredfold safety factor is calculated to be 0.009 mg/kg/day. The theoretical maximum residue contribution (TMRC) from these tolerances is calculated to be 0.000216 mg/kg body weight/day, which occupies approximately 2.4 percent of the ADI. There are no other published tolerances for this chemical.

4. Summary of Regulatory Position and Rationale

The available data submitted to the Agency provide sufficient information to support a conditional registration provided that the following studies are repeated: acute toxicity to freshwater invertebrates and a confined rotational crop study. Therefore, the Agency has accepted the use of quizalofop ethyl on soybeans.

5. Summary of Data Gaps   Due Dates

Environmental Fate Data   24 Months
Acute Toxicity to Freshwater Invertebrates   12 Months

6. Contact person at EPA:

Robert J. Taylor
Product Manager 25
Fungicide-Herbicide Branch
Registration Division (TS-767C)
Office of Pesticide Programs
Environmental Protection Agency
401 M Street SW.
Washington, DC 20460
Phone: (703) 557-1800

DISCLAIMER: The information in this Pesticide Fact Sheet is a summary only and may not be used to fulfill data requirements for pesticide registration and reregistration.

RESMETHRIN

Reason for Issuance: Registration Standard
Date Issued: December 1988
Fact Sheet Number: 193

1. DESCRIPTION OF CHEMICAL

   Generic Name: [5-(phenylmethyl)-3-furanyl] methyl 2,2-dimethyl-
   3-(2-methyl-1-propenyl) cyclopropanecarboxylate

   Common Name: Resmethrin

   Trade and Other Names: Synthrin, SPB-1382, Pynosect, Chrysron

   EPA Shaughnessy Codes: 097801

   Chemical Abstracts Service (CAS) Number: 10453-86-8

   Year of Initial Registration: 1967

   Pesticide Type: Insecticide

   Chemical Family: Pyrethroid

   U.S. and Foreign Producers: Fairfield American (US)
                               Mitchell Cotts Chemical Ltd. (UK)
                               Penick - Bio UCLAF Corp. (US)
                               Sumitomo Chemical Co. (Japan)

2. USE PATTERNS AND FORMULATIONS

   Application: Flying and crawling insect control for household, greenhouse,
   indoor landscaping, mushroom houses, industrial, stored
   product insects, and mosquito control. Resmethrin is also used
   for fabric protection, pet sprays, pet shampoos, and
   application to horses and in horse stables. Resmethrin may be
   used in USDA meat and poultry programs.

   Types of Formulations: Resmethrin is primarily formulated into pressurized
   liquids and ready-to-use solutions. It is also
   formulated as an emulsifiable concentrate, soluble
   concentrate, and on impregnated materials.

3. SCIENCE FINDINGS

Summary Science Statement

Chemical Characteristics of the Technical Material

Physical State: Waxy solid

Color: Off-white to tan

Odor: Chrysanthemate

Molecular Weight and Formula: 338.4

Boiling Point: 180°C at 0.1 mm Hg

Vapor Pressure: 2.58 mm Hg, 200°C

Density: 8.70 lb/gal

Solubility in various solvents: Insoluble in water, 10% in kerosene, and
very soluble in xylene, methylene chloride,
isopropyl alcohol and aromatic hydrocarbons.

Toxicology Characteristics

Acute Oral: LD$_{50}$ - 750 to 4500 mg/kg

Acute Dermal: LD$_{50}$ - 2500 mg/kg

Primary Dermal Irritation: Not considered to be a dermal irritant.

Primary Eye Irritation: Not considered to be an eye irritant.

Dermal Sensitization: Not considered tOH 1.

Toxicology Characteristics

Acute Oral: LD_{50} - 750 to 4500 mg/kg

Acute Dermal: LD_{50} - 2500 mg/kg

Primary Dermal Irritation: Not considered to be a dermal irritant.

Primary Eye Irritation: Not considered to be an eye irritant.

Dermal Sensitization: Not considered to be a skin sensitizer.

Acute Inhalation: LC_{50} >9.49 mg/L (4 hr. exposure)

Major routes of exposure: dermal, eye, inhalation

Delayed neurotoxicity: Studies not required because resmethrin is not an
                         organophosphate.

Subchronic toxicity: A 90-Day subchronic inhalation study in rats did not
                       demonstrate a definite NOEL. At the lowest dose tested,
                       0.1 mg/L, there was evidence of behavioral reactions to
                       treatment, decreased blood levels of glucose in males,
                       decreased body weights and increased serum urea levels
                       in females. 90-Day subchronic dermal studies are
                       required because currently registered use patterns (i.e.
                       veterinary use only animal shampoos) could result in
                       prolonged or repeated exposure to humans.

Chronic Feeding: A rodent (rat) study did not establish a NOEL. There
                     was some evidence liver enlargement at the lowest dose
                     tested (500 ppm or 25mg/kg/day). At the LEL of 2500 ppm
                     or 125 mg/kg/day, increased liver weigth and
                     pathological lesions in this organ were observed. At the
                     HDT of 5000 ppm, or 250 mg/kg/day, additional dose
                     related increases in liver weight, liver pathology, and
                     also increases in thyroid weight and incidences of
                     thyroid cysts were observed.

                     A dog chronic chronic study demonstrated a NOEL of 10
                     mg/kg/day.

Oncogenicity: Resmethrin was determined not to be oncogenic for rats at
dosage levels up to and including 5000 ppm, and for mice at
dosage levels up to and including 1000 ppm.

Teratogenicity: The teratogenic NOEL for rats was 40 mg/kg/day based on
slight increase in aberrations in skeletal findings. There
was an indefinite maternal NOEL with some decrease in food
consumptions at all dose levels tested and a body weight
decrease at the 80 mg/kg/day level. The teratogenic, maternal
and fetotoxic NOELs in the rabbit were greater than 100
mg/kg/day (HDT).

Reproduction: A definite NOEL was not established since there were
marginally statistically significant increases in the number
of pups born dead and decreases in pup weight at weaning at
the lowest dosage tested (500 ppm or 25 mg/kg/day). The
observed effects were judged to be minimal when the low dose
groups were compared with the controls.

Environmental Characteristics

The potential for resmethrin to contaminate groundwater is
unknown. However based on the non-leaching characteristics of
related pyrethriod compounds and the fact that more than one-
half of the 30,000 lbs. used annually is used indoors, it is
not likely that use of this chemical will pose appreciable
risks to groundwater.

Ecological Characteristics

Avian acute toxicity: Resmethrin is slightly toxic to birds, the LD_{50} >
2000 mg/kg - California quail.

Freshwater fish acute toxicity: Resmethrin is very highly toxic to fish
(Bluegill LC_{50} - 0.75 - 2.6 ug/L and
Rainbow trout LC_{50} - 0.28 - 2.4 ug/L).

Freshwater and marine invertebrate toxicity:

The Agency does not have any data on invertebrates species,
but based on the characteristics of related pyrethroid
compounds, resmethrin is believed to be very highly toxic to
invertebrates.

Beneficial insects: Resmethrin is highly toxic to honey bees on contact
(LD$_{50}$ - 0.063 ug/bee).

TOLERANCE ASSESSMENT

No tolerances have been established for residues of resmethrin in or on
any plant or animal commodities. A food additive tolerance of 3 ppm (40 CFR
185.5300) has been established for residues of resmethrin in food items
resulting from treatment of food handling establishments and storage areas.

No Codex MRLS have been established for residues of resmethrin in or on
food or feed.

4.   Required Unique Labeling
All products with mosquito abatement and pest control at aquatic sites
are classified Restricted Use and must contain the following restricted use
classification statement.

RESTRICTED USE CLASSIFICATION
Due to acute fish toxicity

For retail sale to and use only by certified
applicators or persons under their direct
supervision and only for those uses covered by
the Certified Applicator's Certification.

5.   Summary of Regulatory Positions

·   The Agency is not placing resmethrin into Special Review. However
because of its high acute toxicity to fish, the Agency has serious
concerns about resmethrin's toxicity to aquatic invertebrate organisms,
particularly in view of the fact that structurally similar pyrethroid
compounds are highly toxic to these organisms. The Agency will make a
regulatory decision on this issue after completing the review of
outstanding aquatic data. The Agency will impose the restricted use
classification for certain use patterns.

·   The Agency is imposing the restricted use classification for all
products containing resmethrin which bear use directions for mosquito
abatement and insect control uses at aquatic sites.

- Based on the available toxicity data, the Agency is not requiring the use of personal protective clothing or reentry intervals.

- Because of its structural similarity to other pyrethroid compounds which are not known to leach, resmethrin is not expected to contaminate groundwater.

- The available residue chemistry data are insufficient to permit the Agency to conduct a tolerance reassessment.

6.   Summary of Major Data Gaps

Toxicology

Acute Oral
Acute Dermal
Subchronic dermal (21-Day and 90-Day)
Mutagenicity and Genetic Toxicity (full battery of studies including gene
     mutation, chromosome aberration, and other mechanisms of
     mutagenicity).
General metabolism and pharmacokinetics

Residue Chemistry

Nature of the residue (metabolism)
                         - Plants
                         - Livestock
Residue analytical methods
Storage stability
Magnitude of residue -- Meat/milk/poultry/eggs
Food handling establishments

Environmental Fate

Hydrolysis
Photodegradation
               - water
Metabolism studies (Lab)
               - aerobic soil
               - anaerobic aquatic
               - aerobic aquatic

Mobility
      - Leaching and adsorption/desorption
      - Volatility (Lab)
Dissipation (Field)
      - soil
      - aquatic (sediment)

Ecological Effects

Avian and Mammalian Testing
            - Avian Dietary LC_{50}
            - Avian Reproduction

Aquatic Testing
            - Fish early life stage
            - Freshwater fish LC_{50}
            - Acute LC_{50} freshwater invertebrates
            - Acute LC_{50} estuarine and marine organisms
            - Aquatic invertebrate life cycle

Nontarget Area Phytotoxicity

                  Tier I
            - Seedling germination/seedling emergence
            - Vegetation vigor
            - Aquatic plant growth

Nontarget Insects
            - Honeybee: toxicity of residues on foliage

7.   CONTACT PERSON AT EPA

Philip Hutton
Product Manager 17
Insecticide-Rodenticide Branch
Registration Division (TS-767C)
Office of Pesticide Programs
Environmental Protection Agency
401 M Street, S. W.
Washington, D. C.   20460

Office location and telephone number:

Room 214, Crystal Mall #2
1921 Jefferson Davis Highway
Arlington, VA   22202
(703) 557-2690

DISCLAIMER: The information in this Pesticide Fact Sheet
is a summary only and is not to be used to satisfy data
requirements for pesticide registration and reregistration.
The complete Registration Standard for the pesticide may be
obtained from the National Technical Information Service.
Contact the Product Manager listed above for further
information.

ROTENONE

Reason for Issuance: Registration Standard
Date Issued: October 7, 1988
Fact Sheet Number: 198

1. DESCRIPTION OF CHEMICAL

    Common name: Rotenone (and associated resins)

    Chemical Name: (2R, 6aS, 12aS)-1,2,6,6a,12,12a-hexahydro-
                        2-isopropenyl-8,9-dimethoxychromeno[3,4-b]
                        furo[2,3-h]chromen-6-one.

    Other Chemical
    nomenclature:
                        (R)-1,2-dihydro-8,9-dimethoxy-2-
                        (1-methylethenyl)[1]benzopyrano[3,4-b]
                        furo[2,3-h][1]benzopyran-6,12-dione
                        (9th Collective Index); 1,2,12,12a alpha-
                        tetrahydro-2a-isopropenyl-8,9-dimethoxy[1]
                        benzopyranol[3,4-b]furo[2,3-h][1]benzopyran-6
                        (6aH)-one (8th Collective Index); 1,2,12,12a-
                        tetrahydro-8,9-dimethoxy-2-(1-methylethenyl)-
                        [2R-(2a,6a,alpha, 12a alpha)]-(1)benzopyrano
                        (3,4-b)furo(2,3-h)-(1)benzopyran-6 (6aH)-one.

    Other Names: aker- root; Chem Fish; Derris; derris root;
                        Nicouline; Rotacide; Protex; Tubatoxin; and
                        tuba-root; ENT-133; Barbasco; Cube; Haiari;
                        Nekos; and Timbo.

    Chemical Abstracts Service (CAS) Number: 83-79-4

    EPA Pesticide Chemical Code
        (Shaughnessy Number):   71003 - rotenone; 71004 - cube'
                                     resins other than rotenone;
                                     71001 - derris resins.

    Year of Initial Registration: 1947

Pesticide Type: Botanical

Producers: Penick-Bio UCLAF Corporation and Prentiss
   Drug and Chemical Company

2. USE PATTERNS AND FORMULATION

Registered Uses:

Terrestrial Food Crops: Rotenone is registered for foliar
   preharvest application to vegetables, berries, tree fruit,
   nuts, forage crops, and sugarcane.

   In addition to foliar applications, delayed dormant
   applications are made to deciduous tree fruit.

   Soil applications may be made to vegetable crops,
   berries and tree nuts.

Terrestrial Nonfood Crop: Ornamentals, turf, shade trees,
   and tobacco.

Greenhouse Food Crop: Vegetables.

Greenhouse Nonfood: Ornamentals.

Aquatic Non-Food Crop: Fish.

Domesticated Pets and Their Man-made Premises: Cats and Dogs.

Livestock: Cattle (Beef and Dairy), Goats, Horses, Sheep,
   and Swine.

Household: Flying and crawling insects.

Commercial and Industrial Uses: Flying and crawling insects.

Methods of Application: Dusts, sprays, dips, and pumping
   liquid formulations into bodies of water.

Annual Usage: 50,000 to 120,000 lb

Predominant Usage: Aquatic (piscicide); agriculture
   (potatoes, tomatoes, pears, apples); livestock, pets, and
   household

Formulations: The following formulations are registered
0.4-5% dusts; up to 5% wettable powders (WP); up to 0.55
lb/gal emulsifiable concentrate (EC); up to 1.25% soluble
concentrate/liquid (SC/L); up to 5%; wettable powder/dust
(WP/D) up to 2.5%;emulsifiable concentrate (EC); up to
1.5%; soluble concentrate/lquid (SC/L); ready-to-use (RTU)
at 0.1%or less; and up to 0.1% pressurized liquid (PrL).

3.   SCIENCE FINDINGS

Chemical/Physical Characteristics of Rotenone (purity 99.5%
or unspecified) and its associated resins:

Empirical Formula: $C_{23}H_{22}O_6$

Molecular Weight: 394.4

Color: colorless (rotenone, purity 99.5%)

Physical State: Crystalline solid

Specific Gravity: 1.27 at 20 °C

Melting Point: 165-166 °C

Solubility: Water 0.00002 g/100 ml at 20 °C; Ethyl alcohol
0.2 g/100 ml at 20 °C; Carbon tetrachloride 0.6 g/100 ml
at 20°C; Amyl acetate 1.6 g/100 ml at 20°C; Xylene 3.4
g/100 ml at 20 °C; Ethylene dichloride 33.0 g/100 ml at 20
°C; Chloroform 47.2 g/100 ml at 20 °C; Acetone 6.6 g/100
ml at 20 °C; Benzene 8.0 g/100 ml at 20 °C; Chlorobenzene
13.5 g/100 ml at 20°C.

Stability: Decomposes rapidly in organic solvents exposed to
          light and air.

Toxicology Characteristics

* Acute oral LD (rat): LD_{50} = 39.5 +/- 2.21 mg/kg
  for female rats and 102 +/- 12.6 mg/kg for male
  rats.

* 6 Month Feeding (dog): NOEL = 0.4 mg/kg/day;
  LEL = 2 mg/kg/day; levels tested 0, 0.4, 2, and
  10 mg/kg/day.

- 2 Year Feeding (rat): NOEL = 7.5 ppm; LEL = 37.5 ppm; levels tested 0, 7.5, 37.5, and 75 ppm.

- 2 Year Oncogenicity (rat): Negative for oncogenic effects.

- 2 Year Oncogenicity (mouse): Negative for oncogenic effects at doses below the MTD.

- Teratology (rat and mouse): No fetal effects at rates below NOEL for adults.

- Reproduction Toxicity (rat): Reproductive NOEL = 37.5 ppm; Maternal Toxicity NOEL = 7.5 ppm; levels tested 0, 7.5, 37.5, and 75 ppm. No reproductive effects were noted.

- Gene Mutation: Rotenone did not induce gene mutations in bacteria or yeast, but it increased the frequency of gene mutations in mouse lymphoma cells in vitro without metabolic activation. No mutations were induced in mice in vivo by rotenone. No chromosomal effects were observed in rats or mice treated with rotenone, and yeast treated with rotenone did not show increased mitotic recombination or mitotic gene conversion.

Physiological and Biochemical Characteristics

Mechanism of pesticidal action: Rotenone is an inhibitor of oxidative phosphorylation.

Environmental Characteristics

Rotenone is rapidly degraded in soil and water with a half-life of 1-3 days for both aerobic aquatic and anaerobic aquatic soils.

Ecological Characteristics:

- Avian Acute Oral LC$_{50}$

Mallard	2200mg/kg
Pheasant	1680mg/kg

° Fish Acute LC_{50}

Rainbow Trout	22.5 ppb
Channel Catfish	2.6-2.8 ppb
Bluegill	22.5 ppb

° Aquatic Invertebrate EC_{50}

Daphnia magna (Water fleas) 2.1 ug/l
(21 Day EC_{50})

TOLERANCES

Currently, use of rotenone or derris or cube' roots
on growing crops is exempt from the requirement of tolerences,
and neither tolerances nor exemptions exist for rotenone
residues in animal commodities.

4. SUMMARY OF REGULATORY POSITION

Rotenone is not being referred to Special Review because
none of criteria for special review were met.

Cranberry and piscicide uses are classified for restricted
use. Users must consult their State Fish and Game Agency
before applying rotenone products. Because of significant data
gaps on the effects of rotenone in upland game birds, waterfowl,
estuarine and marine organisms, a statement is required on all
end-use products giving directions for use for killing fish.
The statement requires mandatory consultation with state or
federal fish and wildlife agencies and will minimize effects on
non-target organisms.

The Agency is concerned about residues and metabolites of
rotenone and associated resins in/on plants and animals. Data on
the metabolism and nature of rotenone residues are needed to
reevaluate the tolerance exemption.

Water treated with rotenone must not be used as potable
water or for irrigation of crops. A restriction against use of
rotenone within one half mile of potable water or irrigation
intakes must be included on labels of products intended for
use in cranberry bogs and for control of fish. This statement
will be an interim precaution until data on residues in water
are available.

5. SUMMARY OF MAJOR DATA GAPS

Toxicology: Acute dermal toxicity, primary eye and skin
irritation studies, dermal sensitization study, histological
examination of the low and mid dose groups in the rat
chronic feeding study.

Environmental Fate: Hydrolysis, photodegradation in water
and on soil, aerobic soil and aquatic metabolism, leaching,
terrestrial and aquatic field dissipation, confined rotational
crop study, and irrigated crop accumulation.

Environmental Safety: Effects on upland game birds, waterfowl,
and estuarine and marine organisms.

Residue Chemistry: Plant, livestock, and fish metabolism;
method validation, residue data for each registered commodity
and their processed products; food handling establishment
residue data, and all product chemistry data.

6. LABELING REQUIREMENTS

Products containing rotenone intended for terrestrial (food
and non-food) and domestic outdoor uses must include the following statement on the label:
"This pesticide is toxic to fish. Do not apply directly to
water or wetlands (swamps, bogs, marshes, potholes). Runoff and
drift from treated areas may be hazardous to aquatic organisms in
adjacent sites. Do not contaminate untreated water when disposing
of equipment washwater."

Products containing rotenone intended for use in cranberry
bogs must include the following statement on the label:
This pesticide is toxic to fish. Runoff and drift from treated
areas may be hazarddous to aquatic organisns in adjacent sites. Do
not contaminate untreated water when disposing of equipment washwaters.

Products containing rotenone intended for use to kill fish
fish must include the following statement on the label:
"This pesticide is toxic to fish. Fishkills are expected at
recomended rates. Consult your State Fish and Game Agency before
applying this product to public waters to determine if a permit
is needed for such an application. Do not contaminate untreated
water when disposing of equipment washwaters.

CONTACT PERSON AT EPA

Phillip Hutton, PM 17
Insecticide—Rodenticide Branch
Registration Division (TS-767C)
Washington, DC 20460
(703) 557 2690

or

Paul Schroeder
PM Team 17
Insecticide-Rodenticide Branch
Registration Division (TS-767C)
Washington, DC 20460
(703) 557-2602

DISCLAIMER: The information presented in this Chemical
Information Fact Sheet is a summary only and may not be used to fulfill
data requirements for pesticide registration and reregistration.

SODIUM FLUOROACETATE (COMPOUND 1080)

Reason for Issuance: Cancellation/Denial/Suspension/Data Call-In
Date Issued: October 1988
Fact Sheet Number: 174

1. DESCRIPTION OF CHEMICAL

   Generic Name: Sodium Fluoroacetate
                    $CH_2F-COON$

   Common Name: 1080

   Trade and Other Names: Compound 1080, Sodium monofluoroacetate,
                             Ratsbane

   EPA Shaughnessy Code: 075003-4

   Chemical Abstracts Service (CAS) Number: 62-74-8

   Pesticide Type: Vertebrate Pesticide

   U.S. Producers: Only one manufacturer, Tull Chemical Company,
   Inc., sells compound 1080 manufacturing use (technical) product
   in the United States.

2. USE PATTERNS AND FORMULATIONS:

   The principle use of compound 1080 is to control field rodents.
   Thirty nine rodenticide end-use products are currently author-
   ized for use: 35 with California counties, 2 with the Colorado
   Department of Agriculture, and 1 with Klamath County, Oregon.
   The California and Colorado products have intrastate registrations:
   Klamath County has a "special local need" registration.

   Montana, Wyoming, Ranchers Supply of Alpine, Texas, and
   the U.S. Department of Agriculture currently have Federal Registrations for 1080 Toxic Collars. This use will be affected as
   the Agency will require data to support a registration for a
   1080 technical product to be used only in formulation of livestock protection collars.

Formulation Type: Compound 1080 rodenticide is formulàted into grain baits at 0.2% to 0.11%. The registration for 1080 water baits was cancelled in 1986.

3. AGENCY ACTION:

EPA is taking a number of regulatory actions involving 1080 registrations and applications for registrations, since data required to support the registration of 1080 rodenticide products has not been submitted in accordance with Agency requirements. The Agency is issuing:

1) A notice of intent to deny the registration of intrastate products (20 California counties; Colorado Department of Agriculture) for which a complete application for Federal registration was submitted prior to July 31, 1988, in accordance with 40 CFR 152.230.

2) A notice of intent to cancel the registration of the one technical grade compound 1080 product (Tull Chemical Co.).

3) A notice of intent to suspend the one end-use federally registered 1080 rodenticides (Klamath County, Oregon).

4) A data call-in notice [Federal Insecticide, Fungicide, and Rodenticide Act section 3(c)(2)(b)] requiring product chemistry data to support a technical grade compound 1080 product to be used only in livestock protection collars (USDA, Montana Department of Livestock, Wyoming Department of Agriculture, Ranchers Supply Co.).

No actions, other than requests for minimal product chemistry data, are being taken with respect to the 1080 livestock protection collar registrations.

4. BACKGROUND:

The principal use of compound 1080 is to control field rodents. It was available for this use in California, Colorado, Nevada, and Oregon because these states had valid "intrastate" registrations. Other states which had used 1080 for field rodent control relied on the U.S. Department of the Interior registration, which was withdrawn following a 1972 Executive Order prohibiting the use of compound 1080 on federal lands. Prior to 1972, 1080 was also used for predator control, principally coyotes.

California uses over 80 percent of the compound 1080 as a field rodenticide; the bulk is used to control ground squirrels. The only Federal Registration for field rodent use is a "Special Local Needs" registration granted to Klamath County, Oregon.

On July 31, 1985 the Agency concluded a Special Review on
Sodium Fluoroacetate (1080) with the finding that: 1) use of
compound 1080 to control field rodents may have adverse effects on
nontarget wildlife; 2) these risks could be reduced by modifying
the labeling of the products and, in certain cases, by reducing
the concentration of compound 1080 in rodenticide baits. However,
the Agency was unable to determine whether additional regulatory
restrictions were needed due to the lack of critial data. Accordingly, the Agency conditioned any future use of the 1080 rodenticide
on the immediate adoption of certain risk reduction measures and
on the submission of the full complement of data required for
Federal Registration.

The Agency's current action on compound 1080 registrations
relates to the failure of registrants to satisfactorily respond to
two DCI Notices. In November, 1985, a Notice Requiring Submission
of Full Applications for Federal Registration for all intrastate
products was sent to states with intrastate registrations (40 CFR
162.17). A DCI Notice was also issued to California, Colorado and
Nevada requiring the submission of data to support compound 1080
Federal Registrations. Similar DCIs were also issued to Tull
Chemical Co., Inc. and Klamath County. Tull Chemical Co. declined
to submit data, but the California Department of Food and Agriculture (CDFA) agreed to provide the data for the company.

Prior to issuing the November 1985 DCI, the Agency met with
registrants and interested user groups to explain, among other
things, why data were required and how to generate the data.

Thirty six applications for Federal Registrations were submitted by CDFA, and two were submitted by the Colorado Department
of Food and Agriculture. The Nevada Department of Agriculture
failed to submit Federal Registration applications for its three
intrastate products.

In December 1986, the Agency received data from CDFA and the
Colorado Department of Agriculture to support the compound 1080
rodenticide use for California, Colorado, Klamath County, and for
the technical 1080 registration for Tull Chemical Co., Inc.
However, upon review of the data, the Agency determined that the
submissions by California and Colorado are unsatisfactory for
the following reasons:

1. Inadequate safety data to determine hazards to nontarget
    fish, birds, and mammals.

2. Lack of specific directions for 1080 use to enable the
    Agency to determine the food/feed sites for which coumpound
    1080 is/has been used so applicable data requirements can be
    determined; and

3. No validated analytical method with detection limits low enough to determine concentrations of compound 1080 at the level of concern.

A second public meeting was held in October 1987 where the Agency again explained why the data were required and clarified procedures for data development.

On December 15, 1987 another DCI Notice was issued to California, Colorado, Tull Chemical Co., Inc., and Klamath County. This Notice required the submission of four additional environmental fate studies.

On December 17, 1987 the compound 1080 registrants were notified that the Agency would extend the data due dates in the 1985 DCI Notice if the registrants submitted the following items within 30 days: 1) a commitment to fulfill the data requirements, 2) legible draft revised labels, with the use of sites clearly defined; and 3) A Confidential Statement of Formula (EPA Form 8570-4). In addition, progress reports were required for many long-term studies required by the DCI.

Additional data were submitted by CDFA in May 1988, to support compound 1080 use in California, Colorado, Klamath County, and the technical registration for Tull Chemical Co., Inc. Upon Agency review, the data were again found to be unsatisfactory. When CDFA also requested waivers of many of the data requirements, the Agency denied the request.

The Agency is taking action against all rodenticide compound 1080 registrations because of the lack of progress toward completing the DCI long-term data requirements, failure to submit the DCI short-term requirements by the due dates, unacceptable qualifications to the commitments to satisfy data requirements, and failure to submit administrative forms.

The Agency is requiring the four compound 1080 toxic collar registrants to register a compound 1080 manufacturing-use product (technical) to be used only in the formulation of livestock protection collars. The data required to support this registration will involve only product chemistry requirements.

The Agency also will be taking action with regard to the registrations of strychnine products, another vertebrate control pesticide. The Agency is issuing a rescissioin of the authority to sell or distribute intrastate products, and initiating a process to suspend the remaining strychnine registrations until all data required by three Data Call-In Notices are submitted to the Agency and, upon Agency review, are found to be acceptable. (These actions against strychnine are independent of the temporary cancellation action required by the April 11, 1988 order of the United States District Court for the District of Minnesota in the case of Defenders of Wildife v. Administrator.

STREPTOMYCIN

Reason for Issuance: Registration Standard
Date Issued: September 1988
Fact Sheet Number: 186

1. DESCRIPTION OF CHEMICAL

   Chemical Name: O-2-Deoxy-2-(methylamino)-d-L-glucopyranosyl-(1->2)-O-5-
       deoxy-3-C-formyl-x-L-lyxofuranosyl-(1->4)-N,N'-bis
       (aminoiminomethyl)-D-streptomine

   ANSI Common Name: Streptomycin

   Other Common Names: streptamine

   Principal Trade Names: Agri-Mycin 17®, Agri-Step®, Plantomycin®

   EPA (Shaughnessy) Code: 006306, 006310 (streptomycin sulfate)

   Chemical Abstracts Service (CAS) Number: 57-92-1, 3810-74-0 (streptomycin
       sulfate)

   Year of Initial Registration: 1958

   Pesticide Type(s): Fungicide

   Chemical Family: Aminoglycoside antibiotic isolated from the bacterium
       Streptomyces griseus

   U.S. and Foreign Producers: U.S. Merck & Co., Inc. and Pfizer, Inc.

2. USE PATTERNS AND FORMULATIONS

   Registered uses: Terrestrial food crop uses on fruit and vegetables,
       terrestrial nonfood crop uses on tobacco and ornamentals,
       and greenhouse nonfood crop use on ornamentals

   Uses: Ninety-eight percent of annual production is used on apples, pears,
       and tomatoes

Pests controlled (in general): Fungal diseases of selected fruit, vegetables, seed, and ornamental crops

Types of Formulations: Dust, wettable powder, wettable powder/dust, emulsifiable concentrate, pelleted/tablets, and liquid ready-to-use

Types and Method of Application: Foliar application by ground equipment, such as airblast. Other methods include: aircraft, duster attachment mounted over conveyor belt, hand-held or motor driven sprayers, dip treatment, tree injection treatment, slurry seed treatment.

Application Rates: Terrestrial food crop - 25 to 200 ppm
Terrestrial nonfood crop - 50 to 200 ppm

3. SCIENCE FINDINGS

Chemical Characteristics:

Physical state: powder
Color: pink to tan
Odor: burned sugar
Molecular Weight: 581.58; 1457.40 (streptomycin sulfate)
Empirical Formula: $C_{21}H_{39}N_7O_{12}$; $C_{42}H_{84}N_{14}O_{36}S_3$ (Streptomycin sulfate)

Toxicology Characteristics

Streptomycin has been used since the late 1940's to treat bacterial infections in humans. As a result of this use as a human drug, there is an extensive body of toxicological data available on streptomycin. Thus, all toxicological data requirements have been waived.

Physiological and Biochemical Characteristics

Metabolism and Persistence in Plants and Animals: The Agency has determined that plant and animal metabolism data were not needed for the following reasons:

- metabolism of streptomycin in mammals has already been traced in connection with its use in humans;

- residues are non-detectable (< 0.5 ppm) in or on crops when treated according to label use rates and directions;

- large amounts of toxicological data exist;

- most crops are treated at or before transplanting (celery, peppers, potatoes and tomatoes) and pome fruits are treated foliarly but with a 30-day PHI for pears and a 50-day PHI for apples and crabapples

- potential daily exposure to streptomycin as a pesticide is < 0.01% of the daily clinical dosage (1-4 grams/day)

4. TOLERANCE ASSESSMENT

Tolerances have been established for residues of streptomycin in a variety of raw agricultural commodities at 0.25 ppm (40 CFR 180.245). These tolerances are supported by the available data. Tolerances must be proposed for residues of streptomycin in or on beans (succulent and dried), bean vines and bean hay to reflect the registered use on beans.

5. SUMMARY OF REGULATORY POSITIONS

- At this time, none of the risk criteria prescribing a Special Review have been met.

- While data gaps are being filled, currently registered manufacturing-use products and end-use products containing streptomycin as the sole active ingredient may be sold, distributed, formulated, and used, subject to the conditions of this Standard. In order to maintain existing registrations, registrants must provide or agree to develop additional data specified in the Data Appendices.

- No significant human dietary exposure is anticipated due to the lack of detectable residues and the long PHIs.

- Because of the large amount of existing human data, all toxicology requirements have been waived by the Agency for this Standard.

- The Agency is deferring to the Food and Drug Administration on the issue of development of streptomycin resistant microorganisms.

6. LABELING REQUIREMENTS

All streptomycin products must bear appropriate labeling as specified in 40 CFR 156.10. Appendix II of the Standard contains information on label requirements.

In order to remain in compliance with FIFRA, no pesticide product containing streptomycin may be released for shipment by the registrant after 12 months from receipt of the Guidance Document, unless the product bears amended labeling which complies with the specifications in the Standard.

In order to remain in compliance with FIFRA, no pesticide product containing streptomycin may be distributed, sold, offered for sale, held for sale, shipped, delivered for shipment, or received and (having been so received) delivered or offered to be delivered by any person after 24 months from receipt of the Guidance Document, unless the product bears amended labeling which complies with the specifications of the Standard.

Under the Precautionary Statements section of the label, the following statements must appear:

May cause allergic skin reactions. Do not breathe dust or spray mist. Wear dust mask and rubber gloves. Wash thoroughly after handling. This material is not to be used for medical, veterinary, or human purposes.

Do not apply this product in a way that will contact unprotected workers, either directly or through drift. Only protected handlers may be in the area during application. Do not enter or allow entry into treated areas until (sprays have dried/dusts have settled/vapors have dispersed, as applicable) to perform hand labor tasks.

<u>Decontamination</u>

If the pesticide comes in contact with skin, wash off with soap and water. Always wash hands, face, and arms with soap and water before smoking, eating, drinking or toileting. Before removing gloves, wash them with soap and water.

7. SUMMARY OF DATA GAPS

Product Chemistry
(must be resubmitted)

Environmental Fate

Hydrolysis
Photodegradation
Aerobic and Anaerobic Metabolism
Soil Dissipation
Fish Accumulation
Adsorption/Desorption

Ecological Effects

Avian Acute Oral
Avian Subacute Dietary
Honey Bee Acute LD_{50}

8. <u>CONTACT PERSON AT EPA</u>

Lois A. Rossi
Product manager (21)
Fungicides-Herbicides Branch
Registration Division (TS-767C)
Office of Pesticide Programs
Environmental Protection Agency
401 M St., SW
Washington, D.C. 20460

Office location and phone number:

Room 227, Crystal Mall #2
1921 Jefferson Davis Highway
Arlington, VA
(703) 557-1900

9. <u>DISCLAIMER</u>: The information in this Pesticide Fact Sheet is a summary only and may not be used to satisfy data requirements for pesticide registration and reregistration. The complete Registration Standard for the pesticide may be obtained from the National Technical Information Service.

Contact the Product Manager listed above for further information.

STRYCHNINE ALKALOID AND SULFATE

Reason for Issuance: Suspension and Rescission of Sales
Date Issued: October 1988
Fact Sheet Number: 175

1. DESCRIPTION OF CHEMICAL

> Generic Name: Strychnine Alkaloid $C_{21}H_{22}N_2O_2$
> (chemical)   Strychnine Sulfate $(C_{21}H_{22}N_2O_2)H_2SO_4 + 5H_2O$
>
> Common Name: Strychnine
>
> Trade and
> Other Names: Nux Vomica
>
> EPA Shaughnessy Code: Strychnine Alkaloid 076901-8
>   Strychnine Sulfate 076902-6
>
> Chemical Abstracts Service (CAS) Number:
>
> > Strychnine Alkaloid 57-24-9
> > Strychnine Sulfate 60-41-3
>
> Pesticide Type: Vertebrate pesticide.
>
> U.S. and Foreign Producers: All manufacturers import the technical material. Importers include H. Interdonati, Inc.; H. R. Harkins, Inc. and Noris Chemical Co. There are approximately 100 registrants of end-use products. The California Department of Food and Agriculture and California counties are registrants of 37 end use products. The U.S. Department of Agriculture has seven strychnine registrations: 6 end use products and a technical product which is a repackage of an importer's product.

2. USE PATTERNS AND FORMULATIONS:

There are 383 products; 194 are registered for use above ground and 189 are registered for use below ground. The approximately 200 use sites include rangelands, pastures, many crops, forests, and below ground application for pocket gophers and moles.

Formulation Types: Strychnine is usually formulated in grain baits at 0.2% to 2.63% but is also incorporated into a salt block at 5.79%.

3. AGENCY ACTION:

EPA is taking a number of regulatory actions involving strychnine registrations and applications for registrations, since data required to support the registration of strychnine products has not been submitted in accordance with Agency requirements. The Agency is issuing:

1.) A notification of rescission of authority to sell or distribute intrastate products for (CDFA and numerous California counties) which a complete application for Federal registration was not submitted prior to July 31, 1988 in accordance with 40 CFR 152.230. Further, regarding all strychnine applications which are not associated with intrastate products, registration will be withheld pending the receipt and acceptance of required data.

2.) A notice of intent to suspend all registrations which are not in compliance with the submission schedule for data and other information (Colorado, Department of Agriculture; CDFA and numerous California counties; South Dakota Department of Agriculture; Klamath County, Oregon; Nevada Department of Agriculture; USDA; and non-governement registrants for both end-use and technical grade strychnine products.)

These actions against strychnine are independent of the temporary cancellation action required by the April 11, 1988 order of the United States District Court for the District of Minnesota in the case of Defenders of Wildife v. Administrator.

4. BACKGROUND

In March 1972, an Executive Order was issued prohibiting the use of all toxicants, including strychnine, for control of predators on federal lands or in federal programs. In the same year, the Environmental Protection Agency (EPA) cancelled all registrations of strychnine for predator control. In addition, in February 1978, EPA restricted the use of strychnine formulations with concentrations greater than 0.50% to use only by certified applicators. The restriction was based on acute oral toxicity, hazards to nontarget species, and use and accident history.

In 1976, EPA initiated a Special Review (formerly Rebuttable Presumption Against Registration [RPAR]) of the above-ground uses of strychnine. The Special Review criteria that were met or exceeded for above-ground uses are: 1) acute toxicity to mammals and birds; and 2) significant reduction in populations of nontarget organisms and fatalities to members of endangered species.

The Special Review was concluded in October 1983, with publication of a Notice of Intent to Cancel (48 FR 48522). The Notice

allowed continued registration of strychnine for certain above-
ground uses with specific label modifications. The Notice also
required full cancellation of the uses for the control of prairie
dogs, deer mice, meadow mice, chipmunks, marmots/woodchucks on
rangeland/ pastures and cropland; all rodents and small mammals
(except ground squirrels, marmots/woodchucks) around rock piles
and lava outcrops, jackrabbits around airports, and porcupines on
nonagricultural sites.

The Agency also required label modifications for uses involving ground squirrels, jackrabbits, kangaroo rats, and cotton
rats on rangeland/pastures and cropland; ground squirrels, marmots/
woodchucks, jackrabbits, and porcupines on nonagricultural sites;
and birds on cropland and nonagricultural sites. An administrative
hearing was convened for the prairie dog, ground squirrel and
meadow mouse uses.

In March 1987, the Agency published a Federal Register notice
that withdrew the 1983 Notice of Intent to Cancel product registrations for control of prairie dogs, meadow mice, and ground
squirrels and allowed these uses to continue if certain protective
measures were implemented. This action reflected the terms of the
settlement of the administrative hearing. The Agency was sued in
1986 in the U.S. District Court in Minnesota regarding the terms
of the settlement agreement by the Defenders of Wildlife and the
Sierra Club, who alleged that continued use of strychnine for
above-ground use would result in takings of protected wildlife
under the Endangered Species Act, the Migratory Bird Treaty Act,
and the Bald and Golden Eagle Protection Act. The court ordered
EPA to temporarily cancel all registrations of strychnine for
above-ground use to protect nontarget species. EPA will issue a
Federal Register notice, currently scheduled for mid-September to
implement the court's decision.

The Special Review also concluded that efficacy/safety data
for ground squirrels should be developed in order to determine
the lowest efficacious dosage. These data were required through
a Data Call-In (DCI) Notice which was issued in 1984. The Agency's
current action with regard to the withdrawal/suspension of strychnine registrations relates to a total of three DCI notices which
have been issued:

1. The August 1984 Notice requiring ground squirrel efficacy/
   safety data pursuant to the Special Review;

2. An October 1986 Notice requiring the submission of residue
   chemistry, toxicology, environmental fate, environmental
   safety and efficacy data for all strychnine products, as

well as draft revised labels and data compensation forms; and

3. A December 1987, DCI Notice requiring the submission of four additional environmental fate studies.

In October 1987, the Agency met with registrants and interested user groups to explain, among other things, why data were required and how to generate the data.

All strychnine registrants were notified by a December 1987 letter that the Agency would extend the data due dates in the DCIs if the registrants submitted the following items within 30 days: 1) a commitment to fulfill the data requirements contained in the three DCI notices; 2) legible draft revised labels, with the use sites clearly defined; and 3) a Confidential Statement of Formula (EPA Form 8570-4). In addition, schedules were established for progress reports on long-term studies.

The Agency is taking action against all strychnine products because of the lack of progress toward completing the long-term data requirements required in the DCIs, unacceptable qualifications to the commitments to satisfy data requirements, and failure to submit administrative forms.

The Agency is also taking action with regard to the registrations of compound 1080, which is a vertebrate pesticide used principally to control field rodents. The Agency is denying intrastate product applications for registration, cancelling the one technical manufacturer's registration, and suspending the "special local need" registration. A data call-in requiring minimal product chemistry data to support the registtation.

These actions are being taken because of the failure of registrants to satisfactorily respond to DCI requirements concerning compound 1080.

STRYCHNINE UPDATE

Reason for Issuance: Temporary Cancellation Above Ground Uses
Date Issued: September 28, 1988
Fact Sheet Number: 178

1. DESCRIPTION OF CHEMICAL

   Generic Name: Strychnine Alkaloid $C_{21}H_{22}N_2O_2$
   (chemical)   Strychnine Sulfate $C_{21}H_{22}N_2O_2)H_2SO_4+SH_2O$

   Common Name: Strychnine

   Trade and
   Other Names: Nux Vomica

   EPA Shaughnessy Code: Strychnine Alkaloid 076901-8
                           Strychnine Sulfate 076902-6

   Chemical Abstracts Service (CAS) Number:

        Strychnine Alkaloid 57-24-0

   Pesticide Type: Vertebrate pesticide.

   U.S. and Foreign Producers: All manufacturers import the technical
grade material.   Importers include H. Interdonati, Inc.; H. R. Harkins,
Inc.; Noris Chemical Co.; and the U.S. Department of Agriculture (USDA).
There are approximately 100 registrants of end-use products (37 regis-
trants include the California Department of Food and Agriculture, and
California counties).

2. USE PATTERNS AND FORMULATIONS:

   There are 383 products; 194 are registered for use above ground and
   189 are registered for use below ground.   The approximately 200 use
   sites include rangelands, pastures, many crops, forests, and below
   ground application for pocket gophers and moles.

   Formulation Types: Strychnine is usually formulated in grain baits
   at 0.2% to 0.5% but is also incorporated into a salt block at 5.79%.

3. AGENCY ACTION:

On August 23, 1988, the Eighth Circuit Court of Appeals denied
EPA's motion for a stay of an April 11, 1988, order of the U.S.
District Court for the District of Minnesota requiring EPA to
temporarily cancel all strychnine registrations for above-ground
use to protect non-target species. Accordingly, the Agency is
issuing a Federal Register (FR) notice to implement the court's
order. Since temporary cancellation is not a remedy available
under FIFRA, the FR notice identifies the court as the authority
for the temporary cancellation action. The District Court has
reviewed the FR Notice and has no objection to the issuance of
the notice. The FR notice was signed on September 28, 1988 by the
Aministrator and it is to be issued in the Federal Register on
Wednesday, October 5, 1988.

4. BACKGROUND

In 1976, EPA initiated a Special Review [formerly Rebuttable
Presumption Against Registration (RPAR)] of the above-ground uses
of the pesticide strychnine in which the risks and benefits of
strychnine use were examined in detail. This review process was
concluded on October 19, 1983, with the publication of a Notice
of Intent to Cancel. That Notice allowed for continued registration
of certain above-ground uses of strychnine with certain label modifications and required full cancellation of other uses. The primary
concern of the Agency in making its risk/benefit determinations
was risk to non-target, endangered, and threatened species. Prior
to issuance of the Notice, EPA had consulted with the Fish and
Wildlife Service under section 7 of the Endangered Species Act.

An administrative hearing was requested for the uses to control
prairie dogs, ground squirrels and meadow mice. The remaining uses
addressed in the Notice of Intent to Cancel were either amended to
comply with the terms of the Notice or were cancelled by operation
of law. The cancellations did not become effective for the three
uses involved in the hearing.

After the administrative hearing began, settlement discussions
occurred. In the summer of 1986, a settlement was reached allowing
for the continued (with certain label modifications to protect
non-target species) registration of strychnine for above-ground use
to control prairie dogs, ground squirrels and meadow mice. All
parties agreed to the settlement except the Defenders of Wildlife
and the Sierra Club. The settlement was based on a new (1984)
biological opinion issued by the Fish and Wildlife Service in which
it was determined that aboveground strychnine use could continue
without jeopardizing the blackfooted ferret if certain protective
measures were taken. The blackfooted ferret was the species of most
concern in regard to strychnine use.

In August 1986, the Defenders of Wildlife and the Sierra Club filed suit against EPA in the United States District Court for the District of Minnesota. In the suit they alleged that continued registration of strychnine for above-ground use would result in takings of protected wildlife under the Endangered Species Act, the Migratory Bird Treaty Act, and the Bald and Golden Eagle Protection Act. In January 1987, Defenders of Wildlife provided certain information to the Agency regarding deaths of non-target wildlife from strychnine use (the "non-target kill book"). In March 1987, EPA issued a new Notice of Intent to Cancel which reflected the terms of the settlement agreement and allowed the continued use of strychnine to control prairie dogs, ground squirrels, and meadow mice under specific terms and conditions. The Agency informed the Defenders of Wildlife that it would review the information regarding non-target deaths, but did not want to hold up the issuance of the March 1987 Notice because it contained significant measures to protect non-target species.

The "non-target kill book" information was referred to the Fish and Wildlife Service and EPA requested a new biological opinion regarding the above-ground use of strychnine. In May and June, 1988 the Fish and Wildlife Service issued new biological opinions on above-ground use of strychnine. These opinions currently are being reviewed by EPA as part of a reassessment of the risks and benefits posed by above-ground use of strychnine.

In 1984, 1986 and 1987 the Agency sent Data Call-In (DCI) notices to the registrants of strychnine. The information required to be submitted to the Agency includes residue chemistry, toxicology, environmental fate, environmental safety, and efficacy data. The deadlines were extended in December, 1987. The Agency is reviewing progress made toward fulfilling the DCI requirements to determine what future regulatory action may be appropriate.

SULFLURAMID

Reason for Issuance: Conditional Registration—New Chemical
Date Issued: March 23, 1989
Fact Sheet Number: 205

1. DESCRIPTION OF THE CHEMICAL

    Generic Name: N-Ethyl Perfluorooctanesulfonamide

    Common Name: Sulfluramid

    Trade Name: GX-071

    EPA Shaughnessy Code: 128992
    (OPP Chemical Codes)

    Chemical Abstracts Service (CAS) Number: 4151-50-2

    Year of Initial Registration: 1989

    Pesticide Type: Insecticide

    Chemical Family: Fluorinated Sulfonamide

    Producer: Griffin Corporation of Valdosta, GA

2. USE PATTERNS AND FORMULATIONS

Target Pests: Roaches, Ants

Registered Uses: Indoor roach trap, indoor ant trap, technical
                  manufacturing-use product.

Methods of Application: Child resistant bait station.

Formulations: Technical, bait station.

3. SCIENCE FINDINGS

Chemical Characteristics:

Physical State: Solid Crystals

Color: White

Odor: None

Melting Point: 96 degrees C

Vapor Pressure: 4.3×10^{-7} mm Hg at 25 degrees C
9.99×10^{-7} mm HG at 34 degrees C

Density: 0.1485 g/ml at 25 degrees C

Storage Stability: Stable for a minimum for 1 year when stored in a polyethylene bag under ambient conditions.

Octanol/Water Partition Coefficient: $P_{ow} = 3.10$

Flammability: NR

Solubility: The Technical Grade of the Active Ingredient does not dissolve or disperse in water.

Toxicology Characteristics:

Technical Formulation:

GX-071 Technical
EPA File Symbol 1812-327

Acute Oral Toxicity-Rat : $LD_{50} = 500$ mg/kg
Toxicity Category IV

Single Dose Dermal Toxicity-Rabbits : $LD_{50} > 2000$ mg/kg
Toxicity Category III

Primary Eye Irritation-Rabbits
Not considered an eye irritant.

Primary Dermal Irritation-Rabbits
    Mild skin irritant.   Score of 0.13.

Delayed Contact Sensitivity - Guinea Pigs
    Data gap.

Salmonella/Mammalian Activation Gene Mutation Assay
    Negative up to the limits of solubility (624 ug/plate and above)
    and   dosing (tested up to 10,000 ug/plate)

Sister Chromatid Exchange in Chinese Hamster Ovary Cells
    Negative for inducing sister chromatid exchange in vitro.

Rat Primary Hepatocyte Unscheduled DNA Synthesis Assay
    Data gap.

End-Use Formulations

Raid Roach Controller II
EPA Registration Number 1812-329
Acute Oral Toxicity Study - Rat
  No deaths occured following oral exposure to 5 grams/kg.

Raid Ant Controller II
EPA Registration Number 1812-330
Acute Oral Toxicity Study - Rat
  No deaths occured following oral exposure to 5 grams/kg.

Ecological Characteristics:

Avian Dietary LC_{50}
    Mallard Duck --> LC_{50} = 165 ppm
    Bobwhite Quail --> LC_{50} = 460 ppm

Avian Oral LD_{50}
    Bobwhite Quail --> LC_{50} = 473 mg/kg

Freshwater Fish LC_{50}
   Rainbow Trout --> Normal $LC_{50} > 10$ ppm
                     Measured $LC_{50} > 2$ ppm

Acute LC_{50} Freshwater Invertebrate
   Daphnia pulex --> Normal $LC_{50} > 10$ ppm
                     Measured $LC_{50} > 0.21$ ppm

4.   Summary of Regulatory Position and Rationale

   - The Agency has determined that it should allow the conditional
     registration of sulfluramid for indoor nonfood use in bait
     stations to control ants and roaches. Adequate data are
     available to assess the acute toxicological effects of
     sulfluramid to humans.

   - Data Gaps: Primary Hepatocyte Unscheduled DNA Synthesis
                 Dermal Sensitization

   - This registration is conditionally approved with an expiration
     date of September 15, 1990.

7.   CONTACT PERSON AT EPA

   Phillip O. Hutton
   Product Manager (17)
   Insecticide Rodenticide Branch
   Registration Division (H7505C)
   Office of Pesticide Programs
   Environmental Protection Agency
   401 M Street, S. W.
   Washington, D. C. 20460

   Office location and telephone number:

   Room 207, Crystal Mall #2
   1921 Jefferson Davis Highway
   Arlington, VA 22202
   (703) 557-2690

DISCLAIMER: The information in this Pesticide Fact Sheet
is a summary only and is not to be used to satisfy data
requirements for pesticide registration and reregistration.
Contact the Product Manager listed above for further
information.

SULFOTEPP

Reason for Issuance: Registration Standard
Date Issued: September 30, 1988
Fact Sheet Number: 185

1. DESCRIPTION OF CHEMICAL

Chemical Name: O,O,O',O'-tetraethyl dithiopyrophosphate (International Union for Pure and Applied Chemistry)

ANSI Common Name: Not applicable

Other Common Names: Sulfotep (British Standards Institution and International Organization for Standardization); tetraethyl thiodiphosphate (9th Collective Index); tetraethyl thiopyrophosphate (8th Collective Index); thiodiphosphoric acid tetraethyl ester; sulfotepp (Entomological Society of America)

Principal Trade Names: Bladafume; Dithio; Dithione; Plantfume

EPA Pesticide Chemical (Shaughnessy) Number: 079501

Chemical Abstracts Service (CAS) Number: 3689-24-5

Year of Initial Registration: 1951

Pesticide Type(s): Insecticide/acaricide

Chemical Family: Organophosphate

U.S. Registrants: Centerchem, Inc.; Fuller Systems, Inc.; Plant Products Corp.

2. USE PATTERNS AND FORMULATIONS

Registered uses: Greenhouse ornamentals (non-food crop)

Predominant uses: Azaleas, carnations, chrysanthemums, geraniums, roses, snapdragons

Pests controlled: Aphids, spider mites, whiteflies, mealybugs, scales, thrips

Types of Formulations: Ready-to-use liquid; impregnated materials

587

Types and Method of Application: Fogging with liquid spray, smoking
with impregnated materials

Application Rates: 1.75 fl. oz. 4.5%/5,000 ft^3 (liquid fog);
1.75 oz. 15%/5,000 ft^3 (smoke generator)

3. SCIENCE FINDINGS

Chemical Characteristics

Physical state: liquid

Color: pale yellow

Odor: Unknown

Molecular Weight & Formula: 322.3, $C_8H_{20}O_5P_2S_2$

Boiling Point: 136–139°C at 2 mm Hg

Vapor Pressure: 1.7 x 10^{-4} mm Hg or 22.6 mPa at 20°C

Specific Gravity: 1.196 d25/4°C, where 25°C refers to temperature at
which density (d) of sulfotepp measured and 4°C
temperature at which density of H_2O measured

Solubility in various solvents: 25 mg/l water at room temperature;
completely miscible with chloro-
methane and most organic solvents

Toxicology Characteristics

Acute Oral: Data gap. An acute oral LD$_{50}$ study in the rat is required.

Acute Dermal: Data gap. An acute dermal LD$_{50}$ study in the rabbit
is required.

Acute Inhalation: Data gap. An acute inhalation LC$_{50}$ study in the
rat is required.

Primary Dermal Irritation: Data gap. A primary dermal study in
the rabbit is required.

Primary Eye Irritation: Data gap. A primary eye study in the
rabbit is required.

Dermal Sensitization: Data gap. A study in the guinea pig is
required.

Delayed Neurotoxicity: Data gap. An acute study in the hen is
required.

Major Routes of Exposure: Not well understood.   Inhalation of fumes and
dermal exposure to treated ornamentals.

Subchronic Toxicity: Data gaps.   A 21-day dermal study in the
rabbit is required.   A 90-day inhalation study
is required.
Note: The 90-day feeding studies in rodent and
non-rodent are not required for the registered
use patterns.

Oncogenicity: Not required for the registered use patterns.

Chronic feeding: Not required for the registered use patterns.

Metabolism: Not required for the registered use patterns.

Reproduction: Not required for the registered use patterns.

Teratogenicity: Data gap.   A study in either a rodent or non-rodent
is required.

Mutagenicity: Data gaps.   The gene mutation (Ames Test) and the
chromosomal aberration studies are required, as is
the test for other mechanisms of mutagenicity.

Physiological and Biochemical Characteristics

Mechanism of Pesticidal Action: Neurotoxin

Metabolism and Persistence in Plants and Animals: Metabolism
not understood; short residual period on plant foliage.

Environmental Characteristics: There are no available environmental
fate studies on sulfotepp.   Therefore, groundwater contamination
potential cannot be assessed at this time.   Because of toxicological concerns on the acute hazards (see above), an interim
11-hour minimum reentry interval and 2 hours of ventilation are
being imposed for the uses of sulfotepp until adequate data have
been submitted and evaluated.   The following list summarizes the
environmental fate data requirements for sulfotepp.

Degradation Studies: Data gaps.   Laboratory studies on hydrolysis and photodegradation in air are
required.

Metabolism Studies: Data gap.   A laboratory study on aerobic
soil metabolism is required.

Mobility Studies: Data gaps.   Laboratory studies on adsorption/-
desorption (batch equilibrium study preferred)
and volatility are required.   A field volatility study is reserved until the results of
laboratory studies are known.

Reentry Protection: Data gap. An inhalation exposure study is
required. The registrant is required to propose acceptable
reentry labeling based upon airborne residue levels measured after observing the proposed label conditions, on estimated human exposure to those residues, and on toxicity
of sulfotepp. Because of the highly acutely toxic nature
of sulfotepp, <u>no actual human inhalation exposure monitoring
data should be gathered</u>.

Ecological Characteristics: No data are available for any terrestrial species. The available data indicate that sulfotepp is
"highly toxic" to the rainbow trout and bluegill sunfish. No
data are available on effects on freshwater invertebrates. The
following list summarizes the ecological effects data requirements for sulfotepp.

Avian acute toxicity: Data gap. A single-dose LD_{50} study in
the bobwhite quail is required.

Avian dietary toxicity: Data gap. A subacute dietary LC_{50}
study in the bobwhite quail is required.

Freshwater fish acute toxicity: Rainbow trout: LC_{50} = 1.0 (0.8-
1.3) ppm; bluegill sunfish: LC_{50} = 0.36 (0.27-0.46) ppm.
Available fish studies only partially fulfill the requirements, but may be upgraded if additional data concerning
the studies are available and are submitted.

Marine fish acute toxicity: Not required for the registered use
patterns.

Freshwater invertebrate toxicity: Data gap. An acute LC_{50}
study on aquatic invertebrates is required.

Marine invertebrate toxicity: Not required for the registered
use patterns.

4. TOLERANCE ASSESSMENT

There are no approved tolerances for residues of sulfotepp and no
uses on food crops. Therefore, a tolerance assessment is not required.

5. SUMMARY OF REGULATORY POSITIONS

- Sulfotepp is not a candidate for Special Review at this time.

- Sulfotepp meets the criteria for restricted use classification
because of highly acute inhalation toxicity to humans.

- Groundwater contamination concerns for sulfotepp cannot be assessed until basic environmental fate data requirements are met.

- An interim reentry interval based on standards proposed in Title
40, Code of Federal Regulations, Part 170, Subpart F, Special
Standards for Workers in Greenhouses, Section 66, is being imposed

for the uses of sulfotepp pending submission and evaluation of
data on exposure to airborne residues.

- Protective clothing requirements are being imposed as labeling
amendments for all registered sulfotepp products.

6. LABELING REQUIREMENTS

Statements applicable to all products:

GENERAL WARNINGS AND LIMITATIONS: Sulfotepp is classified as a
RESTRICTED USE PESTICIDE by Title 40, Code of Federal Regulations, Part 162, Section 31, on the basis of its acute inhalation hazard to humans.

Statements for Manufacturing-Use Products:

ENVIRONMENTAL HAZARDS:
This pesticide is toxic to fish. Do not discharge effluent
containing this product directly into lakes, streams, ponds,
estuaries, oceans, wetlands or public waters unless this product is specifically identified and addressed in a NPDES permit.
Do not discharge effluent containing this product to sewer systems without previously notifying the sewage treatment plant
authority. For guidance, contact your State Water Board or
Regional Office of the Environmental Protection Agency.

Statements for End-Use Products:

General Warnings and Limitations:

Worker Protection Statement: WEAR THE FOLLOWING PROTECTIVE
CLOTHING DURING LOADING, APPLICATION, EQUIPMENT REPAIR, EQUIP-
MENT CLEANING, EARLY REENTRY TO TREATED AREAS, AND DISPOSAL
OF THE PESTICIDE. Wear a protective suit of one or two pieces
that covers all parts of the body except the head, hands, and
feet. Wear chemical-resistant gloves and chemical-resistant
shoes, shoe coverings, or boots. Wear goggles and a pesticide
respirator approved by the National Institute for Occupational
Safety and Health (NIOSH) and the Mine Safety and Health Administration (MSHA) at all times during application and early
reentry to treated areas.

Reentry Statement: Reentry after applying is restricted until
one of the following intervals has elapsed:

(1) Two hours of ventilation using fans or other mechanical
ventilation systems.

(2) Four hours of ventilation using vents, windows or other
passive ventilation systems.

(3) Eleven hours with no ventilation, followed by one hour of
mechanical ventilation.

(4) Eleven hours with no ventilation, followed by two hours of passive ventilation.

If necessary to reenter the greenhouse for any reason during the specified intervals after application, protective clothing described in the Worker Protection Statement must be worn. All greenhouses must be posted during the exposure period and until safe to reenter.

Disposal Statement: Pesticide wastes are acutely hazardous. Improper disposal of excess pesticide, spray mixture or rinsate is a violation of Federal Law. If these wastes cannot be disposed of by use according to label instructions, contact your State Pesticide or Environmental Control Agency or the Hazardous Waste representative at the nearest EPA Regional Office for guidance. In addition, interested parties may call the RCRA/Superfund Hotline toll free (1-800-424-9346) for information on Resource Conservation and Recovery Act requirements.

7. SUMMARY OF DATA GAPS

Toxicology   All the acute toxicity studies are required, including the acute delayed neurotoxicity study in the hen. The subchronic feeding studies in the rodent and nonrodent are not required due to lack of oral exposure in the registered use patterns. The 21-day dermal and 90-day inhalation studies are required. The 90-day neurotoxicity study is reserved depending upon the results from the acute delayed neurotoxicity study. A teratology study in one species is required and all the mutagenicity studies are required.

Environmental Fate/Exposure   The laboratory degradation studies on hydrolysis and photodegradation in air are required. The aerobic soil metabolism study is required. The laboratory mobility studies on adsorption/desorption (for which the batch equilibrium study is preferred) and volatility are required. The field volatility study is reserved pending results from the laboratory studies. The reentry protection study on inhalation exposure is required, but there is to be no actual monitoring of human inhalation exposure.

Fish and Wildlife   An avian single-dose acute oral LD_{50} study and a subacute dietary LC_{50} study using the bobwhite quail are required. Additional data are required on the freshwater fish toxicity study already submitted. A study on the acute LC_{50} to aquatic invertebrates is required.

Product Chemistry   All the applicable data on product identity, analysis and certification of product ingredients, and physical and chemical properties are required.

8. CONTACT PERSON AT EPA

William H. Miller
Product Manager 16
Insecticide-Rodenticide Branch
Registration Division (TS-767C)
Office of Pesticide Programs
Environmental Protection Agency
401 M St., SW
Washington, D.C.   20460

Office location and phone number:

Room 211, Crystal Mall #2
1921 Jefferson Davis Highway
Arlington, VA
(703) 557-2600

9. DISCLAIMER: The information in this Pesticide Fact Sheet is a summary only and may not be used to satisfy data requirements for pesticide registration and reregistration. The complete Registration Standard for the pesticide may be obtained from the contact person listed above.

TEFLUTHRIN

Reason for Issuance: Conditional Registration—New Chemical
Date Issued: February 3, 1989
Fact Sheet Number: 190

1. Description of Chemical

    Generic Name: 2,3,5,6-tetrafluoro-4-methylphenyl)methyl-(1a,3a)-(Z)-

    (+)-3-(2-chloro-3,3,3-trifluoro-1-propenyl)-2,2-dimethyl-

    cyclopropane carboxylate

    Common Name: Tefluthrin

    Trade Name: Force®

    Other Proposed Names: N/A

    Code Number: 1ClA 0993

    EPA Shaughnessy Code: 128912

    Chemical Abstracts Service (CAS) Number: 79-538-32-2

    Year of Initial Registration: 1989

    Pesticide Type: Insecticide

    Chemical Family: Pyrethroid

    U.S. and Foreign Producers: ICI Americas, Inc.

2. Use Patterns and Formulations

    Application Sites: Cornseeds

    Types and Methods of Application: Ground Application: In-band
    treatment - incorporated into the top 1 inch of soil during corn planting
    operations.

Application Rates: Applied to corn seeds at a rate of up to 0.163 pounds active ingredient per acre overall and 0.7 lbs ai/A in the band. The product is applied once per season.

Types of Formulations: 1.5% granular and 89% technical.

Limitations:

o   Registration is being approved with an expiration date of July 31, 1993. Tolerances expire July 31, 1994.

o   RESTRICTED USE PESTICIDE. Toxic to fish and aquatic organisms. For retail sale to and use only by Certified Applicators, or persons under their direct supervision, and only for those uses covered by the Certified Applicator's certification.

o   CROP ROTATION RESTRICTION: Do not rotate to crops other than corn.

o   ENDANGERED SPECIES RESTRICTION: For ground application, do not apply this product within 20 yards of water (ponds, streams, or lakes).

3.   Science Findings

Summary Science Statement: Tefluthrin is a new synthetic pyrethroid. Technical Tefluthrin exhibits high mammalian toxicity by the oral, dermal, and inhalation route of exposure. It is not considered to be mutagenic, carcinogenic, nor teratogenic in test animals. It is readily absorbed by mammals, and the majority of the residue is largely excreted in the feces and urine by 48 hours. The results of the acute toxicty on the end-use formulation (Force) indicates the product is of moderate to low toxicity. The end use product is irritating to the eyes and can cause eye injury. Goggles or face shield are required when handling the product.

Sufficient data are available to characterize tefluthrin from an environmental and ecological effects standpoint. The results of acute oral and sub-acute dietary studies indicates that tefluthrin is practically non-toxic or slightly toxic to birds. Avian reproduction data indicates tefluthrin has no adverse effects on reproduction in birds. The results of acute toxicity studies indicate that tefluthrin is extremely toxic to fish and other aquatic organisms. Based upon the high toxicity to aquatic organisms from laboratory tests chronic fish, aquatic invertebrate and exposure data (aquatic residue monitoring study) are being required to assess potential hazards to aquatic organisms in the environment.

Tefluthrin is very water soluble (low mobility, low runoff, and low leaching), highly lipophilic (strongly binds or adsorbs to soil organic matter) and is a very stable and persistent compound. Estimated Environmental Concentration in water is expected to be extremely low from this use but the chemical will accumulate and persist in sediment. Based upon its low mobility it is not expected to leach into groundwater.

Tefluthrin may pose a risk to endangered aquatic species. Pending a formal consultation with the Fish and Wildlife Service to determine use limitations with respect to these species, the product label consists of language which will minimize the risk to endangered species.

Chemical Characteristics:

Physical State: Crystalline solid

Color: Off white

Odor: None

Boiling Point: Decomposes at 295 °C

Melting Point: 44.6 °C (Pure), 39.4 to 43.2 °C (Technical)

Vapor Pressure: 8×10^{-6} KPa at 20 °C

Density: 1.48 g/cm^3 at 25 °C

Storage Stability: 9 months at ambient temperature

Octanol/Water Partition Coefficient: log K_O/w = 6.5 at 20 °C

Flammability: Flashpoint 124 °C

Solubility: Water – 0.02 ppm; Methanol – 263 g/L; Acetone, Toluene,

Ethyl Acetate, Hexane, and Dicloromethane >500 g/L

Toxicology Characteristics:

Technical Formulation:

Acute Oral Toxicity-Rat: LD$_{50}$ = 21.8 mg/kg(males); 34.6 mg/kg (females)

Toxicity category I.

Acute Dermal Toxicity-Rat: LD$_{50}$ = 316.0 mg/kg(males); 177 mg/kg (females)

Toxicity category I.

Acute Inhalation Toxicity-Rat: LC$_{50}$ = 49.1 mg/m^3(males); 37.1 mg/m^3 (females)

Toxicity category I.

Primary Dermal Irritation-Rabbit: Slightly irritating,

Toxicity category IV.

Primary Eye Irritation-Rabbit: (not available).

Acute Delayed Neurotoxicity – Hen: No signs of delayed neurotoxicity.

90–Day Feeding Study – Rat: NOEL = 50 ppm; LEL = 150 ppm;

  Alterations in liver weight, hemoglobin and cholesterol.

90–Day Oral Dosing Study – Dog: NOEL = 0.5 mg/kg; LEL = 1.5 mg/kg;

  Increased triglycerides and AST (=SGOT).

2–Year Chronic/Oncogenicity Study – Mouse: NOEL = 3.4 mg/kg;

  LEL = 13.5 mg/kg; Hemangiomatous uteri and liver necrosis.

  Not oncogenic at 54.4 mg/kg (MTD).

1–Year Oral Dosing Study – Dog: NOEL = 0.5 mg/kg; LEL = 2 mg/kg;

  Ataxia in both sexes.

Teratogenicity – Rat: Maternal NOEL = 1 mg/kg; LEL = 3 mg/kg;

  Decreased body weight at 3 mg/kg and pyrethroid toxicity at 5 mg/kg.

  Developmental NOEL = 3 mg/kg; LEL = 5 mg/kg; Decreased ossifications.

Teratogenicity – Rabbit: Maternal NOEL <3 mg/kg (LDT).

  Pyrethroid signs. Developmental NOEL >12 mg/kg (HDT).

Multigeneration Reproduction Study – Rat: Parental NOEL = 50 ppm;

  LEL = 250 ppm; Body weight effects. Reproductive NOEL = 50 ppm;

  LEL = 250 ppm; Pup weight effects.

Mutagenicity:

  Reverse mutation (Salmonella [in vitro]): Not mutagenic at

  5000 ug/plate (precip) in S. typhimurium strains with/without S-9.

  TK Locus in L5178Y Mouse Lymphoma Cells (in vitro): Not mutagenic

  up to 4000 ug/mL (cytotox).

Bone Marrow Cytogenics (Rat [in vivo]): No chromosome damage
up to 12 mg/kg (cytotox).

Micronucleus (Mouse [in vivo]): No micronuclei at 50 mg/kg
(single IP dose at MTD).

Dominant Lethal (Mouse [in vivo]): No dominant lethals at 10
mg/kg (MTD).

Unscheduled DNA Synthesis (Rat Hepatocytes): Absence of unscheduled
DNA synthesis up to 10^{-2} M (cytotox).

End Use Formulation

The stated results for the following acute studies are for the
1.67% formulation: oral (rat), dermal (rat), inhalation (rat), primary
dermal irritation (rabbit) and primary eye irritation (rabbit).

Acute Oral Toxicity – Rat: LD_{50} >2940 mg/kg (males),
Approx. = 1550 mg/kg (females). Toxicity Category III.

Acute Dermal Toxicity – Rat: LD_{50} >2000 mg/kg (males and females).
Toxicity Category IV.

Acute Inhalation Toxicity – Rat: 4-hour LC_{50} = 2304 mg/m^3 (females),
>3929 mg/m^3 (males); Toxicity Category III.

Primary Dermal Irritation – Rabbit: Slightly irritating;
Toxicity Category IV.

Primary Eye Irritation – Rabbit: Unwashed eyes showed corneal opacity,
chemosis, and conjunctival discharge clearing in 4 days;
Toxicity Category II.

Dermal Sensitization – Guinea Pig: Not a sensitizer.

Physiological and Biochemical Characteristics:

Foliar Absorption: N/A

Translocation: Not translocated.

Mechanism of Pesticidal Action: Neurotoxicity characteristic of
pyrethroid insecticides - contact action.

Environmental Characteristics:

The environmental fate data indicate that tefluthrin and its soil-aged
residues have very low vertical mobility.   Tefluthrin has extremely
high Kd adsorption coefficient values and 30-day aged residues in
loamy sand and sandy loam soils did not move significantly
beyond the top 5 inches in 35-inch soil columns.   When leached with
the equivalent of 66 cm of rainfall and only 0.3 percent of the
applied material was found in the leachate.   Based on the mobility
data, tefluthrin and its degradates are not likely to leach and
contaminate ground water.   However, Tefluthrin is stable in water
at pH 5 and 7 and stable with respect to degradation under sunlight
in water and on soil with isomerization to its trans isomer being
the major transformation.   Also, field dissipation data indicate
that tefluthrin is quite persistent with a half-life of 92 to 124
days and is even more stable under anaerobic soil conditions.
Therefore, it cannot be excluded that under year by year usage
and over a long period of time leaching might be observed.   Field
dissipation studies conducted for periods of 1 year did not indicate
movement below the 10 cm depth under actual use conditions.   In summary,
although groundwater contamination is not likely to occur, it cannot be

totally excluded with continuous year by year usage. Fish accumulation
data indicate that tefluthrin and its degradates/metabolites are not
likely to accumulate significantly in fish. The confined rotation
crop studies showed accumulation of ^{14}C-tefluthrin residues occurred in
all rotated crops up to 410 days post-treatment with up to 0.75 lb ai/acre
(not confirmed).

Reentry and spray drift data are not required since tefluthrin is applied
at planting times as a band treatment and then covered with soil.

Ecological Characteristics:

Avian Oral Toxicity: Mallard Duck LD_{50} = 4190 mg/kg

Avian Dietary Toxicity: Bobwhite Quail LC_{50} = 15,000 ppm
 (8 days)
                         Mallard Duck LC_{50} = 2317 ppm

Avian Reproduction: Dietary administration at 5 ppm and 25 ppm for 20 weeks

                     had no adverse effects on reproduction in birds.

                     (NOEL – 25 ppm).

Freshwater Fish Acute Toxicity: Bluegill LC_{50} = 130 parts per trillion (ppt)
(96-hr LC_{50} – tech. grade)
                                 Rainbow Trout = 60 ppt

Freshwater Fish Acute Toxicity: Bluegill LC_{50} = 120 ppt
(96-hr LC_{50} – end-use product)
                                 Rainbow Trout = 127 ppt

Freshwater Invertebrate Acute Toxicity: Daphnia = 70 ppt
(48-hr LC_{50} – tech. grade)

Freshwater Invertebrate Acute Toxicity: Daphnia = 185 ppt
(48-hr LC_{50} – end-use product)

Marine Fish & Invertebrate Toxicity: Sheepshead Minnow = 130 ppt
(96-hr LC_{50} tech. grade)
                                       Mysid Shrimp = 53 ppt

                                       Pacific Oyster = >1 ppm

Tolerance Assessment:

Tolerances have been established for residues of tefluthrin in/or on the following agricultural commodities (40 CFR 180.440). These tolerances are due to expire July 31, 1994.

Commodities	Part Per Million
Corn, grain, field and pop	0.06
Corn, forage and fodder, field and pop	0.06

The provisional acceptable daily intake (PADI), based on a NOEL of 0.75 mg/kg/day from a multigeneration reproduction study and a safety factor of 1000, is 0.00075 mg/kg body weight/day. The theoretical maximum residue contribution from the proposed tolerances is 0.00001 mg/kg body weight/day. This is equivalent to about 1.4 percent of the PADI.

Reported Pesticide Incidents: None

4.   Summary of Regulatory Position and Rationale

· The Agency has determined that it should allow the conditional registration of tefluthrin for agricultural use to control insects in/on corn. Adequate data are available to assess the acute and chronic toxicological effects of tefluthrin to humans.

· Since certain long-term fish, aquatic invertebrate, aquatic exposure, and rotational crop data are missing and required, the registration is being conditionally approved with a expiration date of July 31, 1993, which coincides with the date for submission of the data required to satisfy the remaining data gaps listed below. Similarly, the tolerances have been established with an expiration date of July 31, 1994.

* In view of the high toxicity of tefluthrin to aquatic organisms (invertebrates and fish) and the potential hazard associated with exposure to this product, the Agency is concerned about exposure which may result from improper application or use and so is restricting use of this pesticide.

* The Agency has determined that endangered species labeling restrictions are necessary to protect endangered species and is requiring specific limitations on use of this product to prevent or mitigate exposure.

5.   Summary of Data Gaps

	Guidelines	
Name of Study	Reference No.	Date Due
21-Day Dermal	82-2	April 1989
21-Day Feeding	82-2	April 1989
Aquatic Invertebrate Life-Cycle Test	72-4	August 1989
Aquatic Residue Level Monitoring	70-1	March 1991
Fish Life Cycle	72-5	August 1990
Rotational Crop -Field	165-2	March 1993

6. <u>Contact Person at EPA</u>

George T. LaRocca

Product Manager (15)

Insecticide-Rodenticide Branch

Registration Division (TS-767C)

Office of Pesticide Programs

Environmental Protection Agency

401 M Street SW.

Washington, DC 20460

Office location and telephone number:

Room 211, Crystal Mall #2

1921 Jefferson Davis Highway

Arlington, VA 22202

Phone: (703) 557-2400

DISCLAIMER: The information presented in this Pesticide Fact Sheet is for informational purposes only and may not be used to fulfill data requirements for pesticide registration and reregistration.

TERBACIL

Reason for Issuance: Registration Standard (SRR)
Date Issued: August 1989
Fact Sheet Number: 206

1. DESCRIPTION OF CHEMICAL

Generic Name: 3-tert-butyl-5-chloro-6-methyluracil

Common Name: Terbacil

Trade and Other Names: 5-chloro-3-(1,1-dimethylethyl)-6-methyl-2,4(1H,3H)-pyrimidinedione, Sinbar[R], DuPont Herbicide 732, and Geonter.

EPA Shaughnessy Codes: 012701

Chemical Abstracts Service (CAS) Number: 5902-51-2

Year of Initial Registration: 1966

Pesticide Type: Herbicide

Chemical Family: Uracils

U.S. Producers: E.I. duPont de Nemours Company, Inc.

2. USE PATTERNS AND FORMULATIONS

Application sites: Terrestrial food crops.

Types and Methods of Application: Tractor mounted spray boom to soil surfaces and small emerged weeds. Aerial application.

Application Rates: 0.4 to 8.0 lb active ingredient per acre

Annual Usage: 290,000 - 610,000 pounds

Types of Formulation: Wettable powder

3.   SCIENCE FINDINGS

Chemical Characteristics of the Technical Material

Physical State: Crystalline solid

Color: White

Odor: None

Molecular Weight and Formula: 216.7

Boiling Point: Terbacil is a solid at room temperature

Vapor Pressure: 4.7×10^{-7} mm Hg at $29.5°C$

Density: 1.34

Solubility in various solvents: Water, xylene, and dimethylformamide

Toxicology Characteristics

-Acute Oral: 1082 mg/kg (rat) Toxicity Category III
-Acute Dermal: >5000 mg/kg (rabbit) Toxicity Category III
-Primary Dermal Irritation: No irritation demonstrated
-Primary Eye Irritation: slight eye irritation (conjunctiva)
-Dermal Sensitization: Not a dermal sensitizer
-Acute Inhalation: >4.4 mg/liter/4 hours (Toxicity Category III)
-Subchronic dermal (21-day): > 5000 mg/kg. No toxic signs.
-Subchronic Oral (90-day): No Observed Effect Level (NOEL) of
 5 mg/kg/day.
-Chronic Oral dog (2-year): NOEL of 1.25 mg/kg/day.
-Oncogenicity: Data Gap.
-Teratogenicity (rat): Not teratogenic. Maternal NOEL was 62.5
 mg/kg/day. Teratogenicity NOEL > 250 mg/kg/day.
-Teratogenicity (rabbit): Not teratogenic. Maternal and embryo-
 toxicity NOEL was 200 mg/kg/day. Teratogenicity NOEL was 600
 600 mg/kg/day (highest dose tested).
-Reproduction: No observed effects with a NOEL of 62.5 mg/kg/day
 (highest dose tested).
-Mutagenicity: Non-mutagenic in assays tested.
-Metabolism: Data Gap.

-Major route of exposure: Dermal

Physiological and Biochemical Characteristics

-Mechanism of Pesticidal Action: Stops photosynthesis
-Metabolism and Persistence in Plants and Animals: The
 available data are not adequate to assess the nature of
 terbacil in plants or in animals.

Environmental Characteristics

-Terbacil is stable to hydrolysis and photodegrades slowly in
 water.
-It has a potential to contaminate ground water particularly in
 areas with sandy soil surfaces.
-It leaches slower in fine textured soils and also in soils
 having higher organic contents.
-It does not accumulate to significant levels in bluegill
 sunfish.
-Residues resulting from multiple applications of terbacil
 persisted for 2 years following the final application.
-Preliminary data indicate that terbacil is extremely persistent
 with half-lives of 520 days aerobically and 178 days
 anaerobically.

Ecological Characteristics

-Avian acute toxicity:
     >2250 mg/kg (Quail). Practically Non-toxic.
-Avian dietary toxicity:
     >5000 ppm (Mallard duck,). Practically Non-toxic.
-Freshwater fish acute toxicity:
     102.9 ppm (Bluegill sunfish)   Practically Non toxic.
     46.2 ppm (Rainbow trout)   Slightly Toxic
-Marine fish acute toxicity: Data Gap
-Freshwater invertebrate toxicity:
     65 ppm (Daphnia). Slightly toxic.
-Marine invertebrate toxicity:
     >4.9 ppm (Oyster). Moderately toxic
     49 ppm (Shrimp).   Slightly toxic

TOLERANCE ASSESSMENT

-Tolerances have been established for residues of Terbacil in a
 variety of raw agricultural commodities (Refer to 40 CFR
 180.209 for listing of tolerances). Terbacil's tolerances have
 been reassessed using the Tolerance Assessment System
 (TAS). The TAS chronic exposure analysis estimates average
 daily exposure for the overall U.S. population and each of
 the 22 populations subgroups and compares these estimates
 to the acceptable daily intake (ADI) calculated for
 terbacil.

The Theoretical Maximum Residue Contribution (TMRC) for the overall U.S. population is estimated to be 0.001594 mg/kg/day, which occupies approximately 12% of the ADI. The two most highly exposed subgroups are non-nursing infants, less than 1 year old (TMRC= 0.008122 mg/kg/day or 62% of the ADI), and children , 1- 6 years of age (TMRC= 0.004361 mg/kg/day or 34% of the ADI).

4. Summary of Regulatory Positions

This review of terbacil is the second intensive evaluation of the compound. In its original Registration Standard, issued in 1982, the Agency summarized the available data supporting the registration of terbacil and concluded that additional data were needed to fully evaluate the pesticide.

The Agency has since received and reviewed the data and has revised its scientific and regulatory conclusions relative to these data.

o Terbacil will not be placed in Special Review at this time.
o The Agency will impose a ground water contamination advisory statement to reduce point source contamination.
o The Agency is requiring the following testing on all terbacil end-use products: acute oral, acute dermal, primary eye, primary skin, dermal sensitization, and acute inhalation if appropriate.
o The Agency will not establish a reentry interval for terbacil beyond the minimum reentry interval (sprays have dried, dusts have settled and vapors have dispersed).
o The Agency will require updated worker safety and protective equipment statements for end-use products containing terbacil.
o The Agency is requiring a rotational crop statement on all terbacil labels which may involve rotation to crops other than those currently registered (refer to section IV.D for specific wording). In addition residue data are required for representative crops from any crop group which could be rotated from alfalfa, sainfoin, and mint.
o Existing Tolerances for Terbacil per se should be amended to include metabolites A,B, and C, as specified in 40 CFR 180.209 (b).
o The Agency will revoke tolerances associated with the commodities sainfoin (hay), sainfoin (forage), and pears.
o The Agency will revise the tolerance for residues in or on peaches to 0.2 ppm for combined residues of terbacil and its metabolites.

o The Agency will impose a label restrictions against the feeding and grazing of sugarcane forage and spent hay to livestock or require the development and submission of data in support of tolerances for residues in or on sugarcane forage and spent hay.
o The Agency has determined that grasses grown for seed is a food use. Therefore, data depicting residues of terbacil in or on members of the grass forage, fodder and hay group are required.
o The Agency will not propose group tolerances for terbacil.
o The Agency will not grant any significant new food or feed uses for terbacil until the required residue chemistry and toxicology studies have been submitted and reviewed.
o The Agency has identified certain data that will receive priority review when submitted;

Section 158.340 Toxicology
83-1 Chronic Oral Feeding (Rat)
83-2 Oncogenicity (Rat and Mouse)

Section 158.290 Environmental Fate
162-1 Aerobic Soil Metabolism
162-2 Anaerobic Soil Metabolism
163-1 Leaching Adsorption/Desorption
164-5 Soil, Long Term (field)

Section 158.490 Wildlife and Aquatic Organisms
72-3 Estuarine and Marine Testing (Fish)
122- Tier I Nontarget Area Phytotoxicity

Section 158.240 Residue Chemistry
171-4 Metabolism in Plants & Livestock

5.   Required Unique Labeling

A. Groundwater Advisory Statements
B. Environmental Hazards Statement
D. Reentry Statement
E. Feeding and/or Grazing Restrictions
F. Rotational Crop Statement

6.   Summary of Major Data Gaps   Timeframe Ranges

Toxicology   12-48 Months
Environmental Fate/Exposure   12-48 Months
Ecological Effects   24 Months
Residue Chemistry   24-48 Months
Product Chemistry   12-24 Months

7.   CONTACT PERSONS AT EPA

Product Specific Inquiries:
Robert J. Taylor
Product Manager (25)
Fungicide Herbicide Branch
Registration Division (H-7505C)
Office of Pesticide Programs
Environmental Protection Agency
401 M Street, S. W.
Washington, D. C.   20460

Office location and telephone number:
Room 245, Crystal Mall #2
1921 Jefferson Davis Highway
Arlington, VA   22202
(703) 557-1800

Reregistration Document Inquiries:
Donna M. Williams
Review Manager
Reregistration Branch
Special Review and Reregistration Division (H-7508C)
Environmental Protection Agency
401 M Street, S. W.
Washington, D. C.   20460

Office location and telephone number:
Room 1124, Crystal Mall #2
1921 Jefferson Davis Highway
Arlington, VA   22202
(703) 557-0639

DISCLAIMER: The information in this Pesticide
Fact Sheet is a summary only and is not to be used
to satisfy data requirements for pesticide
registration and reregistration.   The complete
Registration Standard for the pesticide may be
obtained from the National Technical Information
Service.   Contact the Review Manager listed above
for further information.

TERBUFOS

Reason for Issuance: Revised Registration Standard
Date Issued: September 9, 1988
Fact Sheet Number: 5.2

1. Description of Chemical

   Common Name: Terbufos
   Chemical Name: S-[[(1,1-dimethylethyl)thio]methyl]O,O-diethyl
   phosphorodithioate
   Other Chemical Nomenclature: S-([tert-butylthio)methyl]O,O-diethyl
   phosphorodithioate (IUPAC); S-(t-butylthio) methyl
   O,O-diethyl-phosphorodithioate (CA, 8th Collective
   Index); S-tert-butylmercaptomethyl O,O-diethyl
   dithiophosphate
   Trade Names: Contraven; Counter; AC 92,100; CL 92,100; and ST-100
   Chemical Abstracts Service (CAS) Number: 13071-79-9
   EPA Shaughnessy Code: 105001
   Year of Initial Registration: 1974
   Pesticide Type: Insecticide-Nematicide
   Chemical Family: Organophosphate
   U.S. Registrant: American Cyanamid Company

2. Use Patterns and Formulations

   Application Sites: Terrestrial food crop use on corn; grain
   sorghum; and sugar beets.

   Formulations: Granular

   Methods of Application: Broadcast (nonsoil-incorporated) with air or
   ground equipment; and soil-incorporated with ground equipment.

3. Science Findings

   Summary Science Statement: Technical terbufos is highly acutely
   toxic by the oral, dermal, and inhalation routes of exposure
   (Toxicity Category I for all three routes). Terbufos does not
   demonstrate an acute neurotoxic, oncogenic, mutagenic,
   reproductive, or teratogenic potential. Animal studies have shown
   that the chemical is a cholinesterase inhibitor reducing plasma,
   brain, and red blood cell cholinesterase activity. The use of

terbufos poses a potential risk to loaders and applicators and to
persons reentering treated fields following nonsoil-incorporated
broadcast application of the chemical. This is due to the high
acute toxicity and the cholinesterase inhibiting properties of the
chemical.

Based on the plasma cholinesterase inhibition no-effect-level of
0.00125 mg/kg/day as defined in a 4-week dog study and, using a
safety factor of 10, the acceptable daily dietary intake for
humans in 0.000125 mg/kg/day. The theoretical maximum residue
contribution from the established tolerances is estimated to be
0.000052 mg/kg/day. This is equivalent to 42 percent of the
acceptable daily intake for the average U.S. population. Due to
the numerous gaps in residue chemistry data, the Agency is unable
to complete a tolerance reassessment of terbufos.

Terbufos is highly toxic to birds, fish, and aquatic invertebrates.
The acceptable short-term field study shows significant acute
mortalities of birds, mammals, reptiles, and fish resulting from
broadcast application of terbufos to corn fields at 1 pound active
ingredient per acre (1 lb ai/A). In the same study, the application
of terbufos as a soil-incorporated treatment to corn fields at 2
lb ai resulted in acute mortalities to birds and reptiles.

The limited environmental fate data are not sufficient to assess
the mobility and leaching properties of tebufos. Terbufos residues
were reported to occur in well water sampling in Iowa and Minnesota.
These reports, however have not been confirmed in the laboratory
and a resampling of the same Iowa wells a year later, in 1986,
showed no detections for terbufos or its degradates. Terbufos is
one of the pesticides the Agency is sampling for in the National
Well Water Survey.

Chemical/Physical Characteristics of the Technical Material:

    Color: Clear, brownish
    Physical State: Liquid
    Odor: Mercaptan-like
    Boiling Point: 55 °C at 0.02 mmHg
    Stability: Relatively stable in water under neutral or slightly
                acidic conditions but is subject to hydrolysis under
                alkaline conditions.

Toxicology Characteristics

o Acute Oral: Toxicity Category I (1.6 and 1.3 mg/kg for male and
        female rats, respectively).

o Acute Dermal: Toxicity Category I (0.81 and 0.93 mg/kg for male
        and female rabbits, respectively).

o Acute Inhalation: Toxicity Category I ($\leq$ 0.2 mg/L).

o Delayed Neurotoxicity: No evidence of acute delayed neurotoxicity
   at the 40 mg/kg dosage level tested in hens.

o Subchronic Feeding: The NOEL for both systemic effects and
   cholinesterase inhibition in a rat subchronic study is 0.25 ppm.

o Subchronic Dermal: The NOEL for systemic effects in a 30-day
   rabbit study is 0.020 mg/kg.

o Mutagenicity: Terbufos did not exhibit mutagenic potential in the
   Ames assay, the in vivo cytogenetic assay, and the dominant lethal
   test.

o Teratogenicity: The NOEL for developmental toxicity in a rat
   teratology study is 0.1 mg/kg/day.

o Reproduction: The NOEL for reproductive effects in a three-
   generation rat reproduction study is 0.25 ppm.

o Oncogenicity: No oncogenic effects observed in an 18-month mouse
   study and a 2-year rat study at doses up to and including 12.0
   ppm (1.8 mg/kg/day) and 8.0 ppm (0.4 mg/kg/day), respectively.

o Chronic Toxicity: The NOEL for plasma cholinesterase (ChE)
   inhibition from a 4-week dog feeding study is 0.00125 mg/kg/day;
   the NOEL for brain/red blood cell ChE from a 1-year dog study
   is 0.060 mg/kg/day. The NOEL for plasma and brain ChE from a
   1-year rat feeding study is 0.5 ppm.

o Metabolism: Terbufos was rapidly excreted as the diethyl
   phosphoric acid and other polar metabolites (83%) in urine
   within 168 hours of administration to male rats. Terbufos
   and its metabolites were not noted to accumulate in tissues.

Ecological Characteristics:

Based on acceptable laboratory data, technical terbufos is highly
   toxic to birds, fish, and aquatic invertebrates.

Acute Avian Toxicity: 28.6 mg/kg (bobwhite).
Dietary Avian Toxicity: 143 and 157 ppm (from two bobwhite studies).
Avian Reproduction: Terbufos was not considered to produce avian
   reproductive effects based on results of a bobwhite quail study
   and a mallard duck study.
Freshwater Fish Acute Toxicity: Ranges from 0.77 to 20.00 ppb.
Freshwater Invertebrate Acute Toxicity: 0.31 ppb for Daphnia magna.
Marine/Estuarine Fish Acute Toxicity: Data gap.
Marine/Estuarine Invertebrate Toxicity: Data gap.
Marine/Estuarine Mollusk Toxicity: Data gap.

Terrestrial Field Study (Level 1): Both soil-incorporated (2 lb
    ai/A) and nonsoil-incorporated (1 lb ai/A) resulted in nontarget
    mortalities, with the latter application much more severe in its
    effects.

Based on the adverse effects observed in the level 1 field study
    described above, level 2 terrestrial field studies are required to
    assess the potential effects on populations of birds, mammals, and
    reptiles. Based on the high acute toxicity to aquatic organisms
    and results of initial modeling[1] conducted by the Agency for the
    1983 Terbufos Registration Standard, the estimated environmental
    concentration (EEC) of terbufos residues likely to occur in the
    aquatic environment may pose an acute hazard for freshwater and
    marine/estuarine species. This modeling was conducted for the
    soil-incorporated application of terbufos.

Potentially greater hazards are likely for aerial applications of
    terbufos granules since soil-incorporated applications typically
    provide less exposure than aerial broadcast applications.

Although these theoretical calculations and modeling indicate that
    the use of terbufos may result in significant adverse effects to
    aquatic species, actual field monitoring data are not available
    to support this finding. Moreover, the environmental fate charac-
    teristics of terbufos are not accurately defined by available
    data. Thus, the models can be used only on a limited basis.

Aquatic residues monitoring studies are required to determine actual
    residues in aquatic systems exposed to runoff and spray drift.
    Although these studies were previously requested in the 1983
    Terbufos Registration Standard, their initiation was delayed
    pending the Agency's recalculation of the EECs. Prior to the
    completion of this task, reports of fish kill incidents demonstrating
    the potential exposure to aquatic organisms under actual field use
    conditions became available. These fish kills reportedly resulted
    from aerial applications of terbufos to corn fields during conduct
    of the level 1 terrestrial field study. In addition, several
    environmental fate studies previously found acceptable do not meet
    current guideline requirements and need to be repeated.

Terbufos has been identified by the Office of Endangered Species
    (OES), U.S. Fish and Wildlife Service (USFWS), as being likely to
    jeopardize the continued existence of certain endangered species

----

[1]The Agency used computer models (SWRRB) and EXAMS) to simulate
runoff from granular application of terbufos and to predict aquatic
concentrations of the chemical. SWRRB is a hydrology model combined
with a pesticide runoff model. EXAMS is a hydrologic model to predict "steady-state" and "pulseload" behavior of organic toxicants in
aquatic ecosystems.

when used on corn and sorghum. Based on this determination, OES
specified reasonable and prudent alternatives to avoid jeopardizing
the continued existence of the identified species. EPA is working
with USFWS and other Federal and State agencies to implement the
alternatives in a technically sound manner.

Formal consultation will be initiated with OES under Section 7 of the
Endangered Species Act regarding the potential exposure to
endangered species resulting from the registered use of terbufos
on sugar beets.

Environmental Characteristics:

Results of an acceptable hydrolysis study indicate that terbufos
hydrolyzes at pH 5, 7, and 9 with a half-life of 2.2 weeks.
Formaldehyde was the major degradate detected in this study.
Results of an acceptable aerobic soil metabolism study indicate
that terbufos degrades in silt loam soil with a half-life of
26.7 days. The major degradates detected in this study included
carbon dioxide, terbufos sulfoxide, and terbufos sulfone.

Results of a field dissipation study, classified as supplementary,
indicate that terbufos residues have a half-life of less than
40 days in field plots of loam soil located near Arcola, Illinois,
and sandy loam soil located near Greeley, Colorado treated with
a 15 percent granular formulation at an application rate of 1 lb
ai/A. The sampling protocol was inadequate to accurately assess
the dissipation of terbufos residues in field soil and a new study
is required.

The available data reviewed by the Agency are not sufficient to
fulfill data requirements nor to assess the environmental fate of
terbufos. Four studies previously reviewed and found acceptable
under the 1983 Terbufos Registration Standard do not meet the
requirements of the Agency's current guidelines and new studies
are required. These are: anaerobic soil metabolism, leaching,
fish accumulation, and field dissipation.

In addition, several new studies are now required due to the
additional method of nonsoil-incorporated, broadcast (air or
ground equipment) application which was not registered at the
time of the 1983 Terbufos Registration Standard.

Tolerance Assessment:

Tolerances for combined residues of terbufos and its ChE-inhibiting
metabolites in or on food commodities are published under Section
180.352 of Title 40 of the Code of Federal Regulations (40 CFR
180.352). These tolerances range from 0.05 to 0.5 ppm.

Residue Data:

In the Terbufos Registration Standard dated June 1983, no outstanding
data gaps were identified for residue chemistry. However, subsequent amendments to registered uses for terbufos and addenda to
the Pesticide Assessment Guidelines (Subdivision O) for Residue
Chemistry have made it necessary to reevaluate portions of the
data base previously reviewed under the June 1983 Standard. As a
result, some of the original conclusions regarding adequacy of the
data and support for tolerances have been modified in this revised
Registration Standard.

Based on the available plant metabolism studies, the nature of
residues in plants is adequately understood. Of the phosphorylated
metabolites, terbufoxon sulfoxide and terbufos sulfoxide comprised
$\leq$ 30 percent of the residues, and terbufos sulfone and terbufoxon
sulfone comprised $\leq$ 7 percent. The major nonphosphorylated metabolite which comprised $\leq$ 30 percent of the organosoluble residues was
nonphosphorylated terbufoxon sulfone. The available poultry and
ruminant feeding studies do not meet current Guideline requirements
for data depicting the metabolism of terbufos in livestock and new
studies are required. The basic GLC analytical procedure published
as Method I in PAM Vol. II is adequate for collection of data pertaining to the combined residues of terbufos and its ChE-inhibiting
metabolites on commodities with established tolerances. Method
validation data pertaining to recovery of individual metabolites
from representative plant commodities are being required. The adequacy of the available methods for detection of terbufos residues
of concern in animal products will be evaluated upon receipt of the
required animal metabolism data. Field trial studies are required
for all crops for which there are terbufos tolerances. Processing
studies are also required in addition to storage stability residue
data.

The available poultry and ruminant feeding studies show that no
detectable residues occur in eggs, chicken tissues, milk, or
cattle tissues from animals fed exaggerated dietary levels of
terbufos and its ChE-inhibiting metabolites. However, additional
animal metabolism data are required and a determination regarding
the need for and nature of tolerances for residues in meat, milk,
poultry, and eggs will be made upon receipt and evaluation of
these data.

The established tolerances for terbufos are presently expressed in
terms of terbufos and its ChE-inhibiting metabolites without
specifying the latter as phosphorylated metabolites. The Agency

will propose revising 40 CFR 180.352 by changing the wording to
read:

". . . terbufos . . . and its phosphorylated (cholinesterase-
inhibiting) metabolites:

o   Phosphorothioic acid, S-(t-butyl-thio) methyl O,O-diethyl ester.

o   Phosphorothioic acid, S-(t-butyl-sulfinyl) methyl O,O-diethyl
    ester.

o   Phosphorothioic acid, S-(t-butyl-sulfonyl) methyl O,O-diethyl
    ester.

o   Phosphorodithioic acid, S-(t-butyl-sulfinyl) methyl O,O-diethyl
    ester.

o   Phosphorodithioic acid, S-(t-butyl-sulfonyl) methyl O,O-diethyl
    ester."

Acceptable Daily Intake:

Based on the plasma ChE inhibition NOEL as defined in a 4-week dog
study (0.00125 mg/kg/day) and using a safety factor of 10, the
acceptable daily intake or reference dose (RfD) for humans is
0.000125 mg/kg/day.

4.   Summary of Regulatory Positions and Rationales

Terbufos is not being placed in special review at this time. Field
studies are needed to completely assess the potential risk to
wildlife, including endangered species. The Agency is conducting
a comparative avian risk assessment of various granular pesticides,
including terbufos. When this assessment is completed, further
regulatory action may be taken.

The restricted use classification of the 15 percent granular end-use
product based on the high acute oral and dermal toxicity to humans
is being retained.

A level II terrestrial field study; monitoring studies in soil,
water, sediment, and fish; and aquatic organism field studies are
being required for the completion of the Agency's assessment of
the potential risk to both avian and aquatic species.

A special 21-day dermal study in rats and a dislodgeable residue
study are being required for the completion of the Agency's
assessment of the potential risk to workers reentering corn fields
following a nonsoil-incorporated application of terbufos.

Due to the lack of pertinent environmental fate data, no conclusions regarding the potential for terbufos to contaminate ground water can be made.

EPA is developing a program to reduce or eliminate exposure to endangered species to a point where use does not result in jeopardy, and will issue notice of any necessary labeling revisions when the program is developed. No additional labeling is required at this time. Labeling requirements issued in PR Notices 87-4 and 87-5 have been withdrawn pending reissuance.

5. Summary of Required Label Modifications

An updated Environmental Hazard statement is required.

Updated worker safety rules and protective clothing statements are required.

A 7-day reentry interval statement is required for use of terbufos as a broadcast application to corn.

A label statement prohibiting use of terbufos as a broadcast application to seed corn prior to detasseling activities is required.

6. Summary of Outstanding Data Requirements

Data	Due Date[1]
Toxicology	
Special 21-day rat dermal	12 Months
Rabbit teratology	15 Months
Fish & Wildlife	
Acute toxicity to estuarine and marine organisms	12 Months
Fish early life stage	15 Months
Aquatic organism accumulation	12 Months

---

[1]/Due date is measured from the date of receipt of the Standard by the registrant unless otherwise specified.

Data	Due Date[1]

Fish & Wildlife (cont'd)

Terrestrial field study	[2]
Aquatic organism field study	6 Months for protocol

Reentry

Dislodgeable residue study	27 Months

Environmental Fate

Photodegradation in water	9 Months
Photodegradation on soil	9 Months
Photodegradation in air	9 Months
Anaerobic soil metabolism	27 Months
Leaching and absorption/desorption	12 Months
Lab volatility	12 Months
Soil field dissipation	27 Months
Rotational crop (field)	50 Months
Fish accumulation	12 Months
Monitoring study (soil, water, sediment, fish)	6 Months (protocol)

Residue Chemistry

Crop field trials	18 Months
Processing studies	24 Months
Storage stability	15 Months
Ruminant and poultry metabolism	18 Months
Residue analytical methodology data	15 Months

Product Chemistry

Majority of data	6-15 Months

---

[1]/Due date is measured from the date of receipt of the Standard by the registrant unless otherwise specified.

[2]/1st Annual Report . . . . December 31, 1989
2nd Annual Report* . . . . December 31, 1990
3rd Annual Report* . . . . December 31, 1991
Final Report** . . . . December 31, 1992

*A determination may be made at this time to conclude the study, in which case a final report will be due 3 months after notification.
**The due date applies if the study has not been determined to be concluded by earlier reviews.

7.   Contact person at EPA

William H. Miller
Product Manager (16)
Insecticide-Rodenticide Branch
Registration Division (TS-767C)
Environmental Protection Agency
Washington, DC 20460

Phone: (703) 557-2600.

DISCLAIMER: The information presented in this Chemical Information Fact
Sheet is for informational purposes only and may not be used to fulfill
data requirements for pesticide registration and reregistration.

TETRACHLORVINPHOS

Reason for Issuance: Registration Standard
Date Issued: October 14, 1988
Fact Sheet Number: 184

1.   DESCRIPTION OF CHEMICAL

      Generic Name: (Z)-2-chloro-1-(2,4,5-trichlorophenyl) vinyl
                     dimethyl phosphate

      Common Name: Tetrachlorvinphos

      Trade and Other Names: Rabon, Stirofos, Gardona, Gardcide

      EPA Chemical (Shaughnessy) Codes: 083701

      Chemical Abstracts Service (CAS) Number: 22248-79-9 [(Z)-isomer]
                                               22350-76-1 [(E)-isomer]
                                               961-11-5 (mixed isomers)

      Year of Initial Registration: 1966

      Pesticide Type: Insecticde

      Chemical Family: Organophosphate

      U.S. and Foreign Producers: E.I. Dupont de Nemours & Co.
                                  Fermenta Animal Health Co.

2.   USE PATTERNS AND FORMULATIONS

      Application: Control of filth or manure flies associated with
      livestock and poultry as a feed additive, in mineral blocks;
      indoor on animal premises, direct animal treatment, garbage dumps.

Types and Methods of Application: Premise treatment by conventional hydraulic sprayer or low pressure knapsack; animal treatment by hand (shaker can, hand duster, grooming brush, dust mitt), pressurized spray, flea collars, ear tags, oral feed additive, dust bags, and oral mineral block

Types of Formulations: Dust, granular, pelleted/tableted, wettable powder, wettable powder/dust, impregnated material, emulsifiable concentrate, ready-to-use and pressurized liquid.

3.   SCIENCE FINDINGS

Summary Science Statement

Technical tetrachlorvinphos is not very toxic by the inhalation route of exposure, and does not produce clinical signs of neurotoxicity. No data are available to assess the toxicity of tetrachlorvinphos by the acute oral and dermal routes of exposure and studies are being required in the Standard. Tetrachlorvinphos is not teratogenic in rabbits but the effects in rats have not yet been assessed. Rat teratology data received in response to a Comprehensive Data Call-In (1984) were not received in time to be included in the reviews for this Standard. Available data indicate that the potential effects of tetrachlorvinphos on reproduction are minimal; the only compound-related effect observed in the rat was an increase in liver size in the F3b weanlings in the highest dose tested (HDT). There are no acceptable mutagenicity studies on tetrachlorvinphos. Based on a chronic feeding study in dogs, cholinesterase inhibition and reduced weight gain are the significant toxic effects resulting from chronic exposure to tetrachlorvinphos.

Based on the Agency's Guidelines for Carcinogen Risk Assessment, the Agency has classified tetrachlorvinphos as a Group C (possible human) carcinogen with a cancer potency estimate (Q_1^*) of 3.14×10^{-3} $(mg/kg/day)^{-1}$.

Available data indicate that tetrachlorvinphos is practically non-toxic to birds, but is highly toxic to fish and honey bees (contact). Available data are insufficient to characterize the environmental fate of tetrachlorvinphos. Preliminary findings indicate that 14^C tetrachlorvinphos degrades with a half-life of <8 days in medium loam soil.

Chemical Characteristics of the Technical Material

Physical State: solid

Color: tan to brown

Odor: mild chemical

Molecular Weight and Formula: 366.0; $C_{10}H_9Cl_4O_4P$

Boiling Point: N/A solid at room temperature

Vapor Pressure: no data submitted

Density: (bulk density) 50-55 lb/cu ft

Solubility in various solvents: at 0 C: 40 ppm in chloroform,
40 ppm in dichloromethane,
20 ppm in acetone,
8 ppm in xylene,
15 ppm in water at 24 C

Toxicology Characteristics

Acute Oral: No available studies

Acute Dermal: No available studies

Primary Dermal Irritation: No available studies

Primary Eye Irritation: No available studies

Dermal Sensitization: No available studies

Acute Inhalation: 3.61 mg/L   Tox. Category III

Major routes of exposure: dermal and inhalation

Delayed neurotoxicity: negative

Subchronic toxicity: normocytic anemia and plasma cholinesterase inhibition

Oncogenicity: Group C carcinogen; (mouse) Sys. NOEL= 240 mg/kg/day

(rat) Sys. NOEL= 6.25 mg/kg/day

Chronic Feeding: (Dog) Sys. NOEL= 3.13 mg/kg/day
                      Sys. LEL= 50 mg/kg/day

Metabolism: (dog and rat) completely metabolized in both; primarily excreted
              in urine

Teratogenicity: (Rabbits) Maternal NOEL= 375 mg/kg
                          Fetotoxic NOEL=150 mg/kg
                          Teratogenic NOEL= >750 mg/kg

Reproduction: (Rat) NOEL=16.65 mg/kg/day
                      LEL= 50.0 mg/kg/day

Mutagenicity: No acceptable studies

Physiological and Biochemical Characteristics

Mechanism of Pesticidal Action: Cholinesterase inhibition following ingestion
                                of tetrachlorvinphos

Metabolism and Persistence
   in Plants and Animals: Currently, there are no end-use products
                           registered for use on any plant commodity. The
                           nature of the residue in animals is not adequately
                           understood. However, the available animal
                           metabolism data indicate that residues of tetra-
                           chlorvinphos will occur in the tissues of ruminants
                           and poultry following ingestion of feeds containing
                           tetrachlorvinphos.

Environmental Characteristics

     Available data are insufficient to characterize the environmental fate
of tetrachlorvinphos.

Ecological Characteristics

Avian acute toxicity: Mallard Duck LD_{50} > 2,000 mg/kg

Avian dietary toxicity: Mallard Duck LD_{50} > 5,000mg/kg

Freshwater fish acute toxicity: Bluegill LC_{50} = 0.53 ppm
                                Rainbow Trout LC^{5}_{0} = 0.43 ppm

Freshwater invertebrate toxicity: No available studies

Marine invertebrate toxicity: No available studies

4. TOLERANCE ASSESSMENT

Tolerances have been established for residues of tetrachlorvinphos in/on a variety of raw agricultural commodities (40 CFR 180.252). A feed additive regulation has also been established for tetrachlorvinphos (21 CFR 186.950). Since there are no currently registered products bearing labeling for use on food crops tolerance revocation procedures (40CFR 180.7) will be initiated for apples, field corn, sweet corn and popcorn, fresh corn forage and fodder, alfalfa, cherries, cranberries, peaches, pears, and tomatoes. The Agency has evaluated the residue and toxicology data supporting the tolerances on animal commodities and has determined that it does not have sufficient data to support the currently established tolerances residues of tetrachlorvinphos. Because of the extensive residue chemistry and toxicology data gaps, no significant new tolerances or new food uses will be granted until the Agency has received data sufficient to evaluate the dietary exposure of tetrachlorvinphos.

There are no Canadian or Mexican tolerances and no Codex Maximum Residue Limits (MRLs) have been established for tetrachlorvinphos residues in poultry fat and eggs. Therefore, no compatibility questions exist with respect to the Codex MRLs.

5. SUMMARY OF REGULATORY POSITIONS

The Agency is not initiating a special review for tetrachlorvinphos at this time. Although tetrachlorvinphos has been classified as a Group C oncogen (possible human), anticipated dietary exposure to humans and the associated risk is not high (10-6).

The Agency has determined that based upon currently available data, tetrachlorvinphos does not meet the risk criteria (40 CFR 152.170) for restricted use, and thus, products containing tetrachlorvinphos do not warrant restricted use classification at this time.

The Agency will reevaluate the safety and efficacy data for various formulations of tetrachlorvinphos to domestic animals (dogs,cats,hoses)in order to determine if further regulatory action is warranted.

The Agency is requiring protective clothing for workers or other users who mix, load, and/or apply tetrachlorvinphos.

The Agency is requiring data to show the concentration and degradation products of tetrachlorvinphos in fresh manure at the time it is used on land as fertilizer.

The Agency will not grant any tolerances for significant new food uses[1] until sufficient data (residue chemistry and toxicology) are submitted for the Agency to perform a tolerance reassessment for tetrachlorvinphos.

6. LABELING REQUIREMENTS

All products allowing for outdoor use must bear the following environmental hazards statement:

"This is toxic to fish. Do not contaminate water when disposing of equipment washwaters."

The following protective clothing statements must appear on all products allowing for use on livestock, agricultural premises, feed additives: For sprays and dusts:

"Wear long-sleeved shirt and pants; chemical resistant gloves; shoes and socks." For ear tags: "Wear gloves when applying tags".

The following statements must appear on all products allowing for domestic use:

"Wear long-sleeved shirt and pants, shoes and socks."

All products allowing for use as a feed additive and/or use on manure (droppings) must bear the following statement:

"Do not use manure/droppings on land where food crops are to be grown during the current growing season."

7.   SUMMARY OF DATA GAPS

DATA	DUE DATE From DATE OF STANDARD
Product Chemistry- All	9-15 Months
Environmental Fate- Hydrolysis Photodegradation (in water) Metabolism Studies Mobility Studies Dissipation Studies (soil) Dissipation Study (excrement)	9-27 Months
Toxicology- Acute Oral (rat) Acute Dermal (rat) Eye Irritation (rabbit) Dermal Irritation (rabbit) Dermal Sensitization (guinea pig) Mutagenicity 21-Day Dermal	9-12 Months
Oncogenicity (rat) Teratology (rat)	50 Months

Ecological Effects- Acute Toxicity to   9 Months
    Freshwater Invertebrates

Residue Chemistry- Livestock Metabolism
                   Analytical Methods
                   Storage Stability   18-24 Months
                   Magnitude of residues
                   in meat, milk, eggs

8. CONTACT PERSON AT EPA

George T. LaRocca
Product Manager (15)
Insecticide-Rodenticide Branch
Registration Division (TS-767C)
Office of Pesticide Programs
Environmental Protection Agency
401 M Street, S. W.
Washington, D. C. 20460

Office location and telephone number: Room 204; (703) 557-2400

Crystal Mall #2
1921 Jefferson Davis Highway
Arlington, VA 22202
(703) 557-2600

DISCLAIMER: The information in this Pesticide Fact Sheet
is a summary only and is not to be used to satisfy data
requirements for pesticide registration and reregistration.
The complete Registration Standard for the pesticide may be
obtained from the National Technical Information Service.
Contact the Product Manager listed above for further
information.

TRIADIMENOL

Reason for Issuance: New Chemical Registration
Date Issued: July 1989
Fact Sheet Number: 204

DESCRIPTION OF CHEMICAL

Generic Name: beta-(4-chlorophenoxy)-alpha-(1,1-dimethylethyl)-
               1H-1,2,4-triazole-1-ethanol and its metabolites
               containing chlorophenoxy and triazole moieties.

Common Name: triadimenol

Trade Name: Baytan

EPA Shaughnessy Codes: 127201

Chemical Abstracts Service (CAS) Number: 5219-65-3

Year of Initial Registration: 1989

Pesticide Type: Fungicide

U.S. and Foreign Producers: Mobay Corporation

USE PATTERNS AND FORMULATIONS

APPLICATION SITES: Seeds of barley, corn, oats, rye,
    sorghum and wheat to control seed- and soil-borne
    diseases and to provide early season control of foliar
    diseases.

METHOD OF APPLICATION: Application will be made as a water-
    based slurry through standard slurry or mist type
    commercial seed treatment equipment.

TYPES OF FORMULATION: 25% dry flowable end-use product
    and 90% technical powder for formulating use.

APPLICATION RATES: For barley, oats, rye and wheat, apply
    0.25-0.5 oz. ai./100 lbs of seed; for sorghum, apply
    0.5 oz. ai./100 lbs of seed; and for corn, apply 1.0 oz.-
    ai./100 lbs of seed.

USUAL CARRIER: water.

SCIENCE FINDINGS

Summary Science Statement
    Available acute toxicity studies indicate that triadimenol
is in toxicity category II (warning) based on an acute inhalation
toxicity study with rats.
    Chronic feeding/oncogenicity studies were conducted in both
the rat and mouse. Clinical chemistry findings in the chronic
feeding study in the rat suggests that the target organ for
toxicity may be the liver. Although there was an accompaying
small increase in liver weight in the females of the high
dose group, there were no histopathologic charges in the
liver in either sex.
    In the chronic feeding study in mice, the results of
blood chemistry, organ weights and gross and histological
examinations, again indicated the liver as the target organ.
    Triadimenol did not induce either genotoxic effects or
chromosomal aberrations in a series of mutagenicity studies.
In addition, no strong structural activity correlation to
other carcinogens has been found. Triadimenol was also found
not to be teratogenic in either the rat or rabbit.
    Environmental fate data indicates that triadimenol is stable
to hydrolysis and appears to be stable to photolysis on the
soil surface. In addition, based on low adsorption coefficients,
triadimenol will have a low potential to leach in soil.
However, triadimenol may have a moderate potential to leach
in some Western soils.
    Additional studies indicate that due to the manufacturing
process, triadimenol should have no adverse effects on non-target
organisms provided waste is disposed of properly. An overview
of the toxicity test results suggests that triadimenol is
practically non-toxic to birds, slightly toxic to fish, and
moderately toxic to aquatic invertebrates. It is also unlikely
that this registration would affect endangered species because
of its relatively low use rates, agricultural techniques
which involve drill planting of most small grains and corn,
and the low toxicity of triadimenol to all animals.

TOXICOLOGICAL CHARACTERISTICS

Acute oral toxicity in rats:
    LD_{50} 689 mg/kg in males
            752 mg/kg in females
    Toxicity category III

Acute dermal toxicity in rats:
    LD_{50} >5000 mg/kg
    Toxicity category III

Acute inhalation toxicity in rats
    LC_{50} >1.56 mg/L
    Toxicity category III

Primary eye irritation in rabbit:
    slight irritation

Primary dermal irritation in rabbit:
    Toxicity category IV

Dermal sensitization in guinea pigs:
    core minimum; no effect

Chronic Studies: Triadimenol has been evaluated in the following
                    studies.

°Rodent Feeding/Oncogenicity

    1. A 2-year feeding/oncogenicity study with rats using
dietary concentrations of 0, 125, 500, and 2000 parts per
million (ppm) equivalent to 0, 6.25, 25.0, and 100 mg/kg bwt/day
in males and females. Clinical chemistry findings suggest
that the target organ for toxicity may be the liver. The
levels of SGOT and SGPT enzymes were consistently higher at
2000 ppm in males and females when compared to controls, and
some increase in these two parameters was also observed at
500 ppm. Although there was an accompanying small increase
in liver weight in 2000 ppm females, there were no accompanying
increases in histopathologic changes of the liver in either
sex. There were only marginal effects seen on other clinical
chemistry parameters, and no effect of test compound on
clinically observed signs of toxicity, food consumption,
hematologic, or urinalysis parameters. The systemic NOEL
(no-observed effect level) is 125 ppm (6.25 mg/kg/day for
males and females) based on the increase in liver enzymes
(SGOT and SGPT). The systemic LEL (lowest effect level) was
500 ppm (25 mg/kg/day for males and for females).
    2. A 2-year chronic feeding/oncogenicity study in mice
using dietary concentrations of 0, 125, 500, and 2000 ppm
(equivalent to doses of 0, 18, 72, and 285 mg/kg/day for males
and females). The results of blood chemistry, organ weights,
and gross and histological examinations indicated the liver to
be the target organ. There were time- and dose-related increases
in SAP (serum alkaline phosphatase), SGOT and SGPT activities
in both male and female animals receiving 500 and 2,000 ppm of
the test material.
    In addition, increased incidence of enlarged livers,
hyperplastic nodules and increased liver weights in both
male and female animals receiving 2,000 ppm of test material
were detected at necropsy. Female animals receiving 2000 ppm
exhibited a significant increase in the incidences of liver
adenomas only, a compound-related oncogenic effect. In

males, there were no differences in the incidences of these
lesions in treated and control males, and the incidences of
liver adenomas were similar to those observed in historical
controls.

Based on this evidence the Agency classified triadimenol
as a Category C (possible human carcinogen) in accordance
with the EPA Guidelines for Carcinogen Risk Assessment (September
24, 1986, 51 FR 33992). This evaluation was confirmed by the
Agency's Scientific Advisory Panel on December 15, 1987. However, it was also concluded that this evidence of carcinogenicity
did not warrant a low dose extrapolation of risks since the
tumors were only benign, were observed in only one sex, and
only at the highest dose tested. Moreover, the chemical was
negative in the genotoxic assay battery.

Based on blood chemistry findings, the systemic NOEL and
the LEL are 125 ppm and 500 ppm respectively (equivalent to
18 and 72 mg/kg/day for males and females).

3. A 3-month rat feeding study using doses of 0, 150,
and 600 ppm (equivalent to 0, 7.5, and 30 mg/kg bwt/day for
males and females) demonstrated a decrease in body weight,
decrease in hematocrit values, eosinophil count and medium
cell hemoglobin and increase in the high dose group and dose-
related increase in liver weight. The NOEL is 150 ppm and
the LEL is 600 ppm.

°Non-Rodent Feeding Study

1. A 2-year male and female dog feeding study using doses
of 0, 150, 600 and 2400 ppm (equivalent to 0, 3.75, 15, and
60 mg/kg bwt/day for males and females). The NOEL is 150 ppm
based on changes in enzyme levels (equivalent to 3.75 mg/kg
bwt/day for males and females). The LEL is 600 ppm. Although
there were significant decreases in mean body weights in males
receiving 150 and 2400 ppm and in females receiving 600 and 2400
ppm, the biological significance of these changes could not be
assessed. There were noted increases in alkaline phosphatase
N-demethylase, and cytochrome P-450 in males receiving 2400 ppm
and significant increases in N-demethylase in females receiving
600 and 2400 ppm and in cytochrome P-450 in females receiving
2400 ppm when compared to controls.

2. A 6-month dog feeding study using doses of 0, 10, 30,
and 100 ppm (equivalent to 0, 0.25, 0.75, 2.5 mg/kg bwt/day
for males and females). The NOEL was demonstrated at doses
up to 100 ppm, the highest dose level tested.

3. A 3-month dog feeding study using doses of 0, 150, 600
and 2400 ppm (equivalent to 0, 3.75, 15, and 60 mg/kg bwt/day
for males and females). Weight gain in all male groups and
in the highest dose female group was significantly less than
the control. Alkaline phosphatase in males and females showed
a dose-related negative trend. There was no gross pathological
changes. Effects at 600 ppm included an increase in serum
cholesterol level in males. Although the NOEL appeared to be

less than 150 ppm based on reduced body weight and decreased
alkaline phosphatase in males, the Agency has concluded that effects
below 600 ppm in the 2-year dog study were not biologically
significant and the longer-term study supercedes the 90-day dog
study.   Therefore, the NOEL remains at 150 ppm.

°Teratology

    1. A rabbit teratology study with a NOEL for maternal
toxicity of 8 mg/kg. The maternal LEL was 40 mg/kg based on
decreased body weight gains and food consumption. The developmental NOEL and LEL were 40 mg/kg and 200 mg/kg respectively.
This study has to be resubmitted with all the findings
statistically analyzed on a per litter and per fetus basis in order
to be upgraded from its current classification as core supplementary.
    2. A rat teratology study using dose levels 0, 30, 60,
and 120 mg/kg/day was determined to be core supplementary
because the NOEL for developmental toxicity (supernumerary ribs)
was not definitively established. The NOEL and LOEL for maternal
toxicity for this study are 30 and 60 mg/kg/day, respectively,
based on decreases in maternal body weight, body weight gain,
and food consumption at 60 and 120 mg/kg/day.   Furthermore,
increased embryolethality (embryotoxicity) was only observed
at the highest dose level tested (120 mg/kg/day). This study
must be repeated to clearly define a NOEL for developmental
toxicity.
    The above rat study indicated that triadimenol caused a
dose-dependent, statistically significant increase in the
incidence of rudimentary supernumerary ribs. Although the
effect at the low dose level was not statistically significant,
it was considered to be treatment related because of the dose-
related trend.
    The biological significance of the manifestation of
supernumerary ribs is subject to scientific debate, especially
if the ribs are not fully developed (rudimentary). Nonetheless,
the margin of safety (MOS) for this effect must be taken into
consideration. The MOS is the ratio between the NOEL for the
effect and the acute exposure in mg/kg/day. A NOEL for
developmental toxicity could not be defined in the rat teratology
study but it is unlikely to be far below the threshold (LEL)
of 30 mg/kg/day observed in the current study.
    Based on worker exposure information and an estimation of
the NOEL at about 15 mg/kg/day for developmental toxicity
(rudimentary supernumerary ribs in rats) and assuming a maximum
dermal penetration of about 10%, a margin of safety was calculated
to be >100 for factory workers involved in seed treatments using
a closed system. Because of possible developmental toxicity
and the lack of a will defined NOEL for this effect, the product
label must include a recommendation for the use of protective
clothing by factory workers involved in the treatment of seeds
and for farm workers handling the treated seed.

°Reproduction.

A rat multigeneration reproduction study using doses of
0, 20, 100, and 500 ppm (equivalent to 0, 1, 5, and 25 mg/kg
bwt/day for males and females) indicated that the NOEL and
LOEL for both parental and pup toxicity are 100 and 500 ppm,
respectively, based on significant body weight and organ weight
changes.   The NOEL for reproductive toxicity is 500 ppm, highest
dose level tested.

°Mutagenicity:

A reverse mutation assay (AMES), a dominant lethal
test in mice, DNA damage/repair, unscheduled DNA synthesis,
in vitro and in vivo (rat) cytogenic assays, and a forward
mutation in mice, all of which were negative for mutagenic
effects.

ENVIRONMENTAL FATE

Hydrolysis: STABLE. Triadimenol in sterile aqueous buffer
              solutions showed no apparent degradation
              at either temperature or pH tested.
              Recovery was 97% greater after 32 days of
              incubation.

Soil Surface Photolysis: STABLE. Triadimenol appears to
              be stable to photolysis on the soil surface.
              Studies indicate that triadimenol photodegrades
              with a half-life of 36 hours in distilled
              water and 17 hours in a photo-sensitized
              (acetone) solution.

Aerobic Soil Metabolism: STABLE. Studies indicate that
              triadimenol has an estimated aerobic half-life
              of 8 to 9 months. Triadimenol reached a maximum
              level of 68% of that applied at 14C in 71 days
              and declined slightly to 45.2% by day 238.
              Consequently, the anaerobic half-life is
              considerably greater than 8-9 months.

Adsorption/Desorption: Because of its low adsorption
              coefficients, triadimenol is shown to have a low
              to moderate potential to bind to soil particles.
              Studies indicate that the adsorption coefficient,
              k, for triadimenol ranged from 2.37 to 5.26.
              The k values for desorption ranged from 1.49
              in a silty clay soil (0.49 ppm) to 9.12 in a
              loam soil (9.57 ppm). Consequently, there
              is no correlation between adsorption and
              soil organic matter content. The highest

degree of adsorption was observed with the
loam soil, intermediate in organic matter
content.

Environmental fate data requirements have been satisfied
with the exception of a field dissipation study. The company will
be required to submit results of this study by July 1990.

ECOLOGICAL CHARACTERISTICS

Studies submitted show that this chemical is practically
non-toxic to birds, slightly toxic to fish and moderately
toxic to aquatic invertebrates. It is unlikely that the seed
treatment use of triadimenol will affect any terrestrial or
aquatic animals. Chronic effects are unlikely due to the low
use rates and because the seed treatment use requires incorporation
of seeds into the soil. For the above reasons it is also
unlikely that this use will affect any endangered species.

BENEFITS

This chemical has been shown to be environmentally safe,
is used at low rates and has a broad biological spectrum.
Triadimenol controls seed-, soil-, and wind-borne pathogens of
wheat, barley, oats, rye, corn and sorghum. Crops may be
grazed 40 days after seeding. The chemical improves winter
survival and drought tolerance of cereals, lowers the inoculum
levels for overwintering foliar diseases and may eliminate
the need for early season foliar sprays.

TOLERANCE ASSESSMENT:

Tolerances are established for the fungicide triadimenol
and its butanediol metabolite (calculated as triadimenol) in or
on the following commodities: 2.5 ppm for green forage of
barley, oats, rye and wheat; 0.1 ppm for straw of barley, oats,
rye and wheat; 0.05 ppm for grains of barley, oats, rye and wheat,
corn fodder, fresh corn (including sweet), corn forage, corn
grain, and green forage of sorghum; and 0.01 ppm for sorghum grain and
sorghum fodder. Tolerances are established for the fungicide
triadimenol and its metabolites containing the chlorophenoxy
moiety (calculated as triadimenol) in or on the following
commodities: 0.1 ppm for fat, meat and meat by-products of cattle,
goats, hogs, horses, and sheep; and 1.01 ppm for eggs, milk, and fat,
meat and meat by-products of poultry.

Where tolerances are established for residues of both 1-(4-
chlorophenoxy)-3,3-dimethyl-1-(1H-1,2,4-triazol-1-yl)-2-butanone
(triadimefon) and triadimenol, including its butanediol metabolite,

in or on the same raw agricultural commodity and its products
thereof, the total amount of such residues shall not yield more
residue than that permitted by the higher of the two tolerances.
The nature of the residue is adequately understood and the
Agency concluded that the pesticide is useful for the purposes
for which tolerances are sought and that the establishment of
the tolerances will protect the public health.

SUMMARY OF MAJOR DATA GAPS:

The Agency concurs with conditional registration of this
chemical for use as a seed treatment fungicide pending submission
of a field dissipation study by July 1990.

CONTACT PERSON AT EPA

Susan T. Lewis,
Acting Product Manager (PM) 21,
Registration Division (H-7505C),
Environmental Protection Agency,
401 M St., SW.,
Washington, DC   20460

Office location and telephone number:
Rm. 227, CM#2,
1921 Jefferson Davis Highway,
Arlington, VA   22202
(703) 557-1900

DISCLAIMER: The information in this Pesticide Fact Sheet is
a summary only and is not to be used to satisfy data
requirements for pesticide registration and reregistration.

TRIBUTYLTIN

Reason for Issuance: Preliminary Determination
Date Issued: October 1, 1987
Fact Sheet Number: 143

1. Description of chemicals

Chemical Name	Common Name	Chemical Abstract Service Number	EPA Shaughnessy Code
bis(tributyltin) adipate	none	7437-35-6	083117
bis(tributyltin) dodecenyl succinate	none	12379-54-3	083101
bis(tributyltin) oxide	TBTO	56-35-9	083001
bis(tributyltin) sulfide	none	4804-30-4	083113
tributyltin acetate	none	56-36-0	083105
tributyltin acrylate	none	13331-52-7	083121
tributyltin fluoride	TBTF	1983-10-4	083112
tributyltin methacrylate and	TBTM	2155-70-6,	083120
copolymer	TBTM	26345-187	083119
tributyltin resinate	none	none assigned	083114

Chemical family: Organotins

Pesticide type: biocide, antifoulant, and disinfectant. The Special Review is being conducted for the use of these chemicals in antifoulant paint registrations. Twenty TBT compounds are registered as pesticidal active ingredients and nine of the compounds are registered for use in antifouling paints. The major TBT pesticide is tributyltin oxide.

Registrations: Antifoulant paints containing tributyltin (TBT)
compounds were initially registered in the early 1960's. At
the initiation of the Special Review there were 364 TBT antifouling paint formulations and 20 formulating intermediates
with a total of 61 manufacturers. Since January, 1986, 162
products have been voluntarily cancelled and some companies have
merged such that there are now 210 registered antifouling
paint formulations and 12 formulating intermediates with
34 manufacturers, although nearly half of the paint formulations
have been suspended for non-compliance with a Data Call In
Notice issued in July, 1986.

2. Use patterns and formulations

Application sites: TBT's are used in antifoulant paints
applied to ship and boat hulls as well as buoys, crab pots,
fish nets, etc. TBT's are also registered as wood preservatives, disinfectants, and biocides for use in cooling towers,
pulp and paper mills, breweries, leather processing facilities,
and textile mills.

Paint formulations: TBT antifouling paints may be classified
into three categories according to the way the TBT moiety is
incorporated into the paint coating and subsequently released.

  °   Free association paints: In these conventional
      coatings the TBT is physically incorporated into
      the paint matrix (which contains the pigment, water-
      soluble resins, and inert substances). The TBT
      leaches from the paint surface by diffusion. Gradually,
      the paint matrix becomes clogged with insoluble materials
      trapping some of the toxicant while leaving the surface
      unprotected.

  °   Copolymer paints: In this category the TBT moiety
      is chemically bonded to a polymer matrix. The
      biocide is released only by chemical hydrolysis
      of the TBT itself. These paints are characterized
      by slow dissolution from ship hulls and thus achieve
      a constant but prolonged, release of antifoulant
      toxicant.

  °   Ablative paint: These paints have characteristics of
      both of the other two types of paint. The TBT is not
      bound to a polymer, but is incorporated into the paint
      matrix. Ablative paints are soft paint films with the
      rosin portion of the paint slightly water soluble so
      that the surface slowly sloughs or ablates away as the
      painted vessel moves through the water. This allows
      new toxicant layers to be exposed and prevents the
      buildup of insoluble materials.

A TBT antifouling paint formulation can have a single TBT
active ingredient, can be combined with one or more of the
other eight TBT antifoulants, can be combined alone with
copper compounds (especially cuprous oxide), can be combined
with triphenyltin fluoride (another organotin antifoulant), or
can be combined with copper and other organotin compounds.
Products are formulated with 0.5 to 41 percent active ingredient
TBT. Application rates are commonly from 150 to 400 square
feet per gallon of paint.

3. Science Findings

Chemical characteristics: Tributyltin compounds are chemically
characterized by a tin (Sn) atom covalently bonded to three
butyl (C_4H_9-) moieties. When released from the paint matrix or
polymer into the aqueous environment, TBT exists mainly as a
mixture of TBT hydroxide, TBT chloride, and TBT carbonate
species from reaction with carbonates in seawater.

Environmental fate: The environmental chemistry and fate of
tributyltin in aquatic environments are complex and not completely understood. Studies indicate that photolysis and
microbial action are potential mechanisms of degradation from
tri- to di- to monobutyltin and finally to inorganic tin.
Studies indicate the half-life of TBT may be 116 days in aerobic
soils, 815 days in anaerobic soils, 6 to 12 days in sea water,
and up to 238 days in fresh water. TBT accumulates in sediment
at levels that are one to four orders of magnitude greater than
the concentration found in the respective water column. This
amassing of toxicant can have serious consequences for organisms
living and feeding in the benthos.

Low concentrations of elemental or inorganic forms of tin
appear to cause negligible toxicological effects in man or wildlife. However, when carbon groups, such as butyl units, are
added to the tin, there is an increase in fat solubility,
ability to penetrate biological membranes, and consequently,
toxicity. As the number of butyl groups is increased from one
to three, there is a corresponding increase in lipophilicity
and toxicity to aquatic organisms. However, the addition of a
fourth butyl group decreases the toxicity of the molecule.

Ecological characteristics: The TBT compounds are toxic
to aquatic organisms at the low parts per billion (ppb) level.
A summary of aquatic TBT toxicity values are presented below:

° Fish

Acute Toxicity:	0.96 - 24.0 ppb
Chronic Toxicity:	> 0.2 ppb
Bioaccumulation:	200- to 4300-fold
Behavioral Toxicity:	Avoidance occurred at 1.0 to 24.0 ppb. Fish may not detect harmful sublethal levels.

° Bivalves

Acute Toxicity:	0.9 to 2.3 ppb
Chronic Toxicity:	0.02 to 0.05 ppb
Bioaccumulation:	2000- to 6000-fold
Bioavailable:	Yes, even with high silt loads.

° Gastropods

Acute Toxicity:	> 0.01 ppb
Chronic Toxicity:	0.002 to 0.02 ppb
Bioavailable:	Yes, was promoted as a molluscicide against schisto- somiasis because it readily adsorbed to organic matter; snails preferentially ingest organic matter. (Lowest value is an extrapolation.)

° Crustaceans

Acute Toxicity:	0.42 to 2.2 ppb
Chronic Toxicity:	> 0.09 ppb
Bioaccumulation:	4400-fold
Bioavailable:	Yes, more from food than from water.
Behavioral Toxicity:	0.5 ppb caused positive photo- taxis in daphnids.

° Algae

Acute and Chronic Toxicity:	growth inhibition at 0.1 to 0.35 ppb
Bioaccumulation:	800- to 30,000-fold
Bioavailable:	Yes, since filter feeders readily consume algae it can be assumed that phytoplankton laden with TBT can be con- summed by aquatic organisms.

TBT concentrations are reported to be highest in areas of heavy boating and shipping activity. Before recoating, old paint containing the remaining TBT residue is scraped from the vessel hull and sometimes the scrapings are washed into the water adjacent the boat or shipyard (despite TBT labels prohibiting this practice). TBT has been measured in marine and fresh water environments at levels indicated below. Note that ND means non-detectable or below the level of detection of the analytical method used.

Chesapeake Bay:	ND to 0.8 ppb
San Diego Bay:	ND to 1.0 ppb
San Francisco Bay:	ND to 0.16 ppb
Honolulu Harbor	0.045 to 0.27 ppb
Los Angeles/	
Long Beach Harbor:	ND to 0.12 ppb
Narragansett Bay:	ND to 0.13 ppb
Thames River (CT):	ND to 0.009 ppb
Mayport Florida:	ND to 0.016 ppb
Lake Superior:	0.02 ppb
Lake Ontario:	0.05 to 0.84 ppb

Population Effects: In France, a correlation has been found between TBT in the water column of certain estuaries and gross malformations in Pacific oysters grown in commercial oyster beds in and around areas of heavy boating activity. Following a ban on TBT antifouling paints on vessels less than 25 meters in length, the degree of shell deformities has decreased and the regeneration rate of juvenile oysters (spat) has improved. In England TBT has been reported causing similar shell deformities in Pacific oysters and reproductive abnormalities (imposex) in dogwelk snails. The Department of Fish and Wildlife of Oregon recently have found shell deformities in commercial Pacific oyster beds in Coos Bay which are near a small shipyard applying and removing TBT antifouling paints.

4. TBT Release Rates

The Tributyltin Data Call In Notice (TBT DCI) required all registrants of TBT antifouling paints to measure TBT release from registered paints following a test method developed in cooperation with the American Society for Testing and Materials (ASTM). In addition, each laboratory conducting the TBT release test was required to test a standard copolymer test paint.

Release rate data were submitted for 96 TBT antifouling paint products by July 1, 1987. From a review of these data, it was determined that at least 57 of the tests were conducted satis-factorily and all data were normalized according to adjustments made using the standard test paint results. Two release rate

values were determined for each product: 1) a short term
cumulative release measured over the first 14 days of the test
period and 2) an average daily release rate (average of the
daily release over weeks 3 to 5 of the test).   Details of
release rate data are available in the Tributyltin Technical
Support Document.

Generally, the release rates start high (short term cumulative
release ranged from 1 to 1128 ug/cm^2) and gradually decrease
over the course of the test period (average daily release rate
ranged from 0.02 to 21.53 $ug/cm^2/day$). Some paints with a high
percentage of TBT have a much higher short term cumulative
release than do other paints tested. It was concluded that
while there was a strong statistical correlation between the
percent active ingredient and the average daily release rate,
the data points were too scattered for the percent active
ingredient alone to be useful for regulating TBT paints. The
scattered data points indicated that other factors, such as the
type and quality of the inert ingredients (resins, rosins,
binders, etc.) may be important in determining the release rate
of a paint and that regulating on percent active ingredient
would not necessarily reduce environmental loading.

Results of the release rate tests also showed that some TBT
ablative and free association paints have lower release rates
than many copolymer paints. The free association paints with
release rates lower than copolymer paints generally had a low
percentage of TBT in their formulations.

5.   Summary of regulatory position and rationale

The Agency initiated a Special Review of TBT products used as
antifoulants in January, 1986, based on concern of possible
adverse effects of TBT to nontarget aquatic organisms. The
Agency recognized that additional data were required and issued
a Data Call In Notice (DCI) for TBT products registered for use
as antifoulants or registered as formulating intermediates used
to produce antifoulant products. The DCI required information
on product chemistry data, TBT release rate data, usage data,
worker exposure data, ecological effects data, and environmental
fate data. The DCI also required submission of any available
efficacy data. For many products, product chemistry data, TBT
release rate data, and usage data have been received. Although
the Agency will not have the ecological and environmental fate
data from the DCI for another one to four years, the Agency
believes sufficient data are available to propose a set of
regulatory actions. In 1985, the Agency issued a DCI on
tributyltin oxide requiring data on chronic toxicity to mammals.
These data are not due into the Agency until 1990.

The Agency examined a range of regulatory options to reduce TBT
loading into the environment. The Agency considered: 1) cancelling all TBT antifouling paint registrations, 2) proposing
a restriction on the maximum permitted percent TBT active
ingredient in registered products, 3) regulating the type of
paint formulation, 4) regulating the release rate, 5) restricting the size of vessel treated, and 6) classifying TBT antifouling paints as restricted use pesticides and requiring additional
wording on the label giving directions concerning application,
removal, and disposal of TBT paints to reduce the amount of TBT
entering the aquatic environment from these activities.

The Agency is proposing continued registration of TBT antifouling
paint products with certain regulatory restrictions: 1) limit
the maximum organotin release rate from paint formulations,
2) prohibit use of TBT on non-aluminum vessels under 65 feet in
length, and 3) classify TBT antifouling paints as restricted
use pesticides and require additional wording on the label
regarding application, removal, and disposal of TBT paints to
prevent introduction of TBT paint wastes into the aquatic
environment. Specifically, the Agency is proposing a maximum
short term cumulative release (days 1 to 14 of test period) of
168 ug organotin (calculated as TBT cation)/cm^2 and an average
daily release rate (averaged over weeks 3 to 5 of the test) of
4.0 ug organotin (calculated as TBT cation)/cm^2/day. The
Agency believes that the proposed release rate restrictions would
reduce loading five-fold from its estimated average daily
release rate of 20 ug/cm^2/day before initiation of the Special
Review (calculated for all TBT paint formulations from the
submitted release rate data). The prohibition of use on vessels
under 65 feet should reduce by 37 percent (the estimated volume
of TBT paint used on this size class by a boat and shipyard
survey) the total amount of TBT antifoulant currently used and
potentially available for environmental contamination. This
limitation in use will result in a reduction in TBT concentration
primarily in estuarine and fresh water areas where these small
vessels are used and moored and where the risk from TBT effects
is the greatest. The restricted use classification and additional
wording on the label regarding size restriction and directions for
applying, removing, and disposing of TBT paints without introducing
paint wastes into the water will help ensure that TBT paints
are not used on vessels under 65 feet and that TBT paints will
be applied and disposed in a manner which will reduce the risk
of inadvertent aquatic contamination. Monitoring and efficacy
data will be required separately which will be used to evaluate
the effectiveness of these proposals.

     Upon receipt and evaluation of additional data, the Agency
may determine that further regulatory action is warranted.

6.   Contact person at EPA

Dr. Janet L. Andersen
Environmental Protection Agency
Office of Pesticide Programs
Registration Division (TS-767C)
401 M Street, S.W.
Washington, D.C.   20460

TRICHODERMA HARZIANUM and *TRICHODERMA POLYSPORUM*

Reason for Issuance: Registration of New Biological Pesticide
Date Issued: July 1989
Fact Sheet Number: 203

1. Description

Generic Names:	*Trichoderma harzianum* (ATTC 20476) and *Trichoderma polysporum* (ATTC 20475)
Trade Name:	Binab™ T
EPA Shaughnessy Codes:	128903 and 128902, respectively
Year of Initial Registration:	1989
Pesticide Type:	Biofungicide
U.S. and Foreign Producers:	Binab™ USA, Inc. c/o E.R. Butts International, Inc. 555 Clinton Avenue P.O. Box 3337 Bridgeport, CT 06605-0337

2. Use Patterns and Formulations

Application sites: Mixtures of *Trichoderma harzianum* and *T. polysporum* are proposed for use in the control of internal decay of wood utility poles, playground structures, and fence posts and for use in the control of decay of pruning wounds of ornamental, shade, and forest trees.

Types of formulations: 28% pelleted end-use product (14% each active ingredient) and 33.2% wettable powder end-use product (16.6% each active ingredient).

Types and methods of application: Pellets are placed in holes drilled into wooden members followed by sealing of the holes with a vented plastic plug. Wettable powder formulation is mixed with water (1:2 v/v) and applied to pruning wounds of trees with a paint brush, followed by application of a wound sealant.

Application rates: For pellets: Three pellets/hole. Holes
are 3/8" diameter, 4 1/2" deep, spaced 4" apart and placed
2-4" above ground level. Product contains approximately
1,400 pellets/pound. For wettable powder: Apply 1:2 mixture
of product with water (v/v) to cover pruning wound.

3. Science Findings

Summary Science Statement:

The toxicological data which were submitted for these active
ingredients included reports of acute oral and hypersensitivity
studies and a request for waiver of all other toxicological
data requirements was made based on the contention that
Trichoderma species are widespread in the environment and
are innocuous. The acute oral toxicity/infectivity study and
hypersensitivity study were classified as Core-minimum and as
such support registration of the active ingredients.

The species Trichoderma does not grow at temperatures above
28°C and is not capable of growth in warm blooded animals or
birds. The proposed use patterns for the products would not
expose aquatic wildlife to the fungi. The fungi are not
pathogenic to plants or insects. Trichoderma species are
naturally occurring in soils throughout the world. The
proposed application sites and application rates would not
cause a detectable increase over naturally occurring background
levels. There would be no increased exposure to any non-target
wildlife of ecological concern. The proposed uses do not pose
a "may effect" situation to any endangered or threatened animal
or plant species.

Data for environmental fate are not triggered under current
requirements for the proposed products since the organisms are
naturally occurring species and the results of initial (Tier I)
tests did not trigger the need for additional testing.

Chemical Characteristics:

Color:	greyish beige
Physical State:	powdered solid or pellet
Odor:	moldy flour
Density:	171 grams/liter (WP);
	650 grams/liter (Pellets)
pH:	5.4

Toxicological Characteristics:

Acute effects:

Acute oral toxicity (mice):
The acute oral LD_{50} toxicity of TUF (an extract from the Trichoderma strains) was greater than 4,000 mg/kg after 1 and 14 days. The acute oral and subcutaneous LD_{50} toxicities of trichodermin (antibiotic) in mice were greater than 1,000 mg/kg and 500-1,000 mg/kg respectively.

Hypersensitivity study (Guinea pig):
No sensitization occurred when Guinea pigs were dermally exposed to Trichoderma spores.

Other Toxicity Testing:
Additional toxicity testing was waived based upon submission of data or information:
1. which indicated that the fungi do not grow at or near the body temperatures of mammals or birds.
2. which demonstrated that no toxins or antibiotics were produced in simulated use situations.
3. which showed that exposure to Trichoderma strains has occurred in personnel working with the fungi for times of up to 18 years with no adverse toxicological effects.
4. which provides the criteria used to determine the extent to which formulated preparations are free from contaminating microorganisms.
5. which confirmed the exempt status of certain inert ingredients.

Ecological Characteristics

Data on the ecological characteristics of the products were waived based upon the fact that the fungi will not grow at or near body temperatures of mammals or birds, that the fungi are ubiquitous in nature and that the use patterns are unlikely to result in additional exposure of aquatic organisms or other non-target wildlife of ecological concern.

4. Benefits

The use of Binab" T pellets will control internal decay of wood in utility poles, playground structures, and fence posts and may replace chemical treatments for these uses to some extent.   The use of Binab" T Wettable Powder will control decay of pruning wounds of ornamental, shade, and forest trees aiding in the maintenance of plant health and esthetic value of trees.

5. Tolerance Assessment

    The products will not be used in situations where tolerances
    are required. Trees to be treated are limited to those which
    will not be used for food or feed production.

6. Summary of Major Data Gaps

    No major data gaps exist for these active ingredients.

    Contact Person at EPA

    Susan T. Lewis
    Acting Product Manager (21)
    Fungicide-Herbicide Branch
    Registration Division (H-7505C)
    Environmental Protection Agency
    401 M St., SW.,
    Washington, D.C. 20460

DISCLAIMER: The information presented in this Pesticide Fact Sheet
is for informational purposes only and may not be used to fulfill
data requirements for pesticide registration and reregistration.

Glossary

ADI	acceptable daily intake
a.i.	active ingredient
EP	end-use product
EUP	end-use product
LDT	lowest dose tested
LEL	lowest effects level
LOEL	lowest observed effect level
MATC	maximum acceptable toxic concentration
mbyp	meat byproduct
MP	manufacturing-use product
MPI	maximum permitted intake
NOEL	no observed effect level
NPDES	National Pollution Discharge Elimination System
PADI	provisional acceptable daily intake
PAI	pure active ingredient
PGAI	pure grade active ingredient
PGI	pre-gazing interval
PHI	pre-harvest interval
PIMS	Pesticide Incident Monitoring System
PIS	primary irritation score
PLD	provisional listing dose
ppb	parts per billion
ppm	parts per million
PSI	primary skin irritant
TMRC	theoretical maximum residue concentration or contribution

Numerical List of Pesticide Fact Sheets
Volume 2

*Indicates revised fact sheet was issued.

Number	Chemical Name
2.1	CRYOLITE*
3.2	ETHOPROP*
5.2	TERBUFOS*
19.1	ALDICARB*
34.2	PHORATE*
48.1	PICLORAM*
82.1	CHLORIMURON ETHYL* (CLASSIC)
89.1	AVERMECTIN B$_1$* (AFFIRM)
93	*BACILLUS THURINGIENSIS*
94.2	2,4-DICHLOROPHENOXYACETIC ACID* (2,4-D)
96.1	DIAZINON*
97.1	ALACHLOR
102.1	CARBON TETRACHLORIDE (AND OTHERS)*
103.1	CADMIUM SALTS*
107.2	HEPTACHLOR*
118	ALUMINUM AND MAGNESIUM PHOSPHIDE
134	DICHLORVOS
140	ACEPHATE
141	DALAPON
142	FENITROTHION
143	TRIBUTYLTIN
144	OXYDEMETON-METHYL
145	FENVALERATE
146	BROMINE CHLORIDE
147	AMITRAZ
148	PHOSALONE
149	PROPANIL

649

Number	Chemical Name
150	CHLORPROPHAM
151	MONOCARBAMIDE DIHYDROGEN SULFATE
152	MALATHION
153	ASULAM
154	PHOSPHAMIDON
155.1	METALAXYL
156	MEVINPHOS
157	FENOXAPROP-ETHYL
158	ALLETHRIN STEREOISOMERS
159	ASSERT
160	*PSEUDOMONAS FLUORESCENS* EG-1053
161	COAL TAR/CREOSOTE
162	HARMONY 75 DF
163	LACTIC ACID
164	CYFLUTHRIN
165	CHLORINATED ISOCYANURATES
166	DCPA
167	FORMALDEHYDE AND PARAFORMALDE-HYDE
168	QUIZALOFOP ETHYL
169	FENTHION
170	MALEIC HYDRAZIDE
171	KARATE (PP321)
172	PHENMEDIPHAM
173	GLYPHOSATE
174	SODIUM FLUOROACETATE
175	STRYCHNINE ALKALOID AND SULFATE
176	ETHEPHON
177	BIFENTHRIN
178	STRYCHNINE UPDATE
179	2,4-DB
180	2-(2,4-DICHLOROPHENOXY)PROPIONIC ACID
181	METIRAM
182	MANEB
183	HEXAZINONE
184	TETRACHLORVINPHOS
185	SULFOTEPP
186	STREPTOMYCIN
187	METHOXYCHLOR
188	OXYTETRACYCLINE
189	PROPAZINE
190	TEFLUTHRIN
191	METALDEHYDE
192	MECOPROP
193	RESMETHRIN
194	DIFENZOQUAT

Number	Chemical Name
195	BENDIOCARB
196	IMAZETHAPYR
197	CARBOFURAN
198	ROTENONE
199	CYPERMETHRIN
200	HEXYTHIAZOX
201	METHOMYL
202	FLURPRIMIDOL
203	*TRICHODERMA HARZIANUM* and *TRICHODERMA POLYSPORUM*
204	TRIADIMENOL
205	SULFLURAMID
206	TERBACIL
207	COUMAPHOS
208	MCPA
209	ETHION

Common Name Index
Volume 2

Generic Name Index
Volume 2

Trade Name Index
Volume 2

Includes all names listed on each individual fact sheet under the headings *Trade Names* or *Other Names*.

AAstar - 492

Abamectin - 55

AC-47,300 - 252

AC 92,100 - 611

Acclaim - 258

Accothion - 252

ACL 70 - 108

Acme MCPA Amine 4 - 401

Affirm - 55

Agricultural Terramycin - 478

Agrimee - 55

Agri-Mycin 17 - 571

Agri-Step - 571

Agritox - 401

Agro One - 401

Agrothion - 252

Aker-root - 560

Alanex - 11

American Cyanamid Co. Code
  No. EI4049 - 374

Ammo - 148

Antimilace - 430

Apamidon - 507

Apavinphos - 455

Apron - 422

Assert Herbicide - 40

Assure - 545

Asulox - 47

Asuntol - 128

Atgard - 203

Attack -65

Avenge - 213

Avid - 55

Azonfene - 500

Baam - 35

Bactimos - 65

Bactospeine - 65

Barbasco - 560

Barricade - 148

Basfapon - 155

Basfapon B - 155

Basfapon N - 155

Bay 21/199 - 128

Bay 21097 - 470

Bay 30130 - 527

Baycid - 264

Bayer S-1102A - 252

Bayer S-1752 - 264

Bayer S-5660 - 252

Bayer 29493 - 264

Bayer 41831 - 252

Baymix - 128

Baytan - 628

Baytex - 264

Beet-Kleen - 114

Belmark - 274

Benzinoform - 98

Benzofos - 500

Berkmycin - 478

Betanal - 485

Meta - 430
Metason - 430
Metasystemox - 470
Metasystox-R - 470
Metathion E-50 - 252
Methoxcide - 442
Methoxone - 401
Metilmerkaptofosoksiol - 470
Mevinox - 455
Mevinphos - 455
Milogard - 536
Milo-Pro - 536
MK-936 - 55
Mocap - 241
M-one - 65
Monourea sulfuric acid adduct -
   465
Morton EP-452 - 485
Multamat - 73
Muscatox - 128
Mycoshield - 478
Namekil - 430
NC6897 - 73
Negashunt - 128
Nekos - 560
Nicouline - 560
Niomil - 73
Nogos - 203
No-pest - 203
Novathion - 252
Novobac - 65
NRDC 149 - 148
Nudrin - 433
Nuvanol - 252
Nux Vomica - 576, 580
Oko - 203
Orthene - 1
OS-2046 - 455
Oxacycline - 478
Oxyatets - 478
Paraformaldehyde - 292
Phenmediphame - 485
Phomene - 401
Phorate - 492
Phosdrin - 455
Phosfene - 455
Phostoxin - 30
Pillarzo - 11

Plantfume - 583
Plantomycin - 571
Polado - 301
Polyram - 449
Polyram-Combi - 449
Pramitol - 536
Prentox - 442
Prentox Malathion 95% Spray - 374
Prep - 219
Prokil - 136
Propanex - 527
Propanilo - 527
Propel - 369
Proprop - 155
Protex - 560
Proturf - 422
Prozinex - 536
Pursuit - 354
Pydrin - 274
Pynamin - 22
Pynamin Forte - 23
Pynosect - 552
Quick Phos - 30
R2170 - 470
Rabon - 621
Radapon - 155
Rampart - 492
Ratsbane - 567
Ravap - 203
Regulox W - 384
Regulox 50W - 384
Resitox - 128
Retard - 384
Revenge - 155
Rhodiacide - 229
Rhodocide - 229
Rhonox - 401
Ridomil - 422
Riomitsin - 478
Ripcord - 148
Riselect - 527
Riton - 203
Rodeo - 301
Rogue - 527
Rotacide - 560
Roundup - 301
Roundup L&G - 301
Royal MH-30 - 384

Alphabetical List of Pesticide Fact Sheets
Volume 1

EPN
EPTC
Ethalfluralin
Ethoprop
Ethylenethiourea (ETU)
Fenaminosulf
Fenbutatin-Oxide
Fenoxycarb
Fensulfothion
Fluchloralin
Fluometuron
Fluridone
Fluvalinate
Fonofos
Formetanate Hydrochloride
Glycoserve
Heliothis NPV
Heptachlor
Hybrex
Imazaquin
Isazophos
Isomate-M
Lactofen
Lead Arsenate
Linalool
Lindane
Linuron
Mancozeb
Methiocarb
Methyl Bromide
Methyl Parathion
Metolachlor
Metribuzin
Metsulfuron Methyl
Monocrotophos
Nabam
Naled

Naptalam
Nitrapyrin
Norflurazon
Oxamyl
Paraquat
Parathion
Pendimethalin
Perfluidone
Phorate
Phosmet
Picloram
Potassium Bromide
Potassium Permanganate
Prometryn
Pronamide
Propachlor
Propargite
Propham
Simazine
Sodium and Calcium Hypochlorites
Sodium Arsenate
Sodium Arsenite
Sodium Omadine
Sodium Salt of Fomesafen
Sulfuryl Fluoride
Tebuthiuron
Terbufos
Terbutryn
Thiodicarb
Thiophanate Ethyl
Thiophanate Methyl
Thiram
TPTH
Trimethacarb
Vitamin D_3
Wood Preservatives

Numerical List of Pesticide Fact Sheets
Volume 1

Number	Chemical Name	Number	Chemical Name
1	Aliette	29	Thiram
2	Cryolite	30	Thrichlorfon*
3.1	Ethoprop	31	Wood Preservatives
4	Naled	32	Bronopol
5.1	Terbufos	33	Dantochlor
6	EPTC	34	Phorate
7	Butylate	35	Captafol
8	Dicamba	36	Chlorothalonil
9	Diuron	37	Chlorpyrifos
10	Fenaminosulf	38	Potassium Bromide
11	Formetanate HCL	39	TPTH
12	Anilazine	40	Alachlor*
13	DCNA	41	Cyanazine
14.1	Fensulfothion	42	Vitamin D_3
15	Chlorobenzilate	43	Disulfoton
16	Dicofol	44	Propachlor
17	Arosurf	45	Demeton
18	Thiodicarb	46	Fenarimol*
19	Aldicarb	47	Glycoserve
20	Amitrole	48	Picloram
21	Carbaryl	49	Naptalam
22.1	Fonofos	50	Pendimethalin
23	Simazine	51	Sulfuryl Fluoride
24	Carbofuran	52	Fluchloralin
25	Carbophenothion	53	Metribuzin
26	Daminozide	54	Nitrapyrin
27	Heliothis NPV	55	Dipropetryn
28	Linuron	56	Cyhexatin

Number	Chemical Name	Number	Chemical Name
57	Chlorpyrifos Methyl	98	Methyl Bromide
58	Ethalfluralin	99	Propargite
59	Actellic	100	Azinphos-Methyl
60	Norflurazon	101	Phosmet
61	Sodium Omadine	102	Carbon Tetrachlor-
62	Clipper (Paclo-		ide
	butrazol)	103	Cadmium
63	Arsenal	104	Terbutryn
64	Bentazon and	105	Cyromazine
	Sodium Bentazon	106	Metolachlor
65.1	Dinocap	107.1	Heptachlor
66**		108	Aldrin
67	3,5-Dibromo	109	Chlordane
68.1	Diflubenzuron	110	Arsenic Trioxide
69**		111	Calcium Arsenate
70	Pronamide	112	Lead Arsenate
71	Metsulfuron Methyl	113	Sodium Arsenite
72	Monocrotophos	114	Sodium Arsenate
73	Lindane	115	Aldoxycarb
74	Perfluidone	116	Parathion
75	Captan	117	Methyl Parathion
76	Trimethacarb	118	Aluminum and Mag-
77	Linalool		nesium Phosphides*
78	Fenoxycarb	119	Fenbutatin-Oxide
79	Sodium and Calcium	120	Methiocarb
	Hypochlorites	121	Prometryn
80	Potassium	122	Dichlobenil
	Permanganate	123	Propham
81	Fluridone	124	Nabam
82	Chlorimuron Ethyl	125	Mancozeb
83	Imazaquin	126	Isomate-M
84	Thiophanate Ethyl	127	EPN
85	Hybrex	128	Lactofen
86	Fluvalinate	129	Oxamyl
87	Copper Sulfate	130	Dinoseb
88	Fluometuron	131	Paraquat
89	Avermectin	132	Sodium Salt of
90	Command		Fomesafen
91	Arsenic Acid	133	Brominated Salicyl-
92	Thiophanate Methyl		anilide
93	Bacillus Thuring-	134*	
	iensis*	135	Dodine
94.1	2,4-Dichlorophen-	136	Diphenamid
	oxyacetic acid	137	Tebuthiuron
95	1,3-Dichloropropene	138	Isazophos
96	Diazinon	139	Ethylenethiourea
97*			(ETU)

*Fact Sheet not currently available.   **Number not in use.

Other Noyes Publications

PESTICIDE WASTE DISPOSAL TECHNOLOGY

Edited by

James S. Bridges and Clyde R. Dempsey

U.S. Environmental Protection Agency

Pollution Technology Review No. 148

This book attempts to define practical solutions to pesticide users' disposal problems.

A major agreement must be reached on what can be done, legally and technically, to deal with the difficulties of proper pesticide-related waste disposal, and who should share in the cost of a clean environment. Pesticide commerce and use are regulated under the Federal Insecticide, Fungicide and Rodenticide Act and by state laws and rules. However, once applications of pesticides are completed, any excess pesticide concentrate, unapplied diluted pesticide, and discarded pesticide containers may be regulated as wastes, some of which may be considered hazardous. Although past disposal problems and future policy changes are of major importance, the primary focus must be the solutions to the existing problems facing the pesticide user industry today.

The book is presented in three parts. Part I covers disposal needs; federal/state regulatory requirements; pesticide degradation properties; disposal technology options, including physical treatment, biological treatment, chemical treatment, land application and incineration options; storage, handling, and shipments of pesticide wastes; and empty pesticide container disposal programs.

Part II addresses issues regarding the effectiveness of current state-of-the-art capabilities, identifies emerging techniques or technologies that may be applicable along with technologies being applied in other areas, and describes the need for research efforts capable of providing results in a three-to-five year time frame as they pertain to the treatment/disposal of dilute pesticide wastewater. Twelve technologies are discussed in some detail.

Part III includes industry's role in users' waste disposal, on-site demonstration projects, users' waste minimization/reuse and users' treatment/storage/disposal.

The condensed contents given below lists **parts and selected chapter titles.**

ISBN 0-8155-1157-4 (1988)　　　　　7"x10"　　　　331 pages

PESTICIDE FACT HANDBOOK
Volume 1

U.S. Environmental Protection Agency

This book contains 130 currently available Pesticide Fact Sheets issued by the U.S. Environmental Protection Agency. Each listing includes a description of the chemical use patterns and formulations, scientific findings, a summary of the Agency's regulatory position/rationale, toxicology, and a summary of major data gaps. The Fact Sheets cover more than 550 trade-named pesticides.

Fact Sheets are issued if one of the following regulatory actions occurs: (1) a Registration Standard has been issued, (2) a significantly different use pattern has been registered, (3) a new chemical is registered, or (4) a Special Review determination document has been issued. Fact Sheets have been prepared for Registration Standards issued since June 1982 and for new chemicals and for chemicals with significantly changed use patterns registered since January 1984. They have also been issued for Special Review final determinations since June 1983.

Noyes has republished these Fact Sheets and bound them in a durable, hard cover edition at $96, a fraction of their cost if purchased separately ($11.00 **per** fact sheet, or $1430).

The Fact Sheets are listed below. A Glossary and a Numerical List, as well as **Indexes of Common Names, Generic Names, and Trade Names,** are also included.

Actellic	Command	Fensulfothion	Paraquat
Aldicarb	Copper Sulfate	Fluchloralin	Parathion
Aldoxycarb	Cryolite	Fluometuron	Pendimethalin
Aldrin	Cyanazine	Fluridone	Perfluidone
Aliette	Cyhexatin	Fluvalinate	Phorate
Amitrole	Cyromazine	Fonofos	Phosmet
Anilazine	Daminozide	Formetanate Hydro-	Picloram
Arosurf	Dantochlor	chloride	Potassium Bromide
Arsenal	DCNA	Glycoserve	Potassium Perman-
Arsenic Acid	Demeton	Heliothis NPV	ganate
Arsenic Trioxide	Diazinon	Heptachlor	Prometryn
Avermectin	3,5-Dibromo	Hybrex	Pronamide
Azinphos-Methyl	Dicamba	Imazaquin	Propachlor
Bentazon and	Dichlobenil	Isazophos	Propargite
Sodium Bentazon	2,4-Dichlorophenoxy-	Isomate-M	Propham
Brominated Salicyl-	acetic Acid	Lactofen	Simazine
anilide	1,3-Dichloropropene	Lead Arsenate	Sodium and Calcium
Bronopol	Dicofol	Linalool	Hypochlorites
Butylate	Diflubenzuron	Lindane	Sodium Arsenate
Cadmium	Dinocap	Linuron	Sodium Arsenite
Calcium Arsenate	Dinoseb	Mancozeb	Sodium Omadine
Captafol	Diphenamid	Methiocarb	Sodium Salt of
Captan	Dipropetryn	Methyl Bromide	Fomesafen
Carbaryl	Disulfoton	Methyl Parathion	Sulfuryl Fluoride
Carbofuran	Diuron	Metolachlor	Tebuthiuron
Carbon Tetrachloride	Dodine	Metribuzin	Terbufos
Carbophenothion	EPN	Metsulfuron Methyl	Terbutryn
Chlordane	EPTC	Monocrotophos	Thiodicarb
Chlorimuron Ethyl	Ethalfluralin	Nabam	Thiophanate Ethyl
Chlorobenzilate	Ethoprop	Naled	Thiophanate Methyl
Chlorothalonil	Ethylenethiourea (ETU)	Naptalam	Thiram
Chlorpyrifos	Fenaminosulf	Nitrapyrin	TPTH
Chlorpyrifos Methyl	Fenbutatin-Oxide	Norflurazon	Trimethacarb
Clipper	Fenoxycarb	Oxamyl	Vitamin D$_3$
			Wood Preservatives

ISBN 0-8155-1145-0 (1988) 6"x9" 827 pages

Other Noyes Publications

PESTICIDE MANUFACTURING AND TOXIC MATERIALS CONTROL ENCYCLOPEDIA 1980

Edited by Marshall Sittig

Chemical Technology Review No. 168
Environmental Health Review No. 3
Pollution Technology Review No. 69

This book contains a total of 514 pesticide materials arranged in an alphabetical and encyclopedic fashion by the common or generic name of each pesticide. It is a thorough revision of our previous *Pesticides Process Encyclopedia* published in 1977, plus additional material relative to toxic materials control.

The data on manufacturing processes were drawn primarily from the patent literature, while the data on product toxicity, emissions and product use came mostly from published and unpublished reports released by the Environmental Protection Agency.

This book is definitely of interest to pesticide manufacturers, chemical raw material suppliers, formulators, growers, farmers and food processors. It should also prove useful to chemists, lawyers, industrial hygienists and environmentalists.

It contains much useful extrinsic information, e.g. *allowable tolerance limits, animal and human toxicities,* and similar data not easily ascertained elsewhere.

The use of pesticides leads to healthier plants and bigger crops, and exports of pesticides could provide fast growth for U.S. producers in the coming years.

An indication of the comprehensive nature of this one-volume encyclopedia is given here:

INTRODUCTION
What Is a Pesticide?
Pesticide Manufacture
Pollution Problems
Pesticide Formulations
  Dusts & Wettable Powders
  Granules
  Liquid Formulations
  Packing & Storage
Pesticide Applications

TOXIC MATERIALS CONTROL
Safe Work Practices
Pollution Control in Manufacture
Restrictions on Exposure & Use
  Concentrations in Air/Water
  Registration
  Residue Tolerances

ENVIRONMENTALLY ACCEPTABLE ALTERNATIVES
Biodegradable Pesticides
Physical Control of Toxic Pesticides

Controlled Release Pesticides
Ultra-Low Volume Application
Undesigned Pesticides
Biological Controls
Pheromones
Integrated Pest Management

DATA ON 514 INDIVIDUAL PESTICIDES:
Acephate
Acrolein
Acrylonitrile
Alachlor
Aldicarb
Aldoxycarb
Aldrin
Allethrin
Allidochlor
Allyl Alcohol
Aluminum Phosphide
Ametryne
Aminocarb
Amitraz
AMS
Ancymidol
Anilazine
Anthraquinone
ANTU
Arsenic Acid
Asulam
Atrazine
Azinphos-Ethyl
Azinphos-Methyl
Aziprotryn
Bacillus Thuringiensis
Barban
Benazolin
Bendiocarb
Benfluralin
Benodalin
Benomyl
Bensulide
Bentazon
Benzene Hexachloride
Benzoximate
Benzoylprop-Ethyl
Benzthiazuron
S-Benzyl Di-sec-butylthiocarbamate
Bifenox
plus 474 other pesticides

RAW MATERIALS INDEX

TRADE NAMES INDEX

ISBN 0-8155-0814-X

810 pages